Ethnobiology

Series editors:

Ethnobiology is the study of the dynamic relationship between plants, animals, people, and the environment. Academic and applied interests include ethnobotany, ethnozoology, linguistics, paleoethnobotany, zooarchaeology, ethnoecology, and many others. The field lies at a dynamic intersection between the social and biological sciences. The major contribution from the biological sciences has come from economic botany, which has a rich historical and scientific tradition. Indeed, the objectives of the colonial enterprise were as much about the quest for "green gold" –herbal medicines, spices, novel cultivars, and others—as it was for precious metals and sources of labor. The view that ethnobiology concerns mostly the discovery of new and useful biota extended into the 20th century. The social sciences have contributed to the field in both descriptive studies but also within quantitative approaches in cognitive anthropology that have led to general principles within ethnobiological classification. Ethnobiological research in recent years has focused increasingly on problem solving and hypothesis testing by means of qualitative and especially quantitative methods. It seeks to understand how culturally relevant biotas are cognitively categorized, ranked, named, and assigned meaning. It investigates the complex strategies employed by traditional societies to manage plant and animal taxa, communities, and landscapes. It explores the degree to which local ecological knowledge promotes or undermines resource conservation, and contributes to the solution of global challenges, such as community health, nutrition, and cultural heritage. It investigates the economic value and environmental sustainability to local communities of non-timber forest products, as well as the strategies through which individual ecological knowledge and practices encourage resilience to change—modernization, climate change, and many others. Most importantly, contemporary ethnobiological research is grounded in respect for all cultures, embracing the principles of prior informed consent, benefit sharing, and general mindfulness.

More information about this series at http://www.springer.com/series/11551

Michelle Cristine Medeiros Jacob
Ulysses Paulino Albuquerque
Editors

Local Food Plants of Brazil

Springer

Editors
Michelle Cristine Medeiros Jacob
Department of Nutrition, Laboratório
Horta Comunitária Nutrir
Federal University of Rio Grande do Norte
Natal, Rio Grande do Norte, Brazil

Ulysses Paulino Albuquerque
Department of Botany, Laboratório
de Ecologia e Evolução de Sistemas
Socioecológicos
Federal University of Pernambuco
Recife, Pernambuco, Brazil

ISSN 2365-7553 ISSN 2365-7561 (electronic)
Ethnobiology
ISBN 978-3-030-69138-7 ISBN 978-3-030-69139-4 (eBook)
https://doi.org/10.1007/978-3-030-69139-4

This Springer imprint is published by the registered company Springer Nature Switzerland AG
The registered company address is: Gewerbestrasse 11, 6330 Cham, Switzerland

Preface

Local and diverse food plants are at the heart of sustainable diets. However, current food systems undermine biodiversity, making diets unsustainable for human and environmental health. For example, in the case of Brazil, it is surprising how little we know about the biodiversity of our food plants. Brazil is a megadiverse country known for its biodiversity and expressive cultural diversity. Nevertheless, we have observed a tremendous dietetic standardization based on a low diversity of plants. In the Caatinga biome, a large region in the dry seasonal forest of northeastern Brazil, for instance, it is commonplace to observe how native foods are neglected and sometimes stigmatized as irrelevant resources. The reasons for this homogenization can be many. One probable hypothesis is that the hegemony of certain exotic foods to the detriment of the locals indicate an outspread tendency of cultural homogenization.

While the design of the food systems is an eminently political and ethical problem, there are still technical and scientific bottlenecks that distance us from sustainable diets. Our book presents some of these obstacles, considering food plants in the Brazilian context. This publication also comprises the register of local food plants, presenting the potential of plants in five of six Brazilian biomes and lacunas and our current state of knowledge. The gaps that we diagnose range from the systematic collection of ethnobiological and ethnoculinary data to the pervasive lack of nutritional composition and consumption data on biodiverse foods.

The initial parts of the book–"Basics on Food Plants and Biodiversity" and "Brazilian Food Plants"–provide a theoretical and situational overview of biodiverse food plants, in addition to reviewing the state of current scientific evidence relating food biodiversity to food and nutrition security. The subjects of food composition and consumption are addressed directly in two parts, "Food Composition Data on Brazilian Edible Plants by Biome" and "Consumption of Brazilian Food Plants." A multidisciplinary approach in research and in extension projects is useful to address the gaps in data on biodiverse foods. To show how this bridge between diverse disciplines offers a broad perspective, we include examples in two parts, "Ethnobotanical Knowledge of Brazilian Food Plants" and "Learning and Teaching Brazilian Food Plants." These 6 parts comprise 21 chapters written by professionals who argue that biodiverse foods are central in promoting sustainable diets and, therefore, planetary health.

With this book, we aim to introduce people to the debate about food biodiversity and food and nutrition security, considering the Brazilian context. Its potential lies in the fact that authors with different perspectives–from several countries, disciplines, theoretical backgrounds, and fields of action–worked on this production, expanding the visualization of problems and solutions from different focal lenses. Our final goal is to help as many people as possible find inspiration and ways to make changes based on the context of their lives and professions, in order to promote diets more sustainable for humans and all other forms of life.

Financial support of the National Council for Scientific and Technological Development, the CNPq (Grant number: 150654/2019-7), allowed us to edit this work. We also had support from the National Institutes of Science and Technology in Ethnobiology, Bioprospecting, and Nature Conservation, certified by CNPq, with financial support from Facepe, the Foundation for Support to Science and Technology of the State of Pernambuco (Grant number: APQ-0562-2.01/17).

Natal, Brazil Michelle Cristine Medeiros Jacob
Recife, Brazil Ulysses Paulino Albuquerque

Contents

Contributors

Vanessa Cristiane Santana Amaral Campus Central, Universidade Estadual de Goiás, Anápolis, GO, Brazil

Patrícia Lima Araújo Education and Health Center, Federal University of Campina Grande, Cuité, Brazil

Elisa Flávia Luiz Cardoso Bailão Campus Central, Universidade Estadual de Goiás, Anápolis, GO, Brazil

Daniel Henrique Bandoni Center of Practices and Research in Nutrition and Collective Food Services/Department of Health, Clinic and Institutions, Universidade Federal de São Paulo (UNIFESP), Santos, SP, Brazil

Déborah Monteiro Barbosa Laboratory of Biocultural Ecology, Conservation and Evolution (LECEB), Campus of Engineering and Agricultural Sciences, Federal University of Alagoas, Rio Largo, AL, Brazil

Fabrícia França Bezerril Departament of Nutrition, Federal University of Paraíba, João Pessoa, Paraíba, Brazil

Leonardo Luiz Borges Campus Central, Universidade Estadual de Goiás, Anápolis, GO, Brazil

Escola de Ciências Médicas, Farmacêuticas e Biomédicas, Pontifícia Universidade Católica de Goiás, Goiânia, GO, Brazil

Ieda Maria Bortolotto Programa de Pós-Graduação em Biologia Vegetal, Instituto de Biociências, Universidade Federal de Mato Grosso do Sul, Campo Grande, MS, Brazil

Maria G. L. Brandão Centro Especializado em Plantas Aromáticas, Medicinais e Tóxicas (Ceplamt), Museu de História Natural e Jardim Botânico, Universidade Federal de Minas Gerais (UFMG), Belo Horizonte, Brazil

Raquel Pires Campos Faculdade de Ciências Farmacêuticas, Alimentos e Nutrição, Universidade Federal de Mato Grosso do Sul, Campo Grande, MS, Brazil

Samantha Salomão Caramori Campus Central, Universidade Estadual de Goiás, Anápolis, GO, Brazil

Mariana Franco Cassino Programa de Pós-Graduação em Botânica, Instituto Nacional de Pesquisas da Amazônia (INPA), Manaus, Brazil

CaSEs – Department of Humanities, Universitat Pompeu Fabra (UPF), Barcelona, Spain

Viviany Moura Chaves Universidade Federal do Rio Grande do Norte, Natal, Brazil

Lee Chen Chen Instituto de Ciências Biológicas I, Universidade Federal de Goiás, Goiânia, GO, Brazil

Charles R. Clement Coordenação de Tecnologia e Inovação, Instituto Nacional de Pesquisas da Amazônia (INPA), Manaus, Brazil

Gabriela Coelho-de-Souza Universidade Federal do Rio Grande do Sul, Porto Alegre, RS, Brazil

Lidio Coradin Biodiversity for Food and Nutrition Project, Brasília, DF, Brazil

Nadja Maria da Costa Melo Biophysics and Pharmacology, Federal University of Rio Grande do Norte, Natal, Rio Grande do Norte, Brazil

Camila Pia Delgado da Silva Graduate Program in Food, Nutrition and Health, Universidade Federal de São Paulo (UNIFESP), Santos, SP, Brazil

Laleska Pâmela Rodrigues da Silva Programa de Pós-Graduação em Saúde e Desenvolvimento na Região Centro-Oeste, Faculdade de Medicina, Universidade Federal de Mato Grosso do Sul, Campo Grande, MS, Brazil

Rafael Ricardo Vasconcelos da Silva Laboratory of Biocultural Ecology, Conservation and Evolution (LECEB), Campus of Engineering and Agricultural Sciences, Federal University of Alagoas, Rio Largo, AL, Brazil

Vanuska Lima da Silva Universidade Federal do Rio Grande do Sul, Porto Alegre, RS, Brazil

Mariana Rodrigues da Silva Lopes Programa de Pós-Graduação em Saúde e Desenvolvimento na Região Centro-Oeste, Faculdade de Medicina, Universidade Federal de Mato Grosso do Sul, Campo Grande, MS, Brazil

Geraldo Alves Damasceno-Junior Instituto de Biociências, Universidade Federal de Mato Grosso do Sul, Campo Grande, MS, Brazil

Luciane Madureira de Almeida Campus Central, Universidade Estadual de Goiás, Anápolis, GO, Brazil

Ana Laura Benevenuto de Amorim Interdisciplinary Graduate Program in Health Sciences, Universidade Federal de São Paulo (UNIFESP), Santos, SP, Brazil

Emanuella de Aragão Tavares Departament of Phamacy, University Federal of Rio Grande Do Norte, Natal, Rio Grande do Norte, Brazil

Joana Yasmin Melo De Araujo Departamento de Nutrição, Universidade Federal do Rio Grande do Norte, Natal, Brazil

Paloma Oliveira Antonino Assis de Carvalho Departamento de Nutrição, Universidade Federal da Paraíba, João Pessoa, Paraíba, Brazil

Francinalva Dantas de Medeiros Education and Health Center, Federal University of Campina Grande, Cuité, Brazil

Patrícia Muniz de Medeiros Laboratory of Biocultural Ecology, Conservation and Evolution (LECEB), Campus of Engineering and Agricultural Sciences, Federal University of Alagoas, Rio Largo, AL, Brazil

Wilson de Melo Cruvinel Escola de Ciências Médicas, Farmacêuticas e Biomédicas, Pontifícia Universidade Católica de Goiás, Goiânia, GO, Brazil

Elias Jacob de Menezes Neto Universidade Federal do Rio Grande do Norte, Natal, Brazil

Maria Elieidy Gomes de Oliveira Department of Nutrition, Post-graduation in Nutrition Sciences, Federal University of Paraiba, João Pessoa, Paraíba, Brazil

Matheus Gabriel de Oliveira Faculdade de Farmácia, Universidade Federal de Goiás, Goiânia, GO, Brazil

Daniela Moura de Oliveira Beltrame National Project Coordinator, Biodiversity for Food and Nutrition Project, Brasília, DF, Brazil

Letícia Zenóbia de Oliveira Campos Universidade Federal do Oeste da Bahia – UFOB, Estrada do Barrocão, Barreiras, BA, Brazil

Clélia de Oliveira Lyra Department of Nutrition, Federal University of Rio Grande do Norte, Natal, Brazil

Fillipe de Oliveira Pereira Education and Health Center, Federal University of Campina Grande, Cuité, Brazil

Joelma Abadia Marciano de Paula Campus Central, Universidade Estadual de Goiás, Anápolis, GO, Brazil

Juliana de Paula-Souza Centro Especializado em Plantas Aromáticas, Medicinais e Tóxicas (Ceplamt), Museu de História Natural e Jardim Botânico, Universidade Federal de Minas Gerais (UFMG), Belo Horizonte, Brazil

Departamento de Botânica, Centro de Ciências Biológicas, Universidade Federal de Santa Catarina (UFSC), Florianópolis, Brazil

Veridiana Vera de Rosso Center of Practices and Research in Nutrition and Collective Food Services/Department of Biosciences, Universidade Federal de São Paulo (UNIFESP), Santos, SP, Brazil

Daline Fernandes de Souza Araújo Health Sciences College of Trairi, Federal University of Rio Grande do Norte, Santa Cruz, Rio Grande do Norte, Brazil

Gracy Kelly Vieira de Vasconcelos Medeiros Department of Nutrition, Post-graduation in Nutrition Sciences, Federal University of Paraiba, João Pessoa, Paraíba, Brazil

Rita de Cássia Ramos do Egypto Queiroga Departament of Nutrition, Federal University of Paraíba, João Pessoa, Paraíba, Brazil

Viviany Teixeira do Nascimento Universidade do Estado da Bahia – UNEB, BR 242, s/n. Loteamento Flamengo. Departamento de Ciências Humanas, Barreiras, BA, Brazil

Gabriela Maria Cota dos Santos Laboratory of Biocultural Ecology, Conservation and Evolution (LECEB), Campus of Engineering and Agricultural Sciences, Federal University of Alagoas, Rio Largo, AL, Brazil

Ivanilda Soares Feitosa Universidade Federal de Pernambuco, Laboratório de Ecologia e Evolução de Sistemas Socioecológicos, Recife, Brazil

Almecina B. Ferreira Jr Universidade Federal do Acre, Centro de Ciências Biológicas e da Natureza, Rio Branco, Acre, Brazil

Washington Soares Ferreira Júnior Laboratório de Investigações Bioculturais no Semiárido, Universidade de Pernambuco, Petrolina, PE, Brazil

Laura Pereira Furquim Museu de Arqueologia e Etnologia, Universidade de São Paulo (USP), São Paulo, Brazil

Danúbia Lins Gomes Laboratory of Biocultural Ecology, Conservation and Evolution (LECEB), Campus of Engineering and Agricultural Sciences, Federal University of Alagoas, Rio Largo, AL, Brazil

Sávio Marcelino Gomes Universidade Federal do Rio Grande do Norte, Natal, Brazil

Gabriela G. Gonçalves Colégio da Villa, Jaguariúna, São Paulo, Brazil

Gerlane Coelho Bernardo Guerra Department of Biophysics and Pharmacology, Federal University of Rio Grande do Norte, Natal, Rio Grande do Norte, Brazil

Rita de Cássia Avellaneda Guimarães Programa de Pós-Graduação em Saúde e Desenvolvimento na Região Centro-Oeste, Faculdade de Medicina, Universidade Federal de Mato Grosso do Sul, Campo Grande, MS, Brazil

Priscila Aiko Hiane Faculdade de Ciências Farmacêuticas, Alimentos e Nutrição, Universidade Federal de Mato Grosso do Sul, Campo Grande, MS, Brazil

Michelle Cristine Medeiros Jacob Department of Nutrition, Laboratório Horta Comunitária Nutrir (LabNutrir), Federal University of Rio Grande do Norte, Natal, Brazil

Thiago Perez Jorge Nutrition Department, Federal University of Rio Grande do Norte, Natal, Brazil

Silvana Maria Zucolotto Langassner Departament of Pharmacy, University Federal of Rio Grande Do Norte, Natal, Rio Grande do Norte, Brazil

Djackson Garcia de Lima Universidade Federal do Rio Grande do Norte, Natal, Brazil

Severina Carla Vieira Cunha Lima Department of Nutrition, Federal University of Rio Grande do Norte, Natal, Brazil

Juliana Salles Machado Departamento de História, Universidade Federal de Santa Catarina (UFSC), Florianópolis, Brazil

Tamires Alcântara Dourado Gomes Machado Departamento de Nutrição, Universidade Federal da Paraíba, João Pessoa, Paraíba, Brazil

Marco Madella CaSEs – Department of Humanities, Universitat Pompeu Fabra (UPF), Barcelona, Spain

Dirce Maria Lobo Marchioni Department of Nutrition, Faculty of Public Health, University of São Paulo, São Paulo, Brazil

Ana Cristina Silveira Martins Food of Engineering, Post-graduation in Food Science and Technology, Federal University of Paraíba, João Pessoa, Paraíba, Brazil

Josué Schneider Martins Universidade Federal do Rio Grande do Sul, Porto Alegre, RS, Brazil

Maria Franco Trindade Medeiros Departamento de Botânica, Museu Nacional, Universidade Federal do Rio de Janeiro, Rio de Janeiro, RJ, Brazil

Fernanda L. B. Mügge Centro Especializado em Plantas Aromáticas, Medicinais e Tóxicas (Ceplamt), Museu de História Natural e Jardim Botânico, Universidade Federal de Minas Gerais (UFMG), Belo Horizonte, Brazil

Zentrum für Molekulare Biologie, Universität Heidelberg, Heidelberg, Germany

Camila Neves Soares Oliveira Ministry of the Environment, Brasília, DF, Brazil

Daniela Dias Ortega CaSEs – Department of Humanities, Universitat Pompeu Fabra (UPF), Barcelona, Spain

Laura Brito Porciuncula Nutrition Department, Federal University of Rio Grande do Norte, Natal, Brazil

Arnildo Pott Programa de Pós-Graduação em Biologia Vegetal, Instituto de Biociências, Universidade Federal de Mato Grosso do Sul, Campo Grande, MS, Brazil

Jean Pierre Poulain Université de Toulouse Jean Jaurès, Toulouse, France

Taylor's University, Kuala Lumpur, Malaysia

Sarah M. S. Prates Centro Especializado em Plantas Aromáticas, Medicinais e Tóxicas (Ceplamt), Museu de História Natural e Jardim Botânico, Universidade Federal de Minas Gerais (UFMG), Belo Horizonte, Brazil

Programa de Pós-Graduação em Ciências de Alimentos, Faculdade de Farmácia, Universidade Federal de Minas Gerais (UFMG), Belo Horizonte, Brazil

Mariana Oliveira Ramos Universidade Federal do Rio Grande do Sul, Porto Alegre, RS, Brazil

Neide Rigo Autonomous Researcher, São Paulo, SP, Brazil

Cecília Rocha Ryerson University, Toronto, ON, Canada

Stefany Areva Severo Universidade do Estado do Rio Grande do Sul/Ecos do Pampa, Rio Grande, RS, Brazil

Myrtle P. Shock Programa de Antropologia e Arqueologia, Instituto de Ciências da Sociedade, Universidade Federal do Oeste de Pará (UFOPA), Santarém, Brazil

Rosa Helena Silva Instituto de Biociências, Universidade Federal de Mato Grosso do Sul, Campo Grande, MS, Brazil

Bernardo Tomchinsky Universidade Federal do Sul e Sudeste do Pará, Instituto de Estudos em Saúde e Biologia, Marabá, Paraíba, Brazil

Adriana Carla Dias Trevisan Universidade do Estado do Rio Grande do Sul/Ecos do Pampa, Rio Grande, RS, Brazil

Author's Biography

Michelle Cristine Medeiros Jacob is a professor in the Nutrition Department and the postgraduate program of social sciences at the Federal University of Rio Grande do Norte. She is the founder of LabNutrir (www.nutrir.com.vc), a community garden laboratory in an undergraduate nutrition program that brings together local plants and people. The United Nations in 2018 recognized the LabNutrir as a reference in the application of the Right to Food Guidelines, especially guideline 11, that is, education and awareness-raising. Dr. Jacob's research interests relate to food and nutrition security, sustainable diets, ethnonutrition, and food systems.

Ulysses Paulino Albuquerque received his Ph.D. in biology in 2001 from the Universidade Federal de Pernambuco, Brazil. He is full professor in the Department of Botany at Universidade Federal de Pernambuco, Pernambuco, Brazil. In 2011, he led the founding of a new Ph.D. program in ethnobiology and nature conservation. Professor Albuquerque has published around 316 journal articles, 200 book chapters, and edited or authored 50 books (including new editions and translations). Professor Albuquerque has served as editor of various peer-reviewed journals and in 2011 co-founded the journal *Ethnobiology and Conservation* as co-editor-in-chief.

Part I
Basics on Food Plants and Biodiversity

Biodiversity Towards Sustainable Food Systems: Four Arguments

Michelle Cristine Medeiros Jacob, Viviany Moura Chaves, and Cecília Rocha

Contents

1 Introduction

For several years, scientists have been warning that food systems have become significant drivers of environmental degradation, of various forms of malnutrition, and of food insecurity (Altieri 2004; Swinburn et al. 2019). The pandemic of COVID-19 demonstrates the practical effect of ignoring the evidence in the name of a narrow focus on food production (to see more about the relationship between environmental degradation and SARS-CoV-2 outbreak see Jacob et al. 2020a). We have never been so close to a global shutdown of our economic system, so close to living on a planet where all forms of life are under threat, and so distant from guaranteeing regular access to nutritious foods to households across the globe (IPES-Food 2020). The global food system is ripe for a change.

Food systems are formed by all activities in food production, transformation, distribution, and consumption, including those leading to food losses and waste.

M. C. M. Jacob (✉) · V. M. Chaves
Universidade Federal do Rio Grande do Norte, Natal, Brazil
e-mail: michellejacob@ufrn.edu.br; vivianychaves@ufrn.edu.br

C. Rocha
Ryerson University, Toronto, ON, Canada
e-mail: crocha@ryerson.ca

M. C. M. Jacob, U. P. Albuquerque (eds.), *Local Food Plants of Brazil*, Ethnobiology, https://doi.org/10.1007/978-3-030-69139-4_1

The interaction and interdependence of food systems, human health, and biodiversity are complex. Sustainable food systems are needed for human health, but the sustainability of food systems depends fundamentally on the preservation of their biodiversity. Sustainable food systems promote global outcomes of human and environmental health, social equity, and economic resilience (IPES-Food 2017). The task of transforming food systems to deliver sustainability requires integrated actions in order (1) to conserve biodiversity and to reduce the impacts on the environment; (2) to shift towards sustainable practices in production, processing, and consumption; (3) to improve socioeconomic welfare; and (4) to consider cultural adequacy of food practices (Béné et al. 2019). In this debate, the biodiversity of plants, animals, and micro-organisms used directly or indirectly for food and agriculture has a crucial role in promoting sustainable food systems (Blicharska et al. 2019). The *Convention on Biological Diversity* defines biodiversity as the variability among living organisms from all sources, including terrestrial, aquatic ecosystems, and the ecological complexes (United Nations 2016). Ecosystems, species, and genes are the three critical components in biodiversity, characterized by attributes, such as diversity, abundance, and composition (Kearns 2010).

In this chapter, we present arguments that highlight the role of biodiversity in making food systems more sustainable. In our analysis, we define biodiverse food plants as the plants of extensive use (e.g., beans, rice, corn) and unconventional food plants as usually native, often neglected, and of culturally limited use (Jacob and Albuquerque 2020). We also consider the non-edible biodiversity of agricultural interest, which includes a multitude of species, such as soil microbiomes, insects, birds, and mammals, which work pollinating crops, regulating pests, balancing nutrients in fields, and storing carbon in soils (Willett et al. 2019). This discussion can help inform food system transformation plans and actions.

2 Biodiversity Towards Sustainable Food Systems: Four Major Arguments

In Fig. 1, we summarize the main opportunities and barriers related to the four arguments to mainstream biodiversity into current food systems.

2.1 *Biodiversity Is Central to Food and Nutrition Security*[1]

The most authoritative and widely used definition of food security is that provided in the FAO's 2001 *State of Food Insecurity* report: "Food security exists when all people, at all times, have physical, social and economic access to sufficient, safe and

[1] In our book, we prefer to use "food and nutrition security" (FNS) instead of "food security" for two main reasons. First, FNS is the term used in Brazilian legislation (see Law 11.346/2006).

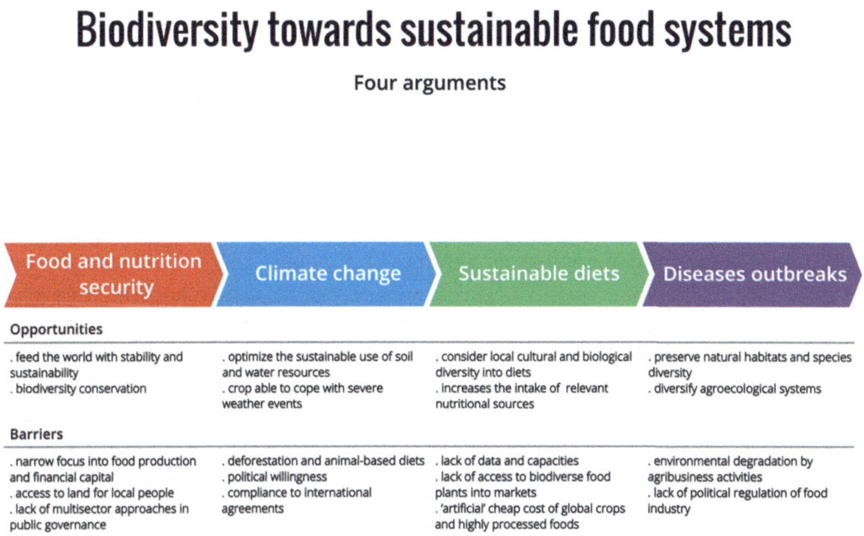

Fig. 1 Four major arguments to mainstream biodiversity into current food systems

nutritious food which meets their dietary needs and food preferences for an active and healthy life" (FAO 2001). From this breathtaking, encompassing definition, many components can be discerned, particularly those addressing the *availability* of and the *access* to food, as well as the *utilization* of food (nutrition uptake), and the *stability* of food availability, access, and utilization. In its 2020 report, the High Level Panel of Experts (HLPE) of the Committee on World Food Security (CFS) proposes amplifying the concept of food security to more explicitly recognize two other dimensions: *sustainability* and *agency* (HLPE 2020).

While the endorsement of the HLPE leads to a more widespread recognition of the importance of sustainability and agency for food and nutrition security (FNS), many scholars and groups working in the area have incorporated these dimensions in their consideration of food security for many years. As an example, the Centre for Studies in Food Security at Ryerson University, Canada, since 2003 has considered the following dimensions of food security (defined collectively as "the 5 As of food security"):[2]

- *Availability*: Sufficient food for all people at all times
- *Accessibility*: Physical and economic access to food for all at all times

---

Second, the term "nutrition" signals that food also needs to offer quality in terms of nutritional health. To read further about this debate, see Ingram (2020). As we show in this section, FAO and the Centre for Studies in Food Security at Ryerson University maintain the original description of the concept (without the "nutrition"). However, the definitions presented by them include the quality component of food embraced by FNS.

[2] See at: https://www.ryerson.ca/foodsecurity/

- *Adequacy*: Access to food that is nutritious and safe and produced in environmentally sustainable ways
- *Acceptability*: Access to culturally acceptable food, which is produced and obtained in ways that do not compromise people's dignity, self-respect, or human rights
- *Agency*: Policies and processes that enable the achievement of food security

A "productionist" view of food security only focuses on the availability dimension, with little regard for issues of poverty and wealth distribution (preventing proper access); nutritional quality of diets and safety of the food made available; the environmental impact of food production; the social and cultural contexts for people accessing food; and little regard for the power dynamics preventing the realization of the human right to adequate food. For years, this "productionist" view has supported the development of an industrial agriculture paradigm, favoring monocultures and emphasizing quantity over quality, to the detriment of biodiversity, the environment, human health, cultures, and social well-being (IPES-Food 2016, 2017). True food and nutrition security depends on a food system that promotes health, fairness, and environmental sustainability.

In the past, using a strict "productionist" approach, biodiversity conservation and food security were often presented as mutually exclusive goals (Sonnino et al. 2014). In a finite resource world, the decisions taken to address one problem were seen to negatively affect the other (Chappell and LaValle 2011). Thus, for example, a conservation focus could limit food production requirements and, as a consequence, increase food insecurity (as if food security depended only on an increase in the absolute quantity of food). However, the practice of converting wildlands to intensive commercial agricultural use, ignoring biodiversity, can produce new challenges to FNS (e.g., pollinator diversity reduction, lower soil fertility). Biodiversity has proven to be central to FNS and vice versa (Sunderland 2011). Presenting these two challenges as an inevitable trade-off is part of a narrative[3] that has proven to be insufficient to analyze the complexity of both. The analysis needed requires a broader focus on food and nutrition security instead of a strict food production approach, with consideration of societal issues such as social justice and governance (Cramer et al. 2017).

However, as shown by Hanspach et al. (2017), a biodiverse environment does not guarantee FNS. They conducted a multivariate analysis of social-ecological data from 110 landscapes in the Global South to study the food-biodiversity nexus. In the landscapes studied, win-win outcomes were associated with high equity, ready

[3]A narrative defines the framings of the stories around the food system, beginning, middle, and end. Three points orient the construction of a food systems failure narrative: what is the failure about, what is threatened and need to be fixed, and where the priorities for action stand. The dominant and narrow narrative is *the food systems failure is their inability to feed the world population. FNS is under threat. The action required is to close the yield gap.* To broaden this narrative, we would prefer to tell the following story*: the food systems failure is their inability to produce equal and equitable benefits. Social justice, democratic process, and small-scale actors are under threat. The action required is decentralization and grassroots autonomy*. See Béné et al. (2019).

access to land for local people, and high human and social capital. On the other hand, trade-offs were related to a narrow focus on financial capital. The authors concluded that avoiding a narrow focus on infrastructure, commercialization, and built capital seems critical for fostering synergies between FNS and biodiversity conservation. It is crucial to broaden the focus by considering strengthening human capital, social capital, and equity to foster win-win relations.

Biodiversity can support FNS in many ways. Blicharska et al. (2019), for example, performed a review to discuss the breadth of ways in which biodiversity can support sustainable development. Analyzing the sustainable development goal 2 (Zero Hunger), they list the direct delivery benefits of biodiversity to FNS (United Nations 2015). Some of them are improving dietary quality; ameliorating soil fertility, structure, quality, and health; providing crop pollination; bearing pest control; expanding agricultural output and future yields; increasing resilience of agricultural systems; providing potential for new crops; and maintaining productivity in marine ecosystems.

Considerations of the synergies between biodiversity preservation and FNS have led to the promotion of agroecology. Diversified agroecological systems offer broader benefits for the environment and society (IPES-Food 2015). With a holistic approach, agroecology considers the sustainable use and management of natural resources and ecosystem services in agriculture. It also explicitly includes social issues into its principles, such as ethical considerations, changes in diet, and social justice (see Altieri 2004). Transitioning towards diversified agroecological systems is more urgent than ever. The COVID-19 outbreak has revealed how intricately linked human, animal, and ecological health are (Altieri and Nicholls 2020). However, a narrow scope without governance arrangements will fail to mainstream biodiversity into global food systems (De Clerck 2016). To prevent future health crises on a global scale, we need to connect agroecological strategies, public policies, and solidarity market arrangements (IPES-Food 2018; Nicholls and Altieri 2018).

2.2 Agricultural Biodiversity Strengthening Resilience to Climate Change

Food production has been a major driver of climate change, being responsible for 26% of all anthropogenic greenhouse gas emissions (Poore and Nemecek 2018). This fact represents a significant concern to FNS since climate change has adverse effects on food production, creating harmful feedback loops in the food-climate nexus (Jacques and Jacques 2012).

Resilience in social-ecological systems is the ability of a given system to sustain itself or recover quickly from difficulties, stresses, and shocks. It comprises three main characteristics: the capacity (1) to absorb shocks, (2) to self-organize, and (3) to learn and adapt. Agricultural biodiversity and associated knowledge strength the resilience to climate change-related stresses. This is the conclusion of a study on global food systems trends that reviewed 172 project reports and case studies from Africa, Central and South America, Asia, and the Pacific (Mijatović et al. 2013). Mijatović and colleagues reported the strategies to strengthen climate change resilience with agricultural diversity by dividing it into three levels. First, at the scale of the landscape, biodiversity protects and restores ecosystems and optimizes the sustainable use of soil and water resources. Second, at the scale of the farming system, biodiversity contributes to the diversification of crops, agroforestry, allowing various adjustments in practices (e.g., soil fertility, rainwater harvesting). Third, at the level of the species or variety, biodiversity improves the stress tolerance through selection and breeding techniques, amplifying the use of resistant species, varieties, and breeds. One clear example provided by the authors at the farming scale is that in agroforestry systems, trees, and shrubs regulate soil moisture and temperature.

Despite the scientific evidence that relates biodiversity and climate change mitigation, current food systems are in the opposite direction, facing an increasing trend towards homogeneity. For example, Khoury et al. (2014) analyzed changes in the diversity of the portfolio of crop species in the food supplies of 152 countries comprising 98% of the world's population from 1961 to 2009. They concluded that globally, national food supplies have become increasingly similar in composition, precisely 36% more similar over the past five decades. A suite of global crop plants builds these national portfolios: maize (*Zea mays* L.), wheat (*Triticum aestivum* L.), and rice (*Oryza sativa* L.), also known as the "big three" cereals. These crops have been bred for intensive agriculture, and they may not be able to cope with the challenging weather events we are already facing (Massawe et al. 2016).

To cope with climate change is urgent to diversify agricultural biodiversity and supply chains. Although the world counts with at least 50,000 species of plants suitable for human consumption, fewer than 300 species make their way into the market (Jacques and Jacques 2012). Brazil has a big potential of biodiversity with an estimated flora of 49,056 species, including algae, angiosperms, bryophytes, fungi, gymnosperms, ferns, and lycophytes (Jardim Botânico do Rio de Janeiro, 2020). These plants are strategic to a massive social change toward plant-based diets. Plant-based diets have the potential to reduce deforestation and methane production by domesticated ruminants (Wolf et al. 2019). Reductions in global deforestation and ruminant numbers could substantially contribute to climate change mitigation goals (Ripple et al. 2014). As the climate change crisis has global as well as local implications, the actors of the international political arena will need to find common ground to achieve the mitigation goals. For example, a recent study by Rajão et al. (2020) shows that almost 20% of soy exports and at least 17% of beef exports from Brazil to the European Union (EU) are contaminated with illegal deforestation of the Amazon and the Cerrado (Brazilian biomes). Related to this fact, there are two significant concerns. First, the increase of greenhouse gas emissions from deforestation and forest fires in Brazil could cancel out EU climate change mitigation efforts. Second, international consumers are demanding to boycott Brazilian products contaminated by deforestation, impacting the national economy. Thus, political will and compliance to international agreements are necessary and urgent to advance the climate change issue.

2.3 *Biodiversity Fosters Sustainable Diets*

The simplification of global and local agricultural systems decreases the availability of diverse food and its consumption (see Khoury et al. 2014), increasing the risk of various forms of malnutrition, potentially leading to undernutrition as well as overweight and obesity. The FAO's report *The State of Food Security and Nutrition in the World* (FAO 2020) shows that in 2019 almost 750 million people, or 10% of the human population, were exposed to severe food insecurity levels. The same report shows that almost two billion people have no regular access to safe, nutritious, and sufficient food and that obesity is also on the rise in all regions of the world.

The consumption of biodiverse food plants is directly associated with a healthy diet. In a broad review of the contribution of wild and cultivated biodiversity to improve diets, Powell et al. (2015) argue that several studies that looked at nutrient intake found a possible relationship between crop diversity and mean nutrient adequacy (a quality diet indicator) across multiple nutrients. In an international research effort, Lachat et al. (2018) assessed data from 24-hour diet recalls from 6226 participants (women and children) in rural areas from seven low- and middle-income countries, to analyze the relationship between dietary species richness and dietary quality. Their dietary quality analysis comprised the mean adequacies of vitamin A, vitamin C, folate, calcium, iron, and zinc. By dietary species richness, they consider the number of species consumed by each individual. Their results showed a positive association between nutritional and biodiversity indicators (species richness), both in the wet and dry seasons. Concerning specifically unconventional food plants, Powell et al. (2013), studying dietary diversity and wild plants in Tanzania, showed that although these plants contributed only 2% of the total energy in the diet, they provided significant percentages of vitamin A (31%), vitamin C (20%), and iron (19%). Even considering that the analysis of biodiversity in diets of industrialized and urban settings needs to advance, the current state of evidence shows that biodiverse food plants are relevant sources of energy, micronutrients, and bioactive compounds (Penafiel et al. 2011). Thus, the consumption of these plants is at the core of the proposal of sustainable or healthy diets,[4] those consisting of a diversity of plant-based foods, with low amounts of animal source foods and low amounts of highly processed foods (Willett et al. 2019).

Brazil has a strong potential for mainstreaming biodiversity into diets. The national ordinance no. 284/2018 identifies 101 native species with nutritional

[4] We understand sustainable diets as a synonym of healthy diets. For us, as for Willet et al. (2019), to be healthy, in a broad sense and long-term, diets need to protect both the environment and human health. To better understand the distinctions made in some cases among sustainable and healthy diets, see Béné et al. (2019)

potential, fostering their integration into national public policies, such as the Food Acquisition Program (*Programa de Aquisição de Alimentos*-PAA) and the National School Food Program (*Programa Nacional de Alimentação Escolar*-PNAE). Nutritional data of these plants are available through the Information System of Brazilian Biodiversity (*Sistema da Informação da Biodiversidade Brasileira*-SiBBr).[5] Moreover, the Brazilian food guide promotes food diversity, considering its role in contributing to a higher variety of nutrients and in protecting the environment (Brasil 2014). Another public health tool is the report *Alimentos Regionais Brasileiros* (Brazilian Regional Foods), published by the Health Ministry (Brasil, 2015), with the purpose of spreading knowledge of the diversity of Brazilian plants, presenting plant information, culinary uses, recipes, and nutritional information. Finally, one of the food composition tables used in the country, the *Tabela Brasileira de Composição de Alimentos* (TBCA/USP), includes a database (*Biodiversidade e alimentos regionais*[6]), containing a variety of plants consumed in Brazil, their scientific name, and cultivar identification.

However, there are several barriers to overcome in the promotion of diets rich in local plants, integrating them into Brazil's food system (see Box 1). Some of them are related to our current knowledge of biodiverse food plants. Jacob and Albuquerque (2020) present four significant gaps that can help to collectively align the research agenda of scientists interested in the topic. First, there is a need to create better strategies to map the biodiverse food plants available in our territory. The creation of research networks and the development of systematic reviews could be strategic to gather these data on a large-scale (see Jacob et al. 2020b as an example). Second, we need to overcome the lack of culinary data in our studies. Some processing food techniques (e.g., to wash, to heat, to infuse, to germinate, to ferment, to cure) or the combination of different foods can modify the diet food matrix and the bioavailability of certain nutrients or toxins. Third, the scarcity of nutritional composition data puts a real barrier to dietary assessments. Finally, we need to improve our capacity to express the relationship between people, plants, and culture in our research tools and teams. As Powell et al. (2015) affirmed, this understanding of complex and dynamic biocultural food systems and landscapes will require that nutritionists, for example, think about landscapes and biodiversity as more than just calories. There is no doubt that the dialogue between nutrition, ethnobiology, anthropology, and agronomy is strategic for improving our capacity to work with biodiverse food plants.

Finally, it is crucial to highlight the virtuous loop between human and environmental health. Biodiverse diets protect the diversity of life by fostering sustainable agricultural practices. Consequently, agricultural diversity can stimulate productivity, stability, ecosystem services, and food systems' resilience (Frison et al. 2011; Khoury et al. 2014). The connecting point to boost this relationship is in our diets. As individuals and as a society, we need to be aware that in the food systems

[5] See at: https://ferramentas.sibbr.gov.br/ficha/bin/view/FN

[6] See at: http://www.tbca.net.br/base-dados/biodiversidade.php

Box 1: Barriers to Promote Biodiverse Food Plants into Diets

- Disconnect between the biodiversity, agriculture, health, education, and other sectors
- Continued lack of resources to develop research and extension systems focused in biodiversity
- Biodiverse food-based approaches all too often fall outside the traditional scope of clinical nutrition and public health
- Lack of skills and institutional capacity necessary to promote multisector approaches
- Lack of data linking biodiversity to dietary diversity and nutrition outcomes
- Relevant information is highly fragmented, scattered in various publications and reports or not easily accessible databases to policymakers and practitioners
- Poorly developed public policies, infrastructure, and markets for most of the biodiversity for food and nutrition
- Reach and influence of the modern globalized food system and trade policies which impede or undermine promotion and consumption of biodiversity for food and nutrition, favoring the consumption of unhealthy processed foods
- Negative perceptions and attitudes to local, nutritionally rich traditional biodiverse foods
- The "artificial" cheap cost of global crops or imported foods which externalize their health and environmental costs
- Lack of farmers' seed networks that support crop diversity sharing
- Lack of innovative food recipes that involve less cooking time and are more in tune with modern food consumption habits and lifestyles
- Lack of consumer demand, which translates into a lack of product awareness

Adapted from Hunter and Fanzo (2013) and Raneri et al. (2019)

dynamic, changes in our consumption patterns have effects on food production (Lawrence et al. 2015). Healthy diets are unaffordable for more than three billion people in the world (FAO 2020). More research is needed to identify and analyze the hidden costs of unhealthy diets, proposing measures for tackling these costs and investing into food system transformation.

2.4 Boosts Food System Resilience to Disease Outbreaks

The health crisis driven by COVID-19 highlighted the risks, weaknesses, and inequities underlying the global food system. In the recent years, we have had to deal with epidemic zoonoses such as avian influenza (H5N1), Severe Acute Respiratory Syndrome (SARS), Ebola virus, and Middle East Respiratory Syndrome (MERS). Over 70% of infectious diseases that have arisen in humans are related to animals, mostly originated in wildlife (FAO 2017).

The emergence of new pandemic and epidemic outbreaks may become more common in the future, considering that ecological catastrophes and climate change increase the frequency of zoonotic diseases (Patz et al. 2005; Alirol et al. 2011). Diverse high-risk human activities, such as industrial livestock production, intensify viral transmission between animals and people (Johnson et al. 2020). For instance, livestock breeding in tropical forests is related to deforestation, which is associated with an increase in infectious diseases such as dengue, malaria, and yellow fever (Wilcox and Ellis 2006).

Another disturbing loop relates to the destruction of natural habitats, biodiversity loss, viruses transmission, and emerging infectious diseases (Olival et al. 2017). According to the United Nations report on zoonotic diseases (United Nations Environment Programme 2020), more virus transmission events occur within communities that have low species diversity, which is referred as dilution effect. The dilution effect occurs because communities with more diversity of animals dilute transmission events keeping the virus prevalence low, thus reducing the number of susceptible animals. Unfortunately, as humans occupy and transform these environments, they disrupt the ecology of wildlife, altering the ecosystem balance, and increasing the likelihood that viruses will find intermediate hosts (Volpato et al. 2020; Jacob et al. 2020a). The conservation of biodiversity and the ecosystem plays a critical role in protecting humans from emerging infectious diseases.

A paradigm shift from industrial agriculture to diversified agroecological systems is an urgent challenge towards resilience. A resilient global system will help us cope with future shocks, making them less likely and critical (Kahiluoto 2020). Experts identify a series of actions to trigger a shift towards resilient food systems, including redirecting agricultural subsidies and investments into research on agroecology (IPES-Food 2016). In some cases, traditional practices focused on agrobiodiversity exhibit greater productivity than conventional agricultural methods. Prieto et al. (2015) demonstrated that the production of animal fodder in managed pastures reduces environmental stress in diverse systems, with taxonomic (interspecies) and genetic (intraspecies) diversity playing different and complementary roles. For example, a study of 81 smallholder communities in Nicaragua after Hurricane Mitch found that farms that used agroecological methods suffered less soil erosion compared to conventional farms (Holt-Giménez 2002; IPES-Food 2016).

Reconciliation between humanity and nature requires collective action along the entire agri-food chain. Thus, implementing any change towards a more resilient future for food systems requires that key stakeholders, including industry, policymakers, governments, and consumers, all take an active role.

3 Food Sovereignty Is Needed for Biodiversity Preservation and Sustainable Food Systems

According to the *Declaration of Nyéléni* (2007), "food sovereignty is the right of peoples to healthy and culturally appropriate food produced through ecologically sound and sustainable methods, and their right to define their own food and agriculture systems." Food sovereignty is thus the right of peoples to make decisions, democratically, about the management of food resources and policies at global, regional, and local levels (Wittman 2011; Weiler et al. 2015).

Food sovereignty poses challenges in the public and political spheres, as it confronts the concentration of power and the consequent social inequalities in food systems (Plahe et al. 2013; Jacques 2015). Some of the principal demands of the food sovereignty movement include agrarian reform in benefit of landless producers (Rosset 2011), the fight against the control of transgenic seeds by transnational agribusiness companies (Kerr 2013), and the demarcation of indigenous lands and respect for their biocultural heritage (Rocha and Liberato 2013; Queiroz 2015).

Discussions on food sovereignty, biodiversity, and social justice are interconnected. There is evidence that biodiversity loss is associated with a country's economic growth and social inequality (Naidoo and Adamowicz 2001; Mikkelson et al. 2007). Holland et al. (2009), analyzing socioeconomic models and the proportion of threatened species of plants and vertebrates in 50 countries, concluded that inequality is an essential factor in predicting the loss of biodiversity. Thus, the proposal for reform in the use and management of natural resources to meet human and environmental well-being requires an integrative approach.

Agricultural diversity is necessary for sustainable development, FNS, and biodiversity conservation (Zimmerer and De Haan 2017). In the context of food sovereignty, agricultural diversity is the starting point for the construction of food policies focusing on the autonomy of peoples, the resilience of productive systems, the use and conservation of plant and animal genetic resources, and the recognition of the cultural identity and the affirmation of the rights of traditional peoples (see Altieri and Toledo 2011; Lenné and Wood 2011).

In the food sovereignty debate, biodiversity conservation in agricultural systems responds to a cultural need. Native plants have cultural and genetic roles, and preserving them means safeguarding biocultural diversity (Jacob et al. 2020b; UNESCO 2003). The preservation of traditional knowledge associated with plants is crucial to food sovereignty since people select plants in nature, rationally, with precise purposes (Moerman 1979). Rangel-Landa et al. (2017), studying socio-ecological factors that influence the decision to domesticate native species, concluded that uncertainty in resource availability is a major factor motivating the management of edible plants. In the case of medicinal and ceremonial plants, reciprocal interchanges, curiosity, and spiritual values are essential factors. So, the knowledge embedded into the decision to domesticate plants is relevant to safeguard the biodiversity heritage of a people.

Agricultural diversity and biodiversity conservation are also important in the context of agroforestry. FAO defines agroforestry as land-use systems and technologies where woody perennials are cultivated on the same land management units as agricultural crops and animals, in some particular spatial arrangement or temporal sequence (FAO 2013). Ethnoagroforestry analyzes traditional forms of agroforestry management, considering cultural, economic, and environmental factors from local communities (Moreno-Calles et al. 2016). Agroforestry practices connect synergically with the use and conservation of biodiversity, by integrating nature and culture, wild and domestic diversity, and, finally, different scales and forms of land management, providing the basis for food sovereignty and sustainability management of ecosystems (Moreno-Calles et al. 2016). For example, in Mexico, 80% of the forests belong to 30,000 traditional communities (INEGI 2008). Part of the agroforestry systems occurs mainly in indigenous areas. And most of the country's environmental movements are based on the distribution of agroforestry management zones, indicating the active participation of indigenous peoples as forest guardians and promoters of food sovereignty (Toledo et al. 2015). Studies in ethnoagroforestry have shown a relationship between traditional agroforestry and soil fertility (Garcíalicona et al. 2017), biological conservation (Franco-Gaona et al. 2016; Hill et al. 2019), and high diversity (Hoogesteger van Dijk et al. 2017).

There is no food sovereignty without biodiversity. And the preservation of biodiversity depends crucially on local people's rights to manage their natural resources. Therefore, some relevant pillars to food sovereignty policies are (1) genetic resources, ecology, and evolution; (2) governance policy, institutions and legal agreements; (3) food, nutrition, health, and disease; and (4) factors of global change with socio-ecological interactions (Zimmerer et al. 2019).

4 Final Considerations

Our review demonstrates the breadth of ways in which biodiversity supports the transformation of sustainable food systems, with positive outcomes for human and environmental health. We argued that biodiversity contributes to sustainable food systems and human health by supporting food and nutrition security, strengthening resilience to climate change, fostering sustainable diets, and boosting resilience to disease outbreaks. Unfortunately, we face the rapid decline of biodiversity globally, threatening more species with global extinction now than ever before. We will not achieve the goals of conserving biodiversity until 2030 without transformative changes across economic, social, and cultural factors that guide human decisions. These transformative changes demand new forms of food systems governance based on food sovereignty. This complexity of factors must guide our analysis and actions as scientists, politicians, professionals, and citizens towards sustainable food systems.

References

Alirol E, Getaz L, Chappuis F et al (2011) Urbanisation and infectious diseases in a globalised world. Lancet Infect Dis 11(2):131–141
Altieri MA (2004) Linking ecologists and traditional farmers in the search for sustainable agriculture. Front Ecol Environ 2:35–42
Altieri MA, Toledo V (2011) The agroecological revolution in Latin America: rescuing nature, ensuring food sovereignty and empowering peasants. J Peasant Stud 38(3):587–612
Altieri MA, Nicholls CI (2020) Agroecology and the emergence of a post COVID-19 agriculture. Agric Human Values 37:525–526
Béné C, Oosterveer P, Lamotte L et al (2019) When food systems meet sustainability – current narratives and implications for actions. World Dev 113:116–130
Blicharska M, Smithers RJ, Mikusiński G et al (2019) Biodiversity's contributions to sustainable development. Nat Sustain 2:1083–1093
Brasil (2014) Guia Alimentar para a População Brasileira. Brasília: Ministry of Health. http://bvsms.saude.gov.br/bvs/publicacoes/guia_alimentar_populacao_brasileira_2ed.pdf. Acessed 9 Aug 2020
Brasil (2015) Alimentos regionais brasileiros. Brasília: Ministry of Health. http://189.28.128.100/dab/docs/portaldab/publicacoes/livro_alimentos_regionais_brasileiros.pdf. Acessed 9 Aug 2020
Chappell MJ, LaValle LA (2011) Food security and biodiversity: can we have both? An agroecological analysis. Agric Human Values 28:3–26
Cramer W, Egea E, Fischer J et al (2017) Biodiversity and food security: from trade-offs to synergies. Reg Environ Chang 17:1257–1259
Declaration of Nyéléni (2007). Forum Food Sovereignty 23rd. https://nyeleni.org/IMG/pdf/DeclNyeleni-es.pdf. Acessed 9 Aug 2020
De Clerck F (2016) Biodiversity central to food security. Nature 531:305
FAO (2001) The State of Food Insecurity in the World 2001. Rome: FAO. http://www.fao.org/3/a--y1500e.pdf. Acessed 9 Aug 2020
FAO (2013) Advancing Agroforestry on the Policy Agenda: A guide for decision-makers. Agroforestry Working Paper no. 1. Rome: FAO. http://www.fao.org/3/a-i3182e.pdf. Acessed 8 Aug 2020
FAO (2017) The future of food and agriculture: trends and challenges. Rome: FAO. http://www.fao.org/3/a-i6583e.pdf. Acessed 8 Aug 2020
FAO (2020) The State of Food Security and Nutrition in the World 2020. Transforming food systems for affordable healthy diets. Rome: FAO. doi.org/https://doi.org/10.4060/ca9692en. Acessed 8 Aug 2020
Franco-Gaona A, Ramírez-Valverde B, Cruz-León A et al (2016) El ekuaro: un sistema agroforestal tradicional michoacano* The ekuaro: a traditional agroforestry system michoacano. Rev Mex Ciencias Agrícolas 16:3357–3370
Frison EA, Cherfas J, Hodgkin T (2011) Agricultural biodiversity is essential for a sustainable improvement in food and nutrition security. Sustainability 3:238–253
García-licona JB, Maldonado-torres R, Moreno-calles AI et al (2017) Ethnoagroforestry management and soil fertility in the semiarid Tehuacán Valley, México. Ethnobiol Conserv 6:1–16
Hanspach J, Abson DJ, French Collier N et al (2017) From trade-offs to synergies in food security and biodiversity conservation. Front Ecol Environ 15:489–494

Hill R, Nates-Parra G, Quezada-Euán JJG et al (2019) Biocultural approaches to pollinator conservation. Nat Sustain 2:214–222
HLPE (2020) Food Security and Nutrition: Building a Global Narrative towards 2030. A report by the High Level Panel of Experts on Food Security and Nutrition of the Committee on World Food Security, Rome. http://www.fao.org/3/ca9731en/ca9731en.pdf. Acessed 8 Aug 2020
Holland TG, Peterson GD, Gonzalez A (2009) A cross-national analysis of how economic inequality predicts biodiversity loss. Conserv Biol 23:1304–1313
Holt-Giménez E (2002) Measuring farmers' agroecological resistance after Hurricane Mitch in Nicaragua: a case study in participatory, sustainable land management impact monitoring. Agric Ecosyst Environ 93:87–105
Hoogesteger van Dijk VM, Casas A, Moreno-Calles AI (2017) Semiarid ethnoagroforestry management: Tajos in the Sierra Gorda, Guanajuato, Mexico. J Ethnobiol Ethnomed 13(1):34
Hunter D, Fanzo J (2013) Introduction: agricultural biodiversity, diverse diets and improving nutrition. In: Fanzo J et al (eds) Diversifying Food and Diets, 1rd edn. HWA, London, pp 33–46
INEGI (2008) Censo Ejidal 2007. Inst. Nac. Estadística, Geogr. e Informática. https://www.inegi.org.mx/programas/cae/2007/. Acessed 9 Aug 2020
Ingram J (2020) Nutrition security is more than food security. Nat Food 1:2
IPES-Food (2015) The new science of sustainable food systems: Overcoming barriers to food systems reform. First Report of the International Panel of Experts on Sustainable Food Systems. http://www.ipes-food.org/_img/upload/files/NewScienceofSusFood.pdf. Acessed 9 Aug 2020
IPES-Food (2016) From uniformity to diversity: a paradigm shift from industrial agriculture to diversified agroecological systems. International Panel of Experts on Sustainable Food Systems. http://www.ipes-food.org/_img/upload/files/UniformityToDiversity_FULL.pdf. Acessed 9 Aug 2020
IPES-Food (2017) Unravelling the Food–Health Nexus: Addressing practices, political economy, and power relations to build healthier food systems. The Global Alliance for the Future of Food and IPES-Food. http://www.ipes-food.org/_img/upload/files/Health_FullReport(1).pdf. Acessed 9 Aug 2020
IPES-Food (2018) Breaking away from industrial food and farming systems: seven case studies of agroecological transition. http://www.ipes-food.org/_img/upload/files/CS2_web.pdf. Acessed 9 Aug 2020
IPES-Food (2020) COVID-19 and the crisis in food systems: Symptoms, causes, and potential solutions, Communiqué by IPES-Food. http://www.ipes-food.org/_img/upload/files/COVID-19_CommuniqueEN%283%29.pdf. Acessed 9 Aug 2020
Jacob MCM, Albuquerque UP (2020) Biodiverse food plants: which gaps do we need to address to promote sustainable diets? Ethnobiol Conserv 9:1–6
Jacob MCM, Feitosa IS, Albuquerque UP (2020a) Animal-based food systems are unsafe: SARS-CoV-2 fosters the debate on meat consumption. Public Health Nutr:1–16
Jacob MCM, Araújo de Medeiros MF, Albuquerque UP (2020b) Biodiverse food plants in the semiarid region of Brazil have unknown potential: a systematic review. PLoS One 15:1–24
Jacques PJ (2015) Civil society, corporate power, and food security: counter-revolutionary efforts that limit social change. J Environ Stud Sci 5:432–444
Jacques PJ, Jacques JR (2012) Monocropping cultures into ruin: the loss of food varieties and cultural diversity. Sustainability 4:2970–2997
Jardim Botânico do Rio de Janeiro (2020) Catálogos e listas. http://jbrj.gov.br/node/1072. Accessed 11 Aug 2020
Johnson CK, Hitchens PL, Pandit PS et al (2020) Global shifts in mammalian population trends reveal key predictors of virus spillover risk. Proc R Soc B Biol Sci 287(1924):20192736
Kahiluoto H (2020) Food systems for resilient futures. Opinion Piece. In: Food Secur, Springer. https://doi.org/10.1007/s12571-020-01070-7
Kearns CA (2010) Conservation of biodiversity. Nat Educ Knowl 3:7
Kerr RB (2013) Seed struggles and food sovereignty in northern Malawi. J Peasant Stud 40:867–897

Khoury CK, Bjorkman AD, Dempewolf H et al (2014) Increasing homogeneity in global food supplies and the implications for food security. Proc Natl Acad Sci U S A 111:4001–4006
Lachat C, Raneri JE, Smith KW et al (2018) Dietary species richness as a measure of food biodiversity and nutritional quality of diets. Proc Natl Acad Sci 115:127–132
Lawrence MA, Friel S, Wingrove K et al (2015) Formulating policy activities to promote healthy and sustainable diets. Public Health Nutr 18(13):2333–2340
Lenné JM, Wood D (2011) Agrobiodiversity Management for Food Security: a critical review. CAB Intern, Wallingford
Massawe F, Mayes S, Cheng A (2016) Crop diversity: an unexploited treasure Trove for food security. Trends Plant Sci 21:365–368
Mijatović D, Van Oudenhoven F, Eyzaguirre P et al (2013) The role of agricultural biodiversity in strengthening resilience to climate change: towards an analytical framework. Int J Agric Sustain 11:95–107
Mikkelson GM, Gonzalez A, Peterson GD (2007) Economic inequality predicts biodiversity loss. PLoS One 2(5):1–5
Moerman DE (1979) Symbols and selectivity: a statistical analysis of native american medical ethnobotany. J Ethnopharmacol 1:111–119
Moreno-Calles AI, Casas A, Rivero-Romero AD et al (2016) Ethnoagroforestry: integration of biocultural diversity for food sovereignty in Mexico. J Ethnobiol Ethnomed 12:1–16
Naidoo R, Adamowicz WL (2001) Effects of economic prosperity on numbers of threatened species. Conserv Biol 15(4):1021–1029
Nicholls CI, Altieri MA (2018) Pathways for the amplification of agroecology. Agroecol Sustain Food Syst 42:1170–1193
Olival KJ, Hosseini PR, Zambrana-Torrelio C et al (2017) Host and viral traits predict zoonotic spillover from mammals. Nature 546:646–650
Patz JA, Campbell-Lendrum D, Holloway T et al (2005) Impact of regional climate change on human health. Nature 438:310–317
Penafiel D, Lachat C, Espinel R et al (2011) A systematic review on the contributions of edible plant and animal biodiversity to human diets. EcoHealth 8:381–399
Plahe JK, Hawkes S, Ponnamperuma S (2013) The corporate food regime and food sovereignty in the Pacific Islands. Contemp Pac 25:309–338
Poore J, Nemecek T (2018) Reducing food's environmental impacts through producers and consumers. Science 360:987–992
Powell B, Maundu P, Kuhnlein HV et al (2013) Wild foods from farm and forest in the East Usambara Mountains, Tanzania. Ecol Food Nutr 52:451–478
Powell B, Thilsted SH, Ickowitz A et al (2015) Improving diets with wild and cultivated biodiversity from across the landscape. Food Secur 7:535–554
Prieto I, Violle C, Barre P et al (2015) Complementary effects of species and genetic diversity on productivity and stability of sown grasslands. Nat Plants 1:1–5
Queiroz RC (2015) Vigilância e proteção de terras indígenas: Programa de Capacitação em Proteção Territorial. Brasília: FUNAI. http://www.funai.gov.br/arquivos/conteudo/cgmt/pdf/Vigilancia_e_Protecao_de_TIs.pdf. Acessed 9 Aug 2020
Rajão R, Soares-Filho B, Nunes F et al (2020) The rotten apples of Brazil's agribusiness. Science 369:246–248
Raneri JE, Kennedy G, Nguyen T et al (2019) Determining key research areas for healthier diets and sustainable food systems in Viet Nam. The International Food Policy Research Institute, Washington
Rangel-Landa S, Casas A, García-Frapolli E et al (2017) Sociocultural and ecological factors influencing management of edible and non-edible plants: the case of Ixcatlán, Mexico. J Ethnobiol Ethnomed 13:1–49
Ripple WJ, Smith P, Haberl H et al (2014) Ruminants, climate change and climate policy. Nat Clim Chang 4:2–5

Rocha C, Liberato RS (2013) Food sovereignty for cultural food security: the case of an indigenous community in Brazil. Food Cult Soc 16:589–602

Rosset P (2011) Food sovereignty and alternative paradigms to confront land grabbing and the food and climate crises. Development 54:21–30

Sonnino R, Faus MA, Maggio A (2014) Sustainable food security: an emerging research and policy Agenda International Journal of Sociology of Agriculture and Food. Int J Sociol Agric Food 21:173–188

Sunderland T (2011) Food security: why is biodiversity important? Int For Rev 13:265–274

Swinburn BA, Kraak VI, Allender S et al (2019) The global syndemic of obesity, undernutrition, and climate change: the Lancet Commission report. Lancet 393:791–846

Toledo VM, Garrido D, Narciso B-B (2015) The struggle for life: socio-environmental conflicts in Mexico. Lat Am Perspect 42:133–147

UNESCO (2003) Convenção para a salvaguarda do patrimônio cultural imaterial, Paris. http://portal.iphan.gov.br/uploads/ckfinder/arquivos/ConvencaoSalvaguarda.pdf. Acessed 9 Aug 2020

United Nations (2015) Transforming our world: the 2030 Agenda for Sustainable Development. New York. https://sustainabledevelopment.un.org/post2015/transformingourworld/publication. Acessed 9 Aug 2020

United Nations (2016) The Convention on Biological Diversity. https://www.cbd.int/convention/. Acessed 9 Aug 2020

United Nations Environment Programme (2020) Preventing the Next Pandemic: Zoonotic diseases and how to break the chain of transmission. United Nations Environment Programme and International Livestock Research Institute, Nairobi. https://reliefweb.int/sites/reliefweb.int/files/resources/ZP.pdf. Acessed 9 Aug 2020

Volpato G, Fontefrancesco MF, Gruppuso P et al (2020) Baby pangolins on my plate: possible lessons to learn from the COVID-19 pandemic. J Ethnobiol Ethnomed 16(19):1–12

Weiler AM, Hergesheimer C, Brisbois B et al (2015) Food sovereignty, food security and health equity: a meta-narrative mapping exercise. Health Policy Plan 30:1078–1092

Wilcox BA, Ellis B (2006) Forests and emerging infectious diseases of humans. Unasylva 57:11–18

Willett W, Rockström J, Loken B et al (2019) Food in the Anthropocene: the EAT–Lancet Commission on healthy diets from sustainable food systems. Lancet 393:447–492

Wittman H (2011) Food sovereignty: a new rights framework for food and nature? Environ Soc Adv Res 2:87–105

Wolf C, Ripple WJ, Betts MG et al (2019) Eating plants and planting forests for the climate. Glob Chang Biol 25:3995

Zimmerer KS, De Haan S (2017) Agrobiodiversity and a sustainable food future. Nat Plants 3(4):17047

Zimmerer KS, de Haan S, Jones AD et al (2019) The biodiversity of food and agriculture (agrobiodiversity) in the anthropocene: research advances and conceptual framework. Anthropocene 25:2–16

Biodiversity, Ethno-diversity and Food Cultures: Towards More Sustainable Food Systems and Diet

Jean Pierre Poulain

Contents

During the 1990s and the early 2000s, the issue of world hunger faded into the background. Not that the problem had disappeared, but other matters (such as the global obesity epidemic or the political and health issues surrounding the use of GMOs) took centre stage in the media. In June 1998, the sharp rise in the prices of food raw materials, particularly cereals, triggered the so-called hunger riots in many countries. This gave new impetus to an old question: How can we feed a growing human population? For several decades, the stakes seemed to be more on the side of quality than quantity, at least in the Western world. So much so that agricultural Europe set up production quotas for certain products and reintroduced the practice of fallow land.

A growing environmental awareness arose during the same period. It led to biodiversity being seen as a common heritage of humanity, worthy of protection, and subsequently to reconsider the consequences of decisions made at local levels in a self-regulated world. With globalisation, the idea that certain human cultures were disappearing gave way to the concept of ethno-diversity, echoing that of biodiversity, both of which were regarded as a heritage for all humanity. Connected to the question of sustainability, the two concepts allowed for the reorganisation of the

J. P. Poulain (✉)
Université de Toulouse Jean Jaurès, Toulouse, France

Taylor's University, Kuala Lumpur, Malaysia
e-mail: poulain@univ-tlse2.fr

M. C. M. Jacob, U. P. Albuquerque (eds.), *Local Food Plants of Brazil*, Ethnobiology, https://doi.org/10.1007/978-3-030-69139-4_2

Malthusian model and the transformation of the status of diversity of food cultures into resources, to face the challenge of hunger.

1 From Politicization of Hunger to Sustainability

The adequacy of food production to the needs of the world's population is seemingly an issue that is easy to model. Moreover, historically it is the epistemological starting point for modern social sciences: economics, demography and to some extent sociology and anthropology. In 1798, in his "An Essay on the Principle of Population", Thomas Malthus theorized that the world's population grows in geometric progression, while the capacity to produce food evolves according to an arithmetic progression, thus drawing a catastrophic future. He predicted that if care were not taken to reduce population growth, the earth would become the scene of deadly struggles because of hunger. As agronomic and demographic knowledge evolved, the theme of world hunger became clearer. In 1938, Anton Zischka (1938) wrote a book entitled *Brot für 2 Milliarden Menschen* (Bread for Two Billion Men). The contemporary reader will note both the demographic scale of two billion as a warning horizon and the ethnocentrism that posits bread as universal food. A few years later, Maximilien Sorre (1943) developed a theoretical reflection proposing a way of overcoming the two competing paradigms that was exploring the relationship between man and nature in geography: the determinism of "anthropo-geography" that gave priority to the environment, and the "possibilism" of Paul Vidal de La Blache, for whom human action is almost limitless. Sorre promoted a vision focusing on the interactions between man and his environment, which would later give birth to modern ecological anthropology (Steward 1955). In his "Geopolitics of Hunger", Josué de Castro (1952) took the question of hunger out of the charity order to place it in the political order. According to him, the instruments used to fight against hunger were to be found in social and political organization.

The second half of the twentieth century was marked by ecological awareness. In 1974, René Dumont, a leading figure of Third-Worldism, ran for the French presidential election. During the television campaign, he appeared on screen with a glass of water in his hand and made a gloomy prognosis: "If we do nothing, we will not be able to drink any more water in twenty years' time". The French magazine *Le Nouvel Observateur* launched an ecological supplement called *Le Sauvage,* with "Vegetable activists" on the front page of its first issue, and an article announcing without detour that "humanity must prepare to become vegetarian". This was followed by a demonstration highlighting (albeit a little quickly) "that producing 1 kg of animal protein requires as many resources as producing 7 kg of vegetable protein". As the population was growing faster than production capacity, and the cultivated areas would soon reach its limits, there would be no choice but to change eating habits, including reducing, if not abandoning, animal products, in order to cope with the coming crisis.

The 1980s marked a turning point in this debate. In France and other developed countries, the poor were hungry. In 1956, the founder of the Emmaüs movement, Abbé Pierre had taken a strong stance on the poverty issue. However, it was not the lack of food that he highlighted, but the lack of a roof and protection against the cold. But this time the scandal arose from the contrast between the overabundance of agricultural surpluses, the mountains of tomatoes or cauliflowers spilled by angry peasants in front of the gates of sub-prefectures and the carcasses of cattle and the tons of butter withdrawn from the market to support the prices and piled up in the cold rooms of the European community. By founding the "Restos du cœur" in 1985, the humorist Coluche, supported by numerous stars of French show business, initiated both the mobilization of "well-fed" France and the change of the status of food. The right to food began to become a fundamental right. The movement spread throughout Europe, and soon charities working in this sector were recognized as being of public utility. The development of Europe agricultural further encouraged mobilization efforts.

At international level, the 1992 Earth Summit in Rio de Janeiro introduced the notion of ethno-diversity in Article 8 of the "Convention on Biological Diversity" signed on 5 June. Conceived to mirror the concept of biodiversity, it deals with the "conduct of societies" and calls for countries to respect, preserve and maintain the uses and knowledge of indigenous and local communities relevant for the protection of biodiversity. With the signing of the Kyoto Protocol in 1997, followed by the Johannesburg (2002) and Nagoya (2010) Summits – all places where alerts were issued and commitments made – the issue of sustainable development took shape and articulated the economic, environmental and social challenges. However, behind the apparent consensus, there is a misunderstanding: in the expression "sustainable development", the "rich" emphasizes sustainability and the "poor" emphasizes "development" (Brunel 2008).

2 Population Dynamics and Politics

Population dynamics have been the subject of considerable scientific progress, particularly with the "demographic transition" model (Notestein 1944, 1948; McKeown 1976). It attempts to account for the impact of development factors on the demographic structure. The development is accompanied by a population growth that can sometimes be a factor of dynamism since the population is getting younger, with a favourable ratio between active and retired people. But these additional individuals also eat... This explains the frequent "stop and go" phenomena experienced by developing economies.

The definition of nutritional needs is too often based on the Western food model and should be diversified. Périssé (1996) defines six major families of food models according to the staple food (rice, wheat, corn, cassava, yam, etc.). In the field, the anthropology of food provides us with a considerably larger number of consumption models. Changes are also difficult to predict beyond certain summary rules, such as the share of animal products increasing with the increase in purchasing power. Finally, the data available in many countries are to be taken with caution as it is very complicated to ascertain the informal economy. Moreover, in some countries, certain organizations may have an interest in aggravating situations in order to justify their existence or their requests for aid.

The modelling of food production might seem more manageable, but again, the more knowledge progresses, the more complex the issue becomes. The cultivated areas, water resources, sunshine and, more broadly, the climate, the type of seed, cultivation methods, soil amendments and methods of pest control, harvesting and storage are all interconnected variables, which determine productivity and which in turn have an effect on climate, water availability and so on. The models on which the predictions are based were, and will probably still be for a while, a little too simplistic. The gloomy predictions made in the past, be it by Malthus, the Club of Rome in the late 1960s or the ecologists of the 1980s have all been challenged by the facts. Prediction is a difficult art in general, and agronomy and demography are no exception. The present food production is much more than is needed to feed all the people on the planet. Unfortunately, however, there is still famine in one half of the world, while the other half is increasingly drowning in overabundance. The most optimistic will point out that the number of people suffering from hunger has not changed over the last three decades of the twentieth century and has even fallen slightly, while the world population has gone from 4.5 billion in 1975 to 6.1 billion in 2000. But others consider the current situation has worsened this last decade and is even more unbearable as we have the means to feed humanity. Jean Ziegler (2006), United Nations Special Rapporteur on the Right to Food, considers that agriculture today can feed 12 billion human beings (i.e. almost twice the world's population). "The 100,000 people who die of hunger every day are therefore murdered," he writes. With these words, he calls for collective responsibility. Therefore, the current situation is not caused by the incapacity to produce food to feed the planet, but by other political and economic reasons. The problem for the moment, and probably for some time to come, is not a problem of production, but a question of distribution, an economic issue, a political problem (Poulain 2018).

The situation worsened in 2008 as a result of several events: a year 2007 of very poor harvests in different parts of the world, the development of biofuels (especially in the USA from corn and in Brazil from sugar cane) and, finally, speculative movements which are the main cause of the worsening pressure on prices (Ziegler 2006).

3 The Importance of Modelling

Forecast errors should not prevent the interest in predictive models. Moreover, there is the possibility that if gloomy predictions on food and nutrition security have not come true, it could partly be due to the awareness they raised and which contributed to the reorganization of production and distribution systems and the implementation of new agricultural policies. The models are "simple", which is all the more reason to try to improve them. To do so, let us look at the assumptions on which they are based.

Modelling work is located between two epistemological perspectives, between two more or less competing theoretical frameworks: an economic-ecological framework and a socio-anthropological framework (see Fig. 1). The first perspective focuses on rational intergenerational responsibility, i.e. the environmental responsibility of our generation towards future generations. It is embodied in questions such as: "What kind of land will we leave to our children?" or the following one "Can we live on credit on the backs of future generations?". The second focuses on intragenerational responsibility, i.e. the ethical "scandal" of the discrepancies between the situations of individuals living in developed societies and those living in underdeveloped countries. It emphasizes the North-South relations and promotes fair trade and the respect of biodiversity and ethno-diversity (El Bilali 2019).

These two perspectives do not use the same scientific resources. The first takes econometrics, systemic ecology, etc. as a model, adopting an epistemological, objectivist and idealistic stance (seeking the major laws that structure phenomena). The second is enshrined in the logic of socio-anthropology and development disciplines and is subjectivist and empirical (concerned with the field and the vision of the actors). These two frameworks are in conflict with each other: one gives primacy to the ecosystem, the other to humans. The latter criticizes the former for building "off-ground" models unrelated to field data, and the former criticizes the low degree

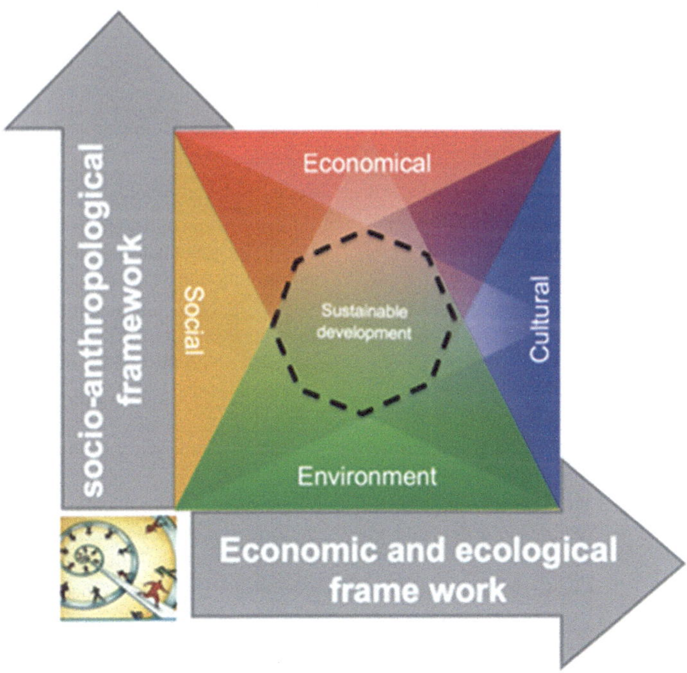

Fig. 1 Introducing cultural heritage into sustainability. Source: Poulain (2018a)

of generalization of the work of the latter. Table 1 summarizes the main tensions between these two perspectives on sustainability.

Predictive models are useful to professionals and politicians alike because they provide a global vision to help guide action. However, their rusticity can be avoided and the processing power available to contemporary research may create the possibility to consider the increasing complexity of these models. The example of the challenges of feeding humanity shows the interest of distinguishing, in the classical model of sustainable development, the social and cultural dimensions. That is to say, issues related to inequalities in the hierarchical scale and those related to ethno-cultural diversities. In the case of food, the firsts correspond to issues of accessibility, while the seconds take into account food cultures, types of food and the ways of preparing them.

Even if the current models are reductionist and unreliable, they still had the merit of sounding the alarm. It is possible to strengthen their predictive capacity by combining demography and nutritional anthropology. This involves replacing the "population" variable in Malthus' model (which more or less assumed that all men eat the same thing) with "food need". This is the result of the population and its transitional dynamics and the food consumption patterns they use. The enrichment of these slightly "off-ground" models therefore involves the introduction of demographic data to take account of population dynamics (stages of the demographic transition process) and anthropo-nutritional data that take account of the diversity of food models used by populations.

In the theory of "convergent modernity", dominant in nutritional epidemiology as well as in business circles, the evolution of food consumption is supposed to be

Table 1 Paradigmatic tension of sustainability considering the economic-ecological and the socio-anthropological perspectives

	Economic and Ecological	Social and Anthropological
Key concepts	Priority to environment Ecological footprint, (km/food), CO_2	Priority to human beings Fair-trade, sustainability, slow food, community-supported agriculture
Orientation and evaluation criteria	Universality of global ecological stakes Intergenerational equity What kind of planet will we leave to our children?	Specificity of social and cultural situations Intragenerational equity Solidarity with the victims of hunger and health scandals
System evaluation modality	Mathematical modelling	Case study, experience feedback
Market relation	The market is set, it is required to understand how it works	The market is a social and political construction. Action is necessary to orient it towards a beneficial situation
Position of consumers	Consumers make choices	Consumers are actors of the systems
Scientific disciplines	Econometrics, ecology, agronomy	Anthropology, sociology, sciences of development

Source: Stassart and Collet (2001)

in line with the Western model (Mahbubanim 2013). That is to say, make the distinction between the differences of practicies linked to the social positions and those related to ethno-cultural belongings. If at the macro scale and in a very simplified conception of the protein transition, using only the opposition animal proteins versus vegetable proteins, this model has some consistency, as soon as the origin of protein sources is introduced in a more diversified way, it loses much of its relevance. Work on protein transition shows that each country follows a singular transitional path. It is because foods that are sources of protein are the subject of many taboos, religions or social and cultural prescriptions (including the issue of animal death for food) that the forms of nutritional and protein transitions are widely determined by cultures (Poulain 2007a, b, Fourat and Lepiller 2017, Drewnowski and Poulain 2019). We can see it, in a multicultural society like Malaysia where the three main communities (Malays, Chinese and Indians) have different forms of proteins transition (Drewnowski et al. 2020).

At the theoretical level, the challenge is to develop sufficient knowledge on food models to identify the room for manoeuver they allow in the variability of needs. At the strategic level, ethno-culinary diversity (culinary systems) and ethno-food diversity must be considered as a resource and not as an obstacle. Taking it into account avoids putting all the eggs in the same basket.

As an extension of these reflections, it is possible to reorganize Malthus' model to better reflect the functioning of the global food system (Poulain 2007a, 2018a, b). It is characterized by food production capacity and food needs to be met. The production capacity is determined by a number of factors: the surface cultivated, the seeds used and the cultivation methods implemented, the inputs and preservatives used for raw products, the methods for preserving and transporting processed products and, lastly, the climate and its evolution. When talking about production, what Malthus called the "population" must be replaced by the "food need" to be satisfied. This need equals the number of mouths to feed (Malthus' population) and their numerical evolution. However not all people eat the same way. While the definition of an adequate diet can be defined in nutritional terms, there is also the matter of culture, tradition, religion, etc. to consider. It is therefore necessary to take into account the food consumption patterns used by various populations and any changes in these patterns (see Fig. 2).

Thus, two complementary perspectives for action emerge. The first calls for action on production capacity. This has been the lever used so far by agricultural and food policies. Great progress has been made, but not without some controversy over its health, environmental and social consequences (Ziegler 2006). There are several conflicting conceptions that can be placed on a continuum ranging on the one hand from the increasingly important and precise use of biotechnologies to act upon the natural processes underlying agriculture and on the other hand to agro-ecology, which is increasingly respectful of these same natural processes. Between these two poles, there is a wide range of positions. Whichever way you look at it, biodiversity is a resource across different philosophies.

On the food quantity side, there is another very important lever: that of reducing losses and wastage, which is estimated at nearly 30% of the quantity produced (FAO

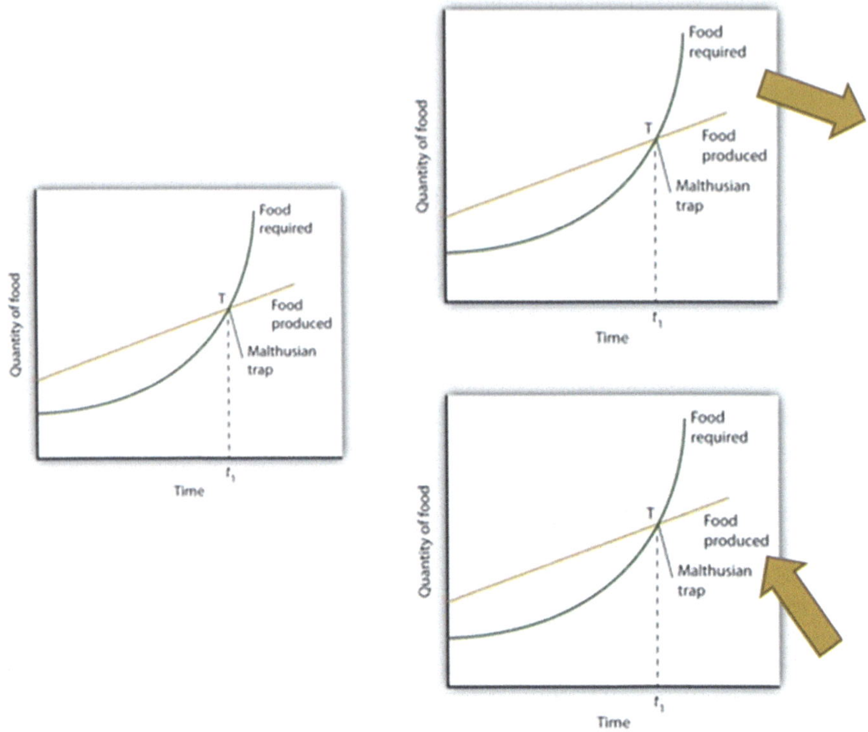

Fig. 2 Back to Malthus' theory: Levers of action. Food required strategy and food produced strategy. Available at: https://bit.ly/3lg8ku5

2013). Actions must be taken at the level of production, processing, and consumption. Such a project could lead to reconsidering animal nutrition models. But it is also possible to modify the expression of the need for food (and at the same time the resources needed to produce the food likely to satisfy this need) by playing on the nature of the food consumed, for example, animal or plant protein sources. Within these two broad categories, the sub-categories do not engage the same amounts of resources. It is therefore possible to act on the foods consumed to reduce the pressure on the ecosystem. However, the way in which animals are fed can considerably vary the environmental impact (Poore and Nemecek 2018) depending on whether they consume products that potentially put them in competition with humans or whether they process non-consumable products into protein. Moreover, this issue needs to be ecologically and culturally contextualized.

Consumption patterns and their transition dynamics vary significantly on a global scale (Drewnowski and Poulain 2019). They result from the use that human communities have been able to make of the resources made available to them in the ecological niche, the resources they have been able to implant and the systems of representations and values that have enabled them to use them in a meaningful way. Food biodiversity and ethno-diversity must therefore be taken into account in order to build a more sustainable food system. It is urgent to start their systematic study (Dernini et al. 2013).

After showing how the interest in biodiversity had made it possible to recognize that of ethno-diversity, we proposed to reformulate the Malthus model. If this has made it possible to launch alerts, does it have poor predictive capacities? This is because it is poorly connected with the empirical data. The putting in relation of the "population" with the "food production" rests on presuppositions on the way in which the men satisfy their need for food which opens the door to ethnocentrisms (in this case of the Western-centrisms) and conceptions and evolutionists.

To get out of these obstacles, we have proposed to make the distinction, within the classical theory of sustainable development, between the "social" and the "cultural". Then two scientific perspectives appear, the economic-ecological approach and the socio-anthropological approach, which make visible two dimensions of sustainability in apparent contradiction. The absence of such a distinction in the classical theory of "sustainable development" has the effect of hypertrophying macroeconomic movements and leaving the cultural variability of food needs in the shade. This theoretical clarification then makes it possible to come back to Malthus's model by taking into account the way in which food needs vary in their modalities (types of food) and manage resource needs and different environmental impacts. By taking note that what people eat is largely defined by societies, the concept of food culture, which is heir to the notions of biodiversity and ethno-diversity, establishes a dialogue with the economy and nutritional epidemiology in the aim to better face the hunger and climate challenges looming in the more or less near future.

References

Brunel S (2008) À qui profite le développement durable. Larousse, Paris

De Castro J (1952). The geography of hunger. Monthly Review Press; Revised edition (1977)

Dernini S, Meybeck A, Burlingame B, Gitz V, Lacirignola C, Debs P, Capone R, El Bilali H (2013) Developing a methodological approach for assessing the sustainability of diets: the Mediterranean diet as a case study. New Medit 12(3):28–36

Drewnowski A, Poulain JP (2019) What lies behind the transition from plant-based to animal protein? Am Med Assoc J Ethics 20:E987–E993

Drewnowski A, Mognard E, Gupta S, Ismail MN, Karim N, Tibère L, Laporte C, Alem Y, Khusun H, Februhartanty J, Anggraini R, Poulain JP (2020) Socio-cultural and economic drivers of plant and animal protein consumption in Malaysia: the SCRiPT study. Nutrients

El Bilali H (2019) Research on agro-food sustainability transitions: where are food security and nutrition? Food Security:1–19

FAO (2013) Food wastage: Key facts and figures, http://www.fao.org/news/story/en/item/196402/icode/

Fourat E, Lepiller O (2017) Forms of food transition: sociocultural factors limiting the diets' animalisation in France and India. Sociol Rural 57(1):41–63

Mahbubani K (2013) The great convergence: Asia, the west and the logic of one world. Public Affairs, New York

McKeown J., (1976) The modern Rise of population, Londres. Edward Arnold

Notestein F. W., (1944) "Problems of policy in relation to areas of heavy population pressure", Milbank Memorial Fund Quarterly, 22, n° 4
Notestein F. W., (1948) "Summary of the demographic background of problems of undeveloped areas", Milbank Memorial Fund Quarterly, 26, n° 3.
Périssé J (1996) Typologie des disponibilités alimentaires par pays en 1990 basée sur un critère énergétique. In Desjeux JF, Hercberg S dir. (1996), La nutrition humaine, la recherche au service de la santé INSERM-Nathan, Paris
Poore J, Nemecek T (2018) Reducing food's environmental impacts through producers and consumers. Science 360(6392):987–992
Poulain JP (2007a) dir L'homme, le mangeur et l'animal, Les Cahiers de l'OCHA Paris
Poulain JP (2007b) L'homme, le mangeur, l'animal : la relation homme-animal à travers les modèles alimentaires. Sciences des Aliments, 27/1,: 3–9
Poulain JP (2018) Sociology of food. Bloomsbury, London
Poulain JP (2018a) Dictionnaire des cultures alimentaires. PUF, Paris (2012)
Poulain JP (2018b) Nourrir l'humanité (modélisation). In Poulain J. P. (dir.) Dictionnaire des cultures alimentaires. PUF, Paris, 928–933
Sorre M (1943) Les fondements biologiques de la géographie humaine. Colin, Paris
Stassart P, Collet E (2001) Construction socio-technique de la qualité. In Mormont M et Van Huylenbroeck (dir.) A la recherche de la qualité, analyses socio-économiques sur les nouvelles filières agro-alimentaires. Editions de l'Université de Liège, Liège pp. 51–86
Steward JH (1955) Theory of culture change, the methodology of multilinear evolution. University of Illinois Press, Urbana
Ziegler J (2006) Le droit à l'alimentation. Fayard, Paris
Zischka A (1938) Brot für 2 Milliarden Menschen, Goldmann, Leipzig

Unconventional Food Plants: Food or Medicine?

Washington Soares Ferreira Júnior, Letícia Zenóbia de Oliveira Campos, and Patrícia Muniz de Medeiros

Contents

1 Introduction

Several human groups, especially in rural environments, depend on local plant species for the treatment of diseases and food (FAO 2010). Some of these species are poorly studied (particularly wild species), and, although they are used by some human groups, they are underused in several other regions (FAO 2010). Considering that in 2019 there were around 690 million people in the world under malnutrition (FAO et al. 2020), the study and popularization of these locally important wild species may promote food security for different human populations.

There are wild or cultivated species that have an important nutritional role as well as good potential for the prevention and treatment of diseases, which are undervalued or used only locally by a few human groups (Leal et al. 2018; Peisino et al. 2019). In this sense, unconventional food species can offer alternatives to strategies

W. S. Ferreira Júnior (✉)
Laboratório de Investigações Bioculturais no Semiárido, Universidade de Pernambuco, Petrolina, PE, Brazil
e-mail: washington.ferreira@upe.br

L. Z. de O. Campos
Universidade Federal do Oeste da Bahia – UFOB, Estrada do Barrocão, Barreiras, BA, Brazil

P. M. de Medeiros
Laboratório de Ecologia, Conservação Evolução Biocultural, Campus de Engenharias e Ciências Agrárias, Universidade Federal de Alagoas, Rio Largo, Alagoas, Brazil
e-mail: patricia.medeiros@ceca.ufal.br

M. C. M. Jacob, U. P. Albuquerque (eds.), *Local Food Plants of Brazil*, Ethnobiology, https://doi.org/10.1007/978-3-030-69139-4_3

that address not only food and nutrition security but also the treatment of diseases in several vulnerable human groups.

In this sense, the term "unconventional food plants" has been widely used in Brazilian scientific literature to refer to plants that have food potential, but are unknown or unused by most of the population. This concept, therefore, considers not only wild plants, but low spread cultivated species, in addition to unconventional parts of conventional plants (Kinupp and Lorenzi 2014). The issue has been gaining strong media appeal (Junqueira and Perline 2019), behaving like an umbrella for the popularization of a series of species, especially short-lived ruderal species (Oliveira and Ranieri 2017). However, outside the Brazilian context, this term has not been used, so ethnobiological studies usually work with wild food plants (Pieroni 1999; Lentini and Venza 2007; Sõukand 2016).

A large number of studies carried out in different regions have shown that some wild plants are used by human groups both for food and for the treatment of diseases. For example, when reviewing the use of wild plants by different human groups in northeastern Portugal, Pinela et al. (2017) observed that 33 species (out of a total of 37) had some medicinal application, in addition to the indications for use as food. Furthermore, the research by Purba et al. (2018) recorded a preparation known as "terites" made by the Batak Karo ethnic group in Indonesia. This preparation is based on a juice that is the product of partially digested food from cattle, with the addition of several plants, being used in food for large festivities and indicated for the treatment of diseases. The study by Yang et al. (2020), when investigating the knowledge about food and medicinal plants in four traditional human groups in India, registered 75 useful species, 19 of which indicated both for food and for the treatment of diseases.

Another example shows that the research by Urso et al. (2016), carried out with different human groups located in southeastern Angola, found that approximately 11% of the plants that are known in the studied communities are indicated as medicine and food. For example, the fruit of the species *Aframomum alboviolaceum* (Ridl.) K.Schum. (Zingiberaceae) was indicated for medicinal use as anthelmintic, being also consumed raw as food. According to the authors, two other species also had the same parts of the plants indicated for both uses (medicine and food). These species can be investigated in relation to their potential both nutritionally and for the treatment of diseases (Urso et al. 2016), which is relevant to indicate a list of species that may favor nutritional security and maintain people's health.

These studies indicate that there is an overlap – at least partial – between uses as food or medicine for a set of species in different human groups. This raises the question of the therapeutic potential of neglected wild edible species, which have also been recorded as medicinal in one or more human groups, in addition to the reasons that lead these important species to be neglected. In this sense, this chapter discusses a set of evidences indicating the potential of many of these plants for the prevention and treatment of diseases. In addition, we evaluate here the main factors that may contribute to the popularization of these species, in order to promote the nutritional security and the health of several vulnerable human groups in the future.

2 What Is the Potential of Unconventional Food Plants (UFPs) in the Medicinal Use?

Traditional health strategies, which incorporate plants such as food, medicine, or both, can play an important role in individual well-being. Tows and Andel (2016) argue that many plants are used historically within a food-medicinal *continuum*. However, studies developed on knowledge and use of plants are characterized by a historical segmentation involving research that has investigated these uses separately. There has been an increase in interest in understanding whether there is a separation threshold between these two categories, thus leading to research on how these uses are related, whether the same plant is used for both purposes, regardless of the part of the plant used, whether the same part of the plant is used, concurrently, for both purposes (see Ferreira Júnior et al. 2015).

Therefore, in order to better understand the different ways that the *continuum* can happen, we will present some examples of ethnobotanical studies that addressed this theme in their research, within the classification proposed by Pieroni and Quave (2006). The authors suggested that uses as food and medicine can be related in three different ways (Fig. 1). In the first, it is suggested that certain plants are indicated as food and medicine, but the parts used or forms of preparation are not related. In this sense, we can cite a study carried out in the Brazilian semiarid region, in which the authors sought to understand the overlap of the food category with other categories of use, especially medicine. The authors found that the species *Hancornia speciosa* Gomes (Apocynaceae), popularly known as mangaba, has its fruits used exclusively as food, while the latex derived from the stem is used as medicine in the treatment and prevention of problems related to the gastrointestinal system (Campos et al.

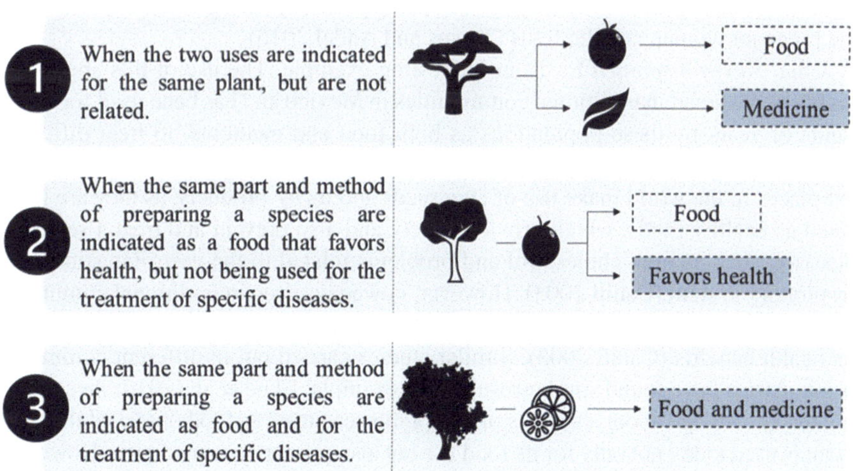

Fig. 1 Different degrees of relationship between the food and medicinal uses of plants, proposed by Pieroni and Quave (2006)

2016). In the study, the participants also reported that green fruits with latex were in other times also used as medicine. However, in the course of time, a similarity between the taste of the latex present in the green fruits and the latex coming from the stem has been recognized, and it started to be used as medicine (Campos et al. 2016). In this case, we can also address other important issues that are mainly related to the role of taste of different parts of the same resource and its availability throughout the year for medicinal use. Thus, considering that the fruit has its presence strongly marked by seasonality, it would not be possible to use them throughout the year for the treatment/prevention of a certain disease, unlike latex from the stem.

Still within the first way in which food and medicinal uses can be related, Rigat et al. (2009) in a study carried out in the Iberian Peninsula, Catalan region, observed that a significant number of plants can be used within a medicinal food *continuum*, although they emphasize that the parts used for each purpose are different. In this study, the authors found that fruits are used more often as food, while flowers are used as medicine. The authors justify their findings by the fact that flowers are richer in products of secondary metabolism, thus having a greater concentration of important components for the treatment of diseases. Additionally, the authors pointed out that many condiment plants are used at the same time in both categories (food and medicine), which is also a trend in other works developed in this perspective. It is suggested that these condiments are classified as functional foods, thus being included in the second classification proposed by Pieroni and Quave (2006) in which they postulate that a plant can be consumed as food, not being indicated for the treatment/prevention of one or more specific diseases, although people recognize its positive impact on health. For example, in West Africa many species are used by women during pregnancy as they are considered to promote strength, ensuring a strong and healthy pregnancy. However, these are not used for a specific system, but guarantee the strengthening action in general, resulting in the well-being of the pregnant woman and the fetus (Towns and Andel 2016).

Chia, *Salvia hispanica* L., is an interesting example. The use of this species is highly widespread in traditional communities in Mexico and has been used for thousands of years by these populations as both food and medicine, to treat different diseases, using both seeds and oil extracted from seeds and leaves. Currently, different places in the world make use of chia seeds and its by-products, as they are indicated as healthy foods, which provide satiety and also prevent and treat a series of diseases, such as high cholesterol and problems related to the gastrointestinal and respiratory system (Cahill 2003). However, nowadays even in traditional communities, satiety is seen as a side effect, and this species is normally ingested because of its health benefits (Cahill 2003). Similar studies carried out in different communities in Africa have found similar results. For example, Ekué et al. (2010) investigating the knowledge about a species of Malpighiaceae used as food observed that this plant is used today not only for its food use but also primarily as medicine. However, the plant is not indicated for a specific medicinal use, as causes a general well-being in the individual.

In this sense, Sõukand (2016) points out that most modern societies no longer need to use wild plants to satisfy the hunger and that, if they continue to be used, it is because they have something beyond the characteristics to supply energy demands and are considered healthy. The following account of the work developed by Sõukand (2016) on the island of Saaremaa, Estonia, northern Europe, can be used as an example in relation to the use of certain species: "I use it because it has vitamin C, it tastes good and you can make tea every day"… "it gives you strength, vitality, but they are not specifically used for healing... it gives you more energy and makes you more intelligent". The discourses brought up in the studies that try to understand how the food and medicinal uses are related express a very interesting discussion about the fact that these uses are so strongly linked that it becomes difficult to draw a line of separation, even if it is not for the treatment/prevention of a specific disease.

Finally, the third classification proposed by Pieroni and Quave (2006) suggests that plants are used as food and also in the prevention and treatment of one or more specific diseases (the same part of the plant used, with the same method of preparation for both purposes). It is very common to find this type of use, especially in indigenous communities. Certain authors have classified these plants in which the same part is used, at the same time, for both purposes, as medicinal foods. They play a very specific medicinal role, such as in the treatment of chronic diseases like diabetes and hypertension. These foods often need to go through processing to remove toxins before being consumed (Jiang and Quave 2013). These medicinal foods can, over time, be classified as preferred species by traditional communities and can even be interpreted as a strategy for optimizing the use of native resources (Johns 1990).

These examples reinforce the ideas proposed by Jiang and Quave (2013). The authors state that diet and health are closely related, especially in traditional communities, where diet is used as a health strategy for the treatment and prevention of a series of diseases. Similarly, Etkin in different studies in African communities found that the use of species that act as food and medicine has been strongly fixed in certain cultures, that there is an immense difficulty in drawing a line of separation between these two categories of use (Etkin 1996; Etkin 2006).

In general, many authors argue that native food species have played an important role in providing nutrients and compounds considered effective for the prevention and treatment of numerous diseases in traditional societies (Johns 1990; Etkin 2006; Leonti 2012). In this regard, Leonti (2012) points out that there are certainly co-evolutionary aspects between plant species and characteristics of human nutrition, and the use of one species in the food-medicine *continuum* has contributed strongly to the development of modern pharmacopoeias. Thus, it can be inferred that selecting a resource that can contribute to both nutrition and disease prevention would have been more advantageous for populations throughout human evolution (Etkin 2006).

2.1 Phytochemical and Pharmacological Evidence of the Medicinal Potential of UFPs

There is a need for interdisciplinary scientific fields, which integrate the pharmacological aspects of biodiversity resources that are used to feed different human groups (Heywood 2011). Chemical compounds ingested in the diet can be important for the prevention or relief of symptoms linked to several diseases, such as diabetes, cardiovascular diseases, cancer, allergy, osteoporosis, and menopause (Heinrich and Prieto 2008). Thus, wild plants that are also indicated for the treatment of diseases and are used in the diet of one or more human groups may indicate both an absence of separation between food and medicinal uses and the need for interdisciplinary studies to assess their impacts on nutritional security and people's health.

Several studies have evaluated the pharmacological activities in vitro and in vivo of edible wild plants, which are indicated as food and medicinal by different human groups. For example, the study by Guarrera and Savo (2013) found 67 wild edible species and 18 cultivated species consumed by informants in Italy and perceived as important for health. When conducting a review of the pharmacological activities of these species in the literature, the authors observed that the species showed the following properties: hypotensive, hypoglycemic, antifungal, inhibition of cancer, anti-inflammatory, hypolipidemic, neuropharmacological (sedative), cancer preventive, antibacterial, immunostimulatory, antimicrobial, and anti-diabetic. These properties have been observed for the treatment of diseases belonging to different body systems, such as digestive system, respiratory system, genital-urinary system, endocrine system, cardiovascular, osteoarticular, and nervous system.

In another example, the species *Leea macrophylla* Roxb. ex Hornem. (Vitaceae), used as food and medicine in several human groups in India, was found to present antioxidant and antimicrobial activity in vitro (Joshi et al. 2016). With these activities, the species can be important for the prevention of diseases linked to oxidative stress and for the treatment of infections (Joshi et al. 2016). Several other studies have observed the potential of edible wild plants for the treatment of diseases (Dewanjee et al. 2013; Zahara et al. 2015; Narzary et al. 2016; Bello et al. 2019) (see Table 1).

Table 1 shows 47 species of wild plants that are indicated by different human groups for both food and medicinal use. These have been investigated in relation to their chemical constituents and pharmacological properties. The table is not exhaustive and considers only some works in the literature that have studied the pharmacological properties of edible wild species in different regions. Based on the studies, we observed a diversity of pharmacological activities presenting antioxidant (44 species), anti-diabetic (9 species), antimicrobial (8 species), and anti-inflammatory (6 species) activities.

Table 1 Phytochemical characterization and pharmacological activities of wild plant species indicated for food and medicinal uses in different human groups from varied regions

Family	Species	Common names in Brazil	Phytochemical characterization	Pharmacological activities	References
Acanthaceae	*Asystasia gangetica* (L.) T. Anderson	Espinafre indiano[a]	Ferulic acid, salicylic acid, myricetin, quercetin, apigenin, kaempferol, catechin	Antioxidant	Datta et al. (2019)
Acanthaceae	*Hygrophila schulli* (Buch.-Ham.) M.R.Almeida & S.M. Almeida	-	Gallic acid, catechin hydrate, vanillic acid, caffeic acid, epicatechin, p-coumaric acid	Antioxidant, anti-diabetic	Alam et al. (2020)
Aizoaceae	*Trianthema portulacastrum* L.	–	Saponins, steroids, alkaloids, flavonoids, terpenes, benzoic acid derivatives, and cinnamic acid derivatives	Antimicrobial activity, anti-inflammatory activity, analgesic and antinociceptive activity, antipiretic activity, antihyperglycemic effects, hepatoprotective effects, cancer preventive and therapeutic effects	Yamaki et al. (2016)
Amaranthaceae	*Achyranthes aspera* L.	Carrapicho[b]	Vanillic acid, ferulic acid, rutin, quercetin, apigenin, kaempferol	antioxidant	Datta et al. (2019)
Amaranthaceae	*Amaranthus viridis* L.	Caruru de mancha, caruru, caruru verde, bredo, bredo verdadeiro, caruru de soldado, caruru de porco[d]	Gallic acid, chlorogenic acid, syringic acid, ferulic acid, ellagic acid, apigenin	Antioxidant	Datta et al. (2019)
Apiaceae	*Eryngium foetidum* L.	Coentro bravo, coentro da colônia, coentro de caboclo[d]	Alkaloids, saponins, cardiac glycoside, steroid, coumarins, phenolic compounds, tannins, flavonoid, lignin, proteins, starch	Antioxidant, anthelmintic	Narzary et al. (2016), Swargiary et al. (2016)

(continued)

Table 1 (continued)

Family	Species	Common names in Brazil	Phytochemical characterization	Pharmacological activities	References
Apocynaceae	*Cryptolepis sinensis* (Lour.) Merr.	–	Alkaloids, saponins, cardiac glycoside, steroid, anthraquinones, coumarins, phenolic compounds, tannins, flavonoid, anthocyanins, phlobatannins, lignin, proteins, starch	Antioxidant	Narzary et al. (2016)
Araceae	*Arum dioscoridis* Sm.	–	Saponins, alkaloids, carbohydrates, phenols, tannin, flavonoid	Antioxidant	Jaradat and Abualhasan (2016)
Araceae	*Arum elongatum* Steven	–	Saponins, alkaloids, carbohydrates, phenols, tannin, flavonoid	Antioxidant	Jaradat and Abualhasan (2016)
Araceae	*Arum hygrophilum* Boiss.	–	Saponins, alkaloids, carbohydrates, phenols, tannin, flavonoid	Antioxidant	Jaradat and Abualhasan (2016)
Araceae	*Arum palaestinum* Boiss.	–	Saponins, alkaloids, carbohydrates, phenols, tannin, flavonoid	Antioxidant	Jaradat and Abualhasan (2016)
Asteraceae	*Tragopogon dubius* Scop.	–	Flavonoids, phenylmethane derivatives	Antimicrobial, antioxidant	See the review of Abdalla and Zidorn (2020)
Asteraceae	*Tragopogon graminifolius* DC.	–	Phenylpropane derivatives	Antimicrobian, healing, antioxidant	See the review of Abdalla and Zidorn (2020)
Asteraceae	*Tragopogon orientalis* L.	–	Phenylmethane derivatives	Antioxidant	See the review of Abdalla and Zidorn (2020)

Family	Species	Common names in Brazil	Phytochemical characterization	Pharmacological activities	References
Asteraceae	*Tragopogon porrifolius* L.	–	Flavonoids, terpenoids	Antitumor, anti-inflammatory, antioxidant	See the review of Abdalla and Zidorn (2020)
Asteraceae	*Tragopogon pratensis* L.	–	Flavonoids, terpenoids, phenylmethane derivatives	Antitumor, antioxidant	See the review of Abdalla and Zidorn (2020)
Asteraceae	*Bidens biternata* (Lour.) Merr. & Sherff	Picão[c]	Alkaloids, glycosides, steroids, tannins	Antioxidant	Zahara et al. (2015)
Asteraceae	*Blumea lanceolaria* (Roxb.) Druce	–	Alkaloids, saponins, cardiac glycoside, steroid, phenolic compounds, tannins, flavonoid, anthocyanins, lignin, proteins	Antioxidant	Narzary et al. (2016)
Asteraceae	*Enhydra fluctuans* Lour.	–	Alkaloids, saponins, cardiac glycoside, steroid, anthraquinones, coumarins, phenolic compounds, tannin, flavonoid, ferulic acid, ellagic acid, quercetin, apigenin, kaempferol	Antioxidant	Narzary et al. (2016), Datta et al. (2019)
Asteraceae	*Gynura bicolor* (Roxb. ex Willd.) DC.	Espinafre de okinawa[e]	Phenolic acids, flavonoids, carotenoids, anthocyanins, essential oils, methoxypyrazines, amino acids, glycosides	Antioxidant property, anti-inflammation, anti-diabetic effects (anti-hyperglycemic effect), anticancer	See the review of Do et al. (2020)
Asteraceae	*Blumea lacera* (Burm. f.) DC.	–	Gallic acid, catechin hydrate, vanillic acid, caffeic acid, epicatechin, p-coumaric acid, myricetin	Antioxidant, anti-diabetic	Alam et al. (2020)

(continued)

Table 1 (continued)

Family	Species	Common names in Brazil	Phytochemical characterization	Pharmacological activities	References
Athyriaceae	*Diplazium esculentum* (Retz.) Sw.	–	Gallic acid, quercetin	Antioxidant, anti-diabetic	Junejo et al. (2018)
Berberidaceae	*Berberis aristata* DC.	–	Epicatechin, p-coumaric acid, quercetin	Antioxidant, anti-diabetic	Alam et al. (2020)
Caryophyllaceae	*Drymaria cordata* (L.) Willd. ex Schult.	cordão-de-sapo, jaboticaá, jaraquicaá, mastruço-do-brejo, agrião-selvagem, erva-tostão[b]	Alkaloids, saponins, cardiac glycoside, steroid, coumarins, phenolic compounds, tannins, flavonoid, proteins	Antioxidant	Narzary et al. (2016)
Cleomaceae	*Cleome gynandra* L.	mussambe [e]	Saponins, reducing compounds, tannins, alkaloids, volatile oils, phenols, anthocyanosides, flavonoids, sterols, and triterpenes, coumarins	Antioxidant	Nabatanzi et al. (2015)
Convolvulaceae	*Ipomoea aquatica* Forssk.	cacon[e]	p-Hydroxy benzoic acid, chlorogenic acid, vanillic acid, syringic acid, p-coumaric acid, sinapic acid, rutin, myricetin	Antioxidant	Datta et al. (2019)
Cucurbitaceae	*Momordica charantia* L.	Melão de são caetano, melãozinho, fruto de cobra[b]	Phenolic acids, triterpenes	Anti-diabetic, anticancer, antioxidant, antimicrobial, analgesic, and neuroprotective	See the review of Nagarani et al. (2014)
Cucurbitaceae	*Momordica cochinchinensis* (Lour.) Spreng	-	Phenolic acids, triterpenes	Anticancer, antimicrobial	See the review of Nagarani et al. (2014)
Cucurbitaceae	*Momordica dioica* Roxb. ex Willd.	melão de são ceaetano[f]	Phenolic acids	Analgesic and neuroprotective activity	See the review of Nagarani et al. (2014)

Family	Species	Common names in Brazil	Phytochemical characterization	Pharmacological activities	References
Cucurbitaceae	*Momordica balsamina* L.	Melão de são caetano[f]	Triterpenes	Antioxidant, antimicrobial	See the review of Nagarani et al. (2014)
Fabaceae	*Sesbania sesban* (L.) Merr.	Cânhamo[e]	Gallic acid, catechin hydrate, vanillic acid, caffeic acid, rutin hydrate, ellagic acid, myricetin, kaempferol, quercetin	Antioxidant, anti-diabetic	Alam et al. (2020)
Fabaceae	*Erythrina variegata* L.	Garra de trigre[e]	Gallic acid, catechin hydrate, vanillic acid, caffeic acid, rutin hydrate, ellagic acid, myricetin, kaempferol, quercetin	Antioxidant, anti-diabetic	Alam et al. (2020)
Lamiaceae	*Clerodendrum viscosum* Vent.	Clerodendrum[d]	Alkaloids, flavonoids, phenol, reducing sugar, saponins, tannins	Antioxidant, anthelmintic	Swargiary et al. (2016)
Malvaceae	*Adansonia digitata* L.	Embondeiro[e]	Tannic acid	Antioxidant, anti-inflammatory	Ayele et al. (2013)
Phyllanthaceae	*Antidesma acidum* Retz.	–	Alkaloids, saponins, cardiac glycoside, steroid, anthraquinones, coumarins, phenolic compounds, tannins, flavonoid, anthocyanins, phlobatannins, proteins	Antioxidant	Narzary et al. (2016)
Rubiaceae	*Oldenlandia corymbosa* L.	Erva-diamante[e]	Vanillic acid, ferulic acid, sinapic acid, ellagic acid, quercetin, apigenin, kaempferol	Antioxidant	Datta et al. (2019)

(continued)

Table 1 (continued)

Family	Species	Common names in Brazil	Phytochemical characterization	Pharmacological activities	References
Rutaceae	*Zanthoxylum acanthopodium* DC.	mamica de cadela; mamica de porca[e]	ascorbic acid, phenolic acids and flavonoids, being ascorbic acid, gallic acid, methyl gallate, caffeic acid, syringic acid, rutin, p-coumaric acid, ferulic acid, quercetin, apigenin, kaempferol	Antioxidant	Seal (2016)
Rutaceae	*Murraya koenigii* (L.) Spreng.	Curry indiano[e]	Flavonoids, phenol, reducing sugar, steroids, tannins	Antioxidant, anthelmintic	Swargiary et al. (2016)
Solanaceae	*Solanum villosum* Forssk.	–	Alkaloids, terpenoids.	Larvicide, antimicrobial, antioxidant	See the review of Zahara et al. (2019)
Solanaceae	*Solanum anguivi* Lam.	–	Saponins, reducing compounds, tannins, alkaloids, phenols, flavonoids, sterols, and triterpenes, coumarins	Antioxidant	Nabatanzi et al. (2015)
Solanaceae	*Solanum nigrum* L.	–	Saponins, reducing compounds, tannins, alkaloids, phenols, flavonoids, sterols, and triterpenes, coumarins	Antioxidant	Nabatanzi et al. (2015)
Solanaceae	*Physalis angulata* L.	Bucho de rã; camapu, camapum, joá, joá de capote, mata fome, balão[c]	Saponins, reducing compounds, alkaloids, anthocyanosides, flavonoids, sterols, and triterpenes, coumarins	Antioxidant	Nabatanzi et al. (2015)
Verbenaceae	*Lippia javanica* (Burm f.) Spreng.	mato-limão[e]	Alkaloids, saponins, cardiac glycoside, steroid, anthraquinones, coumarins, phenolic compounds, flavonoid, anthocyanins, lignin, proteins	Antioxidant, anthelmintic	Narzary et al. (2016), Swargiary et al. (2016)

Family	Species	Common names in Brazil	Phytochemical characterization	Pharmacological activities	References
Vitaceae	*Tetrastigma angustifolium* Planch.	–	Alkaloids, saponins, cardiac glycoside, steroid, phenolic compounds, tannins, flavonoid, anthocyanins, phlobatannins, lignin	Antioxidant	Narzary et al. (2016)
Vitaceae	*Cyphostemma adenocaule* (Steud. ex A. Rich.) Desc. ex Wild & R.B. Drumm.	–	Betulin, betulinic acid, cyphostemmic acid A, cyphostemmic acid B, cyphostemmic acid C, cyphostemmic acid D, epigouanic acid A, lupeol, zizyberanal acid, β-sitosterol and its glucoside, 3 β, 28-dihydroxy-30-norlupan- 20-one	Antioxidant, anti-inflammatory	See the review of Bello et al. (2019)
Vitaceae	*Leea macrophylla* Roxb. ex Hornem.	Léia[d]	Alkaloids, glycosides, flavonoids, steroidal/triterpenes, tannins, saponins, mucilages, proteins, amino acids, and sugars	Antioxidant, antimicrobial, anti-inflammatory	Joshi et al. (2016), Dewanjee et al. (2013)
Zingiberaceae	*Aframomum angustifolium* (Sonn.) K. Schum.	Longoza[e]	Reducing compounds, alkaloids, volatile oils, flavonoids, sterols, and triterpenes, coumarins	Antioxidant	Nabatanzi et al. (2015)

[a]Passos (2019)
[b]Lorenzi (2006)
[c]Lorenzi (2008a)
[d]Lorenzi (2008b)
[e]FAO (2020)
[f]Fernandes et al. (2009)

3 What Factors Influence the Sharing of UFPs in Different Human Groups?

People behave differently to what concerns UFP knowledge and consumption. Such differences may occur between individuals from the same population (intracultural variation) and between populations (intercultural variation). In terms of intracultural variation, literature has searched for the role of socioeconomic variables in the knowledge and/or consumption of wild food plants.

Age is one of the most studied factors, and several investigations have found that the elders either consume, know, or cite more use reports for wild food plants (Ghorbani et al. 2012; Cruz et al. 2013; Bortolotto et al. 2015; Geng et al. 2016). Such a pattern is not restricted to food plants as it was also found for medicinal and other uses. Since knowledge is accumulated with time and, more specifically, local ecological knowledge comes from experience and observation, it is intuitive to imagine that, in most cases, wild food plant knowledge will increase with age.

Gender is another variable widely studied as a possible predictor of wild food plant knowledge and use. However, there is no clear gender pattern in ethnobiological literature. Several studies have found no differences between male and female individuals (Joshi et al. 2015; Ochoa and Ladio 2015; Punchay et al. 2020), and some authors have suggested that this dominium is shared and appropriated by both genders and that female and male learning contexts were also experienced in similar ways (Ochoa and Ladio 2015).

However, to a lesser extent, in some cases, men (Kang et al. 2013) or women (Nascimento et al. 2013) are the most knowledgeable individuals. Such gender differences may have to do with the biased role of men and women in daily activities and wild food collection areas. In a study with Polish migrants in Misiones (Argentina), Kujawska and Luczaj (2015) found that, although no gender differences could be found for species gathered in ruderal areas, men knew more species from primary and secondary forests, which led to a gender bias in the overall number of species.

Other (less studied) predictors of intracultural heterogeneity in wild food plant knowledge and/or consumption include occupation, school education, income, and family structure. Studies have found that married people, parents, and those living in houses with members from more than one generation are more likely to have a greater knowledge of wild food plants (Ochoa and Ladio 2015; Ong and Kim 2016; Punchay et al. 2020).

Less instructed and poorer individuals may also accumulate more knowledge on this use category (Cruz et al. 2013; Ong and Kim 2016). Additionally, in many cases, people with current or past field activities (including agriculture and extractivism) may know or consume more wild food plants than people with other jobs (Cruz et al. 2013). However, some studies were not able to find correlations between some of these variables and wild food plant knowledge (Campos et al. 2015; Punchay et al. 2020).

Even communities in similar social-environmental conditions may present differences in terms of the explanatory power of socioeconomic variables. A study conducted in three neighboring communities from the Brazilian semiarid searched for predictors of wild food plant knowledge and consumption and found that, for all the studied factors, one community behaved differently from the others (Campos et al. 2015). For example, in one of the studied communities, men were found to know more wild food plants than women. However, in the other two communities, no gender biases could be identified. This finding indicates that the processes that generate heterogeneity are too sensitive, and even small social-environmental differences between communities may be responsible for significant alterations in knowledge distribution and sharing.

Although people who experienced hunger may accumulate more knowledge of wild food plants (Ong and Kim 2016), the association between wild foods and poverty/hunger may have led to cultural taboos that can influence cultural transmission. A group of species known as famine foods are accessed by local communities in times of scarcity, when crops and other foods are no longer available (Nascimento et al. 2012). Some of these species are hard to collect and/or require special preparations to eliminate anti-nutritional properties, which makes the knowledge about them an important biocultural inheritance. However, their association with difficult times and unpleasant memories has often made people reluctant to talk about this group of plants, which can compromise cultural transmission and make the youngest generations unaware of their collection and preparation strategies.

To what concerns intercultural variations, ethnicity and migration have found to be important predictors of wild food plant knowledge and/or consumption. People with different histories of relationships with the environment may accumulate a distinct body of knowledge on these species. In the above-cited study with Polish migrants in Argentina, Kujawska and Luczaj (2015) found that knowledge varied depending on the migrants' origin, given that migrants who went to Misiones via Brazil had a more diversified knowledge of edible plants. Other studies that compared different cultural groups in our out the context of migration have also found important differences (Ochoa and Ladio 2015; Pieroni and Cattero 2019).

Additionally, access to the urban-industrial society has also shown to interfere with wild food plant knowledge. In a study with local communities along the Paraguay River (Brazil), Bortolotto et al. (2015) found that people living closer to cities know fewer plants than those far from urban centers. Therefore, besides the well-known effect of urbanization in medicinal plant use, other use categories may also lose importance when people have access to alternative products, and wild food plants seem to be among them.

4 Final Considerations

In this chapter, we indicate evidence demonstrating that a set of unconventional food plants has been indicated in various human groups from different regions in both food and medicinal use. In addition, several of these species have been investigated in relation to phytochemical characterization and pharmacological properties, demonstrating the potential of UFPs for the prevention and treatment of diseases.

However, even with the nutritional and pharmacological potential of these species, a set of socioeconomic and cultural factors affect their popularity within the same human group or between different cultures, in various regions. As several human groups have undergone socioeconomic changes, it is interesting to promote interdisciplinary investigations both to assess the pharmacological potential of locally useful food species and to promote their popularity in different human groups. These studies will contribute to nutritional safety and the health of people in various regions.

References

Abdalla MA, Zidorn C (2020) The genus *Tragopogon* (Asteraceae): a review of its traditional uses, phytochemistry, and pharmacological properties. J Ethnopharmacol 250:112466

Alam MK, Rana ZH, Islam SN et al (2020) Comparative assessment of nutritional composition, polyphenol profile, antidiabetic and antioxidative properties of selected edible wild plant species of Bangladesh. Food Chem 320:126646

Ayele Y, Kim J-A, Park E et al (2013) A methanol extract of Adansonia digitata L. leaves inhibits pro-inflammatory iNOS possibly via the inhibition of NF-κB activation. Biomol Ther 21:146–152

Bello OM, Jagaba SM, Bello OE et al (2019) Phytochemistry, pharmacology and perceived health uses of non-cultivated vegetable *Cyphostemma adenocaule* (Steud. ex A. Rich.)Desc. ex Wild and R.B. Drumm: a review. Sci Afr 2:e00053

Bortolotto IM, Mello Amorozo MC, Neto GG et al (2015) Knowledge and use of wild edible plants in rural communities along Paraguay River, Pantanal, Brazil. J Ethnobiol Ethnomed 11:46

Cahill JP (2003) Ethnobotany of Chia, *Salvia hispanica* L. (Lamiaceae). Econ Bot 57:604–618

Campos LZO, Albuquerque UP, Peroni N et al (2015) Do socioeconomic characteristics explain the knowledge and use of native food plants in semiarid environments in Northeastern Brazil? J Arid Environ 115:53–61

Campos LZO, Nascimento ALB, Albuquerque UP, Araújo EL (2016) Criteria for native food plant collection in northeastern Brazil. Hum Ecol 44:775–782

Cruz MP, Peroni N, Albuquerque UP (2013) Knowledge, use and management of native wild edible plants from a seasonal dry forest (NE, Brazil). J Ethnobiol Ethnomed 9:79

Datta S, Sinha BK, Bhattacharjee S, Seal T (2019) Nutritional composition, mineral content, antioxidant activity and quantitative estimation of water soluble vitamins and phenolics by RP-HPLC in some lesser used wild edible plants. Heliyon 5:e01431

Dewanjee S, Dua TK, Sahu R (2013) Potential anti-inflammatory effect of Leea macrophylla Roxb. leaves: a wild edible plant. Food Chem Toxicol 59:514–520

Do TVT, Suhartini W, Mutabazi F, Mutukumira AN (2020) Gynura bicolor DC. (Okinawa spinach): a comprehensive review on nutritional constituents, phytochemical compounds, utilization, health benefits, and toxicological evaluation. Food Res Int 134:109222

Ekué MRM, Sinsin B, Eyog-Matig O, Finkeldey R (2010) Uses, traditional management, perception of variation and preferences in ackee (Blighia sapida K.D.Koenig) fruit traits in Benin: implications for domestication and conservation. J Ethnobiol Ethnomed 6:12

Etkin N (1996) Medicinal cuisines: diet and ethnopharmacology. Int J Pharmacog 34:313–326

Etkin N (2006) Edible medicines: an ethnopharmacology of food. The University of Arizona Press, Arizona

FAO (2010) The second report on the state of the world's plant genetic resources for food and agriculture. FAO, Rome

FAO, IFAD, UNICEF et al (2020) The state of food security and nutrition in the world 2020. Transforming food systems for affordable healthy diets. FAO, Rome

FAO (2020) Species and cultivars: common names. http://www.fao.org/agriculture/crops/thematic-sitemap/theme/hort-indust-crops/microgardens/speciesandcultivars/commonnames/en/. Accessed 10 Oct 2020

Ferreira Júnior WS, Campos LZO, Pieroni A et al (2015) Biological and cultural bases of the use of medicinal and food plants. In: Albuquerque UP, Medeiros PM, Casas A (eds) Evolutionary Ethnobiology. Springer, New York pp. 175–184

Fernandes CD, Catinda JF, Reis AL et al (2009) Estudo Farmacobotânico das Folhas de *Momordica charantia* L. (Cucurbitaceae). Visão Acadêmica 10:1

Geng Y, Zhang Y, Ranjitkar S et al (2016) Traditional knowledge and its transmission of wild edibles used by the Naxi in Baidi Village, northwest Yunnan province. J Ethnobiol Ethnomed 12:10

Ghorbani A, Langenberger G, Sauerborn J (2012) A comparison of the wild food plant use knowledge of ethnic minorities in Naban River Watershed National Nature Reserve, Yunnan, SW China. J Ethnobiol Ethnomed 8:17

Guarrera PM, Savo V (2013) Perceived health properties of wild and cultivated food plants in local and popular traditions of Italy: a review. J Ethnopharmacol 146:659–680

Heinrich M, Prieto JM (2008) Diet and healthy ageing 2100: will we globalise local knowledge systems? Ageing Res Rev 7:249–274

Heywood VH (2011) Ethnopharmacology, food production, nutrition and biodiversity conservation: towards a sustainable future for indigenous peoples. J Ethnopharmacol 137:1–15

Jaradat NA, Abualhasan M (2016) Comparison of phytoconstituents, total phenol contents and free radical scavenging capacities between four Arum species from Jerusalem and Bethlehem. Pharm Sci 22:120–125

Jiang S, Quave CL (2013) A comparison of traditional food and health strategies among Taiwanese and Chinese immigrants in Atlanta, Georgia, USA. J Ethnobiol Ethnomed 9:61

Johns T (1990) The origins of human diet and medicine. The University of Arizona Press, Arizona

Joshi A, Prasad SK, Joshi VK et al (2016) Phytochemical standardization, antioxidant, and antibacterial evaluations of Leea macrophylla: a wild edible plant. J Food Drug Anal 24:324–331

Joshi N, Siwakoti M, Kehlenbeck K (2015) Wild vegetable species in Makawanpur District, Central Nepal: developing a priority setting approach for domestication to improve food security. Econ Bot 69:161–170

Junejo JA, Gogoi G, Islam J et al (2018) Exploration of antioxidant, antidiabetic and hepatoprotective activity of Diplazium esculentum - a wild edible plant from North Eastern India. Futur J Pharm Sci 4:93–101

Junqueira AH, Perline EA (2019) Gosto, ideologia e consumo alimentar: práticas e mudanças discursivas sobre plantas alimentícias não convencionais - PANC. Cad Ling e Soc 20:17–35

Kang Y, Łuczaj Ł, Kang J et al (2013) Wild food plants and wild edible fungi in two valleys of the Qinling Mountains (Shaanxi, central China). J Ethnobiol Ethnomed 9:26

Kinupp VF, Lorenzi H (2014) Plantas Alimentícias não Convencionais no Brasil: Guia de identificação, Aspectos Nutricionais e Receitas Ilustradas. Instituto Plantarum, São Paulo

Kujawska M, Luczaj L (2015) Wild edible plants used by the Polish Community in Misiones, Argentina. Hum Ecol 43:855–869

Leal ML, Alves RP, Hanazaki N (2018) Knowledge, use, and disuse of unconventional food plants. J Ethnobiol Ethnomed 14:1–9

Leonti M (2012) The co-evolutionary perspective of the food-medicine continuum and wild gathered and cultivated vegetables. Genet Resour Crop Ev 59:1295–1302

Lentini F, Venza F (2007) Wild food plants of popular use in Sicily. J Ethnobiol Ethnomed 3:15

Lorenzi H (2006) Manual de identificação e controle de plantas daninhas: plantio direto e convencional, 6th edn. Instituto Plantarum, São Paulo

Lorenzi H (2008a) Plantas medicinais – Brasil: nativas e exóticas, 2nd edn. Instituto Plantarum, São Paulo
Lorenzi H (2008b) Plantas ornamentais no Brasil: arbustivas, herbáceas e trepadeiras, 4th edn. Instituto Plantarum, São Paulo
Nabatanzi A, Kabasa JD, Nakalembe I (2015) Phytochemicals and antioxidant properties of five wild edible plants consumed by pregnant women in Buikwe district, Uganda. Int J Pharmacogn Phytochem Res 7:1267–1271
Nagarani G, Abirami A, Siddhuraju P (2014) Food prospects and nutraceutical attributes of Momordica species: a potential tropical bioresources – a review. Food Sci Hum Wellness 3:117–126
Narzary H, Islary A, Basumatary S (2016) Phytochemicals and antioxidant properties of eleven wild edible plants from Assam, India. Med J Nutr Metab 9:191–201
Nascimento VT, Lucena RFP, Maciel MIS et al (2013) Knowledge and use of wild food plants in areas of dry seasonal forests in Brazil. Ecol Food Nutr 52:317–343
Nascimento VT, Vasconcelos MAS, Maciel MIS et al (2012) Famine foods of Brazil's seasonal dry forests: ethnobotanical and nutritional aspects. Econ Bot 66:22–34
Ochoa JJ, Ladio AH (2015) Current use of wild plants with edible underground storage organs in a rural population of Patagonia: between tradition and change. J Ethnobiol Ethnomed 11:70
Oliveira BPT, Ranieri GR (2017) Narrativa midiática e difusão sobre Plantas Alimentícias Não Convencionais (PANC): contribuições para avançar no debate. Cad Agroecol 13:7
Ong HG, Kim YD (2016) The role of wild edible plants in household food security among transitioning hunter-gatherers: evidence from the Philippines. Food Secur 9:11–24
Passos MAB (2019) Plantas Alimentícias Não Convencionais (PANC) Ocorrentes em Roraima. Revista Eletrônica Científica Ensino Interdisciplinar 5:14
Peisino MCO, Zouain MS, Christo Scherer MM et al (2019) Health-promoting properties of Brazilian unconventional food plants. Waste Biomass Valori 11:4691–4700
Pieroni A (1999) Gathered wild food plants in the upper valley of the Serchio River (Garfagnana), Central Italy. Econ Bot 53:327–341
Pieroni A, Quave CL (2006) Functional foods or food medicines? On the consumption of wild plant among Albanians and Southern Italians in Lucania. In: Pieroni A, Price L (eds) Eating and healing: traditional food as medicine. Haworth Press, New York, pp 101–129
Pieroni A, Cattero V (2019) Wild vegetables do not lie: comparative gastronomic ethnobotany and ethnolinguistics on the Greek traces of the Mediterranean Diet of southeastern Italy. Acta Bot Brasilica 33:198–211
Pinela J, Carvalho AM, Ferreira ICFR (2017) Wild edible plants: nutritional and toxicological characteristics, retrieval strategies and importance for today's society. Food Chem Toxicol 110:165–188
Punchay K, Inta A, Tiansawat P et al (2020) Traditional knowledge of wild food plants of Thai Karen and Lawa (Thailand). Genet Resour Crop Evol 67:1277–1299
Purba EC, Silalahi M, Nisyawati (2018) Gastronomic ethnobiology of "terites"—a traditional Batak Karo medicinal food: a ruminant's stomach content as a human food resource. J Ethn Foods 5:114–120
Rigat M, Bonet MA, Garcia S et al (2009) Ethnobotany of food plants in the High River ter Valley (Pyrenees, Catalonia, Iberian Peninsula): non-crop food vascular plants and crop food plants with medicinal properties. Eco Food Nutr 48(4):303–326
Seal T (2016) HPLC determination of phenolic acids, flavonoids and ascorbic acid in four different solvent extracts of zanthoxylum acanthopodium, a wild edible plant of Meghalaya state of India. Int J Pharm Pharm Sci 8:103–109
Sõukand R (2016) Perceived reasons for changes in the use of wild food plants in Saaremaa, Estonia. Appetite 107:231–241
Swargiary A, Daimari A, Daimari M et al (2016) Phytochemicals, antioxidant, and anthelmintic activity of selected traditional wild edible plants of lower Assam. Indian J Pharmacol 48:418–423

Towns A, Andel TT (2016) Wild plants, pregnancy, and the food-medicine continuum in the southern regions of Ghana and Benin. J Ethnopharmacol 179:375–382
Urso V, Signorini MA, Tonini M et al (2016) Wild medicinal and food plants used by communities living in Mopane woodlands of southern Angola: results of an ethnobotanical field investigation. J Ethnopharmacol 177:126–139
Yamaki J, Venkata KCN, Mandal A et al (2016) Health-promoting and disease-preventive potential of Trianthema portulacastrum Linn. (Gadabani)—an Indian medicinal and dietary plant. J Integr Med 14:84–99
Yang J, Chen W-Y, Fu Y et al (2020) Medicinal and edible plants used by the Lhoba people in Medog County, Tibet, China. J Ethnopharmacol 249:112430
Zahara K, Ahmad N, Bibi Y et al (2019) An insight to therapeutic potential and phytochemical profile of Solanum villosum (L). Med Drug Discov 2:100007
Zahara K, Bibi Y, Tabassum S et al (2015) A review on pharmacological properties of Bidens biternata: a potential nutraceutical. Asian Pacific J Trop Dis 5:595–599

Part II
Brazilian Food Plants: An Overview

Biodiversity for Food and Nutrition: Promoting Brazilian Underutilized Edible Plants into Food and Nutrition Security National Policies

Daniela Moura de Oliveira Beltrame, Camila Neves Soares Oliveira, and Lidio Coradin

Contents

1 Introduction

Brazil is considered the main country among the 17 megabiodiverse countries, holding 15 to 20% of the world's biodiversity. The country has the greatest wealth of flora in the world, as well as the largest remnants of tropical ecosystems (Ulloa Ulloa et al. 2017). In addition to possessing an immense biodiversity, Brazil also has a rich sociocultural diversity, represented by many traditional peoples and communities, such as indigenous, *quilombolas*, riverine communities, artisanal fisherfolk, *caiçaras*, *sertanejos*, and *pantaneiros*, among others, that have a strong relation with the biodiversity of the environment in which they live and are guardians of a huge body of traditional knowledge (knowledge, beliefs, and customs) about their use and conservation (SEDS 2020; MMA 2018a).

D. Moura de Oliveira Beltrame (✉)
National Project Coordinator, Biodiversity for Food and Nutrition Project, Brasília, DF, Brazil

C. Neves Soares Oliveira
Ministry of the Environment, Brasília, DF, Brazil
e-mail: camila.oliveira@mma.gov.br

L. Coradin
National Project Director, Biodiversity for Food and Nutrition Project, Brasília, DF, Brazil

M. C. M. Jacob, U. P. Albuquerque (eds.), *Local Food Plants of Brazil*, Ethnobiology, https://doi.org/10.1007/978-3-030-69139-4_4

Paradoxically, for decades, because we have not been valuing our natural resources, we are reducing the use of biodiversity and the use of the species that surround us, thereby increasing the fragility of the food system. The country's agriculture and food security are very reliant on exotic or introduced crops or species from other regions of the world, such as sugar cane, soy, orange, and rice from Asia; coffee originating in Africa; or maize from Central America (Cunha Alves & Azevedo 2018). There is a steady expansion of export-oriented agribusiness, characterized by large land concentration, monoculture prevalence, and expansion of livestock production, reinforcing unsuitable and unsustainable land management practices that contribute to the destruction of natural habitats and to the loss of native biodiversity in Brazil (Beltrame et al. 2018).

At the same time, Brazil faces dietary simplification and high levels of malnutrition, even though much of this neglected native biodiversity is highly nutritious. As we use a smaller number of species, the proper consumption of food and nutrients needed to promote the growth, development, and maintenance of our health is put at risk. This was evidenced in the Household Budget Survey (POF 2008–2009) (IBGE 2010): only 10 fruits (orange, banana, apple, papaya, mango, watermelon, tangerine, grape, pineapple, and *açaí*) correspond to 91% of total fruits consumed by Brazilians. Of these, only pineapples and *açaí* are native to Brazil; the other eight species originate from other countries. In addition, only 25% of Brazilians consume the amount of fruits and vegetables recommended by the World Health Organization (400 g per day), that is, foods that provide important nutrients for health, such as vitamins, minerals, and dietary fiber.

It is worth mentioning, however, that markets continue to demand new product options, which is why Brazilian biodiversity presents an enormous potential to satisfy these demands and generate wealth. Exploring these food sources could also help to provide sustainable solutions to diversifying diets, also tackling malnutrition problems, and promoting agricultural development through local food procurement, while also promoting biodiversity conservation and climate change adaptation and resilience.

To this end, the need to develop new public policies that could, on the one hand, contribute to ensuring the protection of native biodiversity and, on the other hand, improve the food security of the population was evident. It is easy to verify, in the different regions of Brazil the existence of a great number of species of current economic value or of potential use, whose use remains restricted to local or regional level. Increasingly, initiatives dedicated to meeting market demands for new products occupy a prominent position. Thus, it was clear that the country needed to intensify investments aimed at increasing knowledge about native species, including better utilization of this rich biodiversity, so that native species could become better known and used by society.

Knowledge and evidence gaps still needed to be addressed to better integrate biodiversity for enhancing food and nutrition security, including scientific data on the nutritional value and composition of native neglected species. To this end, the

Biodiversity for Food and Nutrition Project (BFN) in Brazil worked from 2012 to 2018 to strengthen the conservation and sustainable use of biodiversity by providing evidence of its benefits for nutrition and well-being, mainstreaming biodiversity into national food and nutrition policies and strategies, and increasing awareness of the potentials offered by native biodiversity. The Project generated food composition data, influenced policies, developed recipes, and helped to increase appreciation of the value of underutilized, nutrient-rich biodiversity through strategic research partnerships with universities and research institutes and advocacy and lobbying at policy and regulatory frameworks related to food and nutrition security.

The GEF-funded Biodiversity for Food and Nutrition Project was led by Brazil, Kenya, Sri Lanka, and Turkey and was coordinated by Biodiversity International with implementation support from the United Nations Environment Program (UNEP) and the Food and Agriculture Organization of the United Nations (FAO).

The four countries participating in the project represent some of the world's most megadiverse countries thanks to the extraordinary diversity of ecosystems and species existing within their borders. They each contain unique biological diversity and have associated traditional ecological knowledge that supports a large proportion of the world's food supply in a range of ecosystems that are global priorities for conservation. The project sought to address the issue of diminishing the use of local agrobiodiversity by contributing to the improvement of global knowledge of biodiversity for food and nutrition and, in so doing, enhancing the well-being, livelihoods, and food security of target beneficiaries in the four countries through the conservation and sustainable use of this biodiversity and the identification of best practices for up-scaling.

Globally, the BFN has made a number of highly significant contributions to global policy, including the Voluntary Guidelines for Mainstreaming Biodiversity into Policies, Programmes and National and Regional Plans of Action (FAO 2016) and inputs to the document UNEP/CBD/SBSTTA/19/INF/1: "Strategic Scientific and Technical Issues Related to the Implementation of the Strategic Plan for Biodiversity 2011-2020: Biodiversity, Food Systems and Agriculture" (CBD 2015).

Through the focus on mainstreaming biodiversity in the different sectors, the project contributed to multiple environmental outcomes linked to Sustainable Development Goals (SDGs) and Aichi Biodiversity targets, as well as social and economic benefits. These include limiting biodiversity loss; mitigating climate change; strengthening seed systems to ensure that biodiversity is conserved, available, and accessible; developing markets that ensure that a diversity of foods is available and affordable; reviving traditional knowledge and cultural heritage; supporting rural development; and strengthening local economies. A food system transformed in this way contributes to environmental conservation while providing income generation benefits, especially for the youth and for vulnerable groups, including women, smallholder farmers, indigenous people, and local communities, as a result of diversification options (FAO 2019).

2 Project Components and Key Interventions

2.1 Knowledge Base

The starting point for increasing the knowledge base about Brazilian neglected species was the Plants for the Future, an ongoing initiative from the Ministry of the Environment that aims to prioritize native flora species of current economic value or potential use, as well as to raise the attention of the Brazilian people to the enormous possibilities and opportunities of using these species. The initiative was established in Brazil in the early 2000s and was designed to respond to the concerns and aspirations of the society and to change the current scenario regarding the use, conservation, and benefits that can be generated from native species. This initiative aimed at offering to the society new options to meet the growing demand for new species of food, aromatic, medicinal, and cosmetic interest, among other possible uses. It identified and prioritized native species from the Brazilian flora of current or potential use, allowing new choices to farmers and, in the case of food species, creating new gastronomic possibilities, rescuing traditional knowledge, and diversifying options for the Brazilian cuisine. The initiative was developed through the implementation of five subprojects, one for each of the five major geopolitical regions of the country, in order to value and draw attention to the importance of the native species of each region as well as for regional traditions. The species were prioritized through a careful work conducted in partnership with specialists from all different regions of Brazil, based on predetermined criteria for each of the defined groups of use for each region (food, aromatic, fibrous, forage, medicinal, oleaginous, and ornamental, among others).

Books from the Plants for the Future collection are available in the Ministry of Environment's publications website.[1] From this work, new species of Brazilian flora are being made available to farmers, with special attention to family agriculture, which can find new options to diversify their crops. In the same way, new investment opportunities for the business sector are being generated in the development of new products to help to boost the Brazilian bioeconomy, as well as offering new perspectives and incentives for the use of native species in governmental policies and programs.

The BFN project selected 78 underutilized edible species identified by the Plants for the Future to carry out nutritional composition analysis, in partnership with public universities and research institutes across the country. The food composition work followed methodologies developed by the Food and Agriculture Organization of the United Nations and the International Network for Food Data Systems (FAO/INFOODS). Initially, data was compiled from scientific literature, documents, and reports from local universities and research institutes. Gap analysis revealed lack of reliable data for many nutrients for most fruit species, especially dietary fiber,

[1] Available at: https://www.mma.gov.br/publicacoes/biodiversidade/category/54-agrobiodiversidade

vitamins, and minerals. Therefore, in a second stage, laboratorial analysis was carried out. The results were made available online on the Biodiversity and Nutrition[2] tool (SiBBr 2018), a food composition database developed by the BFN Project in partnership with the Information System on Brazilian Biodiversity (SiBBr) at the Ministry of Science, Technology and Innovation (MCTI).

Numerous students, professors, and researchers were involved in the food composition research and results proved that many of the prioritized native species were richer in nutrients compared with more commonly consumed exotic foods in Brazil (Beltrame et al. 2016). For example, the fruit camu-camu (*Myrciaria dubia* (Kunth) McVaugh) contains 30 times more vitamin C than oranges (1620 mg/100 g versus 53 mg/100 g)' the flour from the babaçu (*Attalea speciosa* Mart. ex Spreng) mesocarp contains 18,3 mg of iron per 100 g, compared to 1 mg in the same amount of wheat flour or 2,3 mg in maize flour, while the baru nut (*Dipteryx alata* Vogel) contains 30 mg of protein/100 g, compared to 14 mg in walnuts and 18 mg in almonds (NEPA 2011; SiBBr 2018). Being non-domesticated wild species, the nutritional value of these underutilized foods can vary greatly depending on the harvest location. For example, samples of camu-camu are reported to contain up to to 6112 mg of vitamin C/100 g of pulp (Yuyama et al. 2002).

Working through regional partners ensured the development of capacities in different regions of the country and scaling up research and extension activities that were already developed regionally. It also helped to raise awareness among students, researchers, and professors about the importance of food composition and the role of Brazilian underutilized edible species. These groups acted as multipliers within education and research institutions, building additional human capacity.

Federal ministries linked to initiatives such as the National School Feeding Program (PNAE), the Food Procurement Program (PAA), and the Minimum Price Guarantee Policy for Sociobiodiversity Products (PGPM-Bio) can benefit from the food composition data to demonstrate the value of native species as sources of nutritious foods and, potentially, sources of income for family farmers and communities. Similarly, the nutrition benefits of native biodiversity to diversify and improve the nutritional content of diets can be demonstrated to consumers.

2.2 *Influencing Policies*

Another key component of the BFN Project and one of the strategies to address the barriers identified by the Project Preparation Phase, such as "poorly developed infrastructure and markets," "trade policies that undermine promotion and consumption of BFN," and "inadequate agricultural and food security policies and strategies," was mainstreaming biodiversity into public policies related to food and nutrition security, agriculture, rural, and social development, education, and health.

[2] Available at: https://ferramentas.sibbr.gov.br/ficha/bin/view/FN

At an early stage of the BFN Project, the partners involved – which included representatives of the ministries of the environment, agriculture, social development, agrarian development, science and technology, education, and health, as part of the federal framework at the time – identified an already established federal multi-sectoral institutional framework and associated federal initiatives and policies as a strategic opportunity to enhance the mainstreaming of biodiversity for improved food and nutrition outcomes. A project governance mechanism – the National Steering Committee – was established to coordinate and manage this partnership transparently and included the abovementioned ministries as well as representatives of the National Supply Company (CONAB), the Brazilian Agricultural Research Corporation (EMBRAPA), and the National Food and Nutrition Security Council (CONSEA).

Through partnership with these institutions, the BFN project team was involved in a series of national activities promoting the importance of biodiversity for food and nutrition within the following federal government initiatives: the Food Procurement Program (PAA), National School Feeding Program (PNAE), Minimum Price Guarantee Policy for Sociobiodiversity Products (PGPM-Bio), National Plan for the Promotion of Sociobiodiversity Value Chains (PNPSB), Development of Organic Agriculture (Pro-Orgânico), National Plan for Agroecology and Organic Production (PLANAPO), and National Food and Nutrition Policy (PNAN). This framework of public policies was not static, so new plans and programs were created or concluded during the lifetime of the BFN Project.

Many of these federal initiatives presented entry points for improving nutrition or livelihoods with links to underutilized species, for example, the PAA and PNAE, two long-lasting and robust food procurement programs: at least 30% of the food purchased through PNAE with federal funds must be bought directly from family farmers, while both the PNAE and the PAA pay a premium of up to 30% in the price of organic or agroecological produce and prioritize purchases from settlers of the agrarian reform and traditional communities, such as *quilombolas* and indigenous communities. These were key incentives and subsidies for procuring native neglected and underutilized food species for public food procurement. In doing so, these programs created unique opportunities for the use of natural resources from the various Brazilian ecosystems and promoted the opportunity for the development of new institutional markets for underutilized plants and their products while provided incentives for the management and sustainable use of Brazilian food and agricultural biodiversity (Beltrame et al. 2016).

A major achievement in Brazil and critical to the success of public policies to increase the attention given to native neglected species was the development of a new supporting policy, the *Official list of native Brazilian sociobiodiversity species of nutritional value* published by Ordinance N°163/2016 (Brazil 2016) and updated by Ordinance N° 284/2018 (Brazil 2018a), which officially defined and recognized over 100 Brazilian underutilized food species. This was the first policy of its kind in the country able to incentivize the production and consumption of nutritious neglected and underutilized species, facilitating greater procurement of these species and their integration into school feeding and other federal initiatives.

The list of sociobiodiversity species was a response to meet the demands by the institutions responsible for the PAA, whose employees struggled to identify neglected and underutilized native species that were valid for incentives reserved for regional food products. Ministries now refer to the ordinance list to monitor PNAE and PAA institutional purchases of native biodiversity species as well as to expand the number of species included in PGPM-Bio, initiatives which support production and marketing (with a fair price) for family farmers. The inclusion of target underutilized species in the ordinance has greatly increased their marketing potential and ultimately, and smallholders are encouraged to conserve, use, produce, and commercialize local native species.

Many steps were taken in this direction, as one example, on the supply side, the ordinance has provided the 25 million communities of gatherers and family farms who continue managing and growing the native species (MMA 2019) with a dependable market outlet for their produce. A derived public policy has created additional market incentives by establishing a sociobiodiversity label for all produce and products in the list of native Brazilian sociobiodiversity species. The label was initially created in 2018 by Ordinance N° 129/2018 and updated by Ordinance N° 161/2019 (Brazil 2018b; Brazil 2019a). The label can be requested by family farmers, cooperatives, and small and medium enterprises with a strong participation in family farming who have registered with the National Program to Strengthen Family Farming (PRONAF). Together, the registration to PRONAF and the sociobiodiversity label open up additional market opportunities for family farmers and communities of gatherers.

The promotion and monitoring of the use of native biodiversity in school procurement has also benefited from the creation of a food quality index called Food and Nutrition Security Quality Index (Índice de Qualidade da Coordenação de Segurança Alimentar e Nutricional – IQ COSAN) (FNDE 2018), a tool developed to guide nutritionists that develop school menus. It allocates point-based ratings to school meals depending on levels of dietary diversity and the absence of unhealthy foods such as sugars, sweets, and processed and fried foods. An additional 2.5 points are allocated if meals include any number of the native species listed by Ordinance N° 284/2018. IQ-COSAN is acting as both a regulatory mechanism for healthy eating in schools, in line with the country's nutrition guidelines, and an incentive for school nutritionists and school managers to incorporate underutilized native species into school meals to increase their ratings (FNDE 2018).

Another significant achievement in the Influencing Policies component was the BFN's contributions to the second edition of the Dietary Guidelines for the Brazilian Population (Fig. 1) (MS 2014), launched by the Ministry of Health and its National Food and Nutrition Policy (PNAN). The Dietary Guidelines highlighted Brazil's rich heritage of neglected and underutilized species, their associated traditional knowledge and food cultures, and also considered that healthy diets derive from socially and environmentally sustainable food systems. Among the supporting documents developed to assist the implementation of the revised dietary guidelines in Brazil was the BFN's contribution to the new edition of Brazilian Regional Foods (Fig. 1) (MS 2015), including a chapter on "Biodiversity for Food and Nutrition."

Fig. 1 Dietary Guidelines for the Brazilian Population and Regional Foods book, publications from the Ministry of Health that promote the use of regional recipes and neglected foods to diversify and improve diets and nutrition

The book combines recipes and nutritional information and is used throughout schools and other institutions and aims to contribute to a revival of culinary skills using regional foods and native food biodiversity.

PNAN's Health in School Program (PSE) was also identified as a strategic entry point for promoting biodiversity in the school environment. This program fostered visits by primary healthcare professionals to public schools and provides capacity building for school teachers and managers for the development of educational activities. Activities were carried out to monitor eating habits and nutrition and to promote healthy eating behaviors consistent with the Brazilian dietary guidelines. New training materials for the implementation of PSE were developed, including booklets targeting teachers and healthcare professionals (MS 2018, 2019). To promote the inclusion of biodiversity, technical contributions were made to ensure that biodiversity was included in all training materials with considerable attention drawn to regional foods, particularly native fruit species and their roles in promoting better health and nutrition.

The BFN also lobbied for the inclusion of food biodiversity in National Council for Food and Nutrition Security (CONSEA) documents. As a result, the Reference Document for the 5th National Conference on Food and Nutrition Security (CNSAN), held in November 2015, incorporated biodiversity as one of the main aspects related to food and nutrition security. CNSAN, held every 4 years, is where guidelines and priorities for food and nutrition security actions were set and communicated to the CONSEA to inform policy making. The main outcome from CNSAN was a Policy Letter, containing several recommendations related to the sustainable use of biodiversity to achieve food sovereignty. Some recommendations focused especially on expanding public policies and actions to ensure self-sufficiency for family farmers through agroecological practices and promotion of biodiversity, such as the creation of organic seed banks and the promotion of value chains for non-conventional vegetables and native fruits.

By interacting with CONSEA and working with partner ministries on government plans such as the National Plan for Food and Nutrition Security (PLANSAN 2016–2019), the National Pact for Healthy Food, and the National Plan for Agroecology and Organic Production (PLANAPO), BFN promoted biodiversity as one of the solutions for combating food and nutrition insecurity, to promote healthy diets and improve producers' livelihoods (Beltrame et al. 2016).

2.3 *Raising Awareness*

Activities to promote and raise awareness of native biodiversity species, as a third component of the BFN Project, included the development of recipes using prioritized neglected and underutilized species. The collection of 335 recipes was made available online in the book Brazilian Biodiversity: Tastes and Flavours[3] (Fig. 2) (Santiago and Coradin 2018). This recipe book was also a result of a participatory

Fig. 2 Book with the collection of 335 recipes using ingredients from neglected species from Brazilian biodiversity published by the BFN Project

[3]Available at: https://www.mma.gov.br/publicacoes/biodiversidade/category/54-agrobiodiversidade.html

process that involved researchers, chefs, and students from partner universities and research institutes across the country.

Sociobiodiversity species were also highlighted in the 2nd Best Recipes from School Meals awards, a competition promoted by PNAE with school cooks, and a chapter about Brazilian biodiversity was included in the book derived from the awards (FNDE 2018). The award saw school cooks from all over Brazil competing in the preparation of the healthiest and most nutritious school meal that is also appealing to students. In 2018, 2252 school cooks from Brazil's 5 regions took part in the competition.

Another example of capacity building and awareness raising was the establishment of partnerships with the Educating through School Gardens and Gastronomy Project (PEHEG) to diversify school curricula using school gardens and gastronomy as educational tools which helped to promote healthy eating habits, appreciation of regional ingredients and recipes, learning of cooking techniques, as well as experiencing flavors, food textures, and aromas of native neglected and underutilized species among schoolchildren. The inclusion of Brazilian underutilized edible plants in school feeding and school curricula through PEHEG represented a cultural appreciation of local resources and food practices, raising awareness and recognition of traditional knowledges and techniques of cultivation, rational extraction, production, and transformation. In addition, it contributed to the food and nutrition security of the communities involved with each school of this Project, as those foods were rich in micro and macronutrients. The PEHEG contributed to the process of educating, empowering, and engaging the educators and public agents, as well as the members of the school communities about those themes, through the implementation of school gardens and activities related to gastronomy as pedagogical tools for environmental and food and nutrition education. In short, this Project helped to create awareness about healthy eating and environmental issues, encouraging the students and communities to promote the necessary transformations to build a more sustainable society (Santos et al. 2020).

As part of the effort to improve the production chain of native biodiversity and add more value to the products, a series of booklets on best practices for the sustainable collection of wild, organic native products was published by the Ministry of Agriculture (MAPA 2014) and the Ministry of the Environment (MMA 2018b) for 21 species, 15 of which are species prioritized and proposed by the BFN project. The booklets were divided in two series, one targeting the producers themselves and the other with more technical information for extension workers who provide assistance to the producers. This was a main achievement to tackle one of the barriers for increasing markets and demand for native biodiversity identified, since most of these foods are not domesticated and cultivated, but collected from the wild, highly perishable and demand special care for their production, processing, transportation, and storage, in order to achieve the desired quality standards demanded by consumers (Beltrame et al. 2016).

To raise awareness about native neglected food species with the general population, the BFN project also organized or participated in a number of cultural exhibitions and gastronomic events in different Brazilian cities (Fig. 3), for example,

Fig. 3 Examples of events and tasting sessions organized by the BFN project to raise awareness and about neglected food species from the Brazilian biodiversity. Photos credits: Shawn Landerz, Itamar Sandoval, Julceia Camillo

cooking demonstrations and tasting, promotion of farmers markets with native species, and campaigns such as the “Organic Food Week” organized by the federal government in different Brazilian cities. To celebrate the conclusion of project’s activities, the Biodiversity for Food and Nutrition International Symposium was held in Brasília in 2017, with the presence of government’s authorities and other partners from federal ministries, researchers, students, and chefs from Brazil and abroad. During the Symposium, the main results of the project were presented (including cooking demonstrations and a tasting session with the recipes developed by the partners for the “Brazilian Biodiversity: Tastes and Flavors” book), and ideas for future actions were discussed.

3 Outcomes

The BFN Project together with the Plants for the Future initiative has contributed to increase the knowledge and awareness about the role of native biodiversity for food and nutrition security in Brazil today so weakened by the restriction and strong dependence on few species. With the inclusion of new species in the agricultural system, the country begins to have new and unusual cultivation options, as well as a better condition to face the global environmental changes, including climate change. In addition to promoting conservation activities, encouraging the use of new options of agricultural interest, and providing information on hundreds of native plants of real or potential value, these actions will certainly contribute to promote the formation of new productive chains. This work is critical to increase information and change society's perception of the strategic importance of conserving biodiversity and native genetic resources.

At the end of the BFN Project, changes in behaviors and attitudes were evident within ministries and federal institution partners of the BFN. *Undoubtedly, one of the great achievements of the Project in Brazil was providing to stakeholders in different levels involved with public policies related to institutional food procurement and school feeding programs an official* list of native species of the Brazilian sociobiodiversity with current or potential nutritional value (Ordinance N° 284/2018) and the BFN Food Composition Database.

Together, these two instruments facilitated the procurement of products from sociobiodiversity species, as well as greater incentives for family farmers to promote the production and sale of these products in the scope of the PAA and PNAE. Likewise, it is expected for a greater number of native food species to be included in the list of PGPM-Bio, as has occurred with buriti (*Mauritia flexuosa* L.f.), that became part of its 2018 subsidy list. This agricultural income policy compensates producers should their sociobiodiversity products not reach the market value established by the National Supply Company.

Increases in institutional expenditures on Brazilian native biodiversity and their products were documented, although investments remain small when compared to overall food purchases. For example, expenditures in sociobiodiversity products by PAA increased from 2,75% of the total expenditure in 2016 to 5% in 2017, and payments made by PGPM-Bio to sociobiodiversity producers grew steadily in recent years, growing from R$ 5 million in 2014 to more than R$ 13 million in 2019 (Brazil 2019b; CONAB 2020). These initiatives can be considered strategic tools to promote the conservation and sustainable use of sociobiodiversity and offer a glimpse of the market potential for expanding the amount spent in these products. This also means that there is a great opportunity to monitor how and if increasing the diversity of foods produced, purchased, and consumed will, in the longer-term, affect the livelihoods of the small-scale producers and promote better food and nutrition security while enabling climate change adaptation and biodiversity conservation.

Despite its importance, these initiatives represent only the beginning of a process. Other similar actions need and should be conducted at a local and regional level, so that it is possible to rescue and broaden the population's interest in the use of Brazilian underutilized edible plants. The continuity of actions of this nature should be developed not only because of the importance and necessity of the conservation of the rich biodiversity existing in the country but also with a view to promote greater participation of native species in our diet, to create new options for the use of these species, to consolidate new productive chains, and specially to show that native species can be much more nutritionally rich than the basic species that currently dominate our dietary habits.

The future development of new market chains and the increase of interest by the private sector are expected, especially now with a new era of bioeconomy in Brazil. In the long term, expanding production and marketing of native nutritious species have also the potential to increase the income of family farmers and gatherers, as well as the diversification and improvement of diets and nutritional status of beneficiaries of programs linked to food and nutrition security and the population in general. Decision-makers, family farmers and gatherers, and the general public will be able to recognize biodiversity's value, thus promoting greater conservation and sustainable use.

References

Beltrame DMO, Oliveira CNS, Borelli T, de Andrade Cardoso Santiago Monego ET, Vera de Rosso V, Coradin L, Hunter D (2016) Diversifying institutional food procurement – opportunities and barriers for integrating biodiversity for food and nutrition in Brazil. Revista Raízes 36(2):55–69

Beltrame DMO, Oliveira CNS, Coradin L (2018) Biodiversidade Brasileira – Novas possibilidades e oportunidades. In: Fundo Nacional para o Desenvolvimento da Educação (ed) Melhores Receitas da Alimentação Escolar, 2nd edn. FNDE, Brasília, pp 15–24

Brazil (2016) Portaria Interministerial no 163, de 11 de maio de 2016. Espécies nativas da sociobiodiversidade brasileira de valor alimentício. Brasília, DF: Diário Oficial da União. Seção 1, 04 de Nov de 2015

Brazil (2018a) Portaria Interministerial MMA e MDS N° 284, de 30 de maio de 2018. Institui a lista de espécies da sociobiodiversidade, para fins de comercialização in natura ou de seus produtos derivados, no âmbito das operações realizadas pelo Programa de Aquisição de Alimentos-PAA. Brasília, DF: Diário Oficial da União. Seção 1, N° 131, 10 de julho de 2018

Brazil (2018b) Portaria N° 129, de 7 de março de 2018. Institui o Selo de Identificação da Participação da Agricultura Familiar – SIPAF. Brasília, DF: Diário Oficial da União. Seção 1, N° 46, 08 de março de 2018

Brazil (2019a) Portaria N° 161, de 9 de agosto de 2019. Institui o Selo Nacional da Agricultura Familiar - SENAF e dispõe sobre os procedimentos relativos à solicitação, renovação e cancelamento do selo, e dá outras providências. Brasília, DF: Diário Oficial da União. Seção 1, N° 156, 09 de agosto de 2019

Brazil (2019b) Ministry of the Environment Brazil: 6th National Report to the convention on biological diversity/Ministry of the Environment. Secretariat of Biodiversity. https://www.cbd.int/doc/nr/nr-06/br-nr-06-en.pdf. Accessed 30 June 2020

Companhia Nacional de Abastecimento (CONAB) (2020) Boletins da Sociobiodiversidade. https://www.conab.gov.br/info-agro/analises-do-mercado-agropecuario-e-extrativista/boletim-da-sociobiodiversidade/boletim-sociobio. Accessed 28 June 2020

Convention on Biological Diversity (CBD) (2015) Strategic Scientific and Technical Issues Related to the Implementation of the Strategic Plan For Biodiversity 2011–2020: Biodiversity, Food Systems And Agriculture. https://www.cbd.int/kb/record/meetingDocument/105616?RecordType=meetingDocument&Event=SBSTTA-19. Accessed 20 July 2020

Cunha Alves AA, Azevedo VCR (2018) Embrapa Network for Brazilian plant genetic resources conservation. Biopreserv Biobank 16(5):350–360

Food and Agriculture Organization (FAO) (2016) Voluntary Guidelines for Mainstreaming Biodiversity into Policies, Programmes and National and Regional Plans of Action on Nutrition. http://www.fao.org/family-farming/detail/en/c/416118/. Accessed 21 July 2020

Food and Agriculture Organization (FAO) (2019) Mainstreaming biodiversity conservation and sustainable use for improved human nutrition and well-being: global project findings and recommendations. FAO, Rome

Fundo Nacional para o Desenvolvimento da Educação (FNDE) (2018) Índice de Qualidade da Coordenação de Segurança Alimentar Nutricional - IQ COSAN http://www.fnde.gov.br/programas/pnae/pnae-area-para-gestores/ferramentas-de-apoio-ao-nutricionista. Accessed 15 July 2020

Instituto Brasileiro de Geografia e Estatística (IBGE) (2010) Pesquisa de Orçamentos Familiares 2008–2009: antropometria e estado nutricional de crianças, adolescentes e adultos no Brasil. IBGE, Rio de Janeiro

Ministério da Agricultura, Pecuária e Abastecimento. MAPA (2014) Cadernos de Boas Práticas de Extrativismo Sustentável Orgânico. https://www.gov.br/agricultura/pt-br/assuntos/sustentabilidade/organicos/arquivos-publicacoes-organicos. Accessed 15 July 2020

Ministério da Saúde (MS) (2014) Dietary Guidelines for the Brazilian Population. http://bvsms.saude.gov.br/bvs/publicacoes/dietary_guidelines_brazilian_population.pdf. Acessed 15 July 2020

Ministério da Saúde (MS) (2015) Alimentos Regionais Brasileiro, 2nd edn. Ministry of Health, Brasília. http://bvsms.saude.gov.br/bvs/publicacoes/alimentos_regionais_brasileiros_2ed.pdf. Accessed 15 July 2020

Ministério da Saúde (MS) (2018) Caderno de atividades: Promoção da Alimentação Adequada e Saudável: Educação Infantil. Ministério da Saúde, Universidade do Estado do Rio de Janeiro, Brasília

Ministério da Saúde (MS) (2019) Caderno de atividades: Promoção da Alimentação Adequada e Saudável: Ensino Fundamental II. Ministério da Saúde, Universidade do Estado do Rio de Janeiro, Brasília

Ministério do Meio Ambiente (2018a) Biodiversidade Brasileira http://www.mma.gov.br/biodiversidade/biodiversidade-brasileira. Accessed 20 July 2020

Ministério do Meio Ambiente (MMA) (2018b). Cadernos de Boas Práticas de Extrativismo Sustentável Orgânico http://www.mma.gov.br/publicacoes/desenvolvimento-rural/category/200-extrativismo-sustent%C3%A1vel Accessed 15 July 2020

Ministério do Meio Ambiente (MMA) (2019). Catálogo de Produtos da Sociobiodiversidade ofertados pelos povos e comunidades tradicionais em Unidades de Conservação Federais. ICMBio, Brasília

Núcleo de Estudos e Pesquisas em Alimentação (NEPA) (2011) Tabela brasileira de composição dos alimentos (TACO) 4th edn. http://www.nepa.unicamp.br/taco/contar/taco_4_edicao_ampliada_e_revisada.pdf?arquivo=taco_4_versao_ampliada_e_revisada.pdf. Accessed 20 July 2020

Santos ARD, Almeida NL, Santos AS, et al (2020) The integration of food biodiversity in school curricula through school gardens and gastronomy in Brazil. In: Hunter D, Monville-Oro, Burgos B, et al (ed) Agrobiodiversity, school gardens and healthy diets promoting biodiversity, food and sustainable nutrition. Routledge, New York, p 283–290

Secretaria Especial do Desenvolvimento Social (SEDS) (2020) Povos e Comunidades Tradicionais http://mds.gov.br/assuntos/seguranca-alimentar/direito-a-alimentacao/povos-e-comunidades-tradicionais. Accessed 25 July 2020

Sistema de Informações sobre a Biodiversidade Brasileira (SiBBr) (2018) Ferramenta Biodiversidade & Nutrição: banco de dados de composição nutricional e receitas https://ferramentas.sibbr.gov.br/ficha/bin/view/FN/. Accessed 20 July 2020

Ulloa Ulloa C, Acevedo-Rodríguez P, Beck S et al (2017) An integrated assessment of the vascular plant species of the Americas. Science 358:1614–1617

Yuyama K, Aguiar JPL, Yuyama LKO (2002) Camu-camu: um fruto fantástico como fonte de vitamina C. Acta Amaz 32(1):169–174

Bioactive Potential of Brazilian Plants Used as Food with Emphasis on Leaves and Roots

Fernanda L. B. Mügge, Sarah M. S. Prates, Juliana de Paula-Souza, and Maria G. L. Brandão

Contents

F. L. B. Mügge
Centro Especializado em Plantas Aromáticas, Medicinais e Tóxicas (Ceplamt), Museu de História Natural e Jardim Botânico, Universidade Federal de Minas Gerais (UFMG), Belo Horizonte, Brazil

Zentrum für Molekulare Biologie, Universität Heidelberg, Heidelberg, Germany

S. M. S. Prates
Centro Especializado em Plantas Aromáticas, Medicinais e Tóxicas (Ceplamt), Museu de História Natural e Jardim Botânico, Universidade Federal de Minas Gerais (UFMG), Belo Horizonte, Brazil

Programa de Pós-Graduação em Ciências de Alimentos, Faculdade de Farmácia, Universidade Federal de Minas Gerais (UFMG), Belo Horizonte, Brazil

J. de Paula-Souza
Centro Especializado em Plantas Aromáticas, Medicinais e Tóxicas (Ceplamt), Museu de História Natural e Jardim Botânico, Universidade Federal de Minas Gerais (UFMG), Belo Horizonte, Brazil

Departamento de Botânica, Centro de Ciências Biológicas, Universidade Federal de Santa Catarina (UFSC), Florianópolis, Brazil

M. G. L. Brandão (✉)
Centro Especializado em Plantas Aromáticas, Medicinais e Tóxicas (Ceplamt), Museu de História Natural e Jardim Botânico, Universidade Federal de Minas Gerais (UFMG), Belo Horizonte, Brazil
e-mail: mbrandao@ufmg.br

M. C. M. Jacob, U. P. Albuquerque (eds.), *Local Food Plants of Brazil*, Ethnobiology, https://doi.org/10.1007/978-3-030-69139-4_5

1 Introduction

Archaeological finds show that plants have been used for millennia as sources of food and medicine in the Americas. Colonizing Europeans indeed learned the use of species such as corn, potatoes, cassava, tomatoes, cocoa, avocado, pineapple, and cashew, from the Amerindians (Maezumi et al. 2018). Native plants continue to be widely used in Brazil, but the intense miscegenation of cultures that occurred over recent centuries has also popularized the use of exotic species (Ferrão et al. 2008). The intense industrialization and urbanization of the country, which started in the 1950s, has deprived the Brazilian population of knowledge about the potential of native plants for use as food and medicine (Domingues et al. 2012; Fernandes 2004). These processes were aggravated by different economic cycles, starting with the exploitation of brazilwood (*Paubrasilia echinata* (Lam.) Gagnon, H.C.Lima & G.P.Lewis, Fabaceae) by the Portuguese. The economy is currently dominated by mining and agribusiness, both of which seriously damage native natural ecosystems (Dean 1996). Today, only 7% of the Atlantic Forest territory remains, while other ecosystems such as the savannas (Cerrado) and dry forests (Caatinga) are rapidly being replaced by monocultures of eucalyptus, sugarcane, soybeans, and livestock (Bockmann et al. 2018; Venâncio et al. 2018). As a consequence, not only knowledge but also the native plant species themselves are being lost (Brandão et al. 2008a, b; Mügge et al. 2016).

In contrast to this scenario, the consumption of healthier food and the transition to plant-based diets are growing in the developed world (Kahleova et al. 2017; Melina et al. 2016). The development of a "bioeconomy", based on the sustainable use of biodiversity with less impact on the environment is also coming into favor (Valli et al. 2018), and traditional knowledge concerning the use of plant species used as food can be very valuable in this context. Information on species used by local populations in the past can reveal the existence of plants with the potential to be used as functional foods or used as a source of purified biological products. To contribute to this aim, during the last decade, our research group has focused on recovering traditional information about useful Brazilian plants from manuscripts and books published prior to 1960. Each work is carefully searched for data regarding the use of plants, and the information is then publicized (Brandão et al. 2011; Brandão et al. 2012; Breitbach et al. 2013; Fagg et al. 2015; Teixeira et al. 2019). Furthermore, the information is entered into an online database named Dataplamt (www.ceplamt.org.br/dataplamt), built to organize and make information readily available. Searches can be done through links referring to popular and scientific names, traditional uses, places where the plant occurs, and full texts extracted from original books or documents. To date, Dataplamt contains information from 66 documents on 3400 species of Brazilian useful plants.

In two previous studies, we have focused on the search for data concerning native food species. Oliveira et al. (2012) studied the native plant foods recorded in field books of 16 European naturalists who traveled through Brazil in the nineteenth century. More recently, Teixeira et al. (2019) showed data on 504 edible fruits recovered from the book "Dicionário das plantas úteis do Brasil e das exóticas cultivadas" (Dictionary of Useful Plants of Brazil and Exotic Cultivated), organized by the Portuguese botanist Manoel Pio Corrêa (1874–1934), in 1926. In this study, we present additional data about native plant species used as food, with emphasis on leaves and roots, recovered from books published by nine authors from the sixteenth to the twentieth centuries.

2 Methods

2.1 Search and Organization of Data

Firstly, we did a search in *Dataplamt* for plants that were classified as edible ("comestível," in Portuguese). Then, we searched the data using the key words: "frutos comestíveis" (edible fruits), "raízes comestíveis" (edible roots/tubers), "folhas comestíveis" (edible leaves), "flores comestíveis" (edible flowers), and " brotos comestíveis" (edible sprouts)". Information from books published by nine authors, living in different regions of Brazil from the sixteenth to the twentieth centuries, were selected. Table 1 shows the names and nationalities of each author, year of publication of their books, the area upon which their studies focused, and the numbers of edible plant species recorded by each one.

Table 1 Authors' nationalities, phytogeographical domains and regions studied in Brazil, years of the studies and number of plants cited in each publication

Abb[a]	Author/nationality	Area of study	Period in Brazil	Number of plants
SO	Gabriel Soares de Sousa/Portuguese	Atlantic Forest (Bahia state)	1575–1592	5
LI	Friar Cristovão de Lisboa/Portuguese	Amazonia (Maranhão state)	1624–1635	5
PI	Guilherme Piso/Dutch	Atlantic Forest and Caatinga (Pernambuco state)	1637–1644	7
VE	Friar Mariano C. Vellozo/Brazilian	Atlantic Forest and Cerrado (Minas Gerais and Rio de Janeiro states)	Natural	18
SP	Richard Spruce/English	Amazonia (Amazonas and Pará state)	1849–1864	4
CH	Pedro Napoleão Chernoviz/Polish	All regions of Brazil	1840–1855	9
PK	Theodor Peckolt/German	All regions of Brazil	1847–1912	30
CR	Manoel Pio Corrêa/Portuguese	All regions of Brazil	18XX-1934	26
CN	Paul Le Cointe/French	Amazonia	1891–1956	96

[a]Abbreviation

To check whether a species was native of Brazil, as well as the current botanic name, was performed using the website "Flora do Brasil".[1] It is often difficult to identify species from books published between the sixteenth and eighteenth centuries, because taxonomy as we currently know it was established only in the late eighteenth century, and the rules of botanical nomenclature did not exist until the beginning of the twentieth century. In particular, imprecise identification of some of the plants referred by Sousa (1587), Piso (1648), and Lisboa might detract from the accuracy of the study. However, reliability is somewhat increased by the fact that the editors of each of these books have added comments and more precise identification of the described species. Data on recent laboratory studies on bioactivity were obtained from PubMed.

The recorded plants were then divided in two tables: one with species submitted to studies that confirmed its health benefits (Table 2) and another table with the other plants that have not been studied (Table 3). Information about which author cited the plants is also included in the Tables (abbreviated names), as well as citations by other authors also present in Dataplamt.

2.2 Species Cited by Other Authors

The results obtained from this study were compared with data recorded by European naturalists who lived or travelled in Brazil in previous centuries. The first search was performed in Oliveira et al. (2012), which describes historical data recovered from the nineteenth century (indicated as "1-167" in Tables 1 and 2). Other data were obtained directly from the field books of the naturalists Auguste de Saint-Hilaire (Brandão et al. 2012), George Gardner (Fagg et al. 2015), and Richard Spruce (Santos-Fonseca et al. 2019). Additional data collected by von Martius in Amazonia in the nineteenth century was recorded by Breitbach et al. (2013) and from Corrêa in the twentieth century in Teixeira et al. (2019). Importantly, these naturalists provide detailed information about the plants and its uses. They also provide an adequate coverage of the country in terms of area and phytogeographic domains: von Martius, for example, described plants from the North region (Amazon); Gardner and Burchell from Midwest region (Caatinga and Cerrado); and Saint-Hilaire described species from Southeast and South regions (Cerrado and Atlantic Forest).

[1] See at: http://floradobrasil.jbrj.gov.br/2010/

Table 2 Food species cited by the authors submitted to bioactivity studies

Family/species/ popular names		SO	LI	PI	VE	SP	CH	PK	CN	CR	Correlated studies on bioactivity
Aizoaceae *Sesuvium portulacastrum* (L.) L./Beldroega miúda	Lvs				x			x		x	Antimicrobial, antioxidant (Al-azzawi et al. 2012; Chandrasekaran et al. 2011)
Alismataceae *Echinodorus macrophyllus* (Kunth) Micheli/ Chapéu de couro[a]	Lvs									x	Anti-inflammatory (Silva et al. 2016), nephroprotective (Portella et al. 2012)
Amaranthaceae *Amaranthus cruentus* L./Caruru	Lvs				x			x			Obesity (Kanikowska et al. 2019), anti-inflammatory (Tang and Tsao 2017)
Amaranthus spinosus L./Bledo	Lvs							x			Antioxidant, nutritional, (Mondal and Maity 2016, Sarker and Oba 2019), diabetes (Bavarva and Narasimhacharya 2013)
Amaranthus viridis L./Bredo americano	Lvs			x	x			x			Antioxidant, nutritional, (Sarker and Oba 2019), Mondal and Maity (2016), nutritional (Silva et al. 2018)
Anacardiaceae *Spondias tuberosa* Arruda/Ambu [1,2]	Rts	x		x			x			x	Antimicrobial (Santos et al. 2019)
Aquifoliaceae *Ilex dumosa* Reissek/Mate[a]	Lvs				x					x	Gan et al. (2018)
Ilex integerrima (Vell.) Reissek/ Mate[a]	Lvs									x	Gan et al. (2018)
Ilex paraguariensis A.St.-Hil./Mate[a]	Lvs						x			x	Gan et al. (2018) + Several other studies
Ilex pseudobuxus Reissek/Caúna[a]	Lvs									x	Gan et al. (2018)
Ilex theezans Mart. ex Reissek/Matea	Lvs									x	Gan et al. (2018)
Araceae *Caladium bicolor* (Aiton) Vent./Ara	Lvs, Rts	x	x				x	x	x	x	Antidiarrheal (Salako et al. 2015)

(continued)

Table 2 (continued)

Family/species/ popular names		SO	LI	PI	VE	SP	CH	PK	CN	CR	Correlated studies on bioactivity
Colocasia esculenta (L.) Schott/Inhame [1-4]	Rts								x	x	Immunomodulator (Pereira et al. 2018), diabetes (Eleazu et al. 2016)
Xanthosoma sagittifolium (L.) Schott/Mangará [3-4]	Lvs, Rts	x	x					x			Antitumor (Caxito et al. 2015), prevent colon cancer (Jackix et al. 2013), antioxidant (Arruda et al. 2005), antifungal (Schmourlo et al. 2005)
Asteraceae *Sonchus oleraceus* L./Serralha lisa	Lvs				x		x		x	x	Antioxidant (Mawalagedera et al. 2016); anti-aging agent (Ou et al. 2015)
Ayapana triplinervis (Vahl) R.M.King & H.Rob./Aiapana[a] [1]	Lvs									x	Antioxidant (Taïlé et al. 2020)
Acmella oleracea (L.) R.K.Jansen/ Agrião do Pará	Lvs								x	x	Chemical, nutritional (Neves et al. 2019), gastroprotective (Maria-Ferreira et al. 2014; Nascimento et al. 2013)
Cactaceae *Pereskia aculeata* Mill./Ora pro nobis	Lvs				**x**					x	Antioxidant, antimicrobial (Garcia et al. 2019; Souza et al. 2016), improve intestinal motility and lipid profile (Barbalho et al. 2016)
Dioscoreaceae *Dioscorea chondrocarpa* Griseb./Cará	Rts									x	Antimicrobial (Barnabé et al. 2014)
Dioscorea trifida L.f./Inhame roxo	Rts		x		x				x	x	Food allergy (Mollica et al. 2013)
Euphorbiaceae *Manihot esculenta* Crantz/Aipim [1-16]	Rts	x	x	x	x	x	x	x	x	x	Several studies
Fabaceae *Neptunia oleracea* Lour./Juquiry manso	Lvs								x	x	Antioxidant, management of diabetes (Lee et al. 2019)
Pachyrhizus erosus (L.) Urb./Jacatupé	Rts						x				Management of blood glucose (Park et al. 2016, Santoso et al. 2019)

(continued)

Table 2 (continued)

Family/species/ popular names		SO	LI	PI	VE	SP	CH	PK	CN	CR	Correlated studies on bioactivity
Marantaceae *Maranta arundinacea* L./ Araruta	Rts					x				x	Antiulcerogenic (Rajashekhara et al. 2014), immunostimulant (Kumalasari et al. 2012)
Nyctaginaceae *Boerhavia diffusa* L./Tangeraca mansa	Lvs							x			Gastrointestinal disorders, anticancer (Mishra et al. 2014)
Boerhavia erecta L./Erva tostão de Minas	Lvs							x			Antioxidant, anti-inflammatory (Compaore et al. 2018)
Oxalidaceae *Oxalis barrelieri* L./Azedinha	Lvs									x	Antidiarrhea (Fokam Tagne et al. 2018)
Piperaceae *Peperomia pellucida* (L.) Kunth/ Jaboti-membeca	Lvs							x		x	Antimicrobial (Alves et al. 2019)
Piper umbellatum L./Mático [1,3,7]	Lvs	x		x	x	x		x		x	Intestinal anti-inflammatory (Arunachalam et al. 2020), anti-gastric ulcer (Silva Junior et al. 2016), anticancer (Iwamoto et al. 2015)
Plumbaginaceae *Plumbago scandens* L./ Caá-jandiwap, loco[1]	Lvs, Rts			x	x		x		x	x	Antimicrobial *Helicobacter pylori* (de Paiva et al. 2003, Wang and Huang 2005)
Polygonaceae *Rumex acetosella* L./Alecrim de São José	Lvs							x		x	Management of diabetes (Özenver et al. 2020)
Polygonum aviculare L./ Sanguinária	Lvs									x	Pancreatic lipase inhibitions (Park et al. 2019), antioxidant, anti-inflammatory (Granica et al. 2013), antiobesity (Sung et al. 2013), antimicrobial (Salama and Marraiki 2010)

(continued)

Table 2 (continued)

Family/species/ popular names		SO	LI	PI	VE	SP	CH	PK	CN	CR	Correlated studies on bioactivity
Talinaceae *Talinum paniculatum* (Jacq.) Gaertn./ Língua de vaca	Lvs								x	x	Antimicrobial (Reis et al. 2015)
Tropaoelaceae *Tropaeolum pentaphyllum* Lam./Capuchinha [a]	Lvs/ Rts							x		x	Nutritional, antioxidant (De Bona et al. 2017), antimicrobial (Cruz et al. 2016)

[a]Used as tea. *Lvs* leaves, *Rts* roots.

3 Results and Discussion

Data on the use of Brazilian native plant food species were recovered from books published by nine authors that lived in Brazil during different times in the past five centuries (Table 1). The earliest was the Portuguese Gabriel Soares de Sousa (1540–1592) who lived in in Bahia for 17 years. In his book "Tratado Descritivo do Brasil em 1587" (Descriptive Treaty of Brazil in 1587), he gave detailed information specifically about plants from the Atlantic Forest used by the Amerindians. Friar Cristovão de Lisboa (1583–1652) was a Portuguese who lived in Brazil from 1624 to 1635 and worked in a religious mission in the state of Maranhão; once back in Portugal his observations were published first in 1627 in the book "History of animals and plants of Maranhão" (História dos animais e plantas do Maranhão). As the title suggests, the book has information about animals and plants from the Amazonian region, accompanied by richly detailed high quality drawings which allow identification of the plants (Marques 1996). Also in the seventeenth century, northeast Brazil was invaded by the Dutch, and Guilherme Piso and Marcgrave lived there for 8 years. Later in 1648, back in the Netherlands, they published the book "Historiae Naturalis & Medicae," in which several native food and medicinal plants used by the Amerindians are described (Piso 1648).

From the Portuguese invasion in 1500 to the next three centuries, Brazilian territory was under strict control in order to prevent natural resources of being exploited by other countries. In 1808, the Portuguese royal family fled to Rio de Janeiro from Portugal after Napoleon Bonaparte's invasion. They lived in Brazil for the next 13 years, a period that was characterized by notable progress in the country's economy, culture and science, such as the creation of the first School of Medicine in Salvador and the Botanical Garden of Rio de Janeiro. Foreigners finally had permission to enter the country and many European naturalists came to Brazil. They explored all regions of the country and identified new plant species and gathered information on their use by the local communities (Brandão et al. 2008a, b; Oliveira et al. 2012). Among these naturalists was the English Richard Spruce (1817–1893), who lived in the Amazon for 17 years. After his death, Wallace edited his "Notes of

Table 3 Brazilian plant foods not studied and authors who cited their uses

Families/species/popular names/parts used	Authors								
	SO	LI	PI	VE	SP	CH	PK	CN	CR
Alstromeriaceae *Alstroemeria caryophyllaea* Jacq./Madressilva/Rts									x
Alstroemeria monticola Mart. ex Schult. & Schult.f./Carajurú/Rts							x		
Bomarea edulis (Tussac) Herb./Jaranganha/Rts									x
Amaranthaceae *Blutaparon vermiculare* (L.) Mears/Caá-ponga/Lvs							x		
Amaryllidaceae *Zephyranthes candida* (Lindl.) Herb./Carapitaia/Rts							x		x
Anacardiaceae *Spondias venulosa* (Engl.) Engl./Acaiá/Rts		x		x					x
Apiaceae *Eryngium nudicaule* Lam./Gravatazinho/Rts									x
Aquifoliaceae *Ilex affinis* Gardner/Congonha[a]/Lvs									x
Ilex brasiliensis (Spreng.) Loes./Erva mate[a]/Lvs									x
Ilex conocarpa Reissek/Congonha[a]/Lvs									x
Ilex chamaedryfolia Reissek/Congonha[a]/Lvs									x
Ilex diuretica Mart. ex Reissek/Congonha[a]/Lvs									x
Ilex paltorioides Reissek/Congonha[a]/Lvs									x
Araceae *Dracontium asperum* K.Koch/Taja de cobra/Rts								x	
Urospatha caudata (Poepp.) Schott/Apê/Rts							x	x	x
Xanthosoma riedelianum (Schott) Schott/Mangarito/Rts									x
Xanthosoma striatipes (Kunth & C.D.Bouché) Madison/Banana do brejo/Rts[3]							x		x
Arecaceae *Euterpe oleracea* Mart./Açaí/Lvs/Spr [6,7,14]									x
Asteraceae *Erechtites valerianifolius* (Wolf) DC./Caruru/Lvs				x					x
Pacourina edulis Aubl./Pacurina/Lvs								x	x
Balanophoraceae *Langsdorffia hypogaea* Mart./Espiga de sangue/Flw							x		
Lophophytum mirabile Schott & Endl./Batata escamas/Rts									x
Ombrophytum peruvianum Poepp. & Endl./Parasita/Rts									x
Bignoniaceae *Dolichandra unguis-cati* (L.) L.G.Lohmann/Poampé/Rts								x	x
Campanulaceae *Centropogon cornutus* (L.) Druce/Crista de peru/Lvs									x

(continued)

Table 3 (continued)

Families/species/popular names/parts used	Authors								
	SO	LI	PI	VE	SP	CH	PK	CN	CR
Cannaceae *Canna paniculata* Ruiz & Pav./Paca vira/Rts							x		x
Canna glauca L./Achira/Rts			x	x		x	x	x	x
Cardiopteridaceae *Citronella paniculata* (Mart.) R.A.Howard/ Congoinha[a]/Lvs[2]									x
Convolvulaceae *Distimake tuberosus* (L.) A.R. Simões & Staples/Jetuca/Rts									x
Dioscoreaceae *Dioscorea bahiensis* Kunth/Cará/Rts									x
Dioscorea cinnamomifolia Hook./Cará assú/Rts				x					x
Dioscorea dodecaneura Vell./Cará barbado/Rts									x
Dioscorea fodinarum Kunth/Cará/Rts									x
Dioscorea glandulosa (Klotzsch ex Griseb.) Kunth/ Cará/Rts									x
Dioscorea hassleriana Chodat/Cará coco/Rts				x					x
Dioscorea heptaneura Vell./Cará branco/Rts									x
Dioscorea lacerdaei Griseb./Cará/Rts									
Dioscorea laxiflora Mart. ex Griseb./Caratinga bravo/ Rts								x	x
Dioscorea leptostachya Gardner/Cará/Rts									x
Dioscorea multiflora Mart. ex Griseb./Cará/Rts									x
Dioscorea ovata Vell./Cará inhame/Rts									x
Dioscorea piperifolia Humb. & Bonpl. ex Willd./ Cará/Rts								x	x
Dioscorea sinuata Vell./Caratinga brava/Rts				x					x
Dioscorea stegelmanniana R.Knuth/Cará/Rts									x
Dioscorea trifoliata Kunth/Caranambú/Rts									x
Euphorbiaceae *Manihot caerulescens* Pohl/Maniçoba/Rts									x
Manihot caerulescens Pohl ssp. *caerulescens*/Maniçoba/Rts									x
Manihot dichotoma Ule/Maniçoba/Rts									x
Manihot glaziovii Müll.Arg./Maniçoba/Rts									x
Manihot violacea Pohl/Maniçoba/Rts									x
Fabaceae *Luetzelburgia auriculata* (Allemão) Ducke/Pau chapada/Rts									x
Heliconiaceae *Heliconia psittacorum* L.f./Pacová catinga/Rts									x
Humiriaceae *Humiria balsamifera* (Aubl.) A.St.-Hil./Umiri/Rts					x			x	

(continued)

Table 3 (continued)

Families/species/popular names/parts used	Authors								
	SO	LI	PI	VE	SP	CH	PK	CN	CR
Sacoglottis guianensis Benth./Achua/Rts								x	
Icacinaceae *Casimirella ampla* (Miers) R.A.Howard/Mairá/Rts								x	x
Leretia cordata Vell./Mata fome/Rts									x
Iridaceae *Phalocallis coelestis* (Lehm.) Ravenna/Bibi/Rts									x
Malvaceae *Abutilon megapotamicum* (Spreng.) A.St.-Hil. & Naudin/Benção de Deus/Lvs									x
Abutilon purpurascens K.Schum./Benção de Deus/ Flw									x
Hibiscus bifurcatus Cav./Algodão do brejo/Rts								x	x
Sphaeralcea bonariensis (Cav.) Griseb./Malvaisco/ Lvs, Flw						x			x
Marantaceae *Goeppertia allouia* (Aubl.) Borchs. & S.Suárez/ Tupinambor/Rts									x
Goeppertia tuberosa (Vell.) Borchs. & S.Suárez/ Caeté/Rts			x	x			x		
Ischnosiphon arouma (Aubl.) Körn./Arumá membeca/ Rts							x	x	x
Ischnosiphon petiolatus (Rudge) L.Andersson/ Arumá/Rts								x	
Thalia geniculata L./Arumarana/Rts							x	x	x
Nyctaginaceae *Neea theifera* Oerst./Caparrosa[a]/Lvs								x	x
Nymphaeaceae *Victoria amazonica* (Poepp.) J.E.Sowerby/Uapé/Rts								x	
Oxalidaceae *Oxalis hirsutissima* Mart. & Zucc./Azedinha/Lvs									x
Phytolaccaceae *Phytolacca rivinoides* Kunth & Bouché/Espinafre Guiana/Lvs							x		x
Phytolacca thyrsiflora Fenzl. ex J.A.Schmidt/Carurú/ Lvs									x
Piperaceae *Peperomia transparens* Miq./Língua de sapo/Lvs							x		
Poaceae *Gynerium sagittatum* (Aubl.) P.Beauv./Arina/Spr							x		
Polygonaceae *Rumex brasiliensis* Link/Labaça/Lvs							x		x
Portulacaceae *Portulaca halimoides* L./Beldroega/Lvs									x
Portulaca mucronata Link/Beldroega/Lvs							x		

(continued)

Table 3 (continued)

Families/species/popular names/parts used	Authors								
	SO	LI	PI	VE	SP	CH	PK	CN	CR
Portulaca pilosa L./Alecrim de São José/Lvs [1,2]									x
Solanaceae *Solanum commersonii* Poir./Batata silvestre/Rts									x
Solanum alternatopinnatum Steud./Caruru de espinho/Lvs				x					x
Talinaceae *Talinum fruticosum* (L.) Juss./Beldroega grande/Lvs									x
Tropaeolaceae *Tropaeolum brasiliense* Casar./Capuchinha do Brasil/ Lvs							x		x
Typhaceae *Typha domingensis* Pers./Tabua/Rts								x	x

[a]Used as tea. *Lvs* leaves, *Rts* roots, *Flw* flowers, *Spr* sprouts.

a botanist in Amazonas and the Andes" (1908), providing them with a bibliographic introduction (Seaward 2000).

It is also important to highlight the efforts of the first Brazilian botanist, Friar José Mariano da Conceição Vellozo. Friar Vellozo was born in 1741 in Minas Gerais and died in Rio de Janeiro, in 1811. From an early age, he was interested in botany. In 1755 he started his religious life in Rio de Janeiro and in 1771 he moved to São Paulo, where he started teaching the indigenous people, simultaneously having the opportunity to learn about the uses and original names of native plants. He was commissioned by the Portuguese Crown to compile an extensive botanical inventory of the Flora in Rio de Janeiro and its surroundings. The result is the work "Florae Fluminensis," completed in 1790, which consists of one volume containing the botanical descriptions of 1639 species of plants, native and exotic, and a further 11 volumes of drawn illustrations. Vellozo was a pioneer in this type of work, and the "Florae" includes valuable information on the use of about 300 plants, including those used as food (Brandão 2019).

The most recent authors cited here studied plants in the nineteenth and early twentieth centuries. The Polish medical doctor P. Chernoviz (1812–1882) and the German pharmacist Theodor Peckolt, (1822–1912), lived in Rio de Janeiro and wrote important books containing information on hundreds of useful and medicinal native plants (Peckolt and Peckolt 2016; Ricardo et al. 2017). In 1947 the French Paul Le Cointe (1870–1956) published the book "Amazônia Brasileira" (Brazilian Amazon), with information on hundreds of plants, their traditional uses, and economic viability. The Portuguese botanist Manoel Pio Corrêa (1874–1934) included description of roots and leaves used as food in his book "Dicionário das plantas úteis do Brasil e das exóticas cultivadas" (Dictionary of Useful Plants of Brazil and Exotic Cultivated), organized in 1926, and his previous work entitled "Flora do Brasil" (Brazilian Flora), published in 1909. Although these authors did their own field work and research, they also declare that they used information present in previous sources. We can consider, therefore, that the presence of plants mentioned by previous authors (e.g., Saint-Hilaire, von Martius) is a validation of their uses as food.

A total of 630 plant species used as food were cited in the studied books. Of these, 514 are fruits, most of them already discussed in Teixeira et al. (2019). Thus, in this work, we emphasize the remaining 116 plants, which have other parts used as food. Table 1 shows that Corrêa cited the highest number of species (96) followed by Peckolt (30) and Le Cointe (26). Among the authors that published their works before or during the eighteenth century, Vellozo cited the greatest number of food plants. A total of 18 species was described in his work, while Chernoviz, Piso, Lisboa, and Sousa cited 9, 7, 5, and 5 plants, respectively. Vellozo was Brazilian, and his familiarity with local customs certainly allowed him to have more proximity to the population in order to acquire traditional knowledge. In total, from the works of the 9 authors, 64 roots/tubercles, 52 leaves, 3 flowers, and 2 sprouts were cited as food (Tables 2 and 3).

Among the roots, the use of *Manihot esculenta* Crantz tubers (aipim, cassava, Euphorbiaceae) was described by all the authors. Cassava tubers are rich in special carbohydrates that release glucose slowly in the body without generating blood glucose spikes (Okafor et al. 2017; Udeme et al. 2015). This composition helps maintain high energy levels for longer, being useful for high performance athletes, and makes this food very suitable for management of diabetes. Cassava is also rich in vitamins A, B1, B2, and C and can be consumed by people and animals. Due to its high benefits, cassava was considered by the Food and Agriculture Organization of the United Nations (FAO/UN) as the most relevant food of the twentieth century (FAO 2013). Other five species of *Manihot* were also recorded by Corrêa (*M. caerulescens* Pohl, *M. caerulescens* Pohl subsp. *caerulescens*, *M. dichotoma* Ule, *M. glaziovii* Müll.Arg., *M. violacea* Pohl.), but their potential has not been verified to date.

Other useful roots are those from *Spondias tuberosa* Arruda (Anacardiaceae), *Caladium bicolor* (Aiton) Vent., and *Xanthosoma sagittifolium* (L.) Schott (Araceae), cited by different authors along the centuries. They have also been submitted to laboratory studies, which confirmed their health benefits (Table 2). These plants showed antioxidant (Arruda et al. 2005), antimicrobial (Santos et al. 2019; Schmourlo et al. 2005), and antitumoral (Caxito et al. 2015) and could also have the capacity to prevent colon cancer (Kackix et al. 2013). Table 2 shows that roots of *Colocasia esculenta* (L.) Schott (Araceae) and *Maranta arundinacea* (Marantaceae) have immunomodulatory activity. Immunomodulators are substances that act directly on the immune system, strengthening its defenses and improving its functions. These effects have been associated to saponins and some types of polysaccharides. The adaptogenic effect of ginsengs roots, for example, is associated with the presence of saponins (Metwaly et al. 2019; Shi et al. 2019), while for mushrooms and seaweed, the effects are associated to saccharides (Ling et al. 2017; Mallard et al. 2019). Saponins and some types of polysaccharides, as starch, are very common in roots, and the effects observed for *C. esculenta* were correlated to the presence of saponins (Azubuike et al. 2018; Eleazu et al. 2016; Pereira et al.

2018). Other less known edible roots, but also cited along the centuries, are *Plumbago scandens* L. (Plumbaginaceae), *Spondias venulosa* (Engl.) Engl. (Anacardiaceae), *Canna glauca* L. (Cannaceae), and *Goeppertia tuberosa* (Vell.) Borchs. & S.Suarez (Marantaceae). The use of roots of *P. scandens* was cited by Piso and Vellozo; *P. scandens* roots were recently shown to have activity against *Helicobacter pylori*, what could help in the management of gastric ulcers and prevent cancer (de Paiva et al. 2003; Wang and Huang 2005).

The tubercles of Dioscoreaceae (yams) have been used in the human diet for millenia. Many species have economic importance, due to the presence of diosgenin, which is used as a prototype for oral contraceptive in the pharmaceutical development (Hata et al. 2003). The most important species cultivated for their edible tubers include *D. alata*, originating from Asia, *D. cayenensis*, and *D. rotundata* from West Africa. The tubercles of these species contain several nutrients, including carbohydrates, amino acids, minerals, thiamine, riboflavin, niacin, and ascorbic acid. Diosgenin and allantoin constitute the main secondary bioactive metabolites (Mollica et al. 2013) and have also been shown to possess anti-inflammatory and immunomodulatory properties (Chen et al. 2015). Sixteen species of *Dioscorea* were cited by Corrêa, and four of them (*D. cinnamomifolia* Hook., *D. hassleriana* Chodat, *D. sinuata* Vell., and *D. trifida* L.f.) were also registered by Vellozo in the eighteenth century. The use of *D. trifida* was also observed by Lisboa and Le Cointe in Amazonia. Extracts from the roots were shown to be active in reducing inflammatory parameters associated with food allergies and were suggested to have the potential to prevent and treat this condition (Mollica et al. 2013).

Among plants with leaves that can be used as food, we present 52 species in Tables 2 and 3, with *Piper umbellatum* L. (Piperaceae) being cited by six among the nine studied authors, including Friar Vellozo. In recent studies this species has shown important results as intestinal anti-inflammatory (Arunachalam et al. 2020), anti-gastric ulcers (Silva Junior et al. 2016), and anticancer (Iwamoto et al. 2015). Other six species were also cited by Vellozo and have shown positive results in studies of bioactivity (Table 2), *Sesuvium portulacastrum* (L.) L. (Aizoaceae), which has antioxidant and antimicrobial activities (Al-azzawi et al. 2012; Chandrasekaran et al. 2011); *Amaranthus cruentus* L. (Amaranthaceae) which is anti-inflammatory (Tang and Tsao 2017) and useful in management of obesity (Kanikowska et al. 2019); *A. viridis* L., which has antioxidant properties and has high nutritional potential (Sarker and Oba 2019; Silva et al. 2018); *Sochus oleraceus* L. (Asteraceae), that also has activity as antioxidant (Mawalagedera et al. 2016) and as anti-aging agent (Ou et al. 2015); and *Pereskia aculeata* Mill. (Cactaceae), that has antioxidant and antimicrobial activities (Garcia et al. 2019; Souza et al. 2016) and improves intestinal motility and lipid profile (Barbalho et al. 2016). All these species have been recently considered as non-conventional food plant species ("plantas alimentícias não convencionais – PANCs," in Portuguese). They are plants of spontaneous development, easily found in gardens, backyards, and even on street sidewalks (Kinupp and Lorenzi 2014). Unfortunately, many of the PANC species are considered by the general population as weeds, which is why they are little used in food. Species of PANCs can have significantly higher contents of minerals, fibers, antioxidants, and proteins when compared to domesticated plants. The greater quantity of PANCs used today in Brazil corresponds to exotic plants, introduced in the country mainly by the Portuguese and Africans, brought as slaves. It is important to highlight that the plants presented in Tables 2 and 3 are native to Brazil. The species cited especially by the authors from the sixteenth to the eighteenth centuries, were most likely originally domesticated by Amerindians. An example is *Pereskia aculeata* Mill. (Ora-pro-nobis, Cactaceae, Table 2). Vellozo wrote that "…Africans use ora-pro-nobis instead of hibiscus…," *P. aculeata* is a native species while many species of hibiscus were introduced in Brazil from Africa. The use of PANCs as food has been highly encouraged in Brazil today and has been adopted by a high percentage of the population, and species present in Tables 1 and 2 are important in this movement due to their historical uses and bioactivities.

Leaves are also frequently used as tea, and the caffeinated tea-like beverages have been very popular all over the world, for centuries. Tea from *Camellia sinensis* L. (Theaceae), for example, has been consumed for millennia in India, China, and other Asian countries, but became very popular in England in the seventeenth century (Gomes 2019). When the Portuguese royal family came to Brazil in the nineteenth century, there was an interest in the introduction of the *C. sinensis*. Negotiations were made with China and the government brought in some tea seedlings and a few hundred Chinese for their cultivation. This attempt, however, was not successful, as the full development of the plant was not achieved, mainly due to the poor choice of planted species. On the other side, we have mate tea, produced from leaves of the tree *Ilex paraguariensis* A.St-Hil. (Aquifoliaceae), which has been consumed for millennia by Amerindians living in Argentina, Brazil, Chile, Paraguay, and Uruguay. The Jesuits cultivated mate tea on their missions in Paraguay in the sixteenth century, but formal taxonomic description of the plant, by the French naturalist Auguste de Saint-Hilaire, occurred only in the nineteenth century (Brandão et al. 2012). The French naturalist Aimé Bonpland, who explored the northern parts of South America with Alexander von Humboldt, studied the culture of mate extensively and described the best methods for cultivation, collection and preparation of tea (Corrêa-Filho 1957; Linhares 1969). Currently, mate tea is consumed worldwide as an alternative to coffee and black tea. Studies have demonstrated that leaves of *C. sinensis* and *I. paraguariensis* have similar chemical characteristics. Both are rich in polyphenols, proteins, enzymes, caffeine, carbohydrates, and inorganics, which provide beneficial properties for health. Several recent studies have confirmed activities as antimicrobial, antioxidant, anti-obesity, antidiabetic, and cardiovascular protective effects (Gan et al. 2018). Data about other ten *Ilex* species were found in the consulted bibliography (marked with [a] in Tables 2 and 3). They show the presence of phenolics and choleretic activity in previous studies (Gan et al. 2018), but have not been investigated in more recent research.

The leaves of *Echinodorus macrophyllus* (Kunth) Micheli (Alismataceae), *Ayapana triplinervis* (Vahl) R.M.King & H.Rob. (Asteraceae), *Citronella paniculata* (Mart.) R.A.Howard (Cardiopteridaceae), and *Neea theifera* Oerst. (Nyctaginaceae) are also used as tea. Only the first two have been studied and found to have antioxidant activities (Taïlé et al. 2020) and to show anti-inflammatory and nephroprotective activities (Portella et al. 2012; Silva et al. 2016). It is interesting to observe that these species belong to different families, and this fact increases the possibility of finding potential bioactive substances with different structures and activities. In addition, three flowers (*Langsdorffia hypogaea* Mart./Balanophoraceae, *Abutilon purpurascens* K.Schum, and *Sphaeralcea bonariensis* (Cav.) Griseb./Malvaceae), and two sprouts (*Gynerium sagittatum* (Aubl.) P.Beauv./Poaceae and *Euterpe oleracea* Mart, Arecaceae) were cited by the authors, but the potentials of none of them have been evaluated to date.

The recovery of old eating habits, using different species as food, is considered a strategy for health promotion (Bataglion et al. 2015; Teixeira et al. 2019). Previous studies have shown that phytochemicals, such as phenolic compounds and alkaloids commonly present in plants, can modulate inflammation and improve health problems associated with many degenerative and/or chronic diseases (Oliveira et al. 2014). This has stimulated a renewed interest in accessing plant foods in biodiversity centers, due to the high levels of bioactive substances, capable of preventing such as obesity, diabetes, cancer, coronary heart disease, and Alzheimer's (Neri-Numa et al. 2018). The polyphenols present in the leaves, for example, act as antioxidants due to their capacity to stabilize free radicals by breaking the chain of oxidative reactions and/or retarding the formation of free radicals in oxidizable matrices. These processes work as a defense mechanism of the human body, controlling lipid peroxidation and preventing the development of the degenerative diseases. Table 4 shows that 12 plants with antioxidant activity which is correlated with the presence of phenolics. Ou et al. (2015), for example, show the activity of an extract from *Sonchus oleraceus* L. (Asteraceae) as anti-aging in vitro. In the same

Table 4 Species whose consumption has been suggested to treat or prevent diseases

Plant with antioxidant effect (leaves)	Correlated diseases
Sonchus oleraceus L.	Anti-ageing effect
Amaranthus cruentus L. *Boerhavia erecta* L. *Echinodorus macrophyllus* (Kunth) Micheli *Polygonum aviculare* L. *Piper umbellatum* L.	Anti-inflammatory (many diseases)
Piper umbellatum L. *Xanthosoma sagittifolium* (L.) Schott	Antitumor/prevent colon cancer
Amaranthus spinosus L. *Colocasia esculenta* (L.) Schott *Neptunia oleracea Lour.*	Diabetes
Pereskia aculeata Mill.	Improves intestinal motility (constipation)
Amaranthus viridis L. *Pereskia aculeata* Mill. *Polygonum aviculare* L.	Lipid metabolism (cardiovascular)
Echinodorus macrophyllus (Kunth) Micheli *Sonchus oleraceus* L.	Nephroprotective
Amaranthus cruentus L. *Polygonum aviculare* L.	Obesity

way, Carey et al. (2017) have demonstrated that extract from açaí (*Euterpe oleracea* Mart., Arecaceae) is rich in phenolics, was also effective in prevent age-related cognitive deficits in aged rats. Studies have also shown a positive correlation in the ingestion of antioxidant substances as antidiabetic/antiobesity agents, and this effect occurs by different mechanisms (Cox et al. 2015). Other studies have shown correlation between anti-obesity effects of capsaicin with the presence of flavonoids and polyphenols. The authors have demonstrated that this effect can occur due to changes in the gut microbiota that led to significant impacts on the host's metabolism (Roopchand et al. 2015; Shen et al. 2017; van Dorsten et al. 2012). We can consider that encouraging the consumption of the plants from Table 2 represent an important tool to prevent these diseases, because of their beneficial use for centuries and studies that confirm their relevance as functional foods. We also argue that species from Table 3, and all other cited in historical documents, should have priority in the studies on bioactivity for developing biological products. Detailed, critical studies are essential in order to validate – or invalidate – claims concerning bioactivity and specificity, so that those plants with beneficial properties can be better exploited.

Brazil has the most diverse flora of the world, with 38,700 species approximately, which corresponds to 26.5% of the total of known species (Ulloa et al. 2017). The plants listed in this study confirm the great variety of native plant foods supplied by the Brazilian flora. The data recovered from this historical review covered four of the six main phytogeographic domains of the country: Atlantic Forest, Cerrado, Caatinga, and Amazonia. This information is also important because it was recorded in a time that Brazil's vegetation was still preserved, and the interactions between the Amerindians and the new inhabitants were frequent. The commercialization and management of these plants can also contribute to biodiversity conservation and improve the income of local populations.

4 Conclusion

In this study we collected written references to the traditional use of leaves and roots of native Brazilian plant species and summarize their suggested bioactive properties. These plants may have a great potential for use as functional foods, despite most of them being unknown to the population and to science. The valorization of biodiversity, through sustainable use of underexploited species, is a way to slow down the erosion of genetic diversity in remote regions, while studies with these plants can lead to development of innovative bio-products with high commercial value. We argue that incentives for better use of the species listed in this study must be considered and they should be included in agricultural programs that are done throughout the country.

Acknowledgments The authors are grateful to Conselho Nacional de Desenvolvimentos Científico e Tecnológico (CNPq) and Fundação de Amparo a Pesquisas de Minas Gerais

(FAPEMIG) for financial support in the development of Dataplamt. MGLB is funded by CNPq with a research fellowship (310389/2017-8). We are also in debt to Prof. Christine Clayton from University of Heidelberg for her help with text review.

References

Al-azzawi A, Alguboori A, Hachim MY et al (2012) Preliminary phytochemical and antibacterial screening of Sesuvium portulacastrum in the United Arab Emirates. Pharm Res 4:219–224. https://doi.org/10.4103/0974-8490.102269

Alves NSF, Setzer WN, da Silva JKR (2019) The chemistry and biological activities of Peperomia pellucida (Piperaceae): a critical review. J Ethnopharmacol 232:90–102. https://doi.org/10.1016/j.jep.2018.12.021

Arruda S, Souza EMT, Siqueira E (2005) Carotenoids from malanga (Xanthosoma Sagittifolium) leaves protect cells against oxidative stress in rats. Int J Vitam Nutr Res 75:161–168. https://doi.org/10.1024/0300-9831.75.2.161

Arunachalam K, Damazo AS, Macho A et al (2020) Piper umbellatum L. (Piperaceae): phytochemical profiles of the hydroethanolic leaf extract and intestinal anti-inflammatory mechanisms on 2,4,6 trinitrobenzene sulfonic acid induced ulcerative colitis in rats. J Ethnopharmacol 254:112707. https://doi.org/10.1016/j.jep.2020.112707

Azubuike NC, Onwukwe OS, Onyemelukwe AO et al (2018) Impact of Colocasia esculenta extract and fractions on high-fat diet-induced changes in body weight, adipose tissue and liver of rats. Pak J Pharm Sci 31:2149–2156

Barbalho SM, Guiguer ÉL, Marinelli PS et al (2016) Pereskia aculeata miller flour: metabolic effects and composition. J Med Food 19:890–894. https://doi.org/10.1089/jmf.2016.0052

Barnabé M, Saraceni CHC, Dutra-Correa M, Suffredini IB (2014) The influence of Brazilian plant extracts on Streptococcus mutans biofilm. J Appl Oral Sci 22:366–372. https://doi.org/10.1590/1678-775720140085

Bataglion GA, da Silva FMA, Eberlin MN, Koolen HHF (2015) Determination of the phenolic composition from Brazilian tropical fruits by UHPLC–MS/MS. Food Chem 180:280–287. https://doi.org/10.1016/j.foodchem.2015.02.059

Bavarva JH, Narasimhacharya AV (2013) Systematic study to evaluate anti-diabetic potential of Amaranthus spinosus on type-1 and type-2 diabetes. Cell Mol Biol (Noisy-le-Grand) 59:OL1818–OL1825

Bockmann FA, Rodrigues MT, Kohsldorf T et al (2018) Brazil's government attacks biodiversity. Science 360:865. https://doi.org/10.1126/science.aat7540

Brandão MGL (2019) Plantas úteis e medicinais nas obras de Frei Vellozo. 3i Editora, Belo Horizonte

Brandão MGL, Zanetti NNS, Oliveira GR et al (2008a) Plantas medicinais da Estrada Real. MG Biota 1:4–15

Brandão MGL, Zanetti NNS, Oliveira P et al (2008b) Brazilian medicinal plants described by 19th century European naturalists and in the Official Pharmacopoeia. J Ethnopharmacol 120:141–148. https://doi.org/10.1016/j.jep.2008.08.004

Brandão MGL, Grael CF, Fagg CW (2011) European naturalists and medicinal plants of Brazil. In: Biological diversity and sustainable resources use. InTech, pp 101–120

Brandão MGL, Pignal M, Romaniuc S et al (2012) Useful Brazilian plants listed in the field books of the French naturalist Auguste de SaintHilaire (1779–1853). J Ethnopharmacol 143:488–500. https://doi.org/10.1016/j.jep.2012.06.052

Breitbach UB, Niehues M, Lopes NP et al (2013) Amazonian Brazilian medicinal plants described by C.F.P. von Martius in the 19th century. J Ethnopharmacol 147:180–189. https://doi.org/10.1016/j.jep.2013.02.030

Carey AN, Miller MG, Fisher DR et al (2017) Dietary supplementation with the polyphenol-rich açaí pulps (Euterpe oleracea Mart. and Euterpe precatoria Mart.) improves cognition in aged rats and attenuates inflammatory signaling in BV-2 microglial cells. Nutr Neurosci 20:238–245. https://doi.org/10.1080/1028415X.2015.1115213

Caxito MLC, Correia RR, Gomes ACC et al (2015) In Vitro Antileukemic Activity of Xanthosoma sagittifolium (Taioba) Leaf Extract. Evid Based Complement Alternat Med 2015:1–10. https://doi.org/10.1155/2015/384267

Chandrasekaran M, Senthilkumar A, Venkatesalu V (2011) Fatty acid methyl esters from the leaves of Sesuvium portulacastrum L. Eur Rev Med Pharmacol 15:775

Chen Y, Tang YM, Yu SL et al (2015) Advances in the pharmacological activities and mechanisms of diosgenin. Chin J Nat Med 13:578–587. https://doi.org/10.1016/S1875-5364(15)30053-4

Compaore M, Bakasso S, Meda R et al (2018) Antioxidant and anti-inflammatory activities of fractions from Bidens engleri O.E. Schulz (Asteraceae) and Boerhavia erecta L. (Nyctaginaceae). Medicines 5:53. https://doi.org/10.3390/medicines5020053

Corrêa-Filho V (1957) Ervais e do Brasil ervateiros. Ministério da Agricultura, Rio de Janeiro

Cox AJ, West NP, Cripps AW (2015) Obesity, inflammation, and the gut microbiota. Lancet Diabetes Endocrinol 3:207–215. https://doi.org/10.1016/S2213-8587(14)70134-2

Cruz R, Denardi L, Mossmann N et al (2016) Antimicrobial activity and chromatographic analysis of extracts from Tropaeolum pentaphyllum lam. Tubers. Molecules 21:566. https://doi.org/10.3390/molecules21050566

de Almeida Jackix E, Monteiro EB, Raposo HF, Vanzela EC, Amaya-Farfán J (2013) Taioba (Xanthosoma sagittifolium) Leaves: Nutrient Composition and Physiological Effects on Healthy Rats. J Food Sci 78:1929–1934. https://doi.org/10.1111/1750-3841.12301

De Bona GS, Boschetti W, Bortolin RC et al (2017) Characterization of dietary constituents and antioxidant capacity of Tropaeolum pentaphyllum Lam. J Food Sci Technol 54:3587–3597. https://doi.org/10.1007/s13197-017-2817-z

de Paiva SR, Figueiredo MR, Aragão TV, Kaplan MAC (2003) Antimicrobial activity in vitro of plumbagin isolated from Plumbago species. Mem Inst Oswaldo Cruz 98:959–961. https://doi.org/10.1590/S0074-02762003000700017

Dean WA (1996) A ferro e fogo: a história da devastação da Mata Atlântica brasileira. Companhia das Letras, Rio de Janeiro, p 484

Domingues HMB, Kleiche-Dray M, Petitjean P (2012) Historia das substâncias naturais: saberes tradicionais e quimica: Amazônia e América Latina. Museu de Astronomia e Ciências Afins/Institut de Recher chepourle Développement, Rio de Janeiro/Paris

Eleazu CO, Eleazu KC, Iroaganachi MA (2016) Effect of cocoyam (*Colocasia esculenta*), unripe plantain (*Musa paradisiaca*) or their combination on glycated hemoglobin, lipogenic enzymes, and lipid metabolism of streptozotocin-induced diabetic rats. Pharm Biol 54(1):91–97. https://doi.org/10.3109/13880209.2015.1016181

Fagg CW, Lughadha EN, Milliken W et al (2015) Useful Brazilian plants listed in the manuscripts and publications of the Scottish medic and naturalist George Gardner (1812–1849). J Ethnopharmacol 161:18–29. https://doi.org/10.1016/j.jep.2014.11.035

FAO (2013) Cassava's huge potential as 21st Century crop. http://www.fao.org/news/story/en/item/176780/icode/. Accessed 10 Aug 2020

Fernandes TM (2004) Plantas Medicinais: memória da ciência no Brasil. Editora Fiocruz, Rio de Janeiro

Ferrão JEM, Caixinhas ML, Liberato MC (2008) A ecologia, as plantas e a interculturalidade. Percursos de Interculturalidade, Portugal

Fokam Tagne MA, Noubissi PA, Fankem GO, Kamgang R (2018) Effects of Oxalis barrelieri L. (Oxalidaceae) aqueous extract on diarrhea induced by Shigella dysenteriae type 1 in rats. Heal Sci Rep 1:e20. https://doi.org/10.1002/hsr2.20

Gan RY, Zhang D, Wang M, Corke H (2018) Health benefits of bioactive compounds from the genus Ilex, a source of traditional caffeinated beverages. Nutrients 10:1682. https://doi.org/10.3390/nu10111682

Garcia JAA, Corrêa RCG, Barros L et al (2019) Phytochemical profile and biological activities of "Ora-pro-nobis" leaves (Pereskia aculeata Miller), an underexploited superfood from the Brazilian Atlantic Forest. Food Chem 294:302–308. https://doi.org/10.1016/j.foodchem.2019.05.074

Gomes L (2019) Escravidão: Do Primeiro Leilão de Cativos em Portugal até a Morte de Zumbi dos Palmares. Globo Livros, Rio de Janeiro

Granica S, Czerwińska ME, Żyżyńska-Granica B, Kiss AK (2013) Antioxidant and anti-inflammatory flavonol glucuronides from Polygonum aviculare L. Fitoterapia 91:180–188. https://doi.org/10.1016/j.fitote.2013.08.026

Hata Y, Reguero MT, Garcia LA, Buitrago G, Alvarez A (2003) Evaluacio´n del contenido de sapogeninas em variedades nativas de inhame (Dioscorea spp.) provenientes de la coleccion de La Universidad del Córdoba. Rev Colomb 32:149–157

Iwamoto LH, Vendramini-Costa DB, Monteiro PA et al (2015) Anticancer and anti-inflammatory activities of a standardized dichloromethane extract from Piper umbellatum L. leaves. Evid Based Complement Alternat Med 2015:1–8. https://doi.org/10.1155/2015/948737

Jackix EA, Monteiro EB, Raposo HF et al (2013) Taioba (Xanthosoma sagittifolium) leaves: nutrient composition and physiological effects on healthy rats. J Food Sci 78:H1929–H1934. https://doi.org/10.1111/1750-3841.12301

Kahleova H, Levin S, Barnard N (2017) Cardio-metabolic benefits of plant-based diets. Nutrients 9:848. https://doi.org/10.3390/nu9080848

Kanikowska D, Kanikowska A, Rutkowski R et al (2019) Amaranth (Amaranthus cruentus L.) and canola (Brassica napus L.) oil impact on the oxidative metabolism of neutrophils in the obese patients. Pharm Biol 57:140–144. https://doi.org/10.1080/13880209.2019.1569696

Kinupp WK, Lorenzi H (2014) Plantas Alimentícias Não Convencionais (PANC) no Brasil. Instituto Plantarum, Nova Odessa

Kumalasari ID, Harmayani E, Lestari LA et al (2012) Evaluation of immunostimulatory effect of the arrowroot (Maranta arundinacea. L) in vitro and in vivo. Cytotechnology 64:131–137. https://doi.org/10.1007/s10616-011-9403-4

Lee SY, Mediani A, Ismail IS et al (2019) Antioxidants and α-glucosidase inhibitors from Neptunia oleracea fractions using 1H NMR-based metabolomics approach and UHPLC-MS/MS analysis. BMC Complement Altern Med 19:7. https://doi.org/10.1186/s12906-018-2413-4

Ling N, Zhou X, Ji Y, Li W, Ji C, Qi Z (2017) Immuno-modulatory and cellular antioxidant activities of κ-selenocarrageenan in combination with Epirubicin in H22 hepatoma-bearing mice. Biomed Pharmacother 91:132–137. https://doi.org/10.1016/j.biopha.2017.04.064

Linhares T (1969) História econômica do mate. Livraria José Olympio, Rio de Janeiro

Maezumi SY, Alves D, Robinson M, Souza JG, Levis G, Barnett RL, Oliveira EA, Urrego D, Schaan D (2018) Iriarte JR (2018) The legacy of 4,500 years polyculture agroforestry in the eastern Amazon. Nature 4:540–547

Mallard B, Leach DN, Wohlmuth H, Tiralongo J (2019) Synergistic immuno-modulatory activity in human macrophages of a medicinal mushroom formulation consisting of Reishi, Shiitake and Maitake. PLoS One 14(11):e0224740. https://doi.org/10.1371/journal.pone.0224740

Maria-Ferreira D, da Silva LM, Mendes DA, Cabrini Dde A, Nascimento AM, Iacomini M, Cipriani TR, Santos AR, Werner MF, Baggio CH (2014) Rhamnogalacturonan from *Acmella oleracea* (L.) R.K. Jansen: gastroprotective and ulcer healing properties in rats. PLoS One 9(1):e84762. https://doi.org/10.1371/journal.pone.0084762

Marques JF (1996) Frei Cristóvão de Lisboa, missionário no Maranhão e Grão Pará (1624-1635) e a defesa dos índios brasileiros. Rev da Fac Let 13:323–351

Mawalagedera SMMR, Ou Z-Q, McDowell A, Gould KS (2016) Effects of boiling and in vitro gastrointestinal digestion on the antioxidant activity of Sonchus oleraceus leaves. Food Funct 7:1515–1522. https://doi.org/10.1039/C5FO01508A

Melina V, Craig W, Levin S (2016) Position of the academy of nutrition and dietetics: vegetarian diets. J Acad Nutr Diet 116:1970–1980. https://doi.org/10.1016/j.jand.2016.09.025

Metwaly AM, Lianlian Z, Luqi H, Deqiang D (2019) Black ginseng and its saponins: preparation, phytochemistry and pharmacological effects. Molecules 24(10):1856. https://doi.org/10.3390/molecules24101856

Mishra S, Aeri V, Gaur PK, Jachak SM (2014) Phytochemical, therapeutic, and ethnopharmacological overview for a traditionally important herb: boerhavia diffusa Linn. Biomed Res Int 2014:1–19. https://doi.org/10.1155/2014/808302

Mollica JQ, Cara DC, D'Auriol M et al (2013) Anti-inflammatory activity of American yam Dioscorea trifida L.f. in food allergy induced by ovalbumin in mice. J Funct Foods 5:1975–1984. https://doi.org/10.1016/j.jff.2013.09.020

Mondal A, Maity TK (2016) Antibacterial activity of a novel fatty acid (14E, 18E, 22E, 26E)-methyl nonacosa-14, 18, 22, 26 tetraenoate isolated from Amaranthus spinosus. Pharm Biol 54:2364–2367. https://doi.org/10.3109/13880209.2016.1155628

Mügge FL, Paula-Souza J, Melo JC, Brandão MG (2016) Native plant species with economic value from Minas Gerais and Goiás: a discussion on the currentness of the data recovered by the French naturalist Auguste de Saint-Hilaire. Hortic Bras 34:455–462. https://doi.org/10.1590/s0102-053620160402

Nascimento AM, de Souza LM, Baggio CH, Werner MF, Maria-Ferreira D, da Silva LM, Sassaki GL, Gorin PA, Iacomini M, Cipriani TR (2013) Gastroprotective effect and structure of a rhamnogalacturonan from *Acmella oleracea*. Phytochemistry 85:137–142. https://doi.org/10.1016/j.phytochem.2012.08.024

Neri-Numa IA, Soriano Sancho RA, Pereira APA, Pastore GM (2018) Small Brazilian wild fruits: nutrients, bioactive compounds, healthpromotion properties and commercial interest. Food Res Int 103:345–360. https://doi.org/10.1016/j.foodres.2017.10.053

Neves DA, Schmiele M, Pallone JAL et al (2019) Chemical and nutritional characterization of raw and hydrothermal processed jambu (Acmella oleracea (L.) R.K. Jansen). Food Res Int 116:1144–1152. https://doi.org/10.1016/j.foodres.2018.09.060

Okafor EN, Erukainure OL, Ozumba AU et al (2017) Cassava flour substitution modulates Glycemic responses and Glycemic index of wheat breads in apparent healthy volunteers. J Diet Suppl 14:446–452. https://doi.org/10.1080/19390211.2016.1267061

Oliveira VB, Yamada LT, Fagg CW, Brandão MGL (2012) Native foods from Brazilian biodiversity as a source of bioactive compounds. Food Res Int 48:170–179. https://doi.org/10.1016/j.foodres.2012.03.011

Oliveira VB, Ferreira AVM, Oliveira MC, Teixeira MM, Brandão MGL (2014) Effects of Xylopia aromatica (Lam.) Mart. fruit on metabolic and inflammatory dysfunction induced by high refined carbohydrate-containing-diet in mice. Food Res Int 62:541–550. https://doi.org/10.1016/j.foodres.2014.03.066

Ou ZQ, Rades T, McDowell A (2015) Anti-ageing effects of Sonchus oleraceus L. (pūhā) leaf extracts on H2O2-induced cell senescence. Molecules 20:4548–4564. https://doi.org/10.3390/molecules20034548

Özenver N, Güvenalp Z, Kuruüzüm-Uz A, Demirezer LÖ (2020) Inhibitory potential on key enzymes relevant to type II diabetes mellitus and antioxidant properties of the various extracts and phytochemical constituents from Rumex acetosella L. J Food Biochem 44(10):e13415. https://doi.org/10.1111/jfbc.13415

Park CJ, Lee H-A, Han J-S (2016) Jicama (Pachyrhizus erosus) extract increases insulin sensitivity and regulates hepatic glucose in C57BL/Ksjdb/db mice. J Clin Biochem Nutr 58:56–63. https://doi.org/10.3164/jcbn.15-59

Park JY, Kim CS, Park K-M, Chang P-S (2019) Inhibitory characteristics of flavonol-3-O-glycosides from Polygonum aviculare L. (common knotgrass) against porcine pancreatic lipase. Sci Rep 9:18080. https://doi.org/10.1038/s41598-019-54546-8

Peckolt T, Peckolt G (2016) História das Plantas Medicinais e Úteis do Brasil. Fino Traço, Belo Horizonte

Pereira PR, Corrêa ACNTF, Vericimo MA, Paschoalin VMF (2018) Tarin, a potential Immunomodulator and COX-inhibitor lectin found in Taro (Colocasia esculenta). Compr Rev Food Sci Food Saf 17:878–891. https://doi.org/10.1111/1541-4337.12358

Piso G (1648) História Natural e Médica da Índia Ocidental. Tradução Mário Lôbo Leal, Rio de Janeiro: Ministério da Educação e Cultura. Instituto Nacional do Livro; 1957. 682p.

Portella VG, Cosenza GP, Diniz LRL et al (2012) Nephroprotective Effect of Echinodorus macrophyllus Micheli on Gentamicin-Induced Nephrotoxicity in Rats. Nephron Extra 2:177–183. https://doi.org/10.1159/000339181

Rajashekhara N, Ashok B, Sharma P, Ravishankar B (2014) Evaluation of acute toxicity and antiulcerogenic study of rhizome starch of two source plants of Tugaksheeree (Curcuma angustifolia Roxb. and Maranta arundinacea Linn.). AYU (An Int Q J Res Ayurveda) 35:433. https://doi.org/10.4103/0974-8520.159013

Reis LFC, Cerdeira CD, Paula BF et al (2015) Chemical characterization and evaluation of antibacterial, antifungal, antimycobacterial, and cytotoxic activities of Talinum paniculatum. Rev Inst Med Trop Sao Paulo 57:397–405. https://doi.org/10.1590/S0036-46652015000500005

Ricardo LM, de Paula-Souza J, Andrade A, Brandão MGL (2017) Plants from the Brazilian Traditional Medicine: species from the books of the Polish physician Piotr Czerniewicz (Pedro Luiz Napoleão Chernoviz, 1812–1881). Rev Bras Farmacogn 27:388–400. https://doi.org/10.1016/j.bjp.2017.01.002

Roopchand DE, Carmody RN, Kuhn P et al (2015) Dietary Polyphenols Promote Growth of the Gut Bacterium Akkermansia muciniphila and Attenuate High-Fat Diet–Induced Metabolic Syndrome. Diabetes 64:2847–2858. https://doi.org/10.2337/db14-1916

Salako OA, Akindele AJ, Shitta OM et al (2015) Antidiarrhoeal activity of aqueous leaf extract of Caladium bicolor (Araceae) and its possible mechanisms of action. J Ethnopharmacol 176:225–231. https://doi.org/10.1016/j.jep.2015.10.035

Salama HMH, Marraiki N (2010) Antimicrobial activity and phytochemical analyses of Polygonum aviculare L. (Polygonaceae), naturally growing in Egypt. Saudi J Biol Sci 17:57–63. https://doi.org/10.1016/j.sjbs.2009.12.009

Santos ATL, Pereira Carneiro JN, Pereira da Cruz R et al (2019) UPLC-MS-ESI-QTOF analysis and antifungal activity of the Spondias tuberosa Arruda leaf and root hydroalcoholic extracts. Antibiotics 8:240. https://doi.org/10.3390/antibiotics8040240

Santos-Fonseca DJ, Coelho-Ferreira M, Fonseca-Kruel V (2019) Useful plants referenced by the naturalista Richar Spruce in the 19th century in the state of Pará, Brazil. Acta Bot Brasilica 33(2):221–231. https://doi.org/10.1590/0102-33062018abb0344

Santoso P, Amelia A, Rahayu R (2019) Jicama (Pachyrhizus erosus) fiber prevents excessive blood glucose and body weight increase without affecting food intake in mice fed with high-sugar diet. J Adv Vet Anim Res 6:222. https://doi.org/10.5455/javar.2019.f336

Sarker U, Oba S (2019) Nutraceuticals, antioxidant pigments, and phytochemicals in the leaves of Amaranthus spinosus and Amaranthus viridis weedy species. Sci Rep 9:20413. https://doi.org/10.1038/s41598-019-50977-5

Schmourlo G, Mendonça-Filho RR, Alviano CS, Costa SS (2005) Screening of antifungal agents using ethanol precipitation and bioautography of medicinal and food plants. J Ethnopharmacol 96:563–568. https://doi.org/10.1016/j.jep.2004.10.007

Seaward MRD (2000) Richard Spruce, botânico e desbravador da América do Sul. História, Ciências, Saúde-Manguinhos 7:379–390. https://doi.org/10.1590/S0104-59702000000300007

Shen W, Shen M, Zhao X et al (2017) Anti-obesity effect of capsaicin in mice fed with high-fat diet is associated with an increase in population of the gut bacterium akkermansia muciniphila. Front Microbiol 8:272. https://doi.org/10.3389/fmicb.2017.00272

Shi ZY, Zeng JZ, Wong (2019) Chemical structures and pharmacological profiles of ginseng saponins. Molecules 24(13):2443. https://doi.org/10.3390/molecules24132443. PMID: 31277214

Silva Junior IF, Balogun SO, de Oliveira RG et al (2016) Piper umbellatum L.: a medicinal plant with gastric-ulcer protective and ulcer healing effects in experimental rodent models. J Ethnopharmacol 192:123–131. https://doi.org/10.1016/j.jep.2016.07.011

Silva GP, Fernandes DC, Vigliano MV et al (2016) Flavonoid-enriched fraction from Echinodorus macrophyllus aqueous extract exhibits high in-vitro and in-vivo anti-inflammatory activity. J Pharm Pharmacol 68:1584–1596. https://doi.org/10.1111/jphp.12620

Silva LFLE, Souza DC, Resende LV et al (2018) Nutritional evaluation of non-conventional vegetables in Brazil. An Acad Bras Cienc 90:1775–1787. https://doi.org/10.1590/0001-3765201820170509

Sousa GS (1587) Tratado Descritivo do Brasil em 1587. São Paulo: Companhia Editora Nacional; 1987. 389p.

Souza L, Caputo L, Inchausti De Barros I et al (2016) Pereskia aculeata Muller (Cactaceae) leaves: chemical composition and biological activities. Int J Mol Sci 17:1478. https://doi.org/10.3390/ijms17091478

Sung YY, Yoon T, Yang WK, Kim SJ, Kim DS, Kim HK (2013) The antiobesity effect of polygonum aviculare l. ethanol extract in high-fat diet-induced obese mice. Evid Based Complement Alternat Med 2013:626397. https://doi.org/10.1155/2013/626397

Taïlé J, Arcambal A, Clerc P et al (2020) Medicinal plant polyphenols attenuate oxidative stress and improve inflammatory and vasoactive markers in cerebral endothelial cells during Hyperglycemic condition. Antioxidants 9:573. https://doi.org/10.3390/antiox9070573

Tang Y, Tsao R (2017) Phytochemicals in quinoa and amaranth grains and their antioxidant, anti-inflammatory, and potential health beneficial effects: a review. Mol Nutr Food Res 61:1600767. https://doi.org/10.1002/mnfr.201600767

Teixeira N, Melo JCS, Batista LF et al (2019) Edible fruits from Brazilian biodiversity: a review on their sensorial characteristics versus bioactivity as tool to select research. Food Res Int 119:325–348. https://doi.org/10.1016/j.foodres.2019.01.058

Udeme N, Okafor P, Eleazu C (2015) The metabolic effects of consumption of yellow cassava (Manihot esculenta Crantz) on some biochemical parameters in experimental rats. Int J Toxicol 34:559–564. https://doi.org/10.1177/1091581815606085

Ulloa UC, Acevedo-Rodríguez P, Beck S et al (2017) An integrated assessment of the vascular plant species of the Americas. Science 358:1614–1617. https://doi.org/10.1126/science.aao0398

Valli M, Russo HM, Bolzani VS (2018) The potential contribution of the natural products from Brazilian biodiversity to bioeconomy. An Acad Bras Cienc 90:763–778. https://doi.org/10.1590/0001-3765201820170653

van Dorsten FA, Peters S, Gross G et al (2012) Gut microbial metabolism of polyphenols from black tea and red wine/grape juice is SourceSpecific and colon-region dependent. J Agric Food Chem 60:11331–11342. https://doi.org/10.1021/jf303165w

Venâncio MD, Pope K, Sieber S (2018) Brazil's new government threatens food security and biodiversity. Nature 564:39–39. https://doi.org/10.1038/d41586-018-07611-7

Wang YC, Huang T-L (2005) Screening of anti-Helicobacter pylori herbs deriving from Taiwanese folk medicinal plants. FEMS Immunol Med Microbiol 43:295–300. https://doi.org/10.1016/j.femsim.2004.09.00

Natural Toxins in Brazilian Unconventional Food Plants: Uses and Safety

Fillipe de Oliveira Pereira, Francinalva Dantas de Medeiros, and Patrícia Lima Araújo

Contents

1 Are Unconventional Food Plants Toxic?

Brazil presents one of the greatest biodiversities in the world. Whether native, cultivated, or naturalized, over 48,572 species are currently recognized in the Brazilian flora. Data provided by the Brazilian Ministry of the Environment reveals more than 200 thousand registered species in the Amazon, Caatinga, Cerrado, Atlantic Forest, Pantanal, and Pampa biomes. This approaches 15% of all plant species on the planet (do Brasil 2020; Flora do Brasil 2020 em construção 2020).

Many of these native Brazilian species can be used as food sources, but their potential is still little known, either due to the lack of common knowledge, to restricted distribution/or use in the country, or to simple negligence. Such plants are currently known as unconventional food plants (UFP). UFP refers to plants that have one or more parts with food potential yet with unusual use; those which are not quickly recognized or acquired by the population of a given community; and plants that generally have no market value, being only sold in small scale, or are discarded, or plants which only possess one quality that is important to the market. These UFP can be described as native or exotic, cultivated or spontaneous, and can be adapted to local environments (Kinupp and de Barros 2007). Taken together, UFP become

F. de Oliveira Pereira (✉) · F. D. de Medeiros · P. L. Araújo
Education and Health Center, Federal University of Campina Grande, Cuité, Brazil
e-mail: fillipe.oliveira@professor.ufcg.edu.br; fdmedeiros@ufcg.edu.br

M. C. M. Jacob, U. P. Albuquerque (eds.), *Local Food Plants of Brazil*, Ethnobiology, https://doi.org/10.1007/978-3-030-69139-4_6

an important strategy towards promoting food and nutritional security, and the sustainable use of such natural resources can guarantee the autonomy of local populations, making it urgent to vindicate and expand our understanding of their potential for rational use (Leal et al. 2018).

In a given population, consumption of certain UFP can occur in times of food scarcity or in emergency situations, when the plants become categorized as "famine food plants." Native plant species are the most often utilized in such situations, and in some cases there are reports after consumption of undesirable effects such as stomach complaints, constipation, diarrhea, and even signs of intoxication (Guinand and Lemessa 2001; Nascimento et al. 2018).

The substances involved in triggering such undesirable effects are derived from the plant's biochemical metabolism. Secondary metabolites, or natural products, are so-called because they are not directly involved in the primary processes of basic growth and development but in mechanisms of plant adaptation and defense. These modulate as a way of responding to environmental stress, attracting pollinators or seed dispersers, and defending against herbivores, insects, fungi, and phytopathogenic bacteria (Bennett and Wallsgrove 1994; Shih and Morgan 2020).

Widely used in pharmaceutical research, these natural products determine vegetable taste and odor and can make them unpalatable, with either beneficial or harmful effects on humans and animal health upon ingestion. As an example, plants rich in cyanogenic glycosides or glycosinolates can cause serious problems with ingestion. A traditional plant, manioc (*Manihot esculenta* Crantz) being native to Brazil, is well-known for presenting cyanogenic glycosides which if consumed form hydrogen cyanide, resulting in food poisoning. However, local traditional culture has developed an adequate preparation technique for this species, thus eliminating the natural toxin (Jackson-Malete et al. 2015).

2 Natural Toxins

Natural toxins present wide structural variety; they are toxic compounds produced by plants and can induce various undesirable effects in human health. The toxins are useful for the plants, as adaptations or in defense systems, but they can be harmful to humans when consumed (Jackson-Malete et al. 2015). These natural toxins can also present anti-nutritional properties which can limit the bioavailability of nutrients present in food (Samtiya et al. 2020). The principal classes of secondary metabolites typically present compounds with the characteristics of natural toxins including alkaloids, cyanogenic glycosides, coumarins, tannins, and terpenes, among others (Yamane et al. 2010).

Alkaloids are a diverse group of secondary metabolites, with more than 12,000 molecules reported from about 150 plant families. Their main function is to defend against predators (from vertebrates to fungi) and from other plants, thus being potentially toxic when consumed by humans. Alkaloids are an intrinsic part of the Western diet, and throughout history there have been many case reports of poisoning, whether accidental or through deliberate food poisonings. The compounds in this class can be derived from ornithine (pyrrolizidine and tropane alkaloids), lysine (piperidine and quinolizidine alkaloids), tyrosine (isoquinoline alkaloids), tryptophan (alkaloids from b-carboline, quinoline, and ergot), nicotinic acid (nicotine and myosmine), adenine/guanine (methylxanthine), or glycoalkaloids. Pyrrolizidine alkaloids can bioactivate certain reactive alkylating intermediates, and various quinolizidine, β-carboline, ergot alkaloids, and steroids act as neurotoxins. Despite this, we still lack safe definitions and/or tolerable values for daily intake of many of the alkaloids present in the modern food chain (Cirlini et al. 2018; Koleva et al. 2012).

It is important to differentiate the use of plants that contain natural toxins from the use of these same substances in isolation, since the concentrations they present in the plant define whether they present a health risk or not. This exists in addition to factors such as interaction with other foods and methods of preparation. An example of this tea from the leaves of *Erytroxylum Coca* Lam., whether as food or medicine, being very common in many countries in South America, contains high concentrations of tropane alkaloids, such as those derived from cocaine, tropinone, and calystegines. But is safe given the low concentrations present in the species (respecting moderate use), and the way in which the tea is prepared, which causes degradation of the active substances (Marín-Sáez et al. 2019).

Non-proteic amino acids are often found in vegetables and seeds and can be toxic because they present similar structures, yet lead to production of unnatural proteins that may not work properly. As examples from the UFP species *Canavalia ensiformis* (L.) DC (pig bean), we find the non-proteic amino acid canavanine, an analog of the amino acid arginine; and in *Beta vulgaris* L. (beet), we find azetidine-2-carboxylic acid, an inferior homolog of proline, associated with disorders in neurodevelopment and autoimmune diseases (Rubenstein 2020; Yamane et al. 2010).

When degraded by plant enzymes after maceration (or in the intestinal microbiota after ingesting plant material), cyanogenic glycosides (mentioned above) release hydrogen cyanide (HCN), with a ketone or aldehyde which are highly toxic. Yet glucosinolates are widely distributed in plant tissues. By themselves, they are not toxic to herbivores, but in the presence of the enzyme myrosinase, and depending on the pH, they can degrade into isothiocyanates or nitriles (Kuti and Konoru 2006; Yamane et al. 2010). Certain diseases, such as konzo, tropical ataxic neuropathy, goiter, and cretinism (due to iodine deficiency), are among others associated with chronic food intake involving cyanogenic glycosides. We note indications that individuals with malnutrition are more susceptible to the effects of cyanogenic glycosides (Cressey and Reeve 2019). Almonds, pomegranates, apple and pear seeds, cassava, flaxseed, spinach, lupine, plum, and apricot kernels among others, all contain cyanogenic glycosides.

Phenolic compounds are a class of natural products with great structural diversity, presenting at least one aromatic ring in which at least one hydrogen is replaced by a hydroxyl group. They are widely distributed in the plant kingdom and in many species used as food. The group subclasses of flavonoids, tannins, coumarins, phenolic acids, and lignans serve as examples.

Flavonoids protect plants against UV radiation and pathogenic microorganisms and are also involved in plant pigmentation (Perez-Vizcaino and Fraga 2018). Flavonoids are currently considered safe for consumption, mainly due to their antioxidant properties, and in protection against chronic and cardiovascular diseases. Yet harmful effects have been seen at high doses, most commonly in food supplements enriched with flavonoid compounds (Kozłowska and Szostak-Węgierek 2018; Perez-Vizcaino and Fraga 2018).

Tannins are polyphenols with molecular weights greater than 500 Da and, for the well-known antioxidant potential of phenolic compounds, are believed to improve the oxidative condition present in *diabetes mellitus* (Kumari and Jain 2012). However, tannins have a strong capacity to form complexes with proteins and digestive enzymes, impairing their activity, and thus human digestion. Depending on the dose, interference in the digestive process can seriously affect absorption of nutrients and minerals such as iron and zinc (Higashijima et al. 2019). For this reason, the group is widely recognized as an anti-nutritional factor. As an example of foods rich in these compounds, we have red wines, teas, green fruits, grapes, apples, cocoa, persimmons, and bananas, among others.

Coumarins are a metabolite class that occurs naturally in many plants and consists of molecules carrying a 1,2-benzopyrone nucleus. *Dipteryx odorata* (Aubl.) Willd. (tonka beans or cumaru) is a significant example of a vegetable source of coumarins. In foods, these metabolites commonly bring strong flavors, yet also important hepatotoxic and carcinogenic properties. Furanocoumarins present phototoxicity (Abraham et al. 2010). Coumarins have an important history in health regulation, and additions to food are prohibited in the United States (21 CFR 189.130). The European Union imposes limits on their intake (European Regulation EC 1334/2008). Information such as this alerts us to inappropriate use of coumarins in food, but may not be applied to all representatives of the coumarin group.

Still, certain phenolic groups such as monophenols and diphenols undergo enzymatic oxidation by polyphenols oxidases and peroxidases, to become reactive molecular species that covalently bind to nutritionally important amino acid residues, such as cysteine and lysine, decreasing their nutritional value.

Terpenes, which are derived from isoprene units, present a wide range of biological structures and activities. Phytoecdysteroids, cardenolides, and iridoid glycosides present toxicity to mammals, and quinoa is a food rich in phytoecdysteroids (Yamane et al. 2010).

Carotenoids are found in many vegetables. They are divided into two main groups: carotenes, which are hydrocarbons – such as β-carotene and lycopene – and xanthophylls, which include oxygen in addition to hydrogen and carbon, such as lutein, zeaxanthin, bixin, and norbixin. Lipophilicity and the presence of conjugated double bonds within the molecule cause carotenoids to slow lipid peroxidation and stabilize cell membranes. Carotenoids are molecules with antioxidant potential (Hammond and Renzi-hammond 2013).

Saponins are widely distributed compounds with great structural variety. They consist of a hydrophobic steroid or triterpene aglycone linked to sugar chains (hydrophilic part) (Oleszek and Oleszek 2020). Plants very rich in saponins have limited use because high concentrations yield a very bitter taste. In addition, it is reported that saponins inhibit certain digestive enzymes. There are also reports in the literature of anti-nutritional effects induced by the formation of iron and zinc complexes which reduce absorption. In addition, by inhibition of digestive enzymes, saponins influence carbohydrate, lipid, and protein bioavailability (Higashijima et al. 2019). For example, saponin extracts from various types of legumes consumed have been reported in China as inhibiting the activity of the pancreatic lipase and α-glycosidase digestive enzymes (Liu and Xu 2015). Depending on consumption, the effects can be harmful since nutrient demand varies by individual; whether one is in post-surgical recovery, a child, or a woman in gestation. Yet, saponins might be interesting in cases of obesity and diabetes. In general, foods rich in saponins present little or low toxicity. The saponin content in foods decreases with cooking; this improves nutritional availability (Samtiya et al. 2020). Examples of plants containing saponins are oats, quinoa, peppers, eggplant, tomato seeds, and yams, among others.

Oxalic acid is considered an anti-nutrient because it prevents calcium absorption and can induce undesirable effects. Reports of excessive intake of foods rich in oxalic acids are relatively uncommon, but their undesirable effects usually involve gastrointestinal effects, hypocalcemia, and renal toxicity (Dassanayake and Gnanathasan 2012). Although the lethal oral dose of oxalic acid is very high (from 5 to 30 g in adults), when absorbed, it can react with calcium in the plasma and form calcium oxalate crystals, inducing renal damage through precipitation in the kidneys (Tsujihata 2008). Oxalic acid can interfere with the absorption of minerals such as magnesium, sodium, and potassium. Examples of oxalic acid-rich food plants are spinach, beets, chard, star fruit, cocoa, sweet potatoes, tomatoes, and black tea, among others (Higashijima et al. 2019).

Nitrites and nitrates interfere with vitamin A metabolism and thyroid gland function. Excessive consumption is associated with hypoxia or acidotic situations, in which nitrite is converted into nitric oxide (NO) (Higashijima et al. 2019). In cell signaling, NO activates soluble guanylyl cyclase to generate the guanosine monophosphate (cGMP) second messenger. Increasing the intracellular concentration of cGMP induces vessel relaxation and better blood distribution in local tissue. cGMP also stimulates mitochondrial biogenesis and oxidation of fatty acids in skeletal muscle; generates energy; and increases metabolic efficiency (Ashmore et al. 2015). These are interesting perspectives for clinical improvement in metabolic syndrome or for aerobic performance in athletes. From a functional perspective, vegetables that are naturally rich in inorganic nitrate positively affect NO biology. A classic example of this is beet juice consumption (Sweazea et al. 2017). Yet, as observed by Santos (2006), cooking of plants in culinary preparations decreases the availability of their nitrates. As studies progress, and when considering the results presented in this chapter (see Table 1), the beneficial and safe effects of nitrate-rich vegetables are better understood.

In a scenario of expanding knowledge concerning the use of UFP in Brazil, the data bring important realizations. Though many foods may contribute to a varied and nutritionally rich diet, preparation for consumption requires special attention to

Table 1 Toxicity reports of unconventional food plants of Brazil

Species (family)	Biomes[a]	Culinary use[b]	Ethnobotanical studies	In vitro studies	In vivo studies	References
Araucaria angustifolia (Bertol.) Kuntze (Araucariaceae).	Atlantic Forest	Seed *in natura*; boiled or roasted seed (cake, flour, pudding, ice cream, souffle)	–	–	–	–
Bixa orellana L. (Bixaceae).	Amazon, Cerrado, Atlantic Forest	Seed (condiment, dye)	–	1. Aqueous, ethanolic extract, and chloroform-ethanol (1:3) from the seeds pulp showed no cytotoxic effect (*Allium cepa*) 2. Aqueous extract from the leaves showed high toxicity and teratogenicity (*Zebrafish embryos*)	3. Diet with waste seeds showed no liver toxicity (rats) 4. Recommendations for safe doses of bixin and norbixin seed extracts	1. (Joseph and Siril 2010) 2. (De Vera et al. 2016) 3. Valério et al. (2015) 4. EFSA Journal (2016)
Byrsonima crassifolia (L.) Kunth (Malpighiaceae)	Amazon, Caatinga, Cerrado, Atlantic Forest, Pantanal	Fruit *in natura*; fruit pulp (cake, jam, yogurt, liquor, sauce, mousse, pudding, puree, soup, ice cream, juice, pie)	–	–	1. Methanol extract from leaves showed low toxicity, induced constipation and decreased locomotor activity at high doses (mice) 2. Hexane, chloroform and methanolic extracts of the fruit showed low acute toxicity (rats)	1. (Herrera-Ruiz et al. 2011) 2. (Perez-Gutierrez et al. 2010)

Species (family)	Biomes[a]	Culinary use[b]	Ethnobotanical studies	In vitro studies	In vivo studies	References
Caryocar brasiliense Cambess (Caryocaraceae)	Amazon, Caatinga, Cerrado, Atlantic Forest	Seed pulp (cake, canned, jam, flour, liquor, oil, ice cream, juice, rice preparations and meat stews)	–	1. Ethanol extract from the fruit showed low cytotoxicity (BALB-C 3T3 mouse fibroblasts) 2. Fruit pulp oil showed no cytotoxic effect (*A. salina*)	2. Fruit pulp oil showed low genotoxic and clastogenic effects (rats) 3. Ethanol extract from the bark (exocarp and external mesocarp) had a low toxic effect on male reproductive functions (Swiss mice) 4. Fruit pulp oil showed low acute and subchronic toxicity (Wistar rats) 5. Fruit pulp oil showed low maternal and embryological toxicity 6. Hydroalcoholic extract of the bran of the fruit peel showed a toxic effect (Swiss mouse)	1. (Roesler et al. 2010) 2. (Traesel et al. 2017b) 3. (Souza et al. 2019) 4. (Traesel et al. 2016) 5. (Traesel et al. 2017a) 6. (Almeida et al. 2010)

(continued)

Table 1 (continued)

Species (family)	Biomes[a]	Culinary use[b]	Ethnobotanical studies	In vitro studies	In vivo studies	References
Genipa americana L. (Rubiaceae)	Amazon, Caatinga, Cerrado, Atlantic Forest, Pantanal	Crystallized fruit; fruit pulp (cake, jam, syrup, liquor, blue bread/cake, ice cream, juice)	–	1. Extract from green genipap fruits showed genotoxic effect (Chinese hamster ovary cells) 2. Ethanol extract from the fruit showed low cytotoxicity (Hep-G2 liver carcinoma cells, normal fetal lung cells MRC-5 and HT-29 colon carcinoma cells)	–	1. (Neri-Numa et al. 2020) 2. (Tauchen et al. 2016)
Hymenaea courbaril L. (Fabaceae).	Amazon, Caatinga, Cerrado, Atlantic Forest, Pantanal	Fruit pulp and seeds (biscuit, cake, cream, sweet, flour, bread, pudding, ice cream, pie)	–	1. Fresh xylem sap showed low cytotoxicity (3T3-A31 mouse fibroblast)	2. Sap showed no toxic effects and genotoxic using the micronucleus test (mice)	1. (da Costa et al. 2014) 2. (Vale et al. 2013)
Mauritia flexuosa L. f. (Arecaceae)	Amazon, Caatinga, Cerrado	Fruit peel (tea); fruit pulp *in natura*; fruit pulp (sweet, flour, jelly, oil, ice cream, juice); stem marrow (starch)	–	–	1. Diet with crude or refined oil inhibited somatic growth and reflex maturation (rats) 2. Diet with crude or refined oil showed low toxicity (rats)	1. (Medeiros et al. 2015) 2. (De Souza Aquino et al. 2015)

Species (family)	Biomes[a]	Culinary use[b]	Ethnobotanical studies	In vitro studies	In vivo studies	References
Pereskia aculeata Mill. (Cactaceae)	Caatinga, Cerrado, Atlantic Forest	Leaves (flour, dough, bread), Leaves/flowers/ young branches (stew, omelet, sauté, salad, pie); Fruit (jam, liqueur, mousse, juice)	–	–	1. Topical application of the hexane fraction from the methanolic extract (leaves) showed low dermal and systemic toxicity in Wistar rats 2. Ethanol extract from leaves showed low acute toxicity in rats	1. (Pinto et al. 2015b) 2. (Silva et al. 2017)
Platonia insignis Mart. (Clusiaceae)	Amazon, Cerrado	Fruit *in natura*; fruit pulp (sweet, jelly, yogurt, liquor, sauce, mousse, ice cream, juice, pie)	–	1. Dichloromethane and ethyl acetate fractions from the ethanolic extract of the seeds showed genotoxic effect and low cytotoxicity (*A. salina and* mammalian V79 cells) 2. Hexane seed extract showed low cytotoxicity (Balb/c murine macrophages and human erythrocytes)	2. Oral administration of the hexane seed extract showed low toxicity in rats	1. (Costa Júnior et al. 2013) 2. (Lustosa et al. 2016)

(continued)

Table 1 (continued)

Species (family)	Biomes[a]	Culinary use[b]	Ethnobotanical studies	In vitro studies	In vivo studies	References
Portulaca oleracea L. (Portulacaceae)	Amazon, Caatinga, Cerrado, Atlantic Forest, Pantanal, Pampa	Young leaves and branches (fresh or cooked – cake, stew, pickles, salad, pie); seed (bread)	–	1. Polysaccharide portions of the aerial parts showed low toxicity in normal mouse cells	2. Chloroform extract from aerial parts showed low toxicity in female albino rats, but with anti-fertility effect 3. Chloroformic extracts and methanol-water from the leaves showed anti-fertility effect in rats 4. Hydroalcoholic extract from the whole plant showed low acute and subacute toxicity in Swiss albino mice 5. Methanolic extract from the whole plant showed toxicity in albino rats 6. Fraction 4 methanolic from the whole plant extract showed toxicity in albino rats	1. (Gatea et al. 2017) 2. (Nayaka et al. 2014) 3. (Obinna et al. 2019) 4. (Shafi and Tabassum 2015) 5. (Oyedeji and Bolarinwa 2012) 6. (Olusina et al. 2015)
Psidium cattleianum Sabine (Myrtaceae).	Caatinga, Cerrado, Atlantic Forest, Pantanal, Pampa	Fruit *in natura*; fruit pulp (sweet, jelly, yogurt, liquor, sauce, mousse, ice cream, juice, pie).	1. Report of heartburn after consumption of fruit *in natura* by users of medicinal plants in Mauritius	2. Hydroalcoholic extract of the leaves showed low cytotoxicity (LLC-MK2 mammalian fibroblast)	–	1. (Mahomoodally et al. 2018) 2. (Alvarenga et al. 2013)

Species (family)	Biomes[a]	Culinary use[b]	Ethnobotanical studies	In vitro studies	In vivo studies	References
Solanum paniculatum L. (Solanaceae).	Amazon, Caatinga, Cerrado, Atlantic Forest	Fruit with savory dishes, canned, cream, omelet	–	1. Ethanol extract from the fruit and leaves showed cytotoxic activity, but not mutagenic (bone marrow cells of Swiss mice) 2. Ethanol extract from the aerial parts showed low toxicity (red blood cells from Wistar rats)	2. Ethanol extract from aerial parts showed low acute toxicity in Swiss mice	1. (Vieira et al. 2010). 2. (Clementino-Neto et al. 2015)
Spondias mombin L. (Anacardiaceae).	Amazon, Cerrado, Atlantic Forest	Fruit *in natura*; fruit pulp (jelly, yogurt, liquor, sauce, ice cream, juice, pie)	–	1. Aqueous and hydromethanol extracts of the leaves did not have a genotoxic effect (*A. cepa*) 2. Aqueous and crude ethanol extracts from the leaves showed a genotoxic effect (*Drosophila melanogaster*) 3. Fixed seed oil showed low cytotoxicity (*A. salina*) 4. Ethanol extract from the bark showed moderate cytotoxicity (*A. salina*)	5. Aqueous and hydromethanolic extracts from the leaves showed mutagenic effect (Swiss albino mice) 6. Aqueous leaf extract showed low acute and subchronic toxicity (rats) 7. Fruit juice showed low toxicity (rats)	1. Oyeyemi and Bakare (2013) 2. (Senes-Lopes et al. 2018) 3. (De Rezende et al. 2018) 4. (Clementino et al. 2018) 5. (Oyeyemi et al. 2015) 6. (Goodies et al. 2015) 7. (Akharaiyi et al. 2018)
Sterculia striata A.St.-Hil. & Naudin (Malvaceae)	Amazon, Caatinga, Cerrado, Atlantic Forest	Almond (*in natura*, cooked, toast, cereal bar, biscuit, oil)	–	–	–	–

[a]Flora do Brasil 2020 em construção (2020)
[b]do Brasil (2018)

those that contain natural toxins. Though the nutritional aspects of UFP are relatively well documented, the presence of natural toxins still requires attention, especially in areas suffering from food insecurity (shortages) in Brazil. This chapter will further explore Brazilian UFP safety and consumption, known natural toxins, and reported toxicities. We reviewed the use of UFP and reports of undesirable effects with consumption, including information on plant toxicity, through extracts or the toxins themselves.

3 Brazilian UFP: Toxic Potential

In this chapter, we seek to answer the following questions: "Are Brazilian UFP safe for human consumption?" and "What are the principal natural toxins found?". Our main concern is not to analyze experimental models from the perspective of reproducibility but to record the presence or absence of information on toxicity: which part of the plant was used or which type of extract or preparation used and which are the possible toxins involved. Determining which plants ought to be considered as UFP is not simple since a plant can be widely known and consumed in one region of the country and not in another. Related to geographic and cultural aspects, a specie may or may not be considered a UFP because different communities or regions in Brazil have different relationships with plants (Jacob 2020).

For this reason, we chose 14 plant species from different botanical genera that fulfilled the role of UFP according to the concept of "relative non-conventionality." The 14 plant species were chosen from an initial list of native (socially and biologically diverse) Brazilian plants of interest for consumption and commercialization that the Ministry of the Environment of Brazil had published through an Interministerial Ordinance No. 284/2018. We chose only species from the native Brazilian flora which present greater distributions in Brazilian biomes, in accordance with the Flora do Brazil 2020 database (Flora do Brasil 2020 em construção 2020). This information can be found under the items "origin" and "phytogeographic domains," and Table 1 summarizes the principal UFP toxicity-related information.

To collect the data, we conducted an updated scientific literature search obeying the following criteria: review articles and original research in Portuguese, English, and Spanish, and (for only current references) those published between 2010 and 2020. We used the following databases for research: Scielo, Embrapa Agricultural Research Databases (BDPA), Web of Science, Medline/PubMed (via the National Library of Medicine), and Scopus. We used as descriptors in the databases the terms "*species name*" and *toxin*, or *toxicity*, or *anti-nutritional*. Initially, the titles and abstracts were submitted to a first screening (paired) in which those which did not meet the eligibility criteria were excluded. The title or abstract needed to cite UFP toxicity in ethnobotanical, in vitro or in vivo studies. Subsequently, a complete reading of the selected articles was carried out to obtain the following information: (1)

parts of the plant used as food, the form of preparation and undesirable effects, and (2) information concerning the toxic effect of the plant or plant extracts. Articles that did not present information on plant toxicity, isolated chemical agents with toxic action, agents stimulating the development and survival of the plant, or pharmacological/therapeutic activity of the plant or its products were excluded.

The toxicity studies were grouped into three types: (1) in vitro assays, using various types of human or other cells in their experimental models; (2) in vivo studies which used mice or rats; and (3) ethnobotanical studies. To evaluate the safety and efficacy of a plant or its products in humans, in vitro and in vivo studies always precede clinical trials. This also applies to research on potential plant food products to obtain more consistent data on food safety for consumption. In addition, ethnobotanical studies help to better understand undesirable effects occurring after consuming the plant or preparations made from them.

In the initial screening stage, the database search resulted in a total of 561 articles (analysis of titles and abstracts) yielding 39 articles for complete reading. In vivo animal studies occupied the largest segment (24), followed by in vitro (17), and ethnobotanical (01) studies. The most studied plants for toxicity were *Caryocar brasiliense* Cambess (pequi), *Portulaca oleracea* L. (beldroega, purslane), and *Spondias mombin* L. (cajá). For the edible plants during the last 10 years, toxicity had not been investigated for *Araucaria angustifolia* (Bertol.) Kuntze (pinhão), *Sterculia striata* A. St.-Hil. & Naudin (chichá), *Hymenaea courbaril* L. (jatobá), and *Platonia insignis* Mart. (bacuri). Below, we provide further details on toxicity for each of the UFP in this chapter.

4 UFP Toxicity

A. angustifolia is an example of gymnosperms present in the Atlantic Forest biome which produce food seeds (popularly known as pinhão). It is widely consumed in Southern Brazil whether baked or roasted, in cakes and flours, in paçoca[1] or puddings, ice creams, or souffle preparations (do Brasil 2018; Kinupp and de Barros 2007). Reports in the literature of its antioxidant properties and on correlated applications for *A. angustifolia* are relatively extensive. However, no reports have been found on undesirable effects with use as food or evidence of toxicity in its products. The few reports that deal with *A. angustifolia* claim toxicity to tumor cells using extracts obtained from non-edible parts of the plant such as the bracts (Souza et al. 2014) or essential oils obtained from leaves of other species of this gender (Elkady and Ayoub 2018). A study with consumers in Curitiba-PR reports that *A. angustifolia* is considered a natural food beneficial to health and that it can be consumed by the whole family (de Godoy et al. 2018).

[1] Paçoca de pinhão: a mixture of different spices with cooked and ground seeds of *A. angustifolia* and cooked and ground meat.

H. courbaril is greatly studied in terms of pharmacological potential, with an emphasis on its bark, leaves, seeds, and roots. However, it is the fruits and seeds that have food purposes, because the seeds generate an edible flour of great nutritional value and the fruit pulp can be consumed fresh (do Brasil 2018; Schwartz 2018). Considering this, we sought reports on phytochemical profiles for the seeds and fruits of *H. courbaril*.

The presence of biscumarins has been demonstrated in the seeds, including ipomopsin and hymenain (Simões et al. 2009). Simões et al. (2009) demonstrated the anti-free radical activity of biscumarins isolated from the seeds of *H. courbaril* and suggest that biscumarins have the function of protecting the seeds of the plant during germination. Even if the fruits are traditionally accepted, new studies providing accurate information on the food safety of *H. courbaril* fruits and seeds are much needed. Fresh xylem sap is rich in physetin (a flavonoid), considered a functional ingredient in food due to its inhibitory effects on glucose metabolism and its anti-obesity activity (Im et al. 2016). Its cytotoxic activity was evaluated, and the physetin, together with fresh *H. courbaril* xylem sap, presented good results in a mouse fibroblast cell line. The path is open for the use of the xylem sap or physetin in future food preparations (da Costa et al. 2014).

The fruits of *P. insignis* are highly appreciated by local populations of the Amazon, Maranhão, and Piauí, where the species is easily encountered. The pulp of the fruit is edible (de Carvalho and Nascimento 2017). In chemical-pharmacological studies, the seeds are the most explored, being justified by the presence of bioactive compounds. Studies have evaluated the toxicity of the seeds, yet not the fruit nor pulp. In short, seed extracts present low toxicity in vitro and in vivo (in rats), as shown in Table 1. Justification for the biological potential of *P. insignis* seeds is based on the presence of several metabolites which have already been identified and studied for toxicity such as polyisoprenylated benzophenones and 1,3-distearyl-2-oleylglycerol which presents neither toxicity nor mutagenicity to *Artemia salina* Leach (do Nascimento Cavalcante et al. 2020). Isolation of the garcinielliptone FC (the polyisoprenylated benzophenone) from *P. insignis* seeds spurred studies which reported: (1) low central nervous system toxicity and no genotoxic effects in mice at 28 days of treatment (Coelho et al. 2018); (2) cytotoxic and mutagenic effects in concentration-dependent human hepatoma cells (HepG2) (da Silva Prado et al. 2017); and (3) low acute oral and intraperitoneal toxicity (in mice) up to 2000 mg/kg (da Silva et al. 2016). This information, when combined with records of use of the edible fruit in Brazil, reinforces the idea its consumption is safe.

In some regions of Brazil, *S. striata* use is restricted to traditional communities, especially in the Cerrado, yet it can be found in other biomes such as the Amazon, the Caatinga, and the Mata Atlântica (Atlantic Forest). Its seeds or almonds are eaten roasted, and from these, other preparations are produced (Bortolotto et al. 2015; do Brasil 2018). *S. striata* seeds are rich in proteins, fibers, and bioactive compounds with antioxidant activity. When compared to other Brazilian seeds and the more popular Cashews or Brazil nuts, a great differential is their low-fat content. Their lipid profile is composed mainly of oleic acid (18: $1^{\Delta 9}$), linoleic acid (18: $2^{\Delta 9,12}$), phytosterols, and tocopherols (Policarpi et al. 2018). Normally, because of high-fat content, it is common to suggest caution in seed consumption because of the risks of intestinal problems and weight gain. This makes *S. striata* seeds different (EMBRAPA 1997). Further, we found no current reports on *S. striata* toxicity, which suggests safe food use.

As will be discussed below, the following species: *Bixa orellana* L. (urucum), *Byrsonima crassifolia* (L.) Kunth (murici), *C. brasiliense*, *Genipa americana* L. (jenipapo), and *S. mombin* presented low toxicity in greatly varied experimental model types (Table 1).

B. orellana seeds are commonly used to produce natural red dyes for application in textile, pharmaceutical, and food products. They are natural dyes and possess recognized antioxidant power, and their main components are bixin (containing a metal ester group) and norbixin carotenoids (containing a carboxyl group) (Nathan et al. 2019). A better analysis of the toxic potential of *B. orellana* and its components has revealed important information. According to the EFSA-European Food Safety Authority (2016), in a scientific safety opinion regarding chemical additives, the use of *B. orellana* seed extract (bixin based and norbixin-based) is safe for intake in general. In Brazil, the National Health Surveillance Agency-ANVISA also recognizes *B. orellana* use as a food additive (INS 160b Urucum, bixin, norbixin, Annatto extract, and Na and K salts). This corroborates studies that compose this chapter (Table 1). Nothing significant has been found related to teratogenicity or embryo toxicity, and only one study reported toxicity for *B. orellana* leaf extract (non-edible) (Table 1). In any case, pregnant women should continue to use the plant with caution. We note that carotenoids originating from *B. orellana* suffer degradation when exposed to light or high temperatures; under these conditions, this helps guarantee their safety in food preparations (Satyanarayana et al. 2006).

B. crassifolia is widely distributed in Central and South America. In North and Northeastern Brazil, its fruits are consumed fresh or in preparations usually sold in markets (Perez-Gutierrez et al. 2010). We note that *B. crassifolia* presents antioxidant and cytoprotective potential because of the carotenoids and flavonoids present in its fruits and other parts (Herrera-Ruiz et al. 2011). *B. crassifolia* fruits are a good source of lutein and zeaxanthin, two important carotenoids for humans for their predominant locations in the macular region of the retina. The protective effects on the retina are believed to be involved with the potent antioxidant activity of these components (Mariutti et al. 2013). In addition to being antioxidants, lutein and zeaxanthin are pro-vitamin A carotenoids. Vitamin A is important in visual processes, regulating and modulating cell growth and differentiation (Yonekura et al. 2016). Phenolic compounds such as the quercetin flavonol, one of the most studied flavonoids in the world, bring recognized antioxidant activity to the fruit of *B. crassifolia*. Due not only to study results, but also to its traditional use, *B. crassifolia* fruit appears to be safe.

The fruit of *C. brasiliense*, being native to the Brazilian Cerrado and found in other biomes such as the Amazon, Caatinga, and Mata Atlântica, is frequently consumed. In the literature, reports concerning the fruit or oil extracts from *C. brasiliense* demonstrate an absence of toxicity. *C. brasiliense* can be considered safe for use. It presents large quantities of antioxidants such as carotenoids and gallic acid

(Rocha et al. 2015; Traesel et al. 2017b). The peel of the fruit is indicated for its nutritional potential and presenting both antioxidant components and many food applications. However, its high tannin concentration requires care concerning the possibility of decreasing nutrient bioavailability (Rocha et al. 2015; Souza et al. 2019). Caution when consuming is suggested, especially in individuals who require greater nutritional demand. To minimize these effects, studies recommend boiling or cooking as viable techniques to control these factors and improve the nutritional value of the food (Samtiya et al. 2020).

The fruits of *G. americana* find culinary use and provide widely known bluish food dyes. It is known in the literature that this coloring potential involves the presence of iridoid components (genipin, genipin 1-β-gentiobioside, geniposide, and genipinic acid). *G. americana* iridoid blue-based pigments are considered a natural alternative additive for food applications (Neri-Numa et al. 2018). Little is known about its properties in food or its undesirable effects on human health. *G. americana* fruit extracts present low in vitro toxicity to normal and tumor cells of humans and mice (Table 1). Though studies evaluating their pharmacological potential have been increasing, safety analyses for the use of iridoid compounds (present in *G. americana*) remain inconclusive. For example, genipin presents low cytotoxicity and genotoxicity when tested in isolation on mouse cells (Tsai et al. 2000). Geniposide, a glycoside of genipin present in the fruits of *G. americana*, presents hepatotoxicity in rats when administered in chronic form in high doses (Tian et al. 2018). However, as with other natural dyes, genipin and iridoid compounds are easily degraded in the presence of light, heat, oxidizing agents, or pH changes; common conditions during culinary preparation (Neri-Numa et al. 2018). Therefore, *G. americana* appears to be safe for culinary use of its fruits.

The research performed for this chapter indicates that consumption of *S. mombin* fruit seems safe, presenting low toxicity in animal models upon administration of fruit juice (Akharaiyi et al. 2018). Yet the in vitro studies are confusing because they point to differing cell toxicities with changes in cell type. Further, no study has studied the principal edible part of the plant (the fruit) to assess cytotoxicity (see Table 1). This is important since fresh *S. mombin* fruit is widely consumed in Brazilian North and Northeastern culinary.

The toxicological aspects of certain of the plants analyzed present relevant questions. More detailed comments concerning *Mauritia flexuosa* L. f. (buriti), *Pereskia aculeata* Mill. (ora-pro-nóbis), *Portulaca oleracea* L., *Psidium cattleianum* Sabine (araçá), and *Solanum paniculatum* L. (jurubeba) are necessary.

The fruits of *M. flexuosa* are sweet to the taste, yet rich in fats. Fatty acid profiles for the mesocarp and seed oils predominantly present palmitic acid (16:0) and oleic acid ($18{:}1^{\Delta 9}$), and *M. flexuosa* also presents carotenoids and phenolic compounds (da Agostini-Costa 2018). Methanol extracts of the fruit present flavonoids (flavones, flavonols, flavononols, flavonones, xanthones, chalcones, catechins, leucoantocyanidins, and aurones), components with outstanding antioxidant activity that bring added economic value (Nobre et al. 2018). In animal models, most current studies have demonstrated *M. flexuosa* oil as safe for food use. However, it is noteworthy that maternal use of the oil in the diet can cause motor development and reflex losses in the offspring (Table 1). Based on the above information, we can infer only that consumption of *M. flexuosa* as a food seems safe, yet that pregnant women should be cautious.

The species *P. aculeata* is popularly known in Brazil as ora-pro-nóbis, the only representative of the Cactaceae family in this chapter. It is part of a family of plants very well adapted to arid and hot regions, such as the Caatinga biome of Brazil, and the plant retains its capacity to accumulate water, and supply minerals, antioxidants, and proteins (Takeiti et al. 2009). *P. aculeata* is a plant with more restricted culinary uses; its leaves and fruits are consumed fresh or in preparations made from its by-products such as flours obtained from its leaves or jellies obtained from its fruits (do Brasil 2018). However, it is the uses of *P. aculeata* in traditional medicine from which studies bring information concerning its safety or undesirable effects. The leaves are traditionally used in Brazil in medicinal preparations, as emollients and for wounds, or inflammations on the skin; the fruits are used as expectorants (Pinto and Scio 2014). An important study reported that the hexane fraction of the methanolic extract of *P. aculeata* leaves did not induce signs of topical or systemic toxicity in dermal irritation tests in rats (Pinto et al. 2015b). Although this is not the usual form of the plant as a food, it is an important indication of its safety when there are no reports of toxicity. The second report found in our search confirmed that ethanolic extracts from the leaves of *P. aculeata* are generally safe because the signs of acute toxicity in animal models (rats) were negligible (Silva et al. 2017).

P. aculeata leaves are known to be rich in proteins, especially in tryptophan (Takeiti et al. 2009). Tryptophan is a precursor amino acid for the indole alkaloids which affect biological systems as bioactive amines, such as in adrenergic, serotonergic, dopaminergic, or cholinergic systems related to pain control. In a study developed by Pinto et al. (2015a), various alkaloids with antinociceptive activity were identified in *P. aculeata* leaves, including tryptamine, abrine, mescaline, hordenine, petunidin, di-tert-butylphenol isomers, and quercetin. The alkaloids present in the leaves of *P. aculeata* may well be involved in its analgesic activity. Due to the toxic potential of alkaloids, it is interesting that they are found in low amounts in food preparations. Foods contaminated with certain classes of alkaloids, such as pyrrolizidine alkaloids, are considered a risk to human health (Knutsen et al. 2017).

It has also been reported that the leaves of *P. aculeata* contain important constituents such as oxalic acid, nitrate, and saponins. Oxalic acid is considered an anti-nutrient that compromises calcium absorption and can induce undesirable effects. However, in practical terms, the species is not expected to present enough oxalic acid to significantly impede the absorption of minerals (de Almeida et al. 2014).

The presence of nitrates in *P. aculeata* needs our attention. Nitrates and their derivatives in the human diet are controversial and it is necessary to be aware of these aspects. Normally, dietary nitrate is reduced to nitrite by bacteria in the oral microbiota. Nitrite, in turn, is absorbed and appears to be responsible for systemic effects in humans (Weitzberg and Lundberg 2013). Although both are considered products with limited carcinogenic potential, nitrite in combination with certain amines or amides can potentially form carcinogenic nitrosamines (Chamandoost et al. 2016). There have always been concerns over consumption of processed animal foods rich in nitrates and its carcinogenic consequences. Also observed in *P. aculeata* is the presence of compounds from the saponin group. High concentrations of saponins give the food a bitter taste, and it is reported that certain digestive enzymes are inhibited in the presence of saponins.

In the context of a varied diet, the consumption of *P. aculeata* is interesting. All these aspects lead us to believe that, even with observed low toxicity concerning the *P. aculeata* leaves, it is necessary, especially for users of centrally acting analgesic drugs, to use caution (in concomitant use of the plant), since analgesic effects may increase.

In the scenario of food preparations with bioactive and functional potential, the species *P. oleracea* is a UFP with increasing prominence. Its stems and leaves are consumed in salads and hot preparations, yet *P. oleracea* also brings applications from traditional medicine, many of them scientifically proven. Its positive effects on the blood biochemistry are emphasized and possibly justified by the high content of omega-3 fatty acids and antioxidant components in its composition (Petropoulos et al. 2016). However, with regard to safety, the studies are sometimes contradictory, and certain reports will be necessary to mention.

In this chapter, all the studies used extracts from edible parts of the plant (aerial parts), but with extractive solvents of different polarities. We noted that the chloroform and hydromethanol extracts induced deleterious effects on the fertility of rats, although being of low toxicity, similar to the hydroalcoholic extracts. The methanolic extracts presented a certain degree of toxicity in rats in a dose dependent manner (see Table 1). Trying to better understand the reason for these effects, we noted that the extractions using the chloroform and chloroform-methanol (v/v) solvents are more efficient and reliably obtain higher concentrations of both phenolic compounds and organic acids (Petropoulos et al. 2016). In addition, it is very common for *P. oleracea* to present in its composition other secondary metabolite classes such as flavonoids and phenolic acids (Gatea et al. 2017). Though many of these components have antioxidant action, depending on the dose consumed, and part of the plant used, they can cause undesirable effects.

In this context, the effects of *P. oleracea* on fertility are more important. In vivo animal studies point to the involvement of flavonoids present in the plant in anti-estrogenic activity and anti-ovulatory activity (Nayaka et al. 2014). Kaempferol, apigenin, luteolin, myricetin, and quercetin are some of the main flavonoids found in *P. oleracea*. The greatest flavonoid content is found in the roots, followed by the stems, and leaves (Gatea et al. 2017; Zhu et al. 2010). Although the leaves are mostly used in cooking, caution is highly suggested when using *P. oleracea* in food preparations for pregnant women.

Gatea et al. (2017) when analyzing the biological potential and phytochemical profile of *P. oleracea*, detected high concentrations of oxalic acid, which deserves attention. This compound acts as a chelator of dietary minerals, especially calcium. Yet even considering this caveat, the use of *P. oleracea* in a balanced and varied diet is strongly suggested, as it provides a variety of bioactive nutrients and is relatively safe. The general toxicity of the plant is not pronounced, but assessing its chronic toxicity is necessary, and regular use in the human diet should be moderated due to its oxalic acid high content.

Our study did not find reports of toxicity for *P. cattleianum*; only one of the studies reported the low cytotoxicity of a hydroalcoholic extract obtained from the leaves and not the fruits, which is the principal form consumed as food (Alvarenga et al. 2013). An ethnobotanical study performed with users of medicinal plants in the Republic of Mauritius demonstrated that when consumed to prevent fever and flu, *P. cattleianum* fruit causes mild heartburn. Further, there are reports of vomiting when its fruits are consumed after using iron supplements (Mahomoodally et al. 2018). However, of the universe of 307 individuals participating in the research, there were few reports, and the amount consumed until these effects appeared was not reported. The phytochemical profile of the fruit, containing vitamin C between 200 mg and 242 mg per gram of fresh fruit justifies the adverse effect (heartburn) (Pereira et al. 2018). Very high daily doses of vitamin C (2000 mg or more) can cause transient gastroenteritis or osmotic diarrhea in some individuals (da Silva et al. 2016). Thus, for those with previous gastric or intestinal problems, it becomes interesting to moderate consumption of citrus fruits such as *P. cattleianum.*

In traditional medicine, the roots, leaves, and fruits of *S. paniculatum* are used for general inflammations, as a diuretic, and against liver diseases (Agra et al. 2007). However, it is their fruits that have food potential. Considering this, we found a single report on the toxicity of an ethanol extract obtained from the fruits of *S. paniculatum* (Vieira et al. 2010). The authors demonstrated the cytotoxicity of the extract when administered to mice in high doses (above 200 mg/kg of weight) yet with an absence of mutagenicity. *S. paniculatum* has many metabolites that seem to be involved in its discrete toxicity, among them are steroidal sapogenins and steroidal glycoalkaloids (solanine, solasonine, and solamargine). Steroidal sapogenins are non-saccharide portions of saponins, and although less common, they are found in the roots, leaves, and vegetable fruits of the genus *Solanum* (Sobolewska et al. 2020; Vieira Júnior et al. 2015). In vivo studies of steroidal sapogenins have demonstrated low acute toxicity in mice (Wei et al. 2014). In general, consumption of foods containing these compounds presents little toxicity and does not appear to be dangerous for consumers (Oleszek and Oleszek 2020).

In turn, steroidal glycoalkaloids are found in large quantities in the fruits of *Solanum* spp., presenting a bitter taste, and undesirable effects in humans (Knuthsen et al. 2009). The principal effects reported include gastrointestinal problems (abdominal pain, vomiting, and diarrhea), sweating, bronchospasm, confusion, hallucination, heart failure, partial paralysis, seizures, coma, and even death (Mensinga et al. 2005; Sucha and Tomsik 2016). The suggested mechanisms of action include induction of apoptosis or necrosis, inhibition of acetylcholinesterase, and interaction with membrane steroids (affecting the function of plasma membranes). We suggest reading Sucha and Tomsik (2016) for further study.

Studies on the toxicity of these compounds in plants of the genus *Solanum spp.* have been increasing. Principally, reports refer to other members of this genus such as potatoes (*Solanum tuberosum* L.), tomatoes (*Solanum lycopersicum* L.), and eggplant (*Solanum melongena* L). This is relevant since vegetables are much more present in the daily world of food than is *S. paniculatum*, yet they share certain glycoalkaloids (Barceloux 2009). We note that glycoalkaloid content varies substantially between different cultivars of the plant and post-harvest conditions (light, mechanical injuries, storage). During cooking, the principal methods used in processing, such as boiling, heating, and frying, do not significantly alter the final glycoalkaloid content of the food (Friedman 2006). Considering the genus Solanum, toxic glycoalkaloids are not generally found in the fruit, but are present in greater quantities in other inedible parts such as the leaves (Knuthsen et al. 2009). In this context, cases of glycoalkaloid poisoning are relatively rare, yet they do occur, chiefly with consumption of little known plants considered wild such as *S. paniculatum* (Barceloux 2009). This may be why the only report we found describing the in vitro toxicity of *S. paniculatum* fruit extract cited high concentrations (Vieira et al. 2010).

Finally, we note that the Brazilian UFP in this chapter present diverse spectra of metabolites considered to be natural toxins and which are often present in more than one plant. In view of this great diversity, Fig. 1 summarizes these toxins and associates them with the UFP in which their occurrence was observed.

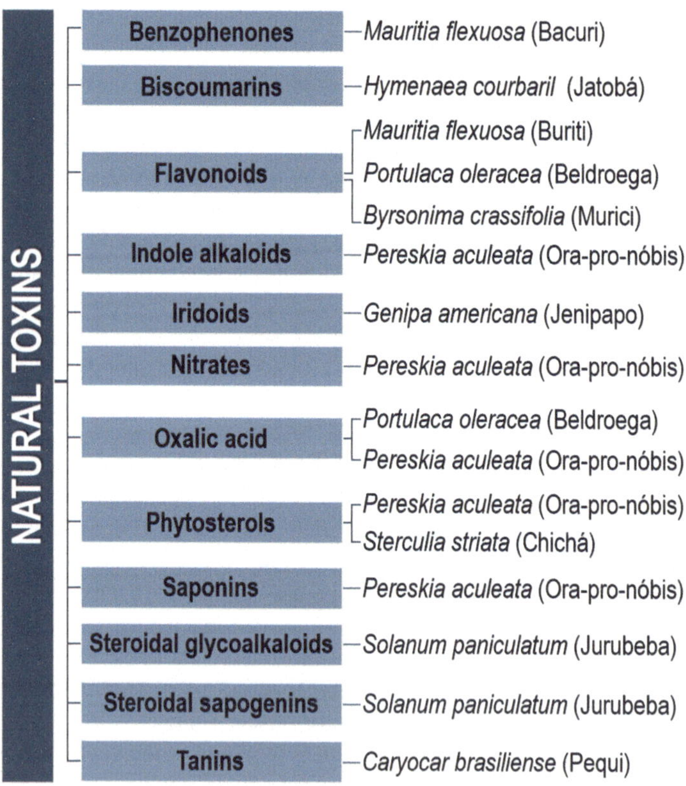

Fig. 1 Distribution of the main groups of natural toxins present in unconventional food plants of Brazil

5 Conclusions

Understanding the toxic properties of certain plants discussed in this chapter raises important questions that go beyond safe consumption. Reflecting critically on these factors can help to better understand the forms of interactions between humans and plants, especially since UFP represent a potentially relevant contribution to a varied and complete diet.

Initially, the studies included in this chapter present a scenario in which practically all UFP's are of great use to popular medicine, while also presenting nutritional potential. Yet often, depending on the part of the plant used, the form of preparation, dosage, and final distribution of the natural toxins involved in these effects, the plants possess toxic or mutagenic properties. In addition, certain metabolites can be considered anti-nutritional, since they interfere in nutrient (food) bioavailability, causing secondary damage to individuals. All of these factors are important and determine the safety of the plant.

Many of the studies used products from plants that are not associated with their principal forms of culinary use in Brazil. Many of these studies presented as their central objective an investigation into the pharmacological potential of the inedible parts of the plants. Ensuring the correct part of the plant is used is essential since there are variations in the phytochemical profiles of the various plant organs. These phytochemical components perform important functions in the plant, and uneven distribution is an important form of adaptation.

The biological properties of natural toxins observed in many studies (given the isolated metabolite) may not represent the toxic activity of the plant as a whole. When consumed, the plant itself presents a phytocomplex system, and interactions between various such compounds during the digestive process can result in reduced bioavailability for any of them. Combinations with other foods can also cause similar results. The amounts of toxin in the ingested plant are in many cases lower, when comparing similar effects from the use of the isolated substance. It is interesting to note that we are cautious with the consumption of UFP associated with drug therapies, as there are real possibilities of drug-food interactions that are not beneficial for humans. Sometimes information concerning natural toxins and how they work in the body remains unclear. This opens up a promising field for future studies.

In general, the information presented in this chapter reveals various UFP species with relatively low toxicity. However, certain UFPs require greater attention to their toxicological aspects: *S. paniculatum* requires more care with its leaves (for containing glycoalkaloids and steroidal saponins) than with its fruits. Due to the presence of flavonoids involved in fertility (for women), and for containing large amounts of oxalic acid, consumption of *P. oleracea* during pregnancy requires caution. Because of potentially negative consequences for the development of newborns, *M. flexuosa* oil should be treated with caution by pregnant women as well. Due to oxalic acid, saponins, and indolic alkaloids, consumption of *P. aculeata* must be moderate since associations of indolic alkaloids with analgesics may be well be harmful. For those who present stomach problems, *P. cattleianum* appears to cause heartburn and moderate use of the fruit should be considered.

Considering our discussions, when there is a need to seek information on food safety in a given UFP, we emphasize the need to ask the following questions: "Is the part of the plant of interest used in this study the edible part?", "What is the predominant phytochemical profile (natural toxins)?", and "Do the current physical processes of culinary practice interfere with either the content or stability of these toxins?" In conclusion, consumption of fresh UFP or derived products, in whole or in part, without adequate knowledge of their botanical, chemical, and toxic characteristics can bring undesirable results. Such knowledge is imperative, especially for plants presenting restricted consumption.

References

Abraham K et al (2010) Toxicology and risk assessment of coumarin: focus on human data. Mol Nutr Food Res 54:228–239

Agra MF et al (2007) Medicinal and poisonous diversity of the flora of "Cariri Paraibano", Brazil. J Ethnopharmacol 111:383–395

Akharaiyi FC et al (2018) Biochemical, hematological and histological effect of spondias mombin l fruit juice on some physiological properties of wistar rats. Int J Pharm Res 10:156–166

Almeida AC et al (2010) Toxicidade aguda dos extratos hidroalcoólicos das folhas de alecrim-pimenta, aroeira e barbatimão e do farelo da casca de pequi administrados por via intraperitoneal. Ciência Rural 40:200–203

Alvarenga FQ et al (2013) In vivo analgesic activity, toxicity and phytochemical screening of the hydroalcoholic extract from the leaves of Psidium cattleianum Sabine. J Ethnopharmacol 150:280–284

Ashmore T et al (2015) Nitrate enhances skeletal muscle fatty acid oxidation via a nitric oxide-cGMP-PPAR-mediated mechanism. BMC Biol 13:110

Barceloux DG (2009) Potatoes, tomatoes, and solanine toxicity (Solanum tuberosum L., Solanum lycopersicum L.). Dis Mon 55:391–402

Bennett RN, Wallsgrove RM (1994) Secondary metabolites in plant defence mechanisms. New Phytol 127:617–633

Bortolotto IM et al (2015) Knowledge and use of wild edible plants in rural communities along Paraguay River, Pantanal, Brazil. J Ethnobiol Ethnomed 11:46

Chamandoost S, Fateh Moradi M, Hosseini M-J (2016) A review of nitrate and nitrite toxicity in foods. J Human Environ Health Promot 1:80–86

Cirlini M et al (2018) Are tropane alkaloids present in organic foods? Detection of scopolamine and atropine in organic buckwheat (Fagopyron esculentum L.) products by UHPLC–MS/MS. Food Chem 239:141–147

Clementino ELC et al (2018) Phytochemical composition and antimicrobial and toxicological activity of Spondias mombin L. (jobo). Rev Cubana de Plantas Med 23

Clementino-Neto J et al (2015) Toxicological, antidiarrheal and spasmolytic activities of solanum paniculatum. Planta Med 82:58–64

Coelho VR et al (2018) A 28-day sub-acute genotoxic and behavioural assessment of garcinielliptone FC. Basic Clin Pharmacol Toxicol 123:207–212

Costa Júnior JS et al (2013) Investigation of biological activities of dichloromethane and ethyl acetate fractions of Platonia insignis Mart. seed. Basic Clin Pharmacol Toxicol 112:34–41

Cressey P, Reeve J (2019) Metabolism of cyanogenic glycosides: a review. Food Chem Toxicol 125:225–232

da Agostini-Costa T (2018) Bioactive compounds and health benefits of some palm species traditionally used in Africa and the Americas – a review. J Ethnopharmacol 224:202–229

da Costa MP et al (2014) Antifungal and cytotoxicity activities of the fresh xylem sap of Hymenaea courbaril L. and its major constituent fisetin. BMC Complement Altern Med 14:245
da Silva, AP dos SCL et al (2016) Pre-clinical toxicology of garcinielliptone FC, a tautomeric pair of polyprenylated benzophenone, isolated from *Platonia insignis* Mart seeds. Phytomedicine 23
da Silva Prado L et al (2017) Evaluation of DNA damage in HepG2 cells and mutagenicity of Garcinielliptone FC, a bioactive benzophenone. Basic Clin Pharmacol Toxicol 120:621–627
da Silva VL et al (2016) Vitamina C (ácido ascórbico). In: Biodisponibilidade de Nutrientes, 5th edn, pp 417–440
Dassanayake U, Gnanathasan CA (2012) Acute renal failure following oxalic acid poisoning: a case report, Journal of Occupational Medicine and Toxicology. BioMed Central 7(1):1–4
de Almeida MEF et al (2014) Caracterização química das hortaliças não-convencionais conhecidas como ora-pro-nobis. Biosci J 30:431–439
de Carvalho EU, Nascimento WMO (2017) Bacuri: platonia insignis. Instituto Interamericano de Cooperación para la Agricultura
de Godoy RCB et al (2018) Consumidor de pinhão: hábitos, atributos de importância e percepção. Braz J Fores Res 38
De Rezende LC et al (2018) New alkyl phenols and fatty acid profile from oils of pulped spondias mombin L. seed wastes. Quim Nova 41:540–543
De Souza Aquino J et al (2015) Effects of dietary brazilian palm oil (Mauritia flexuosa L.) on Cholesterol profile and Vitamin A and e status of rats. Molecules 20:9054–9070
De Vera JS, De Castro MEG, Dulay RMR (2016) Phytochemical constituents and teratogenic effect of lyophilized extracts of *Bixa orellana* L. (Achuete) and *Piper betle* L. (Ikmo) leaves in *Danio rerio* embryos. Der Pharm Chem 8:432–437
do Brasil M (2018) Portaria Interministerial N° 284. https://alimentusconsultoria.com.br/portaria-interministerial-no-284-de-30-de-maio-de-2018-mma. Acessed 28 Mar 2020
do Brasil M (2020) Biodiversidade. https://www.mma.gov.br/biodiversidade.html Accessed 31 Mar 2020
do Nascimento Cavalcante A et al (2020) Toxicity, cytotoxicity, mutagenicity and in vitro antioxidant models of 2-oleyl-1,3-dipalmitoyl-glycerol isolated from the hexane extract of Platonia insignis MART seeds. Toxicol Rep 7:209–216
dos Santos MAT (2006) Efeito do cozimento sobre alguns fatores antinutricionais em folhas de brócoli, couve- flor e couve. Ciência e Agrotecnologia 30:294–301
Elkady WM, Ayoub IM (2018) Chemical profiling and antiproliferative effect of essential oils of two Araucaria species cultivated in Egypt. Ind Crop Prod 118:188–195
EMBRAPA (1997) Centro de Pesquisa Agropecuaria do Meio-Norte Chichá (Sterculia striata St. Hil et Naud): uma nova opção para os mercados nacional e internacional de nozes.
Fernando Aguilar, Riccardo Crebelli, Alessandro Di Domenico, Birgit Dusemund, Maria Jose Frutos, Pierre Galtier, David Gott, Ursula Gundert-Remy, Claude Lambré, Jean-Charles Leblanc, Oliver Lindtner, Peter Moldeus, Alicja Mortensen, Pasquale Mosesso, Dominique Parent-Massin, Agneta Oskarsson, Ivan Stankovic, Ine Waalkens-Berendsen, Rudolf Antonius Woutersen, Matthew Wright and Maged Younes. The safety of annatto extracts (E 160b) as a food aditive. EFSA Panel on Food Additives and Nutrient Sources added to Food (ANS). EFSA Journal 2016;14(8):4544.
Flora do Brasil 2020 em construção (2020) Jardim Botânico do Rio de Janeiro.
Friedman M (2006) Potato Glycoalkaloids and metabolites: roles in the plant and in the diet. J Agric Food Chem 54:8655–8681
Gatea F et al (2017) Chemical constituents and bioactive potential of Portulaca pilosa L vs. Portulaca oleracea L. Med Chem Res 26(7):1516–1527
Goodies ME et al (2015) Antidiabetic activity and toxicity evaluation of aqueous extracts of Spondias mombin and Costus afer on Wistar rats. Br J Pharm Res 6:333–342
Guinand Y, Lemessa D (2001) Wild-food plants in Ethiopia: reflections on the role of wild foods and famine foods at a time of drought. In: Kenyatta C, Henderson A (eds) The potential of indigenous wild foods. UNDP Emergencies Unit for Ethiopia, Ethiopia, pp 31–46
Hammond BR, Renzi-hammond LM (2013) Carotenoids. Adv Nutr 4:474–476

Herrera-Ruiz M et al (2011) Antidepressant effect and pharmacological evaluation of standardized extract of flavonoids from Byrsonima crassifolia. Phytomedicine 18:1255–1261

Higashijima NS et al (2019) Fatores antinutricionais na alimentação humana. Segurança Alimentar e Nutricional 27:e020013

Im WK et al (2016) Fisetin-rich extracts of Rhus verniciflua Stokes improve blood flow rates in mice fed both Normal and high-fat diets. J Med Food 19:120–126

Jackson-Malete J, Blake O, Gordon A (2015) Natural toxins in fruits and vegetables: Blighia sapida and hypoglycin. In: Food safety anld quality systems in developing countries: volume one: export challenges and implementation strategies. Elsevier, pp 17–32

Jacob MM (2020) Biodiversidade de plantas alimentícias não convencionais em uma horta comunitária com fins educativos, DEMETRA: Alimentação. Nutrição & Saúde 15:e44037

Joseph N, Siril EA (2010) Cytotoxic evaluation of annatto (Bixa orellana L.) dye compared with orange red. Cytologia 75:163–167

Kinupp VF, de Barros IBI (2007) Riqueza de Plantas Alimentícias Não-Convencionais na Região Metropolitana de Porto Alegre, Rio Grande do Sul. Rev Bras Biociências 5:63–65

Knuthsen P et al (2009) Glycoalkaloids in potatoes: content of glycoalkaloids in potatoes for consumption. J Food Comp Anal 22:577–581

Knutsen HK et al (2017) Risks for human health related to the presence of pyrrolizidine alkaloids in honey, tea, herbal infusions and food supplements. EFSA J 15:e04908

Koleva II et al (2012) Alkaloids in the human food chain – natural occurrence and possible adverse effects. Mol Nutr Food Res 56:30–52

Kozłowska, A and Szostak-Węgierek, D (2018) n: Kozłowska A, Szostak-Węgierek D. Flavonoids – food sources, health benefits, and mechanisms involved Springer, 1–27.

Kumari M, Jain S (2012) Tannins: an antinutrient with positive effect to manage diabetes.

Kuti JO, Konoru HB (2006) Cyanogenic glycosides content in two edible leaves of tree spinach (Cnidoscolus spp.). J Food Comp Anal 19:556–561

Leal ML, Alves RP, Hanazaki N (2018) Knowledge, use, and disuse of unconventional food plants. J Ethnobiol Ethnomed 14

Liu R, Xu B (2015) Inhibitory effects of Phenolics and Saponins from commonly consumed food legumes in China against digestive enzymes pancreatic lipase and α -glycosidase. Int J Food Prop 18:2246–2255

Lustosa AKMF et al (2016) Immunomodulatory and toxicological evaluation of the fruit seeds from Platonia insignis, a native species from Brazilian Amazon rainforest. Braz J Pharm 26:77–82

Mahomoodally F, Suroowan S, Sreekeessoon U (2018) Adverse reactions of herbal medicine—a quantitative assessment of severity in Mauritius. J Herb Med 12:49–65

Marín-Sáez J, Romero-González R, Garrido Frenich A (2019) Effect of tea making and boiling processes on the degradation of tropane alkaloids in tea and pasta samples contaminated with Solanaceae seeds and coca leaf. Food Chem 287:265–272

Mariutti LRB, Rodrigues E, Mercadante AZ (2013) Carotenoids from *Byrsonima crassifolia*: identification, quantification and in vitro scavenging capacity against peroxyl radicals. J Food Comp Anal 31:155–160

Medeiros MC et al (2015) Buriti oil (*Mauritia flexuosa* L.) negatively impacts somatic growth and reflex maturation and increases retinol deposition in young rats. Int J Dev Neurosci 46:7–13

Mensinga TT et al (2005) Potato glycoalkaloids and adverse effects in humans: an ascending dose study. Regul Toxicol Pharmacol 41:66–72

Nascimento VT, Campos LZ, Albuquerque UP (2018) Plantas alimentícias. In: Albuquerque UP, Alves RR (eds) Introdução à Etnobiologia, 2nd edn. NUPPEA, Recife, pp 139–146

Nathan VK et al (2019) Antioxidant and antimicrobial potential of natural colouring pigment derived from *Bixa orellana* L, Seed Aril. Proc Natl Acad Sci U S A 89:137–143

Nayaka HB, Londonkar RL, Andumesh MK (2014) Evaluation of portulaca oleracea l for antifertility effect in female albino rats. Int J Pharma Pharm Sci 6:86–89

Neri-Numa IA et al (2018) Iridoid blue-based pigments of *Genipa americana* L. (Rubiaceae) extract: influence of pH and temperature on color stability and antioxidant capacity during *in vitro* simulated digestion. Food Chem 263:300–306

Neri-Numa IA et al (2020) *In vitro* bioactivity approach of unripe genipap (*Genipa americana* L., Rubiaceae) fruit extract and its solid lipid microparticle. Food Res Int 127:108720

Nobre CB et al (2018) Antioxidative effect and phytochemical profile of natural products from the fruits of "babaçu" (*Orbignia speciose*) and "buriti" (*Mauritia flexuosa*). Food Chem Toxicol 121:423–429

Obinna VC, Kagbo HD, Agu GO (2019) Investigation of lipophilic and hydrophilic leaf extracts of portulaca oleracea on gestation in albino rats (*Rattus norvegicus*). Thai J Pharm Sci 43:227–231

Oleszek M, Oleszek W (2020) Saponins in food. In: Xiao J, Sarker SD, Asakawa Y (eds) Handbook of dietary phytochemicals, pp 1–40

Olusina OK et al (2015) Effect of fraction 4 of Portulaca oleracea on hematological and plasma biochemical parameters in male Wistar rats. J Chem Pharm Res 7

Oyedeji KO, Bolarinwa AF (2012) Effects of crude extracts of portulaca oleracea on haematological and biochemical parameters in albino rats. Afr J Biomed Res 15

Oyeyemi, IT, Bakare, AA (2013) Genotoxic and anti-genotoxic effect of aqueous extracts of *Spondias mombin* L., *Nymphaea lotus* L. and *Luffa cylindrical* L. on *Allium cepa* root tip cells. Caryologia 66(4):360–367

Oyeyemi IT et al (2015) Genotoxicity and antigenotoxicity study of aqueous and hydro-methanol extracts of *Spondias mombin* L., *Nymphaea lotus* L. and *Luffa cylindrical* L. using animal bioassays. Int Toxicol 8:184–192

Pereira E et al (2018) *Psidium cattleianum* fruits: a review on its composition and bioactivity. Food Chem 258:95–103

Perez-Gutierrez RM et al (2010) Antihyperglycemic, antihyperlipidemic and antiglycation effects of *Byrsonima crassifolia* fruit and seed in normal and streptozotocin-induced diabetic rats. Plant Foods Human Nutr 65:350–357

Perez-Vizcaino F, Fraga CG (2018) Research trends in flavonoids and health. Arch Biochem Biophys 15:107–112

Petropoulos S et al (2016) Phytochemical composition and bioactive compounds of common purslane (*Portulaca oleracea* L.) as affected by crop management practices. Trends Food Sci Technol 55:1–10

Pinto N, Scio E (2014) The biological activities and chemical composition of *Pereskia Species* (Cactaceae)-a review. Plant Foods Hum Nutr 69:189–195

Pinto N, Duque AP et al (2015a) *Pereskia aculeata*: a plant food with antinociceptive activity. Pharm Biol 53:1780–1785

Pinto N, Machado DC et al (2015b) *Pereskia aculeata* miller leaves present in vivo topical anti-inflammatory activity in models of acute and chronic dermatitis. J Ethnopharmacol 173:330–337

Policarpi PB et al (2018) Nutritional potential, chemical profile and antioxidant activity of Chichá (Sterculia striata) nuts and its by-products. Food Res Int 106:736–744

Rocha LB et al (2015) Gallic acid as the major antioxidant in pequi (Caryocar brasiliense Camb.) fruit peel. Rev Bras Plantas Med 17:592–598

Roesler R, Lorencini M, Pastore G (2010) Fontes de antioxidantes do cerrado brasileiro: Citotoxicidade e fototoxicidade *in vitro*. Ciencia e Tecnologia de Alimentos 30:814–821

Rubenstein E (2020) Azetidine-2-carboxylic acid and other nonprotein amino acids in the pathogenesis of neurodevelopmental disorders. Neurodev Dis 2020:27–34

Samtiya M, Aluko RE, Dhewa T (2020) Plant food anti-nutritional factors and their reduction strategies: an overview. Food Prod Proc Nutr 2:1–14

Satyanarayana A et al (2006) Application of annatto dye formulations in different fruit and vegetable products. J Foods Serv 17:1–5

Schwartz G (2018) Jatoba— *Hymenaea courbaril*. In: Rodrigues S, Silva EO, Brito ES (eds) Exotic fruits, pp 257–261

Senes-Lopes TF et al (2018) Genotoxicity of *Turnera subulata* and *Spondias mombin* × *Spondias tuberosa* Extracts from Brazilian Caatinga Biome. J Med Food 21:372–379

Shafi S, Tabassum N (2015) Toxicity evaluation of hydro-alcoholic extract of *Portulaca oleracea* (whole plant) in Swiss albino mice. Int J Pharm Pharm Sci 7:506–510

Shih ML, Morgan JA (2020) Metabolic flux analysis of secondary metabolism in plants. Met Eng Commun 10:e00123

Silva DO et al (2017) Acute toxicity and cytotoxicity of pereskia aculeata, a highly nutritious cactaceae plant. J Med Food 20:403–409
Simões K et al (2009) Ipomopsin and hymenain, two biscoumarins from seeds of *Hymenaea courbaril*. Phytochem Lett 2:59–62
Sobolewska D et al (2020) Saponins as cytotoxic agents: an update (2010–2018). Part I—steroidal saponins. Phytochem Rev 19:139–189
Souza MO et al (2014) Antioxidant and antigenotoxic activities of the Brazilian pine *Araucaria angustifolia* (Bert.) O. Kuntze. Antioxidants 3:24–37
Souza MR et al (2019) Effects of subchronic exposure to *Caryocar brasiliense* peel ethanolic extract on male reproductive functions in *Swiss* mice. Reprod Toxicol 87:118–124
Sucha L, Tomsik P (2016) The steroidal glycoalkaloids from solanaceae: toxic effect, antitumour activity and mechanism of action. Planta Med 82:379–387
Sweazea KL et al (2017) Beet juice energy drink increases nitric oxide bioavailability and lowers blood pressure in healthy men. FASEB J 31:966
Takeiti CY et al (2009) Nutritive evaluation of a non-conventional leafy vegetable (*Pereskia aculeata* miller). Int J Food Sci Nutr 60:148–160
Tauchen J et al (2016) Phenolic composition, antioxidant and anti-proliferative activities of edible and medicinal plants from the peruvian amazon. Braz J Pharm 26:728–737
Tian J et al (2018) Oral chronic toxicity study of geniposide in rats. J Ethnopharmacol 213:166–175
Traesel GK et al (2016) Oral acute and subchronic toxicity studies of the oil extracted from pequi (*Caryocar brasiliense*, Camb.) pulp in rats. Food Chem Toxicol 97:224–231
Traesel GK, de Lima FF et al (2017a) Evaluation of embryotoxic and teratogenic effects of the oil extracted from *Caryocar brasiliense* Cambess pulp in rats. Food Chem Toxicol 110:74–82
Traesel GK, De Araújo FHS et al (2017b) Safety assessment of oil from Pequi (*Caryocar brasiliense* Camb.): evaluation of the potential genotoxic and clastogenic effects. J Med Food 20:804–811
Tsai C et al (2000) *In vitro* evaluation of the genotoxicity of a naturally occurring crosslinking agent (genipin) for biologic tissue fixation. J Biomed Mater Res 52:58–65
Tsujihata M (2008) Mechanism of calcium oxalate renal stone formation and renal tubular cell injury. Int J Urol 15:115–120
Vale CR et al (2013) Assessment of toxic, genotoxic, antigenotoxic, and recombinogenic activities of *Hymenaea courbaril* (Fabaceae) in *Drosophila melanogaster* and mice. Gen Mol Res 12(3):2712–2724
Valério, MA et al (2015) Annatto seed residue (*Bixa orellana* L.): nutritional quality. Food Sci Technol 35(2):326–330
Vieira Júnior GM et al (2015) New steroidal saponins and antiulcer activity from *Solanum paniculatum* L. Food Chem 186:160–167
Vieira PM, Santos SC, Chen-Chen L (2010) Avaliação da mutagenicidade e citotoxicidade de extratos de *Solanum paniculatum* L. usando o teste do micronúcleo *in vivo* em camundongos. Braz J Biol 70:601–606
Wei S et al (2014) Terrestrosin D, a steroidal saponin from *Tribulus terrestris* L., Inhibits growth and angiogenesis of human prostate cancer *in vitro* and *in vivo*. Pathobiology 81:123–132
Weitzberg E, Lundberg JO (2013) Novel aspects of dietary nitrate and human health. Ann Rev Nutr 33:129–159
Yamane H et al (2010) Chemical defence and toxins of plants. In: Comprehensive natural products II: chemistry and biology, vol 4. Elsevier, Amsterdam, p 339
Yonekura L et al (2016) Vitamina A (retinol) e carotenoides. In: Biodisponibilidade de nutrientes, 5th edn, pp 295–340
Zhu H et al (2010) Analysis of flavonoids in *Portulaca oleracea* L. by UV-vis spectrophotometry with comparative study on different extraction technologies. Food Anal Meth 3:90–97

Part III
Ethnobotanical Knowledge of Brazilian Food Plants

Brazilian Food Plants Registered in Historical Documents

Maria Franco Trindade Medeiros

Contents

1 Cultural Memory of Cooking in Brazil

The principle of the process of growing plant species may involve several myths and legends in different human cultures. We have the community with its mythological traditions around the origins of indigenous and local cultures, and, at the same time, we deal with the practice of biological exchange through the circulation of botanical species resulting from intercontinental travels carried out by numerous historical characters of human civilization.

If we consider only two historical landmarks, we could affirm that, since the thirteenth century, we find reports from Marco Polo about his travels and descriptions of an island where there were pepper, nutmeg, as well as other spices that can still be found in the present day around the world (Pereira 1922; Balick and Cox 1996), and when we analyze the *voyages* of the discovery of the Americas, we can also situate over time some facts that marked the way in which humanity was identifying the question of how to feed, what would be the sources of proteins, vitamins, and minerals. Nevertheless, certainly many exchanges between different ethnicities had already occurred in periods prior to Marco Polo's existence and action.

Specifically, talking about the Brazilian territory, it encompasses multiple community experiences woven over the centuries of coexistence between its original peoples and other inhabitants from different regions of origins. Each of its five regions (South, Southeast, Northeast, North, and Midwest) contributes to a specific way in the feeding of Brazilians. From a thorough, detailed research of local

M. F. T. Medeiros (✉)
Laboratório Interativo em Etnobotânica (LINE), Departamento de Botânica, Museu Nacional, Universidade Federal do Rio de Janeiro, Rio de Janeiro, RJ, Brazil

M. C. M. Jacob, U. P. Albuquerque (eds.), *Local Food Plants of Brazil*, Ethnobiology, https://doi.org/10.1007/978-3-030-69139-4_7

traditions related to food and the search for family recipes, we can perceive how Brazilian *cuisine is* varied and rich in flavors.

With ingredients brought from other parts of the world and with the very ones of each Brazilian region, a history of local *cuisine* and, in a larger context, a national food identity were forged. It is a history that was woven with the recombination of ingredients from Europe, where eggs, milk, and wheat flour (*Triticum aestivum* L.) came from, along with what Brazil offered from food species such as corn (*Zea mays* L.), sugar from sugar cane (*Saccharum officinarum* L.) fields, cassava (*Manihot esculenta* Crantz) flour, and countless fruits considered exotic. Thus, in this national *cuisine*, the dishes were commonly composed of a mixture of the creativity of the people dedicated to the preparation of food that elaborated the arrangement of preparations from ingredients not only from Europe and territories with which they maintained commercial relations, such as the Middle East and the Far East, as well as with those of indigenous origin but also composed the culinary adaptations those typical ingredients of the eating habits of the African peoples.

The time of the discovery of Brazil and the following centuries also comprised the period of food exchanges that came and went along the routes of the great navigations (Medeiros 2018). Thus, the cashew of Brazil went to Goa, as well as to Africa the cassava, corn, and peanuts (*Arachis hypogaea* L.). From Africa to Brazil came, for example, okra (*Hibiscus esculentus* L.), yams (*Dioscorea alata* L.), fennel (*Pimpinella anisum* L.), yellow ginger (*Zingiber officinale* Roscoe), and watermelons (*Citrullus lanatus* (Thumb.) Mansf.) (Costa 1983; Baker 1968; Balick and Cox 1996).

In one way or another, it occurred that African food culture has imprinted peculiar traits that are present today. In regions such as Salvador and throughout the region of the *Recôncavo Baiano* (Bahian region located in northeastern Brazil), this influence was affirmed in the tasting appreciation of three ingredients considered basic for this *cuisine* that brings strong African influence. These ingredients are coconut milk (*Cocos nucifera* L.), palm oil (*Elaeis guineensis* Jacq.) and pepper (*Capsicum* spp., *Piper* spp. and *Pimenta* spp.). In the work *Etiópia Oriental* (Eastern Ethiopia), published in 1609, in the city of Évora (Portugal), Friar João dos Santos described the process of obtaining coconut milk saying that it was extracted from fresh coconut, grated and well washed in two or three waters, and squeezed between the hands (Santos 1999 [1609]). In this way the coconut was dry and the water in which it was washed was thick as cow's milk.

Let us take a more detailed look at some of the regions that compound the Brazilian territory in relation to their historical traditions on the feeding modes of local populations.

Vestiges of the food memory of the northeast region can be found in works that Gilberto Freyre (1933) left us. The way of feeding was registered by this sociologist, which leads us to the understanding that in the past and still in the present days the *cuisine* of this region combines extremes in its *menu*. So, this means that the most regional ingredients alongside the international ones are served as elements of the northeastern diet in both the more rustic and refined foods.

In a report about José dos Santos Torres about the menu of his hotel in Pernambuco (northeastern Brazil), Freyre points out that in this space were served various Brazilian and European delicacies. The folklorist Luís da Camara Cascudo (1983) through away the interior of this region and observed the *menu* served in each moment of the day, as did the doctor Antônio da Silva Mello (1946, 1964) who travelled through the interior of the state of Pernambuco during the 1940s for his researches on nutrition.

The brief stay of the French in Maranhão (northeastern Brazil) during the seventeenth century could be registered in detail by the Capuchin priest Claude D'Abbeville, in his work *História da Missão dos padres Capuchinhos na Ilha do Maranhão e Terras Circunvizinhas* (History of the Mission of the capuchin fathers on the Island of Maranhão and Surrounding Lands), of 1614 (D'Abbeville 1874 [1614]). Father D'Abbeville spent only 4 months in Brazil and left us the report on the tasting that had countless fruits and roots of the region. In Maranhão, fruits such as *bacuri* (*Platonia insignis* Mart.), *jacama* (*Annona muricata* L.), *cupuaçu* (*Theobroma grandiflorum* (Willd. ex Spreng.) K. Schum.), and *jenipapo* (*Genipa americana* L.) appear as dessert fruits on the *menus.* Whether *in natura,* or in form of ice cream, candied sweets, candy in syrup or compotes, to this day these culinary finds of D'Abbeville are of popular taste, thus demonstrating that certain "finds" are "eternal".

In northern Brazil, indigenous influence was preponderant. In this region, there is a living past, which is updated in the execution of dishes such as *pato no tucupi* (duck made in *tucupi*, a broth made from manioc root fermentation), *açaí* wines (*Euterpe oleracea* Mart.), and *tacacá* (*tacacá,* a mixture of manioc gum and *tucupi* broth added with seasonings, aromatic herbs, and dry shrimp). The culinary base of Pará (state of this region) was created by indigenous in remote times. Besides, food products from manioc, such as *pirões* (fish porridge) and *beijus* (manioc flour), *açaí, piquiá* (*Caryocar brasiliense* Cambess.), *pupunha* (*Bactris gasipaes* Kunth), and *buriti* (*Mauritia flexuosa* Mart.) integrate the traditional *menu* of this region. In the Amazon, another state of this same region, dozens of native fruits are consumed, and guarana (*Paullinia cupana* Kunth) is highlighted as a drink that prolongs life.

Regarding the southeastern Brazil, there are numerous annotations on the *menus* of the nineteenth century, time when the naturalists August saint-Hilaire and Johann Pohl were in Brazil. Being in the state of Minas Gerais, these naturalists left registered experiences about behaviors and domestic habits of families in this region. It was common practice to have a separate *menu* for men, women, and children. The customs of eating black beans (*Phaseolus vulgaris* L.) with corn flour for lunch and vegetables cooked at supper, commonly kale (*Brassica oleracea* var. *viridis* L.), chicory (*Cichorium endivia* L.), okra, or *almeirão* (*Cichorium intybus* L. var. *intybus*) are described (Saint-Hilaire 1974 [1822]). There are some specialties that remain almost secret. These are stored in the scope of the regional, as is the case of sweet *buriti* or thick syrup of *araçá* (*Psidium guineense* Sw.) and *piquiá*, and in Espírito Santo (also southeastern state), the use of annatto (*Bixa orellana* L.) seeds in the preparation of the *moqueca capixaba* (dish made of tomato sauce, annatto, fish, and seasonings).

It is noteworthy that many of the families of Portuguese origin lent their own names, both family and sugar mill, to baptize a special recipe as a way to honor these people. An example of this tradition is the *Souza Leão* cake, dough made of manioc and coconut milk, as well as the cakes *Cavalcanti, Tia Sinhá*, and *Fonseca Ramos*.

With an extraordinary floristic richness that inspired and gave way to the local culinary creativity of the most varied environments, many recipes have certainly been lost throughout history. The social condition that the *mucamas* (black women servants of the white ladies) and *sinhás* (white ladies) assumed was to be illiterate, and their recipes had as a major form of conservation the immaterial memory and the transmission of their culinary knowledge by the exercise of orality. We believe that this issue constitutes one of the difficulties of working with the theme of food plants of ancient use, because the research of old recipes is compromised to the extent that many of them were not been written. Being usually an activity related to making feminine, the act of cooking was conditioned on women, both of the most humble condition and the most noble, such as the baronesses and viscountesses of the Empire, who were often illiterate.

Foods created over the centuries of creativity in the face of countless ingredients have gained the taste of the Brazilian population, overcoming taboos and prejudices. Wanting to please the palate was a creative effort of generations of cooks who transcended centuries of dedication to the art of dietetics. In fact, we commented only on the actuations of women, but men also integrated this action in the preparation of foods. As Freyre (1933) present us, among the slaves in the domestic service of the *casas-grandes* (the large houses where lived the lords and their families), those who dedicated themselves to the kitchen were within this extreme culinary specialization. Generally, two or three individuals were chosen for the kitchen services. Alongside women, blacks considered incapable of raw service could also be reserved for this service. The note of Freyre about newspaper advertisements from the time of slavery in Brazil that gave visibility to the escape of these black people who worked in the kitchen is interesting. These notes specified that in the act of escape, these blacks were dirty because they were cooks.

Thus, if we can trace a more general panorama, disregarding the infinite specificities of each corner of the Brazilian territory and what was able to mix and create each community experience in its particular living space, this would be the most common scenario of this space so used and so central in the past life of the Colonial and Imperial period of Brazil, that is, the kitchen. *Sinhás* and black people dedicated to the preparation and creation of dishes, sweets, and preserves with fruits and roots of the earth, so amazing in their new flavor, have been gaining the taste of the palate and remain present until the present day in national *cuisine.*

2 Food Species of Brazilian Flora Registered in Past Documents

The food plants and their cultivation, in addition to their multiple uses, can be pointed out as having been one of the first measures thought by those who came to the Brazilian territory with the mission of populating it from the conquest undertaken by the Lusitanians. Since the founding of Colonial Brazil, the newly arrived settlers in the new territory have taken care of securing items for their most pressing survival needs. In this sense, the actions have moved towards the beginning of agriculture and, further forward in history, to the development of some profitable industries. Thus, in this process, the cultivation of cereals and vegetables were the first to be implemented. Then, the fruit species and grasses for the forage of animals gained space in the productive force. On the other hand, the plant species that supplied other genera began to be cultivated later.

In 1551, there is the register made by the Father Manuel da Nóbrega of Pernambuco (northeastern Brazil) that there was the culture of several plants brought from the metropolis. Ciders (*Melissa officinalis* L.), vines (*Vitis vinifera* L.), orange trees (*Citrus* x *sinensis* (L.) Osbeck), lemon trees (*Citrus* x *aurantiifolia* (Christm.) Swingle), and figs (*Ficus carica* L.) were mentioned. Father Nóbrega took care to mention that these cultivars were as good as those in Europe and that they were very productive (Nóbrega 1955 [1551]).

Another register that tell us about this dynamic of food cultivars in the history of occupation of the Brazilian territory is that left by Gândavo, dated 1576. In this document he refers to banana (*Musa paradisiaca* L.) from the Island of São Tomé, ananas (*Ananas comosus* (L.) Merr. var. *comosus*), melons (*Cucumis melo* L.), cucumbers (*Cucumis sativus* L.), pomegranates (*Punica granatum* L.), and figs of many varieties, as well as ciders, lemons, and oranges.

In the Jesuit College located in the city of Olinda, in the former Captaincy of Pernambuco, located in the northeastern region of Brazil, there was the cultivation of several food species, according to a mention made by Father Fernão Cardim (1925 [1584]). On a visit to the College of Olinda in1584, Father Fernão could see that in the Jesuits' backyard there were vines, fig trees, orange trees, melons, cucumbers, and pomegranates. The following year, in 1585, Father José de Anchieta (1933 [1585]) when writing about this same College of Jesuit Fathers still mentions the many coconut trees that existed in that place.

In 1587, in the work *Tratado Descritivo do Brasil* (Descriptive Treaty of Brazil), Gabriel Soares de Sousa also emphasizes the food plants that he found in Brazil. In his register he speaks about the cultivation of several species, being these lime (*Citrus limettioides* Tanaka), French lemon (*Citrus* x *limonia* (L.) Osbeck), and Galician lemon, rice (*Oryza sativa* L.), and yams from Cape Verde and São Tomé, and others such as watermelon, pumpkin (*Cucurbita moschata* Duchesne), mustard (*Brassica* spp.), turnip (*Brassica rapa* L.), cabbage, lettuce (*Lactuca sativa* L.), coriander (*Coriandrum sativum* L.), dill (*Anethum graveolens* L.), parsley (*Petroselinum* spp.), mint (*Mentha* spp.), chives (*Allium schoenoprasum* L.), aubergine (*Solanum melongena* L.), *alfavaca* (*Ocimum* spp.), *bredo* (*Amaranthus* spp.), chicory, carrot (*Daucus carota* L.), spinach (*Tetragonia tetragonioides* (Pall.) Kuntze), and basil (*Ocimum* spp.) (Sousa 1971 [1587]).

Resuming the coconut tree culture, it is noticed that it has gained an expressive development for its variety of uses, and its spread throughout the Brazilian territory was already extensive in 1618, as described in *Diálogos da Grandeza do Brasil,* (Dialogues of the Greatness of Brazil) of Brandão (2010 [1618]). Before the Dutch invasion in northeastern Brazil, in *História da Guerra de Pernambuco* (History of the Pernambuco War), Santiago (1984 [1634]) describes a landscape full of coconut trees and says that there was no residence in which there was at least some individuals of this specie in the city of Olinda, in the former Captaincy of Pernambuco. He also adds that the presence of coconut trees in the backyards of the houses of the city of Olinda gave a special characteristic to that landscape.

It is also been cultivated since the sixteenth century in Brazil the ginger, that was brought from the Island of São Tomé. Its production gained space in Brazilian territory, which contributed to its export in 1575. Thus, as this ginger produced in Brazil was considered of better quality than that of India, it gained the preference of its consumers. For this reason, its culture was forbidden so that trade with the East would not be harmed (Costa 1983). Only on April 24, 1642, this prohibition was repealed and a new provision was enacted, allowing Brazilian people the culture of ginger. This permission was valid for the cultivation of the species on land that was considered unsuitable for sugarcane cultivation. In addition, it had been stipulated that its export would be made available, by the paying the competent duties. Another provision issued on April 10, 1671, changed this scenario. At least in the Captaincy of Pernambuco, the cultivation of ginger was recommended, and its export was allowed without the collection of taxes.

Years later, on March 30, 1678, a royal letter addressed to the governor of Pernambuco communicated that he had ordered the Viceroy of India to consign fruit trees to this captaincy and to the kingdom, for the usefulness that could favor this state, as for the whole Kingdom of Portugal. Later, already on May 20, 1862, there was a new communication from the king issued to the Viceroy with similar orders, and he also ordered that when ships with cargo of useful plants arrived at the captaincy of Pernambuco, the governor should have to look for these plants, and plant them in accordance with India's instructions.

Reaching the years of 1800, on July 7, 1810, the pepper culture of India was recommended, as well as that of other exotic or indigenous plants, as the native species of Brazil were called (Costa 1983). It meant that native plants were now considered for planting. The intention was that uncultivated species, native to Brazil, could constitute new materials suitable for consumption, in addition to becoming export items and moving trade due to their various applications. The cultivators were guaranteed exemption from customs fees by a period of 10 years, awards, honorary medals, and also, exemption from service and recruitment in the military sector. These measures were reiterated in the following year of 1811 by new provisions which were issued in this same sense as the provisions of 1810.

These measures taken point out very clearly how there was a movement of stimulus and protection to the culture of new species that were promising, and, thus, like those mentioned recently, there are many other measures that were issued with the same objective. These documents provide information about the transformation of the landscape of phytogeographic domains existing in Brazil, although they refer more massively to the economic movement determined by the Metropolis to its Colony. The central economies revolved around species not yet mentioned, such as coffee (*Coffea arabica* L.), cotton (*Gossypium* spp.), and sugarcane (Dean 1996). Although there was this centrality in cultivars interesting to the economy, so many other cultivars considered to be peripheral also existed. These peripheral economies were managed by populations that were right on the periphery of the colonial system. These were fields of cultivation for subsistence that were configurated as being relevant to the survival of mestizos, riverside, *quilombolas* (black slaves who formed community enclaves reactionaries to the slavery system), *caiçaras* (communities composed of indigenous peoples, Portuguese settlers, and African slaves who supported themselves from agriculture, small-scale fishing and plant extractivism), and many other Brazilian community formations.

The native species managed by these populations served for the most different utilities, such as for medicine or as a source of food. Taking as an example the secondary forests on the landscape of the Atlantic Forest, we can find nowadays introduced exotic species and managed native species (Oliveira and Silva 2011). These are traces of the memory of past local culture. Among these species, we can cite the jack (*Artocarpus heterophyllus* Lam.), whose culture dates back to ancient times, which can be inferred by the denomination of mills and villages known by this term since many years ago. Another species that we can illustrate here in this incorporation into the floristic composition of a secondary forest would be the coffee.

Among the native species that were managed and that nowadays can be found in secondary forests of the Atlantic Forest, these compound a diverse set including pioneers, initial and late secondaries of ecological succession present in areas equivalent to abandoned swideens (Oliveira and Silva 2011). Considering the plasticity of management practices and the differentiated objectives of each of these practices in terms of production and conservation, these may lead to the incorporation or populational prevalence of certain species in the forest environment, depending on the conditions of the environment, cultural, and economic traits of the locality (Noble and Dirzo 1997). A group of plant resources that can receive greater attention from management practices in the forest environment is that of so-called intentionally managed species, which includes native fruits. This set of species has been gaining over time priority in planting on the edges and within the forests. Among these are the *jabuticaba* (*Myrciaria cauliflora* (Mart.) O. Berg.), the chestnut tree (*Bombacopsis glabra* (Pasq.) Robyns), the *cambucá* (*Plinia edulis* (Vell.) Sobral), and the *cajá-mirim* (*Spondias mombin* L.) (Oliveira and Silva 2011).

Thus, the use of food species by past populations passed through a culture within forest areas. Registers in different supports in this regard are not usual to be found, but the work with oral history can provide more information about these past management practices. What we can perceive through the presence of species such as those mentioned is that there is a valuation of the forest environment in past periods, including the intentional planting of plant species elected as being important for the experience of an ancient community. The landscape then gains contours from the human action carried out in the past and that, also, expands in time and continues to bring interferences on ecological and food patterns nowadays. Thus, in the constitution of a list of species suitable for the ingestion, there is an identity character that is configured as being historical.

3 Brief Words of Conclusion

In the history of the use of food plant resources, historical documentation reveals to us with greater intensity data related to the dominant center of the Brazilian colonial economy. The relations of the human population present in the Brazilian territory that were on the fringes, on the periphery of the central economy, left few documented reports. The historiography of the landscape of the Brazilian phytogeographical domains deserves further investigation in each of its representativeness. What we have found so far is a narrative that highlights the bias of actors and cultures of species that played a central role in the economy.

However, we can point out as key species for the future of human food security the acceptance of food plants that transcend centuries through collective dietary memory.

In addition to this primary need, we leave the indication that the behavior is what could be configured as the key to a balanced relation with flora (and nature in its entirety) in order to consider it not only as a source of consumption, but rather a fundamental part of the experiential constitution of humanity.

References

José de Anchieta A (1933 [1585]) Cartas, informações, fragmentos históricos e sermões do Padre Joseph de Anchieta, S.J. (1554-1591). Civilização Brasileira S.A., Rio de Janeiro

Baker HG (1968) Las plantas y la civilización. Herrero Hermanos Sucessores S.A, Ciudad de México

Balick MJ, Cox PA (1996) Plant, people, and culture: the science of etnobotany. Scientific American, New York

Brandão AF (2010 [1618]) Diálogos das grandezas do Brasil. Senado Federal, Conselho Editorial, Brasília

Cardim Pe F (1925 [1584]) Tratados da terra e gente do Brasil. J. Leite & Cia, Rio de Janeiro

Cascudo Luís da C (1983) História da alimentação no Brasil. Itatiaia, Belo horizonte; Editora da Universidade de São Paulo, São Paulo

da Costa FAP (1983) Anais Pernambucanos. FUNDARPE, Recife

D'Abbeville Pe C (1874 [1614]) História da Missão dos Padres Capuchinhos na Ilha do Maranhão e suas Circumvisinhanças. Typ. do Frias, Maranhão

Dean W (1996) A ferro e fogo. A história e a devastação da Mata Atlântica brasileira. Companhia das Letras, São Paulo

Freyre G (1933) Casa-Grande e Senzala. Maia & Schmidt, Rio de Janeiro

de Gândavo PM (1576) Historia da província Santa Cruz a que vulgarmente chamamos Brasil. Officina de Antonio Gonsaluez, Lisboa

Medeiros MFT (2018) Medicinal plants and state policy in South America: the case of colonial Brazil. In: Albuquerque UP, Patil U, Mathe A (orgs.) Medicinal and aromatic plants of south America – Brazil. Springer, Dordrecht, 71–80

da Silva Mello A (1946) Alimentação no Brasil – problemas e soluções. Edições Cruzeiro, São Paulo

da Silva Mello A (1964) Nordeste brasileiro. Estudos e impressões, José Olympio, São Paulo

Noble IR, Dirzo R (1997) Forest as human-dominated ecosystems. Science 277:522–525

da Nóbrega Pe M (1955 [1551]) Cartas do Brasil e mais escritos do Pe. Manuel da Nóbrega (Opera Omnia). Ed. Serafim Leite. Universidade, Coimbra

Oliveira RR, Silva IM (2011) História da paisagem e paisagens sem história: espécies exóticas e nativas manejadas na Mata Atlântica. In: Peixoto AL Silva IM (orgs.) Saberes e usos de plantas: legados de atividades humanas no Rio de Janeiro. Ed. PUC-Rio, Rio de Janeiro, p 69–92

Pereira FE (ed) (1922) O livro de Marco Paulo (1502). Ed. Nacional, Lisboa

de Saint-Hilaire A (1974 [1822]) Segunda viagem do Rio de Janeiro a Minas Gerais e a São Paulo. Itatiaia, Belo Horizonte; Editora da USP, São Paulo

Santiago DL (1984 [1634]) História da guerra de Pernambuco. FUNDARPE, Recife; Diretoria de Assuntos Culturais

dos Santos Frei J (1999 [1609]) Etiópia Oriental. CNCDP, Lisboa

de Sousa GS (1971 [1587]) Tratado descritivo do Brasil em 1587. C.E.N., São Paulo

Archaeobotany of Brazilian Indigenous Peoples and Their Food Plants

Mariana Franco Cassino, Myrtle P. Shock, Laura Pereira Furquim, Daniela Dias Ortega, Juliana Salles Machado, Marco Madella, and Charles R. Clement

Contents

M. F. Cassino (✉)
Programa de Pós-Graduação em Botânica, Instituto Nacional de Pesquisas da Amazônia (INPA), Manaus, Brazil

ICREA - CaSEs – Department of Humanities, Universitat Pompeu Fabra (UPF), Barcelona, Spain

M. P. Shock
Programa de Antropologia e Arqueologia, Instituto de Ciências da Sociedade, Universidade Federal do Oeste de Pará (UFOPA), Santarém, Brazil

L. P. Furquim
Museu de Arqueologia e Etnologia, Universidade de São Paulo (USP), São Paulo, Brazil
e-mail: laura.furquim@usp.br

D. D. Ortega · M. Madella
ICREA - CaSEs – Department of Humanities, Universitat Pompeu Fabra (UPF), Barcelona, Spain
e-mail: daniela.dias01@estudiant.upf.edu; marco.madella@upf.edu

J. S. Machado
Departamento de História, Universidade Federal de Santa Catarina (UFSC), Florianópolis, Brazil
e-mail: juliana.salles.machado@ufsc.br

C. R. Clement
Coordenação de Tecnologia e Inovação, Instituto Nacional de Pesquisas da Amazônia (INPA), Manaus, Brazil
e-mail: cclement@inpa.gov.br

M. C. M. Jacob, U. P. Albuquerque (eds.), *Local Food Plants of Brazil*, Ethnobiology, https://doi.org/10.1007/978-3-030-69139-4_8

1 Introduction

Human history in Brazil began at least 25,000 years ago (Lahaye et al. 2015; Vialou et al. 2017). Through the millennia, human societies organized and reconfigured their social and political organizations. They also had diverse bodies of knowledge that were shared by people within the society, including their experiences and practices related to plant and landscape management. The breadth of plant food use in Brazil is explained not only by the great local biodiversity and the heterogeneity of Brazilian biomes but also by diverse long-term indigenous food choices.

Archaeological sites are places where material evidence of human occupations, such as ceramic and stone artifacts, plant remains, animal and human bones, earthworks, and/or rock art can be found, and their investigation is fundamental to the reconstruction of human histories. Although it is often difficult to link archaeological sites to modern or historically documented indigenous populations, a notion of the cultural diversity that created the archaeological record can be glimpsed through linguistic diversity (Fig. 1). Current linguistic diversity in Brazil includes more than 160 languages from 39 families. However, pre-Columbian linguistic diversity could have been at least six times higher (ISA 2020). It is notoriously difficult to estimate the population of indigenous groups in the Americas prior to European invasion due to the subsequent loss of lives to wars, epidemic diseases, and slavery. Denevan (1992), a geographer, estimated that there were 24.3 million inhabitants in South America, and this number is similar to a more recent review estimating their numbers at 23.6 million (Koch et al. 2019). Across Brazil there were thousands of indigenous communities of different sizes, economic practices, and social and political organizations. European invasion brought to an end, or significantly changed, most of these communities and their practices as they fled colonization and, in some cases, merged following population losses and/or for greater defensive power (Hemming 1995). The losses of human lives and populations led to the disappearance of some traditional ecological knowledge, and many cultivated plants, yet, great agrobiodiversity remains (Clement 1999).

Food consumption permeates practically all daily and ritual activities of human life and thus strongly influences how people interact with their environments and transform them. Over the long term, these varied dynamics contributed to the formation of Brazilian landscapes and their agrobiodiversity. This chapter will explore the diversity and temporality of food plant use documented in the archaeological record, including the expansion of different crops across the continent and how foodways have been, and continue to be, crucial to the creation of current landscapes.

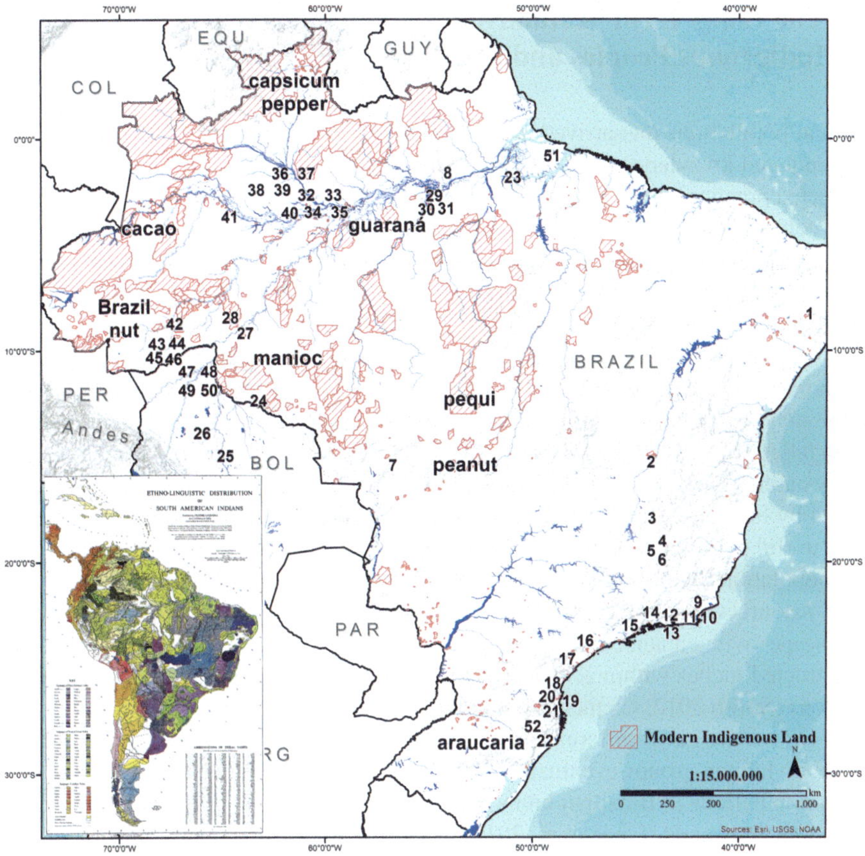

Fig. 1 Locations of archaeological sites (numbered) and the centers of origin for some native crops, by name, within Brazilian territory with respect to the locations of modern indigenous lands (FUNAI 2020). Historical ethnic and linguistic diversity among South American indigenous people is shown below and to the left based on the Ethno-linguistic Distribution of South American Indians, compiled by Čestmir Loukotka (1967) and used with permission from Taylor & Francis. Archaeological sites are: Rock shelters: 1. Furna do Estrago, 2. Santana do Riacho, 3. Lapa do Santo, 4. Lapa dos Bichos, 5. Lapa Pintada, 6. Lapa Grande de Taquaraçu, 7. Santa Elina, 8. Pedra Pintada; Shell mounds: 9. Sambaqui Salinas Peroano, 10. Sambaqui Ponta da Cabeça, 11. Sambaqui Boca da Barra, 12. Sambaqui Beirada, 13. Sambaqui Pontinha, 14. Sambaqui do Meio, 15. Sambaqui do Forte, 16. Moraes, 17. Capelinha, 18. Enseada I, 19. Forte Marechal Luz, 20. Morro do Ouro, 21. Itacoara, 22. Jabuticabeira II, 23. Tucumã, 24. Monte Castelo, 25. Isla del Tesoro, 26. Isla Manechi; Open-air sites: 27. Teotônio, 28. Santa Paula, 29. Serra do Maguari-1, 30. Lago Caranã, 31. Cedro, 32. Hatahara, 33. Açutuba, 34. Lago Grande, 35. Osvaldo, 36. Vila Nova I, 37. Vila Nova II, 38. Floresta, 39. Lago das Pombas, 40. Caldeirão, 41. São João, 42. Cruzeirinho, 43. Fazenda Iquiri II, 44. Sol de Maio, 45. Campo Esperança, 46. Tequinho, 47. El Círculo, 48. Las Palmeras, 49. Chacra Talería, 50. Tumichuchua, 51. Aterro dos Bichos; Pit house: 52. Bonin. Designed by Laura P. Furquim

2 The Long-Term Mutualistic Relationships Between Indigenous Peoples and Plants

Human populations that arrived in lowland South America about 25,000 years ago accumulated knowledge about the flora and fauna of the areas that they settled and perfected practices that aided in the exploitation of these resources. The variation in size, color, sweetness, and flavor in any population of edible plants would have been observed and humans selected their preferred types. Once plant resources are identified as food, or favored types of that food, they can be protected, which can involve the removal of plants competing for light and nutrients. Future harvests are also enhanced when picked fruits, nuts, or tubers are transported back to the settlement for processing or for sharing with other members of the group, as along the trail some may fall from the basket or the seeds of consumed fruits may be discarded. Both fallen fruits and discarded seeds may germinate along the route or in the camps and establish new populations. These may eventually be exploited and also protected by human groups using the trails. Over time this may lead to trails enriched with useful plants and the progressive movement of favored plants from their original populations. Since any settlement is surrounded by trails leading to important food resources, they are central to the development of a mosaic of domesticated landscapes with enhanced availability of different food plants. These are culturally constructed niches within ecosystems that continue to experience natural transformations (Smith 2012; Clement and Cassino 2018).

Over millennia, in different intensities, these transformational dynamics instigated by human groups eventually altered the abundance and distribution of many plant populations in Brazilian biomes, most notably the populations of useful trees that occur in stands (Levis et al. 2018; Balée 1989, 2013). In Amazonia, the most emblematic stands are those of Brazil nut (*Bertholletia excelsa* Bonpl.), the *castanhais* (Shepard Jr. and Ramirez 2011), but there are many others (Levis et al. 2018). Pequi (*Caryocar brasiliense* Cambess.) stands occur across the *Cerrado* (a region with savanna-like vegetation), and those in the upper Xingu River are domesticated (Fig. 1) (Smith and Fausto 2016). In the Atlantic Forest, araucaria (*Araucaria angustifolia* (Bertol.) Kuntze) can occur in immense stands, many of which are directly connected to indigenous activities (Bitencourt and Krauspenhar 2006; Reis et al. 2014). The importance of such tree stands as food resources is huge; for Amazonia, Clement (2019) estimated that the harvest from six species of forest fruit trees that occur in such stands can produce more protein than all the cattle currently in Amazonia (see also Shepard Jr. et al. 2020).

All human settlements have areas where refuse is discarded (called middens by archaeologists) and which often form anthropic soils (Lathrap 1977; Erickson 2003). The by-products from plant processing would normally be deposited in these areas, and discarded seeds or vegetative parts of plants might grow, producing new resources within the settlement. These new plants could be tended and protected, which is the origin of home gardens (Lathrap 1977; Anderson 2005).

As human populations grew during the Holocene, the number and size of settlements increased, many becoming permanent, and each was surrounded by an expanding mosaic of domesticated landscapes. When the food requirements of a community increased, this could be obtained by expanding the gardens further, even beyond the nutrient-rich soils of the middens. This expansion further changed the landscape mosaic around the settlement and some areas were more intensively managed with the production of annual, semi-annual, and perennial food plants (Denevan 2001). These intensively managed areas are called agroforestry systems.

Because plants anywhere in this mosaic are treated by indigenous groups as individuals, in some cases even thought of as "children" (Hastorf 1998; Machado 2012), each is carefully tended and evaluated for future propagation. The result is that some of these plant populations became domesticated over time, sometimes with remarkable differences in size, color, sweetness, and flavor compared to the wild plants of the same species (Clement 1999). The relationship between humans and their plants is mutualistic, where both populations benefit: plants are cared for and propagated, and humans gain their company and nutritional value.

3 Cultural Meals and Cultural Landscapes

Food ingredients are collected, produced, harvested, processed, cooked, and consumed according to the customs and practices of each human culture. The collection of these practices and their organization in specific social, political, and cultural contexts are called foodways (Staller and Carrasco 2010). Consider for a moment one plant, the peanut (the seeds of *Arachis hypogaea* L.), which can be consumed toasted in the shell, or as part of a peanut brittle snack (*pé-de-moleque*), or as peanut butter in a sandwich. While a sandwich might be consumed alone at breakfast or lunch, roasted peanuts might be consumed socially, at soccer games or with beer in a bar, and the peanut brittle is a traditional candy of June festivities in Brazil. Across cultures, people make decisions, both consciously and by habit, as to which plants became parts of their social events and their daily meals. A large number of plants that provided foods of different types – fruits, seeds, roots, spices, leaves, etc. – were the focus of human interventions, and many were domesticated. Foodways are one of the daily expressions of culture, through which consumption practices are linked to choices and knowledge about plant management, production, and processing (Staller and Carrasco 2010; Graff 2018).

The awareness and knowledge of ecological relationships is specific to each ecosystem, or even local landscapes, and changes when human populations move to new ecosystems. Traditional ecological knowledge was, and is, dynamic, as it is learned not only generationally but also from personal experiences, family, and neighbors. It includes behaviors and practices of interactions with plants that favored the creation of cultural niches and landscape mosaics with enhanced food availability. Choices about the adoption of domesticated plants were different among different human cultures. It follows that domesticated plants and domesticated landscapes are cultural artifacts that were molded by each specific culture's knowledge, beliefs, values, and foodways.

The cultivation of plants, through culturally specific ecological knowledge for culturally specific foodways, is the basis for the farming-language dispersal hypothesis (Renfrew and Bellwood 2002). The distribution of Arawak languages may be a lowland South America example of a farming-language dispersal of growing populations (Blench 2012). With information about the variation in words for the same plant among members of the Arawak linguistic dispersal, it is possible to identify a language homeland and even put a date on it. Based on the names for manioc (*Manihot esculenta* Crantz) and hot pepper (*Capsicum* spp.), both of which were domesticated in lowland South America, and maize (*Zea mays* L.) and common bean (*Phaseolus vulgaris* L.), domesticated in Mexico and the central-northern Andes, respectively, Brown et al. (2013a, b, 2014a, b) suggest that the ancestors of today's Arawak speakers had adopted each of these domesticates by 4000 BP in western Amazonia. Thus, meals that contained an abundance of energy (manioc), a diversity of protein (maize and bean, but also fish), spiced with hot pepper may have fueled the Arawak dispersal across lowland South America and into the Caribbean. These four crops are common to other indigenous foodways, both of widely dispersed language families and of languages that did not disperse, reinforcing the observation that choices about plant cultivation are not tied to the plants, but rather to cultures.

Although anthropogenic forests occur all over Brazil, not all groups have adopted the cultivation of herbaceous plants. For example, landscape mosaics enriched with food species are created by indigenous populations with highly mobile lifestyles, such as the Nukak, that do not have cultivated fields (Politis 1996). Also, the deliberate creation of enriched forest spaces has been documented on the Southern edge of the Amazon where the Gorotire Kayapó alter the soil and soil fauna in locations that retain greater amounts of water and, into this improved ecosystem patch, plant a diversity of useful species, including foods and spices (Anderson et al. 1989; Posey 1985). In Southern Brazil, the application of ecological knowledge of araucaria included deliberate seed dispersal and increased burning to promote the expansion of araucaria nut stands (Bitencourt and Krauspenhar 2006; Reis et al. 2014). Each food plant requires local ecological knowledge that directs the management practices central to its production, processing, and consumption. Foodways continuity and variability over time and across space during the Holocene frame the histories of cultural landscapes (Cassino et al. 2019; Ingold 1993).

4 Archaeobotany: Tools for Assessing the Long-Term History of Food Plant Consumption and Management

Archaeobotany is the study of plant remains recovered from archaeological contexts and the human activities that they represent, including cultural foodways (Pearsall 2015; Graff 2018; Hastorf 1999). Macroscopic botanical remains include dried and charred fragments of wood (Fig. 2a, b), fruits (Fig. 2g), seeds (Fig. 2h, i), and tubers.

The majority of archaeological charcoal is from wood and directly related to firewood preferences. Wood used as fuel can indirectly inform about diet, since it is frequently acquired from surrounding vegetation enriched with food plants. By contrast, most of the fragments of tubers, fruits, and seeds are by-products of food processing and were likely deposited in middens or burnt with other waste. For example, the hard seed coats of Brazil nuts are more likely to be found than the edible kernels (seeds). Archaeological charcoal is collected when it is visible during excavation and by sieving or flotation of excavated sediments. The screen size of the sieve determines the size of fragments that are recovered. Given the diminutive size of some seeds, most analysis is carried out on sediment samples processed through fine mesh with openings between 0.5 mm and 2.0 mm (Pearsall 2015).

Plants also leave a number of different microscopic remains. Starch grains (Fig. 2c–e) can adhere to or be incorporated into pieces of ceramics, stone tools, and tartar on human teeth (dental calculus), and, further, they can occur in coprolites (desiccated feces). Starch grains from plant species are identifiable at different taxonomic levels and may also show signs of injury from processing (e.g., grinding) and/or from exposure to heat from cooking (Henry et al. 2009). Phytoliths (Fig. 2f) are silica bodies that form in plant cells and retain the shape of these cells. They are diagnostic at different taxonomic levels as well as characteristic of different plant parts, such as squash (*Cucurbita* spp.) fruit rinds, maize cobs and leaves, palm leaves, or rice chaff (*Oryza* spp.). Phytoliths can be preserved in sediments and also in pottery, grinding stones, dental calculus, and coprolites (Piperno 2006). They have been used to document the use of plants in human diet and also to provide important clues about plant domestication and dispersal (Ball et al. 2015). The enlargement of cells in some plant tissues may accompany increased fruit/seed size that occurs in some domesticated plants (Purugganan and Fuller 2011). If these tissues are phytolith forming, the resulting "domesticated" phytoliths will also have increased in size. It follows that domestication can be tracked in the archaeobotanical record by recording variations in the size of phytoliths. This approach has been used to suggest that squash from ca. 9000 BP in southwestern Ecuador (Piperno and Stothert 2003) and rice from ca. 4000 BP in southwestern Amazonia were undergoing domestication (Hilbert et al. 2017). Beyond direct plant remains, the types of foods that humans consumed can be assessed by the proportions of stable isotopes of nitrogen and carbon in human bones (Lee-Thorp 2008).

Information on how plant management changed plant communities can also be gleaned from the analysis of pollen from lake or waterlogged sediments. Pollen assemblages deposited in such sediments are composed of the pollen from species growing around the sampling location and/or from pollen grains transported in by water or wind. Pollen datasets can document the introduction of exogenous plants, as well as changes in the proportions of different plant taxa in the vegetation that might be due to management (Maezumi et al. 2018; Carson et al. 2014). Further, modern vegetation communities can also be considered cultural artifacts, both from centenary trees and from the offspring of plants that grew in the same area, and they can be tools to assess past vegetation (Cassino et al. 2019).

Fig. 2 Archaeological excavations at the Hatahara (1) open-air site in central Amazonia and the Lapa Pintada (2) rock shelter in the Cerrado. Examples of the diverse plant remains found are *Genipa* sp. (**a**) and *Byrsonima* sp. (**b**) wood charcoals; *Dioscorea* sp. (**c**, **d**) and *Zea mays* (**e**) starch grains, with (**d**) demonstrating the distinctive cross produced in starch by polarized light; *Bactris* sp. (**f**) phytolith; desiccated *Hymenaea* sp. (**g**) fruit rind; and desiccated *Cucurbita* sp. (**h**) and *Passiflora* sp. (**i**) seeds. [Credits: Val Moraes, Central Amazon Project (1), Caroline F Caromano (**a** & **b**), Leandro M Cascon (**c–f**), Myrtle P Shock (2, **g–i**)]

All of these lines of evidence are complementary and refine our understanding of human diet, past, and present. Each of them has particular strengths for documenting different groups of plants: starch grains are useful to document the presence and use as food of carbohydrates-rich plant parts, especially roots, underground organs, and seeds; phytoliths are key for identifying grasses (including maize), palms, and squashes; charcoal frequently preserves identifiable characteristics of nut seed coats, some fruits, and palm seeds; pollen records catch mostly wind pollinated species, like grasses; and current vegetation often reflects plant populations that were increased in frequency and distribution by human groups, but did not become reproductively dependent on cultivation.

5 Food Plant Remains in Brazilian Archaeological Sites

Over the millennia of human occupation of South America, people constructed meaningful places by transforming the environment, building their homes, and attributing memories to the places they lived (Bowser and Zedeño 2009). A great diversity of cultural landscapes was created across the Brazilian biomes by past human occupations employing diverse cultural strategies. Resulting archaeological sites include rock shelters, fluvial and coastal shell mounds, open-air villages, earthworks, such as mounds and geoglyphs, and pit houses (Fig. 1). The archaeological record of foodways is a result of human practices that spanned daily activities of food production and consumption, as well as ritualistic or symbolic practices. As archaeological sites were occupied by different cultures, at different moments in time, the accumulated plant remains may reflect different practices of plant management, social contexts, and symbolic events (Klokler 2012). Further diversity is introduced by the local environment and temporal adoption of different foods. Here, we present archaeobotanical evidence of the use of food plants from sites across Brazil to highlight the diversity of foodways (see Table 1 for species information by site), histories of plant and landscape domestication, and human-mediated plant migrations. In this chapter, the chronological subdivisions of the Holocene follow Walker et al. (2018): Early Holocene ca. 11,700 BP to ca. 8300 BP, Middle Holocene ca. 8300 BP to ca. 4200 BP, and Late Holocene ca. 4200 BP to present.

5.1 Rock Shelters

Across Brazil, and the world, rock shelters often provide privileged records of human plant use due to the exceptional conditions of preservation. These are protected spaces of variable size in rock outcrops that were/are used for daily or specialized activities (Fig. 2.2). They may contain components of cultural communication in the form of rock art and frequently have archaeological deposits with extensive chronologies due to recurrent utilization. Within the geographic lay of the land, rock shelters are permanent and frequently occur on hillsides, such that those people within would have greater visibility of the surrounding area. Brazilian rock shelters that have been studied archaeologically with botanical analyses are located in the savanna (*Cerrado*), the northeastern semi-arid shrubland (*Caatinga*), and eastern Amazonia.

The recurrent use of rock shelters by diverse human populations contributes to the long-term records of plant food use dating back to at least 12,000 years ago in Brazil. From between 12,000 and 9000 BP, people living at the Pedra Pintada rock shelter in eastern Amazonia (Pará state) consumed over a dozen different types of palm and tree fruits, including moriche/buriti (*Mauritia flexuosa* L.f.), tucumã (*Astrocaryum vulgare* Mart.), Brazil nut, nance (*Brysonima* spp.), and jutaí/jatobá (*Hymenaea* spp.) (Shock and Moraes 2019; Roosevelt et al. 1996; Roosevelt 1998). By the Late Holocene, cultivated maize and squash were introduced and the number of tree crops also increased to include cashew (*Anacardium occidentale* L.), hog plum (*Spondias mombin* L.), açaí (*Euterpe oleracea* Mart.), among others (Table 1) (Roosevelt 2000). Resource diversity is not only observed in remains of seeds and fruits. In the Caatinga, at the site of Furna do Estrago (Pernambuco state), human coprolites revealed the presence of phytoliths from a psychoactive plant (*Anadenanthera colubrina* (Vell.) Brenan) and starch from a bromeliad, macambira (*Bromelia laciniosa* Mart. ex Schult. & Schult.f.) (Santos 2014). Current indigenous groups produce flour from the leaves of this bromeliad to make a kind of bread, a practice that seems to have a deep history starting two thousand years ago, alongside the consumption of cultivated maize, manioc, and sweet potato (*Ipomoea batatas* (L.) Lam.).

Exceptional preservation can also lead to the identification of a greater variety of plants, including spices and other non-calorific plant parts, which may have been used less frequently. In the Cerrado of the state of Minas Gerais, dried plant remains from bowl-shaped pits included hundreds of pieces from a diversity of seeds and fruits that are not used in current diets, such as olho-de-boi (*Mucuna sloanei* Fawc. & Rendle), many of which have yet to be identified (Shock 2010). Also in the Cerrado, remains of pollen in coprolites and charcoal of seeds, fruits, and wood from late Holocene occupations in Santa Elina (Mato Grosso state) inform about food diversity in the diet and surrounding vegetation. The occurrence of manioc pollen suggests local plant cultivation alongside the consumption of babassu (*Attalea speciosa* Mart. ex Spreng.) and uricuri (*Attalea phalerata* Mart. ex Spreng.) palm nuts, and fruits like hog plum, jutaí/jatobá, ice-cream-bean (*Inga* spp.), and Annonaceae species (Chaves 2005; Scheel-Ybert and Solari 2005).

Beyond assessing agrobiodiversity, rock shelters with long depositional sequences allow for the tracing of chronological changes in material culture and the use of plants. The introduction of cultivated plants, for instance, may lead to new foods and ways of preparing them, as well as new choices about the management of such plants. These changes can be gradual or abrupt. In the state of Minas Gerais, the remains of seeds and fruits from the sites of Lapa dos Bichos and Lapa Pintada (Fig. 2.2) demonstrate the gradual adoption of cultivated plants arriving from other areas. This process starts with manioc and bottle gourds (*Lagenaria siceraria*

(Molina) Standl.) at ca. 4300 BP, followed by maize at ca. 2000 BP, and peanut, squash, and beans from ca. 750 BP (Shock et al. 2013). In the same region, microbotanical remains from Early Holocene stone tools from Lapa Grande de Taquaraçu and Lapa do Santo attest to the use of stone artifacts for processing plants rich in carbohydrates, even though the taxonomic association of many starch grains is still unclear (Angeles Flores 2015; Angeles Flores et al. 2016; Ortega 2019). At Lapa do Santo, the starch grains are probably from sweet potatoes and yams (*Dioscorea* sp.), accompanied by phytoliths from palms, bamboos, and possibly Zingiberales rhizomes, between 12,700 and 8000 BP (Ortega 2019).

Some of the plants recovered from rock shelters are not the result of food-related daily practices. One such example is *Scleria* spp. in the archaeological sites of Lapa dos Bichos, Lapa Pintada, and Pedra Pintada, where the seeds were used as beads (Shock 2010). Archaeobotanical analyses of some of these types of remains can contribute to understanding plants as representations in exceptional activities. In the state of Minas Gerais, many of the rock shelters have human burials. Early Holocene burials from Santana do Riacho contain food plant remains from jatobá (*Hymenaea courbaril* L.), pequi, and two types of palms that researchers suggest might either have been consumed during rituals in connection with the burial or included as accompaniments for the dead (Resende and Prous 1991).

5.2 Coastal and Fluvial Shell Mounds

Shell mounds are archaeological sites formed mainly through the intentional accumulation of shells and sediments, and they also contain faunal and floral remains, cultural artifacts, hearths, and human burials. They were constructed and occupied by populations that were fishing, hunting, collecting, and cultivating plants and used the mounds for marking their identity on the landscape, for dwelling, camping, and for mortuary practices (Scheel-Ybert et al. 2003; Gaspar et al. 2008; Villagran 2010; Klokler 2012; Figuti et al. 2013; Pezo-Lanfranco et al. 2018).

In Brazil, shell mounds, or *sambaquis*, occur along the entire Atlantic coast, both near the coastline and in lagoons and estuaries, from the south of the state of Rio Grande do Sul to the state of Amapá (Imazio da Silveira and Schaan 2005). There are also fluvial sites, located along rivers in multiple drainage basins, notably from Amazonia and the Atlantic Forest. Choice of site locality was influenced geographically by the seasonality of river floods or tidal ebb. Shell mounds are diverse in relation to their antiquity, location, and material culture: the oldest ones in Brazil date from the transition from the Early to the Middle Holocene (Eggers et al. 2011; Pugliese et al. 2018), but different human groups occupied these landmarks until the colonial period (Bandeira 2013). Along the south coast of Brazil there are monumental shell mounds, e.g., Jabuticabeira II, that measures 400 x 250 m (approximately 9 soccer fields) and is 6 m high (that of a standard two-story home) (Villagran 2010), as well as very low sites. Some shell mounds present artifacts mainly from stone tool industries, while others also have large numbers of ceramic vessels. The inhabitants of the shell mounds had foodways based on a diverse and rich diet (e.g., Colonese et al. 2014).

Analyses of human dental calculus from shell mounds along the northern coast of Santa Catarina (Enseada 1, Itacoara, Morro do Ouro, Forte Marechal Luz, and Jabuticabeira II) revealed evidence for the consumption of maize, sweet potatoes, yams, palm fruits, and araucaria nuts during the late Holocene (Wesolowski et al. 2007; Wesolowski et al. 2010; Boyadjian 2012; Boyadjian et al. 2016). Starch grains from Jabuticabeira II show evidence of damage, supporting the functional identification of polished stone artifacts as tools for macerating plant parts (Boyadjian 2012; Boyadjian et al. 2016). On the coast of the state of Rio de Janeiro, wood charcoal found in middle and late Holocene shell mounds (Forte, Salinas Peroano, Boca da Barra, Ponta da Cabeça, Beirada, and Pontinha) belong to many taxa that produce edible fruits (Table 1), such as *Byrsonima* sp., *Pouteria* sp., and *Mouriri* sp. (Scheel-Ybert 2000; Scheel-Ybert and Gaspar 2014), which suggests that firewood acquired in the surrounding domesticated landscapes reflects plants that were also used as food.

Fluvial shell mounds are constructed by the accumulation of terrestrial mollusks (*Megalobulimus* spp. or *Pomacea* spp.) and are generally of smaller dimensions than the coastal ones. In the state of São Paulo, ca. 60 km from the coast, carbon and nitrogen isotopes extracted from human bones found in burials at the Capelinha and Moraes sites, and dating from the early and middle Holocene, respectively, indicate a diet based on terrestrial fauna, with contributions from plants with a C_3 isotopic signature, which may include tubers, legumes, and many shrubs and trees (Eggers et al. 2011; Colonese et al. 2014). Further analyses at Moraes were carried out on phytoliths and starch grains from dental calculus that indicate the consumption of starchy plants, including yams and sweet potatoes (Boyadjian 2007).

In Amazonia, fluvial shell mounds contain the earliest ceramic remains in South America, anthropic soils, and some of the oldest remains of cultivated plants. They are concentrated in three areas: (i) at the mouth of the Amazon River, on Marajó Island, and in coastal estuaries from present-day French Guiana to the state of Maranhão; (ii) in the region between the floodplains of the Amazon, Tapajós and Xingu rivers; and (iii) in the seasonally flooded savannas of southwestern Amazonia, between the Guaporé wetlands and the Bolivian Llanos de Mojos (Lombardo et al. 2013; Hilbert et al. 2017; Watling et al. 2018; Furquim et al. 2021).

Bolivian shell mounds revealed the early consumption of manioc, leren tubers (*Calathea* sp.), and squash dated to ca. 10,500 BP (Lombardo et al. 2020), which supports genetic studies that posit southwestern Amazonia as a center for early plant domestication and experimentation (Clement et al. 2016). On the Brazilian side of the seasonally flooded savannas, the inhabitants of the Monte Castelo shell mound (middle to late Holocene) managed different ecosystems, such as forests and savannas, which supplied tree species, such as Brazil nut, nance, and cacao relatives (*Theobroma* sp.), as well as palms, such as tucumã (*Astrocaryum* sp.) and moriche/buriti (Furquim et al. 2021). From ca. 5800 BP, they also began to consume squash, maize, and wild rice (*Oryza* sp.). Over time, these groups developed strategies for harvesting and selecting native Amazonian rice, characterizing the beginning of its domestication (Hilbert et al. 2017). The cultivation of such species is influenced by their ecology and the local seasonality: rice could be managed during the flood and harvested as waters began to recede, whereas maize and squash could be grown on mounds or following the receding flood waters.

On Marajó Island, the Tucumã shell mound shows two distinct late Holocene diets, the first associated with shell rich sediments and the second with anthropic soils. The first occupation, dated from 4400 to 1600 BP, and associated with the earliest ceramics of the region, has a phytolith assemblage showing that maize and squash were consumed from the beginning of the occupation (Hilbert 2017). Maize, however, is not present in the later occupation with anthropic soils and dated to 1600 BP. These changes in diet also included a lower presence of squash phytoliths with sizes comparable to domesticated squash and the increasing presence of palm phytoliths. Hilbert (2017) suggests that these changes are connected to a shift in food production strategies and the incorporation of palm species in dietary practices. At Tucumã shell mound, wild rice was present in the landscape, as evidenced by rice phytoliths in natural soil profiles, but contrary to what happened at Monte Castelo, it was not domesticated or incorporated into the local diet (Hilbert 2017).

The diversity in the adoption of plants within regional diets, as exemplified by the adoption of rice at Monte Castelo and not at Tucumã, even when available in the ecological settings, highlights that specific food choices are related to local social practices. Some shell mound occupations present evidence for cultivation and consumption of a select range of species, while others present stronger tendencies for the use of a wide variety of plants.

5.3 Open-Air Sites in Brazilian Amazonia

Over 6500 archaeological sites have been identified in Brazilian Amazonia to date. The first records of human populations in the region date back to 14,000 BP, although the last two millennia showed a considerable increase in human occupations (Tamanaha 2018). Most Brazilian Amazonian sites were open-air villages, some of them including earthworks, such as the geoglyphs of Southwest Amazonia and the *tesos* (earthen mounds) of Marajó Island. More than half of Amazonian sites have anthropic dark earths (ADEs; Fig. 2.1) (Tamanaha 2018), soils formed by the disposal of organic waste in middens, which mark long human permanence and high population density (Arroyo-Kalin 2017; Neves and Petersen 2006). These soils appear in archaeological sites in great frequency starting ca. 3000 BP; however, there is evidence of ADEs from ca. 7500 BP in the upper Madeira River (Zuse 2014).

At the Teotônio site, along the upper Madeira River, where human occupation was continuous for almost 10,000 years, phytoliths from the early and middle Holocene identified the early presence of cultigens, such as manioc, squash, beans, and other roots and tubers (Watling et al. 2018). Dating to the same period, archaeobotanical evidence of edible fruits such as piquiá (*Caryocar* sp.), Brazil nut, guava (*Psidium* sp.), and several palm genera demonstrate that local foodways also depended upon fruit and nut trees (Watling et al. 2018). Since culture is highly dynamic, local plant exploitation strategies and foodways varied during the long-term human occupations of the site, and new crops, such as maize, sweet potato, yam, and arrowroot, were introduced during the late Holocene (Watling et al. 2020).

Crop cultivation was an important source of food in Amazonia, even in interfluvial areas without dark earths. At geoglyphs in the upper Purus River basin, phytolith analysis of sediments from the last 1500 years of pre-Columbian occupation shows that local foodways were based on the consumption of different parts of non-woody plants and shrubs, both cultivated and not, including maize, squash, *Heliconia* sp., and *Celtis* sp. (Watling et al. 2015), as well as fruit trees and palms, such as Brazil nut, hog plum, uricuri, and peach palm (*Bactris gasipaes* Kunth) (Pärssinen et al. 2020). The considerably smaller size of archaeological fruits of Brazil nut and uricuri, compared to modern ones, suggests a continuous selection for larger fruits over time, in a long process of coupled plant and landscape domestication (Pärssinen et al. 2020).

The combination of food production systems and the exploitation of plants along forest trails transformed the Amazonian landscapes (e.g., Watling et al. 2017). Today, the richness and abundance of domesticated trees and palms is higher around Amazonian archaeological sites (Levis et al. 2017). In the lower Tapajós River, for example, agroforestry systems were developed since 4500 BP, combining the cultivation of annual crops, such as maize, sweet potato, manioc, and other roots and tubers, with the progressive enrichment of edible forest species, such as piquiá and many palms (Maezumi et al. 2018; Troufflard and Alves 2019).

Since food consumption and landscape transformations are intimately related, botanical remains from the surroundings of ancient human settlements can provide further information about local foodways. In a late Holocene occupation at the Hatahara site (Fig. 2.1) in Central Amazonia, wood charcoal from several taxa producing edible fruits, including *Spondias* sp., *Genipa* sp., and *Byrsonima* sp., suggest that firewood was being collected in anthropogenic environments around the village that were enriched by plants that were also used as food (Scheel-Ybert et al. 2016; Caromano et al. 2013). In the same archaeological context, phytoliths of many crops and palms, such as moriche/buriti, patauá (cf. *Oenocarpus bataua* Mart.), and açaí were identified (Bozarth et al. 2009; Caromano et al. 2013; Cascon 2010).

At many other late Holocene Amazonian sites, the remains of cultigens, such as maize and tubers, are generally found together with a broad variety of fruit species, especially palms (Silva et al. 2016). In Central Amazonia, at the Caldeirão site, manioc and maize were identified in association with palm remains, notably peach palm. In the middle Solimões River, Brazil nut, cacao (*Theobroma* sp.), tucumã (*Astrocaryum* sp.), and other fruits were found in association with maize cultivation (Cassino 2018). Brazil nut is widespread in the archaeological record, as it is also found in the lower Rio Negro, the middle Rio Unini, and the lower Amazon (Shock et al. 2014). In Eastern Amazonia, on Marajó Island, açaí, tucumã, *Inga* spp., and *Sterculia* spp. were consumed (Roosevelt 1989).

Archaeobotanical data suggest that, since the beginning of human occupation in Amazonia, people started managing native fruit species and, very early on, initiated the dynamics that eventually brought about the domestication of numerous species (Roosevelt et al. 1996; Shock and Moraes 2019; Watling et al. 2018). Although in some periods a given species might have been more important in some local diets, such as manioc in the Xingu basin (Heckenberger 1998) or maize in late occupations along many Amazonian rivers (Watling et al. 2020), archaeobotanical data from Brazilian Amazonia show that pre-colonial indigenous peoples had rich and diversified diets, arising from the association of domesticated crops and a great variety of palms and fruit trees, produced in horticultural and agroforestry systems and from fruit exploitation in the forest (Shepard Jr. et al. 2020). Through thousands of years these food systems have guaranteed long-term food security despite climate and social changes (Maezumi et al. 2018).

5.4 *Pit Houses in the Southern Brazilian Highlands*

Pit houses are part of a wide range of earthwork constructions of different forms and sizes found in the Southern Brazilian Highlands, including enclosures in circular and elliptical shapes formed by causeways, mounds, as well as a variety of subterranean structures used for food processing, storage, and various types of hearths and kilns. The diameters of pit structures vary from 2 to 20 m; however most are less than 5 m across. While they may occur in isolation, pit structures are normally found together in groups, some of which are composed of over a hundred structures (Corteletti et al. 2016). These archaeological sites are located on hill tops or high ridges, at elevations that vary from 600 m to 1800 m. Archaeological data from several excavations in the Brazilian states of Paraná, Rio Grande do Sul, and Santa Catarina, as well as frontier regions with Argentina and Paraguay, dated the appearance of these structures to the late Holocene, ca. 2200 BP, and they were used until the colonial period (Iriarte et al. 2008). There was a peak of these settlements at ca. 950 BP, marked by population growth, more intensive food-production systems, and long-distance human migrations (Iriarte et al. 2008). These diverse archaeological structures have been associated with Southern Jê speaking indigenous populations. Their descendants are the Kaingang and the Laklãnõ Xokleng, who still live in the region today, although in restricted territories.

Sites where earthen enclosures are found are usually associated with central mounds that were used as funerary contexts (Iriarte et al. 2008). These may have been ceremonial spaces due to their formal layout and the lack of domestic remains (Iriarte et al. 2008). Symbolic consumption of a fermented beverage made from maize and honey during mourning was documented among the Kaingang people in historical accounts (Métraux 1946). Cremation practices were common among some of the Jê people in the pre-Columbian period and were historically documented within the Laklãnõ Xokleng before Christianization. Distinct from the mound/enclosure complexes, domestic subterranean structures have circular forms, sometimes presenting "benches" on one or more levels along the walls. Evidence for ceilings is found in postholes that mark where supporting beams entered the floor. Hearths are frequent on the lowest level of the dwellings, along with thin ceramic sherds, some stone tool remains, and charred araucaria nuts.

Archaeobotanical analysis of microscopic remains from ceramic vessels found in such domestic contexts in the Santa Catarina highlands identified the consumption of maize and other domesticated plants, such as beans, squash, and manioc (Corteletti et al. 2016). The authors suggest that the gathering of edible fruit from trees, such as goiabeira-serrana (*Acca sellowiana* (O.Berg) Burret), jabuticaba (*Plinia trunciflora* (O. Berg) Kausel), guabiroba (*Campomanesia* sp.), araçá (*Psidium* sp.), custard apple (*Annona* sp.), queen palm (*Syagrus romanzoffiana* (Cham.) Glassman), jelly palm (*Butia capitata* (Mart.) Becc.), and wolly jelly palm (*Butia eriospatha* (Mart. ex Drude) Becc.), as well as hearts of palm from juçara (*Euterpe edulis* Mart.) could also have had important roles in the local diet. Although these were not identified in the archaeobotanical record, they were reported as food in historical accounts about Kaingang and Laklãnõ Xokleng people (Corteletti et al. 2016). Such fruits and palms have strong cultural and symbolic significance among the Kaingang and the Laklãnõ Xokleng today. They are also indicators of their traditional knowledge of environmental management and landscape domestication. For example, for the Laklãnõ Xokleng, who live in the Upper Itajaí Valley, guava trees are important landscape references. The tree gives its name to their villages, and, in historical times, it marked the rhythm of their mobility as the cycle of fruit production played an important role in attracting game animals to feed on the fruits (Machado 2016). Knowledge about cultivated plants is also very important, as demonstrated among the Kaingang, through the existence and evaluation of origin histories (myths) for maize, beans, and squash (Borba (1908) apud Corteletti et al. 2016).

The strong relationships among indigenous cultures and their forests are well represented by another food, araucaria nuts. For modern and historically documented Southern Jê People, araucaria nuts are not only made into highly nutritious flour, bread, and beverages, but the araucaria tree has strong cultural significance (Noelli 1999; Reis et al. 2014). It is part of the process of transforming the forest in their cultural territory. The distribution of araucaria forests expanded in association with the Southern Jê populations and pit houses in pre-colonial times (Noelli 1999). Paleopalynological analyses indicate the rapid expansion of the Araucaria forests over the high grasslands between 1500 and 800 BP (Reis et al. 2014). As mentioned earlier, this expansion was associated with cultural forest management, through transport, storage, and processing of the seeds (Reis et al. 2014), which increased the availability of this important food for the Jê populations (Iriarte and Behling 2007; Corteletti et al. 2016).

6 Brazilian Food Plants and Their Migrations

In Brazil, migrations and trade among pre-colonial populations are associated with the dispersal of many plant species, such as araucaria (Iriarte and Behling 2007), Brazil nut (Shepard Jr. and Ramirez 2011), pequi (Smith and Fausto 2016), and moriche/buriti (Rull and Montoya 2014). Cacao (*Theobroma cacao* L.), for example, is well-known for its importance in Mesoamerica, where chocolate originated. However, its center of diversity is in western Amazonia (Fig. 1) (Thomas et al. 2012), where it is found in archaeological contexts in Ecuador from ca. 5300 BP (Zarrillo et al. 2018). The first archaeological evidence of early cacao beverages in Mesoamerica dates to before 3000 BP, in the Mayan culture (Henderson et al. 2007). Changes in the forms and decorations of Mesoamerican ceramic vessels across time suggest that beverages from cacao were initially produced by fermenting the sweet pulp surrounding the seeds, similar to the way it was consumed in Amazonia, followed later by the production of nonalcoholic chocolate beverages from fermented cacao seeds (Henderson et al. 2007).

Pre-colonial Native Americans dispersed many domesticated crops very early in time. Manioc was domesticated in southwestern Amazonia (Fig. 1) (Olsen and Schaal 1999), and its early dispersal is shown by archaeobotanical records, which date back to ca. 8000 BP in coastal Peru (Pearsall 1992) and ca. 7000 BP in Panama (Piperno and Holst 1998), and linguistic reconstructions suggest it arrived in Mexico by 6500 BP (Brown et al. 2013b). Likewise, and in the opposite direction, maize, domesticated by 9000 BP in lowland Mexico (Piperno et al. 2009), was identified in archaeobotanical remains as early as ca. 7500 BP in Panama (Piperno et al. 1985) and ca. 6500 BP in Coastal Peru (Grobman et al. 2012) and Amazonia (Brugger et al. 2016). Genetic analyses suggest that this introduced maize further diversified in southwestern Amazonia (Kistler et al. 2018). The complex domestication history of maize is an example of how multiple waves of human-mediated dispersal can be responsible for the diversity and biogeography of many South American domesticated species. This is also the case of peanut, domesticated in Central Brazil (Fig. 1), but identified in the archaeobotanical record of early food production systems in coastal Peru, where peanut populations with a mixture of wild and domesticated features were being cultivated by 8000 BP (Piperno and Dillehay 2008).

After the arrival of European colonizers in the Americas, another series of biocultural exchanges began, leading to at least three centuries of intensive transatlantic plant migrations, which are responsible for the widespread dispersal of South American crops and spices to other continents. This process, known as the Columbian Exchange (Crosby 2003), fueled the dietary enrichment that influenced European demographic expansion, at the expense of the ethnocide of circa 95% of the Amerindian population (Koch et al. 2019). Manioc, already described as a staple food by the first European adventurers, as reported by Gaspar de Carvajal from an expedition in the Amazon River between 1540 and 1542 (Medina 1934), was one of the main cultivars traded overseas. Due to its adaptation to diverse climates, altitudes, and soils, manioc is now cultivated from the Balkans to former European colonies in Asia and Africa (e.g., India, Nigeria, Congo, and Mozambique). Today it is one of the most important sources of energy in many African countries (Lebot 2009).

The spread of the European world system was led by the hunt for valuable spices, foods, and medicines. Numerous Portuguese, Spanish, Dutch, French, and English commercial expeditions of the colonial period searched for the "drugs from the hinterland," American forest spices and other natural products. Among the most pursued species were the Amazonian clove or clove-wood (*Dicypellium caryophyllaceum* (Mart.) Nees), known since the seventeenth century for its similarity with the Indian clove (*Syzygium aromaticum* (L.) Merr. & L.M. Perry); the Amazonian cinnamon or precious-bark (*Aniba canelilla* (Kunth) Mez), whose properties were described by the botanist Alexander von Humboldt in the XVIII century (Donini et al. 2017); and guaraná (*Paullinia cupana* Kunth), a stimulant domesticated by the Sateré-Mawé indigenous people, probably ca. 1000 BP in the Amazon-Tapajós interfluve (Fig. 1) (Atroch et al. 2012; Clement et al. 2010). One of the most important South American spices/vegetables that became globally prominent are the fruits and seeds of the *Capsicum* genus (hot and sweet peppers), including *C. frutescens* (used in Tabasco sauce), *C. chinensis* (used in curry recipes) and *C. baccatum* (popular in southern Brazil), all domesticated in Amazonia (Clement et al. 2010; Bruno 2019; Chiou and Hastorf 2014).

7 Final Considerations

Humans construct, modify and regulate their environments, and their experiences maintain and update traditional knowledge of the ecosystems where they live. Our species, over time, has engaged in wide-ranging environmental modifications and niche construction through our agency and cultural practices. Botanical remains from archaeological contexts across Brazil provide information about the diversity of traditional foodways and their development, which was interwoven by cultural practices into the varying ecological settings of the country. The archaeological examples demonstrate long-term histories of management practices and plant domestication, changes in uses of plant resources over time, how people's cultural choices created diverse cuisines and diets even when groups were living in similar ecosystems, and how some of these diets were based on a great variability of vegetable resources while others were more specialized on a few plants. The consumption of a variety of wild and domesticated plant resources and the processes involved with how they were/are brought into food production systems was/is informed by traditional ecological knowledge. In Brazil, the rich agrobiodiversity demonstrates how many more species were, and are, consumed by indigenous populations than those that can be found today in supermarkets. Some of these foods and spices could be more widely introduced into current foodways to be enjoyed by more people and to encourage healthier diets.

Local food plants of Brazil are a cultural heritage. While some of this heritage is currently shared across Brazilian society, or even across the world, Brazilian indigenous peoples and traditional communities remain the essential guardians of this legacy. These peoples, through their ways of life and the knowledge accumulated over the millennia of occupation of the American continent, maintain the cultural landscapes that support this great diversity of plant foods. Therefore, they are essential to the preservation of food quality and health that are rooted in this diversity. Their agro-ecological practices, which produce agrobiodiversity and rich and resilient foodways, are in stark contrast with modern agribusinesses that are focused on monocultures, threaten biodiversity, and place a multiplicity of cultures and ways of life at great risk.

Today, the breadth and depth of the long-term relationship between indigenous peoples and plants has attracted the attention of scientists and policy makers that are interested in maintaining biodiversity and developing sustainable and resilient food production systems for future human populations. Since this territory, today called Brazil, has been occupied throughout several millennia, archaeology's contribution to a historical perspective is key to interpreting the dynamics underlying the lasting ecological changes of landscape domestication. While the archaeological record will always contain only a fraction of history and the species that were utilized, it helps us to understand long term cultural choices and their effects on the environment and to appreciate the value of producing and maintaining biological and cultural diversity.

Table 1 Food plants found in archaeological sites, ordered by the numbering used in Fig. 1. Analyzed archaeobotanical remains are identified by type: D, dried plant; NWC, non-wood charcoal; PH, phytolith; PO, pollen; S, starch; WC, wood charcoal. Radiocarbon dates are approximate and, when available, preference was given to calibrated dates. Species do not necessarily occur throughout the entire range of dates analyzed by the authors

Archaeological site – type of remain (reference)	Date	Plants
1. Furna do Estrago – PH, PO, S (Santos 2014)	1860–1610 BP	**Herbs:** *Bromelia laciniosa*, *Ipomoea batatas*, *Manihot esculenta*, *Zea mays*, *Solanum* sp., Cucurbitaceae A.Juss., Poaceae Barnhart, Solanaceae A.Juss. **Trees, shrubs, and palms:** *Anadenanthera colubrina*, *Byrsonima* sp., *Eugenia* sp., *Licania* sp., *Matayba* sp., *Neea* sp., *Sebastiania* sp., *Syagrus* sp., Anacardiaceae R.Br., Apocynaceae Juss., Arecaceae Schultz Sch., Combretaceae R.Br., Fabaceae Lindl., Malvaceae Juss., Myrtaceae Juss., Rutaceae A.Juss.
2. Santana do Riacho – D, NWC (Resende and Prous 1991)	4500 BP–modern	**Herbs:** *Lagenaria siceraria* (Molina) Standl., *Zea mays* **Trees, shrubs, and palms:** *Caryocar brasiliense*, *Hymenaea courbaril*, *Oenocarpus bataua*, *Sterculia apetala* (Jacq.) H.Karst., *Syagrus coronata* (Mart.) Becc., *Astrocaryum* sp., Myrtaceae
2. Santana do Riacho – D, NWC (Resende and Prous 1991)	8000–5000 BP	**Trees, shrubs, and palms:** *Sterculia apetala*, *Syagrus coronata*, *Astrocaryum* sp.

(continued)

Table 1 (continued)

Archaeological site – type of remain (reference)	Date	Plants
2. Santana do Riacho – D, NWC (Resende and Prous 1991)	10,000–8000 BP	**Trees, shrubs, and palms:** *Caryocar brasiliense*, *Hymenaea courbaril*, *Syagrus coronata*, *Astrocaryum* sp.
3. Lapa do Santo – PH, S (Ortega 2019)	12,700–8000 BP	**Herbs:** *Ipomoea batatas*, *Dioscorea* sp., Bambusoideae Luerss., Zingiberales. **Trees, shrubs, and palms:** Arecaceae
4. Lapa dos Bichos – D, NWC (Shock 2010; Shock et al. 2013)	750–150 BP	**Herbs:** *Arachis hypogaea*, *Manihot* cf. *esculenta*, *Phaseolus vulgaris*, *Zea mays*, *Chromolaena* sp., *Cucurbita* sp., *Lagenaria* sp., *Lasiacis* sp., *Passiflora* sp., *Trichogonia* sp., Eupatorieae Cass., Gnaphalieae Cass. ex Lecoq & Juill., Vernonieae Cass., Asteraceae Bercht. & J.Presl., Bromeliaceae Juss., Poaceae, Solanaceae **Trees, shrubs, and palms:** *Alibertia edulis* (Rich.) A. Rich., *Bixa orellana* L., *Caryocar brasiliense*, *Guazuma ulmifolia* Lam., *Mucuna sloanei*, *Spondias mombin*, *Spondias tuberosa* Arruda, *Syagrus* cf. *oleracea*, *Alibertia* sp., *Annona* sp., *Byrsonima* sp., *Ficus* sp., *Hymenaea* sp.
4. Lapa dos Bichos – D, NWC (Shock 2010; Shock et al. 2013)	2000–750 BP	**Herbs:** *Manihot* cf. *esculenta*, *Zea mays*, *Lagenaria* sp., *Passiflora* sp., Solanaceae **Trees, shrubs, and palms:** *Spondias mombin*, *Spondias tuberosa*, *Syagrus* cf. *oleracea*, *Hymenaea* sp.
4. Lapa dos Bichos – D, NWC (Shock 2010; Shock et al. 2013)	4250–2000 BP	**Herbs:** cf. *Manihot esculenta*, cf. *Phaseolus lunatus*, *Lagenaria* sp., Solanaceae **Trees, shrubs and palms:** *Caryocar brasiliense*, *Spondias tuberosa*, *Syagrus* cf. *oleracea*, *Hymenaea* sp.
4. Lapa dos Bichos – D, NWC (Shock 2010; Shock et al. 2013)	6500–4250 BP	**Herbs:** Poaceae **Trees, shrubs, and palms:** *Hymenaea* sp., *Syagrus* cf. *oleracea*
5. Lapa Pintada – D, NWC (Shock 2010; Shock et al. 2013)	1200 BP and 800 BP	**Herbs:** *Manihot* cf. *esculenta*, *Zea mays*, *Lagenaria* sp., *Lasiacis* sp., *Passiflora* sp., Vernonieae, Asteraceae, Solanaceae **Trees, shrubs, and palms:** *Caryocar brasiliense*, *Guazuma ulmifolia*, *Syagrus* cf. *oleracea*, *Annona* sp., *Byrsonima* sp., *Hymenaea* sp., *Myrciaria* sp.
5. Lapa Pintada – D, NWC (Shock 2010; Shock et al. 2013)	4400 BP and 4300 BP	**Herbs:** *Manihot* cf. *esculenta*, *Lagenaria* sp., *Passiflora* sp. **Trees, shrubs, and palms:** *Caryocar brasiliense*, *Syagrus* cf. *oleracea*, *Hymenaea* sp.
5. Lapa Pintada – D, NWC (Shock 2010)	7000 BP	**Trees, shrubs, and palms:** *Hymenaea* sp., *Syagrus* cf. *oleracea*
6. Lapa Grande de Taquaraçu – S, NWC (Angeles Flores 2015; Angeles Flores et al. 2016)	11,400–9000 cal BP	**Herbs:** *Dioscorea* sp. **Trees, shrubs, and palms:** Arecaceae

(continued)

Table 1 (continued)

Archaeological site – type of remain (reference)	Date	Plants
7. Santa Elina – PO (Chaves 2005)	400 cal BP	**Herbs:** Asteraceae, Poaceae **Trees, shrubs, and palms:** *Licania* sp., *Sebastiania* sp., Apocynaceae, Combretaceae, Fabaceae – Mimosoideae, Melastomataceae A.Juss., Myrtaceae, Rubiaceae Juss., Rutaceae
7. Santa Elina – PO, WC, NWC (Chaves 2005; Scheel-Ybert and Solari)	4000–2000 cal BP	**Herbs:** *Manihot* sp., Bromeliaceae, Poaceae **Trees, shrubs, and palms:** *Acrocomia aculeata* (Jacq.) Lodd. ex Mart., *Anacardium humile* A.St.-Hil., *Attalea phalerata*, *Attalea vitrivir* Zona, *Caryocar brasiliense*, *Dipteryx alata* Vogel, *Guazuma ulmifolia*, *Guettarda viburnoides* Cham. & Schltdl., *Hymenaea stigonocarpa* Mart. ex Hayne, *Hymenaea courbaril*, *Protium heptaphyllum* (Aubl.) Marchand, *Ficus* sp., *Inga* spp., *Lecythis* sp., *Matayba* sp., *Neea* sp., *Spondias* sp., Anacardiaceae, Annonaceae, Apocynaceae, Arecaceae, Fabaceae, Myrtaceae, Rubiaceae, Sapotaceae Juss.
8. Pedra Pintada – NWC (Roosevelt 1998; Roosevelt et al. 1996; Shock and Moraes 2019)	12,400–9000 cal BP	**Trees, shrubs, and palms:** *Astrocaryum vulgare*, *Attalea microcarpa* Mart., *Attalea spectabilis* Mart., *Bertholletia excelsa*, *Byrsonima crispa* A.Juss., *Hymenaea* c.f. *parvifolia* or *oblongifolia*, *Mauritia flexuosa*, *Mouriri apiranga* Spruce ex Triana, *Oenocarpus* cf. *bacaba*, *Sacoglottis guianensis* Benth., *Talisia esculenta* (Cambess.) Radlk., *Vitex* cf. *cymosa*, *Acrocomia* sp., *Astrocaryum* sp., *Attalea* sp.
8. Pedra Pintada - NWC (Roosevelt 2000)	800–500 BP	**Herbs:** *Casimirella rupestris* (Ducke) R.A.Howard, *Passiflora nitida* Kunth, *Zea mays*, Cucurbitaceae **Trees, shrubs, and palms:** *Acrocomia aculeata*, *Anacardium occidentale* L., *Anacardium giganteum* W.Hancock ex Engl., *Antrocaryon amazonicum* (Ducke) B.L.Burtt & A.W.Hill, *Astrocaryum vulgare*, *Attalea microcarpa*, *Attalea spectabilis*, *Bertholletia excelsa*, *Byrsonima crassifolia* (L.) Kunth, *Byrsonima crispa*, cf. *Endopleura uchi* (Huber) Cuatrec., *Euterpe oleracea*, *Hymenaea courbaril*, *Hymenaea oblongifolia* Huber, *Hymenaea parvifolia* Huber, *Mauritia flexuosa*, *Mauritiella armata* (Mart.) Burret, cf. *Mezilaurus itauba* (Meisn.) Taub. ex Mez, *Mouriri apiranga*, *Moutabea chodatiana* Huber, *Norantea guianensis* Aubl., *Sacoglottis guianensis*, *Spondias mombin*, *Syagrus cocoides* Mart., *Talisia esculenta*, *Vitex cymosa* Bertero ex Spreng., *Coccoloba* sp.
9. Sambaqui Salinas Peroano – WC (Scheel-Ybert 2000; Scheel-Ybert and Gaspar 2014)	4800–1700 cal BP	**Trees, shrubs, and palms:** *Actinostemon* sp., *Condalia* sp., *Cupania* sp., aff. *Duguetia*, *Inga* spp., *Pouteria* sp., cf. *Swartzia* sp., Annonaceae, Arecaceae, Myrtaceae

(continued)

Table 1 (continued)

Archaeological site – type of remain (reference)	Date	Plants
10. Sambaqui da Ponta da Cabeça – WC (Scheel-Ybert 2000; Scheel-Ybert and Gaspar 2014)	3450–2000 cal BP	**Trees, shrubs, and palms:** *Actinostemon* sp., *Byrsonima* sp., *Condalia* sp., *Cupania* sp., aff. *Duguetia*, *Inga* sp., *Pouteria* sp., *Zollernia* sp., Arecaceae, Chrysobalanaceae R.Br., Myrtaceae
11. Sambaqui Boca da Barra – WC (Scheel-Ybert 2000; Scheel-Ybert and Gaspar 2014)	4050–1280 cal BP	**Trees, shrubs, and palms:** *Actinostemon* sp., *Condalia* sp., *Cupania* sp., aff. *Duguetia*, *Inga* spp. *Pouteria* sp., *Zollernia* sp., Annonaceae, Arecaceae, Myrtaceae
12. Sambaqui da Beirada – WC (Scheel-Ybert, 2000; Scheel-Ybert & Gaspar 2014)	4700–3770 cal BP	**Trees, shrubs, and palms:** *Byrsonima* sp., *Condalia* sp., *Cupania* sp., aff. *Duguetia*, *Inga* sp., *Pouteria* sp., Arecaceae, Fabaceae, Myrtaceae
13. Sambaqui da Pontinha – WC (Scheel-Ybert 2000; Scheel-Ybert and Gaspar 2014)	2250–1630 cal BP	**Trees, shrubs, and palms:** *Actinostemon* sp., *Byrsonima* sp., *Condalia* sp., *Cupania* sp., aff. *Duguetia*, *Pouteria* sp., cf. *Swartzia* sp., *Zollernia* sp., Annonaceae, Arecaceae, Fabaceae, Myrtaceae
14. Sambaqui do Meio – WC (Scheel-Ybert 2000; Scheel-Ybert and Gaspar 2014)	5500 cal BP	**Trees, shrubs, and palms:** *Araucaria angustifolia*, *Actinostemon* sp., *Cupania* sp., aff. *Duguetia*, *Zollernia* sp., *Inga* sp., *Pouteria* sp., Fabaceae, Myrtaceae
15. Sambaqui do Forte – WC (Scheel-Ybert 2000; Scheel-Ybert and Gaspar 2014)	5900–1800 cal BP	**Trees, shrubs, and palms:** *Actinostemon* sp., *Byrsonima* sp., *Cupania* sp., *Condalia* sp., aff. *Duguetia*, *Erythroxylum* sp., *Inga* spp., *Mouriri* sp., *Pouteria* sp., cf. *Swartzia* sp., *Zollernia* sp., Annonaceae, Arecaceae, Chrysobalanaceae, Fabaceae, Myrtaceae
16. Moraes – PH, S (Boyadjian 2007)	5900–4500 BP	**Herbs:** *Ipomoea batatas*, *Dioscorea* sp.
18. Enseada I – PH, S (Wesolowski et al. 2007; Wesolowski et al. 2010)	ca. 1390 BP	**Herbs:** *Ipomoea batatas*, *Zea mays*, Poaceae **Trees, shrubs, and palms:** *Araucaria angustifolia*
19. Forte Marechal Luz – PH, S (Wesolowski et al. 2007; Wesolowski et al. 2010)	ca. 1110–850 BP	**Herbs:** *Ipomoea batatas*, Poaceae **Trees, shrubs, and palms:** *Araucaria angustifolia*
20. Morro do Ouro – PH, S (Wesolowski et al. 2007; Wesolowski et al. 2010)	ca. 4030 BP	**Herbs:** *Ipomoea batatas*, *Dioscorea* sp., Poaceae
21. Itacoara – PH, S (Wesolowski et al. 2007; Wesolowski et al. 2010)	ca. 550 BP	**Herbs:** *Ipomoea batatas*, *Zea mays*, Poaceae **Trees, shrubs, and palms:** *Araucaria angustifolia*, Arecaceae

(continued)

Table 1 (continued)

Archaeological site – type of remain (reference)	Date	Plants
22. Jabuticabeira II – PH, S, WC (Boyadjian 2007; Boyadjian 2012; Scheel-Ybert and Gaspar 2014; Boyadjian et al. 2016)	2950–1700 cal BP	**Herbs:** *Ipomoea batatas*, *Zea mays*, *Calathea* sp., *Dioscorea* sp., Poaceae **Trees, shrubs, and palms:** cf. *Eugenia uniflora* L., *Colubrina* sp., *Cupania* sp., *Ficus* sp., *Mouriri* sp., *Pouteria* sp., Anacardiaceae, Annonaceae, Arecaceae, Chrysobalanaceae, Euphorbiaceae Juss., Fabaceae, Myrtaceae, Rubiaceae
23.Tucumã – PH (Hilbert 2017)	ca. 1650 cal BP	**Herbs:** *Cucurbita* sp., *Oryza* sp., Asteraceae, Marantaceae R.Brown, Poaceae **Trees, shrubs, and palms:** *Celtis* sp., Annonaceae, Arecaceae
23.Tucumã – PH (Hilbert 2017)	4300–1650 cal BP	**Herbs:** *Cucurbita* sp., *Oryza* sp., *Zea mays*, Asteraceae, Marantaceae, Poaceae **Trees, shrubs, and palms:** *Celtis* sp., Annonaceae, Arecaceae
24. Monte Castelo PH, NWC (Furquim et al. 2021; Hilbert 2017; Hilbert et al. 2017)	6000–700 cal BP	**Herbs:** *Cucurbita* sp., *Oryza* sp., *Zea mays*, Asteraceae, Marantaceae, Poaceae **Trees, shrubs, and palms:** *Bertholletia excelsa*, *Mauritia flexuosa*, *Astrocaryum* sp., *Byrsonima* sp., *Celtis* sp., *Theobroma* sp., Annonaceae, Arecaceae
25. Isla del Tesoro – PH (Lombardo et al. 2020)	8800–4100 cal BP	**Herbs:** *Zea mays,* c.f. *Calathea* sp., *Cucurbita* sp., *Heliconia* sp., *Manihot* sp., *Oryza* sp., Bambusoideae, Cyperaceae, Marantaceae **Trees, shrubs, and palms:** *Celtis* sp., Arecoideae
26. Isla Manechi – PH (Lombardo et al. 2020)	13,900–5000 cal BP	**Herbs:** *Zea mays, Cucurbita* sp., c.f. *Calathea* sp., *Heliconia* sp., *Manihot* sp., *Oryza* sp., Bambusoideae, Cyperaceae, Marantaceae **Trees, shrubs, and palms:** Arecoideae
27, 28. Teotônio and Santa Paula – PH, S (Watling et al. 2020)	3000–400 cal BP	**Herbs:** *Calathea* cf. *allouia*, *Cucurbita* cf. *moschata*, *Ipomoea batatas*, *Maranta arundinacea*, *Zea mays*, *Dioscorea* sp. **Trees, shrubs, and palms:** *Bactris*/*Astrocaryum* sp., Euterpeae/Attaleinae, Mauritinae/*Euterpe* sp., Fabaceae
27. Teotônio – PH, NWC, S (Watling et al. 2018)	9000–5000 cal. BP	**Herbs:** *Manihot esculenta*, *Calathea* sp., *Cucurbita* sp., *Heliconia* sp., *Phaseolus* sp., Strelitziaceae **Trees, shrubs, and palms:** *Bertholletia excelsa*, *Attalea* sp., *Bactris/Astrocaryum* sp., *Euterpe/Oenocarpus* sp., *Caryocar* sp., *Ficus* sp., *Psidium* sp., Arecaceae
29. Serra do Maguari-1 – PH, S (Maezumi et al. 2018)	ca. 490 cal BP	**Herbs:** *Zea mays*, *Chusquea* sp., *Cucurbita* sp., *Pharus* sp., Olyreae **Trees, shrubs, and palms:** *Astrocaryum* sp., *Attalea* sp., Arecaceae

(continued)

Table 1 (continued)

Archaeological site – type of remain (reference)	Date	Plants
30. Lake Caranã – PO (Maezumi et al. 2018)	4500 cal BP – modern	**Herbs:** *Ipomoea batatas*, *Manihot esculenta*, *Zea mays*, *Cucurbita* sp., *Solanum* sp. **Trees, shrubs, and palms:** *Acrocomia* sp., *Aniba* sp., *Annona* sp., *Astrocaryum* sp., *Attalea* sp., *Bactris* sp., *Brosimum* sp., *Byrsonima* sp., *Caryocar* sp., *Cecropia* sp., *Celtis* sp., *Eschweilera* sp., *Hymenaea* sp., *Jacaratia* sp., *Licania* sp., *Mauritia* sp., *Mauritiella* sp., *Oenocarpus* sp., *Pouteria* sp., *Protium* sp., *Spondias* sp., *Tapirira* sp., *Tetragastris* sp., *Theobroma* sp.
31. Cedro – PH (Troufflard and Alves 2019)	700–200 cal BP	**Herbs:** cf. *Maranta arundinacea* L., *Zea mays*, cf. *Calathea* sp., *Cucurbita* sp., *Oryza* sp. **Trees, shrubs, and palms:** *Celtis* sp., Attaleinae, *Bactris*/*Astrocaryum* sp., Mauritinae/Euterpeae, Annonaceae
32. Hatahara – PH (Bozarth et al. 2009)	ca. 1000 cal BP	**Herbs:** *Zea mays*, *Calathea* sp., *Cucurbita* sp., *Heliconia* sp., *Lagenaria* sp., Cyperaceae Juss., Marantaceae **Trees, shrubs, and palms:** *Bactris* sp.
32. Hatahara – PH, S,WC (Caromano et al. 2013; Scheel-Ybert et al. 2016; Cascon 2010)	ca. 1000 cal BP	**Herbs:** *Zea mays*, *Cyperus* sp., *Dioscorea* sp., Cyperaceae **Trees, shrubs, and palms:** cf. *Attalea maripa* (Aubl.) Mart., cf. *Euterpe oleracea* Mart., cf. *Mauritia flexuosa*; cf. *Oenocarpus bataua*; *Bactris* sp., *Byrsonima* sp., *Cupania* sp., *Diospyros* sp., *Genipa* sp., *Himathantus* sp., *Spondias* sp., *Tapirira* sp., *Vantanea* sp., *Vernonia* sp., Anacardiaceae, Annonaceae, Arecaceae, Chrysobalanaceae, Euphorbiaceae, Fabaceae, Lecythidaceae A.Rich., Melastomataceae, Myrtaceae, Rubiaceae, Sapindaceae Juss., Sapotaceae
33, 34, 35. Açutuba, Lago Grande, Osvaldo – NWC (da Silva et al. 2016)	1750–750 cal BP	**Herbs:** *Zea mays*, tubers **Trees, shrubs, and palms:** Arecaceae
36, 37, 38, 39. Vila Nova I, Vila Nova II, Floresta, Lago das Pombas – NWC (Shock et al. 2014)	2350–450 cal BP	**Trees, shrubs, and palms:** *Bertholletia excelsa* (other taxa were identified but data are not published)
40. Caldeirão – PH, S (unpublished data)	ca. 1300–900 BP	**Herbs:** *Canna indica* L., *Manihot esculenta*, *Zea mays*, Poaceae **Trees, shrubs, and palms:** *Bactris gasipaes*, Arecaceae
41. São João – NWC (Cassino, 2018)	1100–500 BP	**Herbs:** *Zea mays*, Solanaceae **Trees, shrubs and palms:** *Bertholletia excelsa*, *Spondias mombin*, *Astrocaryum* sp., *Byrsonima* sp., *Caryocar* sp., *Oenocarpus* sp., *Theobroma* sp., Annonaceae, Arecaceae, Fabaceae

(continued)

Table 1 (continued)

Archaeological site – type of remain (reference)	Date	Plants
42, 46. Cruzeirinho, Tequinho – D, NWC (Pärssinen et al. 2020)	2000–1300 cal BP	**Trees, shrubs, and palms:** *Attalea maripa* (Aubl.) Mart., *Attalea phalerata*, *Bactris concinna* Mart., *Bactris gasipaes*, *Bertholletia excelsa*, *Oenocarpus mapora* H.Karst., *Spondias mombin*, *Astrocaryum* sp., *Gustavia* sp.
43, 44, 45, 46, 47, 48, 49, 50. Fazenda Iquiri II, Sol de Maio, Campo Esperança, Tequinho, El Círculo, Las Palmeras, Chacra Telería, Tumichuchua – PH (Watling et al. 2015)	2100–150 BP	**Herbs:** *Zea mays*, *Cucurbita* sp., *Heliconia* sp., Chusqueinae, Olyreae, Asteraceae, Marantaceae, Strelitziaceae Hutch. **Trees, shrubs, and palms:** *Celtis* sp., Annonaceae, Arecaceae
51. Aterro dos Bichos – NWC (Roosevelt 1989)	1500–650 BP	**Trees, shrubs, and palms:** *Astrocaryum* sp., *Euterpe* spp., *Inga* spp., *Sterculia* sp.
52. Bonin – PH, S (Cortelleti et al. 2016)	700–600 cal BP	**Herbs:** *Manihot esculenta*, *Zea mays*, *Cucurbita* sp., cf. *Dioscorea* sp., *Phaseolus* sp., Asteraceae, Marantaceae, Poaceae **Trees, shrubs, and palms:** Arecaceae

References

Anderson AB, Posey DA, Balée W (1989) Management of a tropical scrub savanna by the Gorotire Kayapó of Brazil. In: Posey DA, Balée W (eds) Resource management in Amazonia: indigenous and folk strategies, Advances in economic botany, vol 7. New York Botanical Garden, Bronx/New York, pp 159–173

Anderson E (2005) Plants, man and life. Courier Dover Publications, Mineola

Angeles Flores R (2015) Uso de recursos vegetais em Lapa Grande de Taquaraçu: evidencias macro e microscópicas. Master's dissertation, Universidade de São Paulo, São Paulo

Angeles Flores R, de Sousa JCM, de Mello Araujo AG, Ceccantini G (2016) Before Lagoa Santa: micro-remain and technological analysis in a lithic artifact from the Itaparica industry. J Lithic Stud 3(1):6–29. https://doi.org/10.2218/jls.v3i1.1423

Arroyo-Kalin M (2017) Human niche construction and population growth in pre-Columbian Amazonia. Arch Inter 20:122–136. https://doi.org/10.5334/ai-367

Atroch AL, Nascimento Filho FJ, Ângelo PC, Freitas DV, Sousa NR, Clement CR (2012) Domestication and breeding of guarana. In: Borém A, Lopes MTG, Clement CR, Noda H (eds) Domestication and breeding: Amazonian species. Editora da Univ. Fed. Viçosa, Viçosa, pp 333–360

Balée W (1989) The culture of Amazonian forests. In: Posey DA, Balée W (eds) Resource management in Amazonia: indigenous and folk strategies, Advances in economic botany, vol 7. The New York Botanical Garden, Bronx/New York, pp 1–21

Balée W (2013) Cultural forests of the Amazon: a historical ecology of people and their landscapes. The University of Alabama Press, Tuscaloosa

Ball T, Chandler-Ezell K, Dickau R, Duncan N, Hart TC, Iriarte J, Lentfer C, Logan A, Lu H, Madella M, Pearsall DM, Piperno D, Rosen AM, Vrydaghs L, Weisskopf A, Zhang J (2015) Phytoliths as a tool for investigations of agricultural origins and dispersals around the world. J Archaeol Sci 68:32–45. https://doi.org/10.1016/j.jas.2015.08.010

Bandeira AM (2013) Ocupações humanas pré-coloniais na ilha de São Luís-MA: inserção dos sítios arqueológicos na paisagem, cronologia e cultura material cerâmica. Doctoral thesis, Universidade de São Paulo, São Paulo

Bitencourt ALV, Krauspenhar PM (2006) Possible prehistoric anthropogenic effect on *Araucaria angustifolia* (Bert.) O. Kuntze expansion during the late Holocene. Rev Bras Paleont 9(1):109–116. https://doi.org/10.4072/rbp.2006.1.12

Blench R (2012) The role of agriculture in explaining the diversity of Amerindian languages. In: Isendahl C (ed) The past ahead. Language, culture and identity in the Neotropics, Studies in global archaeologyl, vol 18. Acta Universitatis Upsaliensis, Uppsala, pp 13–38

Bowser BJ, Zedeño MN (2009) The archaeology of meaningful places. University of Utah Press, Salt Lake City

Boyadjian CHC (2007) Microfósseis contidos no cálculo dentário como evidência do uso de recursos vegetais nos sambaquis de Jabuticabeira II (SC) e Moraes (SP). Master's dissertation, Universidade de São Paulo, São Paulo

Boyadjian CHC (2012) Análise e identificação de microvestígios vegetais de cálculo dentário para a reconstrução de dieta sambaquieira: estudo de caso de Jabuticabeira II, SC. Doctoral thesis, Universidade de São Paulo, São Paulo

Boyadjian CHC, Eggers S, Reinhard KJ, Scheel-Ybert R (2016) Dieta no sambaqui Jabuticabeira-II (SC): consumo de plantas revelado por microvestígios provenientes de cálculo dentário. Cad LEPAARQ 13(25):131–161. https://doi.org/10.15210/lepaarq.v13i25.7395

Bozarth SR, Price KP, Woods W, Neves EG, Rebellato L (2009) Phytoliths and terra preta: the Hatahara site example. In: Woods W, Teixeira WG, Lehmann J, Steiner C, WinklerPrins AM, Rebellato L (eds) Amazonian dark earths: Wim Sombroek's vision. Springer, Heidelberg, pp 85–98

Brown CH, Clement CR, Epps P, Luedeling E, Wichmann S (2013a) The paleobiolinguistics of domesticated chili pepper (*Capsicum* spp.). Ethnobiol Let 4:1–11. https://doi.org/10.14237/ebl.4.2013.5

Brown CH, Clement CR, Epps P, Luedeling E, Wichmann S (2013b) The paleobiolinguistics of domesticated manioc (*Manihot esculenta*). Ethnobiol Let 4:61–70. https://doi.org/10.14237/ebl.4.2013.2

Brown CH, Clement CR, Epps P, Luedeling E, Wichmann S (2014a) The paleobiolinguistics of maize (*Zea mays* L.). Ethnobiol Let 5(1):52–64. https://doi.org/10.14237/ebl.5.2014.130

Brown CH, Clement CR, Epps P, Luedeling E, Wichmann S (2014b) The paleobiolinguistics of the common bean (*Phaseolus vulgaris* L.). Ethnobiol Let 5 (0):104–115. https://doi.org/10.14237/ebl.5.2014.203

Brugger SO, Gobet E, van Leeuwen JFN, Ledru M-P, Colombaroli D, van der Knaap WO, Lombardo U, Escobar-Torrez K, Finsinger W, Rodrigues L, Giesche A, Zarate M, Veit H, Tinner W (2016) Long-term man–environment interactions in the Bolivian Amazon: 8000 years of vegetation dynamics. Quatern Sci Rev 132:114–128. https://doi.org/10.1016/j.quascirev.2015.11.001

Bruno MC (2019) Indigenous American agricultural contributions to modern global food systems. Oxford Research Encyclopedia of environmental science. Oxford University Press, Oxford. https://doi.org/10.1093/acrefore/9780199389414.013.176

Caromano CF, Cascon LM, Neves EG, Scheel-Ybert R (2013) Revealing fires and rich diets: macro-and micro-archaeobotanical analysis at the Hatahara Site, Central Amazonia. Tipiti 11(2):40–51

Carson JF, Whitney BS, Mayle FE, Iriarte J, Prumers H, Soto JD, Watling J (2014) Environmental impact of geometric earthwork construction in pre-Columbian Amazonia. Proc Natl Acad Sci U S A 111(29):10497–10502. https://doi.org/10.1073/pnas.1321770111

Cascon LM (2010) Alimentação na floresta tropical: um estudo de caso no sítio Hatahara, Amazônia Central, com base em microvestígios botânicos. Master's dissertation, Universidade Federal do Rio de Janeiro, Rio de Janeiro

Cassino MF (2018) Manejo e uso de recursos florísticos em períodos pré-coloniais na Amazônia: um estudo de caso na RDS Amanã. Instituto de Desenvolvimento Sustentável Mamirauá, Tefé

Cassino MF, Alves RP, Levis C, Watling J, Junqueira AB, Shock MP, Ferreira MJ, Andrade VLC, Furquim LP, Coelho SD, Tamanaha EK, Neves EG, Clement CR (2019) Ethnobotany and ethnoecology applied to historical ecology. In: Albuquerque UP, Lucena RFP, Cunha LVFC, Alves RRN (eds) Methods and techniques in ethnobiology and ethnoecology. Springer, New York, pp 187–208. https://doi.org/10.1007/978-1-4939-8919-5_13

Chaves SAdM (2005) Novas perspectivas paleoetnobotânicas e paleovegetacionais do Planalto Central do Brasil: uma proposta palinológica. In: Vialou ÁV (ed) Pré-história do Mato Grosso, vol 1. Santa Elina. EDUSP, São Paulo, pp 149–154

Chiou KL, Hastorf CA (2014) A systematic approach to species–level identification of chile pepper (*Capsicum* spp.) seeds: establishing the groundwork for tracking the domestication and movement of chile peppers through the Americas and beyond. Econ Bot 68(3):316–336. https://doi.org/10.1007/s12231-014-9279-2

Clement CR (1999) 1492 and the loss of Amazonian crop genetic resources. I. The relation between domestication and human population decline. Econ Bot 53(2):188–202. https://doi.org/10.1007/BF02866498

Clement CR (2019) Domesticação da floresta & subdesenvolvimento da Amazônia. In: Mendes dos Santos G (ed) Grupo de Estudos Estratégicos Amazônicos, Caderno de Debates, Tomo XIV. Instituto Nacional de Pesquisas da Amazônia, Manaus, pp 11–52

Clement CR, Cassino MF (2018) Landscape domestication and archaeology. In: Smith C (ed) Encyclopedia of global archaeology. Springer, New York, pp 1–8. https://doi.org/10.1007/978-3-319-51726-1_817-2

Clement CR, Cristo-Araújo M, Coppens d'Eeckenbrugge G, Alves Pereira A, Picanço-Rodrigues D (2010) Origin and domestication of native Amazonian crops. Diversity 2(1):72–106. https://doi.org/10.3390/d2010072

Clement CR, Picanço Rodrigues D, Alves Pereira A, Mühlen GS, Cristo Araújo M, Moreira PA, Lins J, Reis VM (2016) Crop domestication in the upper Madeira River basin. Bol Mus Para Emílio Goeldi Cien Hum 11(1):193–205. https://doi.org/10.1590/1981.81222016000100010

Colonese AC, Collins M, Lucquin A, Eustace M, Hancock Y, Ponzoni RAR, Mora A, Smith C, DeBlasis P, Figuti L (2014) Long-term resilience of late Holocene coastal subsistence system in southeastern South America. PLoS One 9(4):e93854. https://doi.org/10.1371/journal.pone.0093854

Corteletti R, Dickau R, DeBlasis P, Iriarte J (2016) Analises de grãos de amido e fitólitos nas terras altas do sul do Brasil: repensando a economia e mobilidade dos grupos proto-Jê meridionais. Cad LEPAARQ 13(25):162–196. https://doi.org/10.15210/lepaarq.v13i25.7369

Crosby AW (2003) The Columbian exchange: biological and cultural consequences of 1492, 30th Anniversary edn. Greenwood Publishing Group, West Port

Denevan WM (1992) Native American populations in 1492: recent research and a revised hemispheric estimate. In: Denevan WM (ed) The native population of the Americas in 1492, 2nd edn. Univ. of Wisconsin Press, Madison, pp xvii–xxix

Denevan WM (2001) Cultivated landscapes of native Amazonia and the Andes. Oxford geographical and environmental studies. Oxford University Press, Oxford

Donini CVZDS, Fiori MM, dos Santos CFM (2017) Flores e cascas: exploração de cravo, *Dicypellium caryophyllaceum* e *Syzygium aromaticum*, no século XVIII. Paper presented at the VIII Congresso Internacional de História, Maringá, Paraná,

Eggers S, Parks M, Grupe G, Reinhard KJ (2011) Paleoamerican diet, migration and morphology in Brazil: archaeological complexity of the earliest Americans. PLoS One 6(9):e23962. https://doi.org/10.1371/journal.pone.0023962

Erickson CL (2003) Historical ecology and future explorations. In: Lehmann J, Kern DC, Glaser B, Woods WI (eds) Amazonian Dark Earths: origin, properties, management. Kluwer, Dordrecht, pp 455–500

Figuti L, Plens CR, DeBlasis P (2013) Small sambaquis and big chronologies: shellmound building and hunter-gatherers in Neotropical highlands. Radiocarbon 55(2–3):1215–1221. https://doi.org/10.1017/S0033822200048128

FUNAI (2020) Fundação Nacional do Índio. http://www.funai.gov.br/index.php/shape. Accessed July 2020
Furquim LP, Watling J, Hilbert LM, Shock MP, Prestes-Carneiro G, Calo CM, Py-Daniel AR, Brandão K, Pugliese F, Zimpel CA, Silva CA, Neves EG (2021) Facing Change through Diversity: Resilience and Diversification of Plant Management Strategies during the Mid to Late Holocene Transition at the Monte Castelo Shellmound, SW Amazonia. Quaternary 4:8. https://doi.org/10.3390/quat4010008
Gaspar MD, DeBlasis P, Fish SK, Fish PR (2008) Sambaqui (shell mound) societies of coastal Brazil. In: Silverman H, Isbell WH (eds) The handbook of South American archaeology. Springer, pp 319–335. https://doi.org/10.1007/978-0-387-74907-5_18
Graff SR (2018) Archaeological studies of cooking and food preparation. J Arch Res 26(3):305–351. https://doi.org/10.1007/s10814-017-9111-5
Grobman A, Bonavia D, Dillehay TD, Piperno DR, Iriarte J, Holst I (2012) Preceramic maize from Paredones and Huaca Prieta, Peru. Proc Natl Acad Sci U S A 109(5):1755–1759. https://doi.org/10.1073/pnas.1120270109
Hastorf C (1998) The cultural life of early domestic plant use. Antiquity 72(278):773–782. https://doi.org/10.1017/S0003598X00087366
Hastorf CA (1999) Recent research in paleoethnobotany. J Arch Res 7(1):55–103
Heckenberger MJ (1998) Manioc agriculture and sedentism in Amazonia: the Upper Xingu example. Antiquity 72(277):633–648
Hemming J (1995) Red gold: the conquest of the Brazilian Indians, 2nd edn. Papermac, London
Henderson JS, Joyce RA, Hall GR, Hurst WJ, McGovern PE (2007) Chemical and archaeological evidence for the earliest cacao beverages. Proc Natl Acad Sci U S A 104(48):18937–18940. https://doi.org/10.1073/pnas.0708815104
Henry AG, Hudson HF, Piperno DR (2009) Changes in starch grain morphologies from cooking. J Arch Sci 36(3):915–922. https://doi.org/10.1016/j.jas.2008.11.008
Hilbert L, Neves EG, Pugliese F, Whitney BS, Shock M, Veasey E, Zimpel CA, Iriarte J (2017) Evidence for mid-Holocene rice domestication in the Americas. Nat Ecol Evol 1(11):1693–1698. https://doi.org/10.1038/s41559-017-0322-4
Hilbert LM (2017) Investigating plant management in the Monte Castelo (Rondônia-Brazil) and Tucumã (Pará-Brazil) shell mounds using phytolith analysis. PhD Thesis, University of Exeter, Exeter
Imazio da Silveira M, Schaan DP (2005) Onde a Amazônia encontra o mar: estudando os sambaquis do Pará. Rev Arque 18(1):67–79. https://doi.org/10.24885/sab.v18i1.205
Ingold T (1993) The temporality of the landscape. World Arch 25(2):152–174
Iriarte J, Behling H (2007) The expansion of Araucaria forest in the southern Brazilian highlands during the last 4000 years and its implications for the development of the Taquara/Itararé tradition. Environ Arch 12(2):115–127. https://doi.org/10.1179/174963107x226390
Iriarte J, Gillam JC, Marozzi O (2008) Monumental burials and memorial feasting: an example from the southern Brazilian highlands. Antiquity 82(318):947–961. https://doi.org/10.1017/S0003598X00097702
ISA (2020) Povos indígenas no Brasil - Línguas. Instituto Socioambiental. https://pib.socioambiental.org/pt/L%C3%ADnguas. Accessed July 2020
Kistler L, Maezumi SY, De Souza JG, Przelomska NA, Costa FM, Smith O, Loiselle H, Ramos-Madrigal J, Wales N, Ribeiro ER (2018) Multiproxy evidence highlights a complex evolutionary legacy of maize in South America. Science 362(6420):1309–1313. https://doi.org/10.1126/science.aav0207
Klokler DM (2012) Consumo ritual, consumo no ritual: festins funerários e sambaquis. Rev Habitus 10(1):83–104. https://doi.org/10.18224/hab.v10.1.2012.83-104
Koch A, Brierley C, Maslin MM, Lewis SL (2019) Earth system impacts of the European arrival and Great Dying in the Americas after 1492. Quatern Sci Rev 207:13–36. https://doi.org/10.1016/j.quascirev.2018.12.004

Lahaye C, Guérin G, Boëda E, Fontugne M, Hatté C, Frouin M, Clemente-Conted I, Pino M, Felice GD, Guidon N, Lourdeau A, Pagli M, Pessis AM, Da Costa A (2015) New insights into a late-Pleistocene human occupation in America: the Vale da Pedra Furada complete chronological study. Quat Geochronol 30:445–451. https://doi.org/10.1016/j.quageo.2015.03.009

Lathrap DW (1977) Our father the cayman, our mother the gourd: Spinden revisited, or a unitary model for emergence of agriculture in the New World. In: Reed CA (ed) Origins of agriculture. Mouton, The Hague, pp 713–754

Lebot V (2009) Tropical root and tuber crops: Cassava, sweet potato, yams and aroids. CAB International, Oxford, UK

Lee-Thorp JA (2008) On isotopes and old bones. Archaeometry 50(6):925–950. https://doi.org/10.1111/j.1475-4754.2008.00441.x

Levis C, Costa FRC, Bongers F, Peña-Claros M, Clement CR, Junqueira AB, Neves EG, Tamanaha EK, Figueiredo FOG, Salomão RP, Castilho CV, Magnusson WE, Phillips OL, Guevara JE, Sabatier D, Molino J-F, López DC, Mendoza AM, Pitman NCA, Duque A, Vargas PN, Zartman CE, Vasquez R, Andrade A, Camargo JL, Feldpausch TR, Laurance SGW, Laurance WF, Killeen TJ, Nascimento HEM, Montero JC, Mostacedo B, Amaral IL, Guimaraes Vieira IC, Brienen R, Castellanos H, Terborgh J, Vargas Carim MJ, Guimarães JRS, Coelho LS, Matos FDDA, Wittmann F, Mogollón HF, Damasco G, Dávila N, García-Villacorta R, Honorio Coronado EN, Emilio T, DdA LF, Schietti J, Souza P, Targhetta N, Comiskey JA, Marimon BS, Marimon B-H Jr, Neill D, Alonso A, Arroyo L, Carvalho FA, Souza FC, Dallmeier F, Pansonato MP, Duivenvoorden JF, PVA F, Stevenson Diaz PR, Araujo-Murakami A, Aymard CGA, Baraloto C, Amaral DD, Engel J, Henkel TW, Maas P, Petronelli P, JDC R, Stropp J, Daly D, Gribel R, Paredes MR, Silveira M, Thomas-Caesar R, Baker TR, Silva NF, Ferreira LV, Peres CA, Silman MR, Cerón C, Valverde FC, Di Fiore A, Jimenez EM, Peñuela Mora MC, Toledo M, Barbosa EM, Bonates LCM, Arboleda NC, Farias ES, Fuentes A, Guillaumet J-L, Jorgensen PM, Malhi Y, IPA M, Phillips JF, Prieto A, Rudas A, Ruschell AR, Silva N, von Hildebrand P, Vos VA, Zent EL, Zent S, Cintra BBL, Nascimento MT, Oliveira AA, Ramirez-Angulo H, Ramos JF, Rivas-Torres G, Schöngart J, Sierra R, Tirado M, van der Heijden G, Torre EV, Wang O, Young KR, Baider C, Cano A, Farfan-Rios W, Ferreira C, Hoffman B, Mendoza C, Mesones I, Torres-Lezama A, MNU M, van Andel TR, Villarroel D, Zagt R, Alexiades MN, Balslev H, Garcia-Cabrera K, Gonzales T, Hernandez L, Huamantupa-Chuquimaco I, Manzatto AG, Milliken W, Palacios Cuenca W, Pansini S, Pauletto D, Ramirez Arevalo F, Costa Reis NF, Sampaio AF, Urrego Giraldo LE, Sandoval EHV, Gamarra LV, Vela CIA, ter Steege H (2017) Persistent effects of pre-Columbian plant domestication on Amazonian forest composition. Science 355(6328):925–931. https://doi.org/10.1126/science.aal0157

Levis C, Flores BM, Moreira PA, Luize BG, Alves RP, Franco-Moraes J, Lins J, Konings E, Peña-Claros M, Bongers F, Costa FRC, Clement CR (2018) How people domesticated Amazonian forests. Front Ecol Evol 5:171. https://doi.org/10.3389/fevo.2017.00171

Lombardo U, Iriarte J, Hilbert L, Ruiz-Pérez J, Capriles JM, Veit H (2020) Early Holocene crop cultivation and landscape modification in Amazonia. Nature 581(7807):190–193. https://doi.org/10.1038/s41586-020-2162-7

Lombardo U, Szabo K, Capriles JM, May J-H, Amelung W, Hutterer R, Lehndorff E, Plotzki A, Veit H (2013) Early and middle holocene hunter-gatherer occupations in western Amazonia: the hidden shell middens. PLoS One 8(8):e72746. https://doi.org/10.1371/journal.pone.0072746

Loukotka Č (1967) Annals map supplement number eight ethno-linguistic distribution of South American Indians. Ann Am Assoc Geogr 57(2):437–438. https://doi.org/10.1111/j.1467-8306.1967.tb00614.x

Machado JS (2012) Lugares de gente: mulheres, plantas e redes de troca no delta amazônico. Doctoral thesis, Museu Nacional, Universidade Federal de Rio de Janeiro, Rio de Janeiro

Machado JS (2016) Caminhos e paradas. Perspectivas sobre o território Laklãnõ (Xokleng). Rev Mus Arqueo Etnol (27):179–196. https://doi.org/10.11606/issn.2448-1750.revmae.2016.137298

Maezumi SY, Alves D, Robinson M, Souza JG, Levis C, Barnett RL, Almeida de Oliveira E, Urrego D, Schaan D, Iriarte J (2018) The legacy of 4,500 years of polyculture agroforestry in the eastern Amazon. Nat Plants 4(8):540–547. https://doi.org/10.1038/s41477-018-0205-y

Medina JT (ed) (1934) The discovery of the Amazon according to the account of Friar Gaspar de Carvajal and other documents (trans: Heaton HC, Lee BT). The American Geographical Society, New York

Métraux A (1946) The Caingang. In: Steward JH (ed) Handbook of South American Indians, The marginal tribes, vol 1. Smithsonian Institution, Washington, pp 445–477

Neves EG, Petersen JB (2006) The political economy of pre-columbian Amerindians: landscape transformations in Central Amazonia. In: Balée WL, Erickson CL (eds) Time and complexity in historical ecology: studies in the Neotropical lowlands. Columbia University Press, New York, pp 279–310

Noelli FS (1999) A ocupação humana na região sul do Brasil: arqueologia, debates e perspectivas-1872-2000. Revista USP (44):218–269. https://doi.org/10.11606/issn.2316-9036.v0i44p218-269

Olsen KM, Schaal BA (1999) Evidence on the origin of Cassava: Phylogeography of Manihot esculenta. Proc Natl Acad Sci U S A 96:5586–5591

Ortega DD (2019) Microvestígios botânicos em artefatos líticos do sítio Lapa do Santo (Lagoa Santa, Minas Gerais). Master's dissertation, Universidade de São Paulo, São Paulo

Pärssinen M, Ferreira E, Virtanen PK, Ranzi A (2020) Domestication in motion: macrofossils of pre-colonial Brazilian nuts, palms and other Amazonian planted tree species found in the Upper Purus. Environ Arch 25:1–14. https://doi.org/10.1080/14614103.2020.1765295

Pearsall DM (1992) The origins of plant cultivation in South America. In: Cowan CW, Watson PJ (eds) The origins of agriculture: an international perspective. Smithsonian Institution Press, Washington, DC, pp 173–206

Pearsall DM (2015) Paleoethnobotany: a handbook of procedures, 3rd edn. Left Coast Press, Walnut Creek

Pezo-Lanfranco L, Eggers S, Petronilho C, Toso A, da Rocha BD, Von Tersch M, dos Santos AMP, Ramos da Costa B, Meyer R, Colonese AC (2018) Middle Holocene plant cultivation on the Atlantic Forest coast of Brazil? R Soc Open Sci 5(9):180432. https://doi.org/10.1098/rsos.180432

Piperno DR (2006) Phytoliths: a comprehensive guide for archaeologists and paleoecologists. AltaMira Press, Lanham

Piperno DR, Clary KH, Cooke RG, Ranere AJ, Weiland D (1985) Preceramic maize in central Panama: phytolith and pollen evidence. Amer Anthro 87(4):871–878. https://doi.org/10.1525/aa.1985.87.4.02a00090

Piperno DR, Dillehay TD (2008) Starch grains on human teeth reveal early broad crop diet in northern Peru. Proc Natl Acad Sci U S A 105(50):19622–19627. https://doi.org/10.1073/pnas.0808752105

Piperno DR, Holst I (1998) The presence of starch grains on prehistoric stone tools from the humid Neotropics: indications of early tuber use and agriculture in Panama. J Arch Sci 25(8):765–776. https://doi.org/10.1006/jasc.1997.0258

Piperno DR, Ranere AJ, Holst I, Iriarte J, Dickau R (2009) Starch grain and phytolith evidence for early ninth millennium BP maize from the Central Balsas River Valley, Mexico. Proc Natl Acad Sci U S A 106(13):5019–5024. https://doi.org/10.1073/pnas.0812525106

Piperno DR, Stothert KE (2003) Phytolith evidence for early Holocene Cucurbita domestication in Southwest Ecuador. Science 299(5609):1054–1057. https://doi.org/10.1126/science.1080365

Politis GG (1996) Moving to produce: Nukak mobility and settlement patterns in Amazonia. World Arch 27(3):492–511. https://doi.org/10.1080/00438243.1996.9980322

Posey DA (1985) Indigenous management of tropical forest ecosystems: the case of the Kayapo Indians of the Brazilian Amazon. Agrofor Syst 3(2):139–158. https://doi.org/10.1007/BF00122640

Pugliese FA, Zimpel Neto CA, Neves EG (2018) What do Amazonian shellmounds tell us about the long-term indigenous history of South America? Encyclopedia of global archaeology. Springer, Cham. https://doi.org/10.1007/978-3-319-51726-1_3030-2

Purugganan MD, Fuller DQ (2011) Archaeological data reveal slow rates of evolution during plant domestication. Evolution 65(1):171–183. https://doi.org/10.1111/j.1558-5646.2010.01093.x

Reis MS, Ladio A, Peroni N (2014) Landscapes with Araucaria in South America: evidence for a cultural dimension. Ecol Soc 19(2):4. https://doi.org/10.5751/ES-06163-190243

Renfrew C, Bellwood P (2002) Examining the farming/language dispersal hypothesis. McDonald Institute for Archaeological Research, Cambridge

Resende ET, Prous A (1991) Os vestígios vegetais do Grande Abrigo de Santana do Riacho. Arq Mus Hist Nat UFMG 12:87–111

Roosevelt AC (1989) Resource management in Amazonia before the conquest: beyond ethnographic projection. In: Posey DA, Balée W (eds) Resource management in Amazonia: indigenous and folk strategies, Advances in economic botany, vol 7. Bronx/New York, The New York Botanical Garden, pp 30–62

Roosevelt AC (1998) Ancient and modern hunter-gatherers of lowland South America: an evolutionary problem. In: Balée W (ed) Advances in historical ecology. Columbia University Press, New York, pp 190–212

Roosevelt AC (2000) The lower Amazon: a dynamic human habitat. In: Lentz DL (ed) Imperfect balance: landscape transformations in the precolumbian Americas. Columbia University Press, New York, pp 455–479

Roosevelt AC, Lima da Costa M, Lopes Machado C, Michab M, Mercier N, Valladas H, Feathers J, Barnett W, Imazio da Silveira M, Henderson A, Sliva J, Chernoff B, Reese DS, Holman JA, Toth N, Shick K (1996) Paleoindian cave dwellers in the Amazon: the peopling of the Americas. Science 272(5260):373–384. https://doi.org/10.1126/science.272.5260.373

Rull V, Montoya E (2014) *Mauritia flexuosa* palm swamp communities: natural or human-made? A palynological study of the Gran Sabana region (northern South America) within a neotropical context. Quatern Sci Rev 99:17–33. https://doi.org/10.1016/j.quascirev.2014.06.007

Santos IT (2014) Alimentação de grupos humanos pré-históricos do sítio arqueológico Furna do Estrago, Pernambuco, Brasil. Doctoral thesis, Fundação Oswaldo Cruz, Rio de Janeiro

Scheel-Ybert R (2000) Vegetation stability in the Southeastern Brazilian coastal area from 5500 to 1400 14C yr BP deduced from charcoal analysis. Rev Palaeobot Palynol 110(1–2):111–138. https://doi.org/10.1016/s0034-6667(00)00004-x

Scheel-Ybert R, Caromano CF, Azevedo LW (2016) Of forests and gardens: landscape, environment, and cultural choices in Amazonia, southeastern and southern Brazil from c. 3000 to 300 cal yrs BP. Cad LEPAARQ 13(25):425–458. https://doi.org/10.15210/lepaarq.v13i25.7502

Scheel-Ybert R, Eggers S, Wesolowski V, Petronilho CC, Boyadjian CHC, DeBlasis PAD, Barbosa-Guimarães M, Gaspar MD (2003) Novas perspectivas na reconstituição do modo de vida dos sambaquieiros: uma abordagem multidisciplinar. Rev Arque 16(1):109–137. https://doi.org/10.24885/sab.v16i1.182

Scheel-Ybert R, Gaspar MD (2014) Anthracological research on the Brazilian Coast: paleoenvironment and plant exploitation of sambaqui moundbuilders. In: Madella M, Lancelotti C, Savard M (eds) Ancient plants and people: contemporary trends in archaeobotany. The University of Arizona Press, Tucson, pp 237–262

Scheel-Ybert R, Solari ME (2005) Análise dos macrorrestos vegetais do setor oeste: antracologia e carpologia. In: Vialou ÁV (ed) Pré-história do Mato Grosso, vol 1. Santa Elina. EDUSP, São Paulo, pp 139–147

Shepard GH Jr, Clement CR, Lima HP, Mendes dos Santos G, Moraes CP, Neves EG (2020) Ancient and traditional agriculture in South America: tropical lowlands. Encyclopedia of agriculture and the environment. Oxford University Press, New York. https://doi.org/10.1093/acrefore/9780199389414.013.597

Shepard GH Jr, Ramirez H (2011) "Made in Brazil": human dispersal of the Brazil Nut (*Bertholletia excelsa*, Lecythidaceae) in ancient Amazonia. Econ Bot 65(1):44–65. https://doi.org/10.1007/s12231-011-9151-6

Shock M, de Paula Moraes C, da Silva Belletti J, Lima M, da Silva FM, Lima LT, Cassino MF, de Lima AMA (2014) Initial contributions of charred plant remains from archaeological sites in the Amazon to reconstructions of historical ecology. Paper presented at the 3er Encuentro Internacional de Arqueología Amazónica, Quito, Ecuador,

Shock MP (2010) Holocene hunter-gatherer plant use and foraging choice: a test from Minas Gerais, Brazil. PhD thesis, University of California, Santa Barbara

Shock MP, Kipnis R, Bueno L, Silva FM (2013) A chronology of the introduction of domesticated plants in Central Brazil. Tipiti 11(2):52–59

Shock MP, Moraes CP (2019) A floresta é o domus: a importância das evidências arqueobotânicas e arqueológicas das ocupações humanas amazônicas na transição Pleistoceno/Holoceno. Bol Mus Para Emílio Goeldi Cien Hum 14(2):263–289. https://doi.org/10.1590/1981.81222019000200003

Silva FM, Shock MP, Neves EG, Scheel-Ybert R (2016) Vestígios macrobotânicos carbonizados na Amazônia Central: o que eles nos dizem sobre as plantas na pré-história? Cad LEPAARQ 13(25):366–385. https://doi.org/10.15210/lepaarq.v13i25.7369

Smith BD (2012) A cultural niche construction theory of initial domestication. Biol Theory 6(3):260–271. https://doi.org/10.1007/s13752-012-0028-4

Smith M, Fausto C (2016) Socialidade e diversidade de pequis (*Caryocar brasiliense*, Caryocaraceae) entre os Kuikuro do alto rio Xingu (Brasil). Bol Mus Para Emílio Goeldi Cien Hum 11(1):87–113. https://doi.org/10.1590/1981.81222016000100006

Staller JE, Carrasco MD (2010) Pre-Columbian foodways in Mesoamerica. In: Staller JE, Carrasco MD (eds) Pre-Columbian foodways: interdisciplinary approaches to food, culture, and markets in ancient Mesoamerica. Springer, New York, pp 1–20

Tamanaha EK (2018) Um panorama comparativo da Amazônia no ano 1000. Doctoral thesis, Universidade de São Paulo, São Paulo

Thomas E, van Zonneveld M, Loo J, Hodgkin T, Galluzzi G, van Etten J (2012) Present spatial diversity patterns of *Theobroma cacao* L. in the Neotropics reflect genetic differentiation in Pleistocene refugia followed by human-influenced dispersal. PLoS One 7(10):e47676. https://doi.org/10.1371/journal.pone.0047676

Troufflard J, Alves DT (2019) Uma abordagem interdisciplinar do sítio arqueológico Cedro, baixo Amazonas. Bol Mus Para Emílio Goeldi Cien Hum 14(2):553–580. https://doi.org/10.1590/1981.81222019000200015

Vialou D, Benabdelhadi M, Feathers J, Fontugne M, Vialou AV (2017) Peopling South America's centre: the late Pleistocene site of Santa Elina. Antiquity 91(358):865–884. https://doi.org/10.15184/aqy.2017.101

Villagran XS (2010) Geoarqueologia de um sambaqui monumental: estratigrafias que falam. Annablume/Fapesp, São Paulo

Walker M, Head MJ, Berkelhammer M, Björck S, Cheng H, Cwynar L, Fisher D, Gkinis V, Long A, Lowe J (2018) Formal ratification of the subdivision of the Holocene series/Epoch (quaternary system/period): two new Global Boundary Stratotype Sections and Points (GSSPs) and three new stages/subseries. Episodes 41(4):213–223. https://doi.org/10.18814/epiiugs/2018/018016

Watling J, Almeida F, Kater T, Zuse S, Shock MP, Mongeló G, Bespalez E, Santi JR, Neves EG (2020) Arqueobotânica de ocupações ceramistas na Cachoeira do Teotônio. Bol Mus Para Emílio Goeldi Cien Hum 15(2). https://doi.org/10.1590/2178-2547-bgoeldi-2019-0075

Watling J, Iriarte J, Mayle FE, Schaan D, Pessenda LCR, Loader NJ, Street-Perrott FA, Dickau RE, Damasceno A, Ranzi A (2017) Impact of pre-Columbian "geoglyph" builders on Amazonian forests. Proc Natl Acad Sci U S A 114(8):1868–1873. https://doi.org/10.1073/pnas.1614359114

Watling J, Saunaluoma S, Pärssinen M, Schaan D (2015) Subsistence practices among earthwork builders: Phytolith evidence from archaeological sites in the southwest Amazonian interfluves. J Arch Sci Rep 4:541–551. https://doi.org/10.1016/j.jasrep.2015.10.014

Watling J, Shock MP, Mongelo GZ, Almeida FO, Kater T, Oliveira PE, Neves EG (2018) Direct archaeological evidence for southwestern Amazonia as an early plant domestication and food production centre. PLoS One 13(7):e0199868. https://doi.org/10.1371/journal.pone.0199868

Wesolowski V, de Souza SMFM, Reinhard K, Ceccantini G (2007) Grânulos de amido e fitólitos em cálculos dentários humanos: contribuição ao estudo do modo de vida e subsistência de grupos sambaquianos do litoral sul do Brasil. Rev Mus Arqueo Etnol (17):191–210. https://doi.org/10.11606/issn.2448-1750.revmae.2007.89773

Wesolowski V, de Souza SMFM, Reinhard KJ, Ceccantini G (2010) Evaluating microfossil content of dental calculus from Brazilian sambaquis. J Arch Sci 37(6):1326–1338. https://doi.org/10.1016/j.jas.2009.12.037

Zarrillo S, Gaikwad N, Lanaud C, Powis T, Viot C, Lesur I, Fouet O, Argout X, Guichoux E, Salin F, Solorzano RL, Bouchez O, Vignes H, Severts P, Hurtado J, Yepez A, Grivetti L, Blake M, Valdez F (2018) The use and domestication of *Theobroma cacao* during the mid-Holocene in the upper Amazon. Nat Ecol Evol 2:1879–1888. https://doi.org/10.1038/s41559-018-0697-x

Zuse S (2014) Variabilidade cerâmica e diversidade cultural no alto rio Madeira, Rondônia. Doctoral thesis, Universidade de São Paulo, São Paulo

Famine Foods: Thoughts from a Caatinga Research Experience

Viviany Teixeira do Nascimento and Letícia Zenóbia de Oliveira Campos

Contents

1 Introduction

Although eating is a basic human need, food preferences are clearly influenced by the social, political, historical, and economic context in which we live. In addition to these factors, climate issues can temporarily decrease the supply of preferred foods and directly influence what is eaten, causing even the least desirable foods to

V. T. do Nascimento (✉)
Universidade do Estado da Bahia – UNEB, BR 242, s/n. Loteamento Flamengo. Departamento de Ciências Humanas, Barreiras, BA, Brazil
e-mail: vtnascimento@uneb.br

L. Z. O. Campos
Universidade Federal do Oeste da Bahia – UFOB, Estrada do Barrocão, s/n, Morada Nobre, Barreiras, BA, Brazil

M. C. M. Jacob, U. P. Albuquerque (eds.), *Local Food Plants of Brazil*, Ethnobiology, https://doi.org/10.1007/978-3-030-69139-4_9

be consumed (Nascimento et al. 2012). When this occurs, we call these foods "emergency foods" or even "famine foods."

Famine foods are those used only in extreme periods, be they native or exotic. These extreme periods are usually linked to environmental factors such as drought, or periods when there is a significant reduction in agricultural production (Eaton and Konner 1985; Guinand and Lemessa 2000). Regardless of the part used, typical famine foods are those that remain in the environment intact for a long period, being accessed only when all other food sources have been consumed (Irvine 1952).

Although these plants are currently used in extreme situations, they merit emphasis because they were for a long time the main source of food for several human populations, and instead of being considered merely complementary, they were essential for the diet (Etkin and Ross 1982; Eaton and Konner 1985; Johns 1990). Specifically in semiarid regions, many of these species guarantee the sustainability of local diets, providing important nutrients for maintaining the health of populations. In this context, traditional populations stand out, as several studies report that there are foods that are actually used, while others are accessed strictly in times of scarcity. This can generate different implications in the selection of certain species that may be used (Cruz et al. 2014; Bravo-Filho et al. 2018). About the Caatinga specifically, a biome that is characterized by vehemently from drought events and where at certain times famine foods become essential.

Several elements can influence this punctuality in relation to the consumption of famine foods, such as the energy expenditure involved in the collection and preparation of food, in addition to the food taboos linked to the social prejudice toward the practice of consuming famine foods. In view of the problem presented, this chapter will bring considerations about the most striking environmental issues that lead to the use of emergency foods in environments such as the Caatinga, the nutritional characterization of species that have already been studied in this perspective, and finally food taboos linked to the consumption of emergency food.

2 What Are "Famine Foods"?

Wild food plants are those that are not cultivated, that is, they are collected from natural environments with little human interference and are used to feed people in many regions of the planet, although agriculture represents most of what we eat (Nascimento et al. 2013). Within this group, there are those that are available and are consumed and enjoyed in any season or circumstance. Their importance is such that they are commercialized and sometimes even exported, as is the case with the by-products of the *Spondias tuberosa* (Anacardiaceae) fruits, a species native to northeast Brazil (Santos et al. 2011). However, there is also a category or subgroup of "wild food plants" whose consumption is only justifiable in periods when favorite foods are unavailable, and they are called "famine foods" or "emergency foods" (Minnis 1991; Muller and Almedom 2008; Jacob and Albuquerque 2020). This subgroup will be addressed in this chapter.

According to Minnis (1991), plants classified as famine foods must meet two basic assumptions: (1) they must be edible, and (2) they must be available at times when the production and availability of preferred foods is reduced. Therefore, famine foods are usually species well adapted to the most extreme environmental conditions in a given location and resistant to factors such as drought, poor or saline soil, and pests and diseases that can quickly eliminate exotic crops and modern varieties inserted in the local culture by the global market (Minnis 1991; Hughes 2009; Bhansali 2011).

A variety of plant organs can be consumed as famine foods, the most common being the underground organs (roots, rhizomes, tubers, and bulbs), leaves, fruit pulps, seeds, cladodes, or even the entire plant, when it comes to herbs.

In addition to the typical famine foods, occasionally, other sets of plants can gain this status. This is the case of wild plants that have soft parts and do not require a longer time for preparation (usually fruits). They are appreciated and consumed throughout the year, but they have other structures that are usually underground, acting as famine foods. This can also occur with some cultivated plants when parts that are not traditionally consumed are used even if this implies the death of the plant, making it unfeasible for a new harvest cycle (Guinand and Lemessa 2000).

Apart from the types of famine foods explained above, whose parts are only consumed in situations of extreme scarcity, it is noticeable that the literature has associated this name with those wild foods that do not fit as typical famine foods, but that are just not traditionally consumed. That is, they are underutilized foods, but capable of attracting additional consumers in times of scarcity, acting on food and nutrition security, and preventing hunger in times of emergency (Guinand and Lemessa 2000; Hughes 2009).

We must also consider that typical famine foods are often foods that are not actually suitable for food, as they have characteristics that decrease their palatability, such as the presence of thorns. They may taste strange or may be toxic, requiring a prolonged preparation process to reduce their inappropriate properties. Many of these famine foods are low in nutrients, which only help to satiate hunger for a few hours, thanks to their high cellulose content. Finally, they are foods that can be associated with stomach complications, diarrhea, and poisoning. For this set of characteristics, they are commonly considered as a "last resort," used only when all other natural sources of food are exhausted (Castro 1946; Minnis 1991; Guinand and Lemessa 2000).

Although other Brazilian biomes have a number of native and exotic food species that grow spontaneously and can be used as food, they do not meet the basic requirements described above to be categorized as "famine foods." The main characteristic that makes these foods fall into that category is the fact that they are used in periods of extreme drought, where other species are unlikely to be found. For this reason, we selected the Caatinga biome, which has the main characteristics for the consumption of famine foods. After this brief contextualization, we will present below the conditions that lead to the consumption of famine foods by populations in the Brazilian semiarid region, followed by several examples of these resources.

3 What Conditions Favor the Use of Emergency Foods in the Caatinga?

Two ecological characteristics dominate the environmental context of starvation: climatic seasonality and the great dependence of human populations on subsistence and food production systems in the local market. Virtually all naturally occurring famines have occurred in regions that strongly combine seasonal patterns of temperature or humidity with a high degree of variability in these factors from year to year (Cox 1978). In Brazil, the Caatinga is the biome that best fits these conditions.

The Caatinga biome covers the Northeast region of Brazil and some areas of the state of Minas Gerais. It is a region of extreme weather patterns, with high temperatures and high rates of evapotranspiration associated with low relative humidity and precipitation. In this region, a factor that affects the supply of food is the annual distribution of precipitation, as the drought can reach up to 11 months in certain locations (Castro 1946; Nimer 1972). In the longer periods of absence of rain, hunger in the region is widespread, affecting both human and animal populations qualitatively and quantitatively (Castro 1946). This aspect is important because, despite the serious problems in relation to the availability of water and food, the Caatinga is highly populous, with approximately 27 million inhabitants who suffer the effects of the lack of a sustainable food production system (IBGE 2004).[1]

In this semiarid environment of so many uncertainties related to seasonal climatic variability, plants capable of surviving their extreme conditions have adapted. In general, they share the presence of thorns, microphyll, deciduity, and water storage capacity, as observed in cacti and succulents (Prado 2003). This makes the Caatinga region an environment with high rates of floristic endemism that can reach 42% of succulent and woody species (Prado 2003).

Regarding the plants that can be used for human consumption from this extreme environment, there is a diversity of parts that can be used for this purpose, from leaves, cladodes, flowers, and fruits. In addition, this biome has the lowest per capita income rates in Brazil (IBGE 2004), a factor that directly influences the human populations that live there to depend heavily on natural resources. Different studies that seek to understand this human-plant interrelationship have found that there is a variety of species that can be used, with emphasis on those belonging to the Cactaceae family (Nascimento et al. 2012; Lins-Neto et al. 2013; Cruz et al. 2014; Chaves and Barros 2015; Bravo-Filho et al. 2018). However, most references related to food plants from the Caatinga emphasize only those commonly used. For this reason, we carried out a careful bibliographic review in search of mentions, even if discrete, of the foods used only in times of food scarcity caused by low rainfall and which fall within the concept of emergency foods as described earlier in this chapter.

---

[1] It is important to highlight that in recent years, the lack of food for this significant portion of the population has been mitigated by government policies on access to food that favor the acquisition of products through local commerce throughout the year, but which assume an even more important role relevant in combating hunger in the absence of rain.

Below we will briefly present their chemical composition, preparation methods, and some difficulties related to their consumption.

4 Famine Foods of the Caatinga

In Brazil, famine foods have been reported for a long time, although scarcely. In our survey, the oldest yet discreet mentions of this type of food date from the seventeenth century, when the naturalists Piso and Marcgrave prepared a wide inventory of species occurring in northeast Brazil (Piso and Marcgrave 1648). In the past century, a series of works published by the geographer Josué de Castro and collaborators addressed the theme of hunger in Brazil and the food used exclusively in that period (Castro 1946; Castro et al. 1947; Castro 1966). Following these publications, there was an immense silence regarding this type of food, which was only broken in 2012 with the publication of an article with an ethnobotanical focus (Nascimento et al. 2012). It is based on these publications that we will briefly present below the chemical composition, the preparation methods, and some difficulties related to the consumption of certain famine foods. We emphasize that, whenever possible, the species had their Latin names updated according to "Flora 2020". In cases where we did not find the updated Latin name, we left this information clear throughout the text.

4.1 Anacardiaceae

Anacardiaceae is a family with a wide variety of food species. For the Caatinga region, the main species found is *Spondias tuberosa* Arruda (Fig. 1a), a leafy tree native and endemic to Brazil whose fruits and by-products (sweets, jellies, juices, and ice creams) are appreciated inside and outside the northeastern region (Albuquerque and Andrade, 2002; Lins-Neto et al. 2010). For emergency food purposes, the part of this plant consumed is the underground organs. The umbuzeiro root is formed by a spongy parenchyma rich in water (93.3% humidity), and for this reason, it is often mentioned as an important resource for times of drought (Piso and Marcgrave 1648; Castro 1946; Castro et al. 1947; Pickel 2008; Medeiros and Albuquerque 2014). Another nutritional characteristic of this structure is the presence of ascorbic acid, 13.3 milligrams, which can be an alternative vitamin supplement in times when fruits rich in this substance are unavailable (Castro et al. 1947).

4.2 Apocynaceae

Mandevilla tenuifolia (J.C.Mikan) Woodson (Fig. 1b), commonly known as "batata de vaqueiro," is an herbaceous species of the Apocynaceae family, native, but not

Fig. 1 Some famine foods of the Caatinga region: (**a**) *Spondias tuberosa* (plant); (**b**) *Mandevilla tenuifolia* (flower); (**c**) *Encholirium spectabile* (plant); (**d**) *Pilosocereus gounellei* (plant); (**e**) *Opuntia ficus-indica* (plant); (**f**) *Dioscorea coronata* (tuber); (**g**) *Manihot glaziovii* (branches); (**h**) *Dioclea grandiflora* (fruit and seeds)

endemic to Brazil, occurring in different phytogeographic domains including the Caatinga (Koch et al. 2015). In this region, it blooms soon after the first rains, thanks to the presence of an underground organ (xylopodium) that guarantees survival during the dry season. When full of water, the xylopodium becomes edible and is collected and consumed fresh without any kind of preparation similar to a vegetable while the products of the next harvest are not yet available. Thus, the organ is considered an emergency food. Nutritional analysis of xylopodium revealed that it has a low caloric value (63.2 g/100 g) and a large amount of water. Toxic effects related to food consumption of this species are unknown (Nascimento et al. 2012).

4.3 Araceae

Montrichardia linifera (Arruda) Schott commonly known as "aninga" is a non-endemic herbaceous Araceae from Brazil that occurs in the Amazon, Atlantic Forest, and Caatinga (Mayo and Andrade 2020). Its use as a useful food to suppress hunger was recorded on naturalist treaties from the seventeenth century. The edible part is the fruit, which can be eaten fresh or made into flour. According to reports, the ingestion of this species causes everything from simple flatulence to respiratory complications that can lead to death (Piso and Marcgrave 1648; Pickel 2008; Medeiros and Albuquerque 2014). Despite this, it is noticeable for its high caloric value, about 355 kcal/100 g, mainly from carbohydrates 86.5 g/100 g. Minerals such as calcium, iron, and zinc were also recorded in the nutritional analysis of this species (Amarante et al. 2010).

4.4 Arecaceae

Three species of this family have been reported as food in times of scarcity. The oil of the fruit of *Syagrus comosa* (Mart.) Mart. "coco amargoso" (basionym *Cocos comosa* Mart. reported in Castro et al. 1947 and Castro 1966) has a chemical composition (67.3 g/100 g lipids and 9.5 μg/100 g calcium), odor, and taste very similar to *Cocos nucifera* oil (coco da bahia) and can be used as substitute for this. Despite these similarities, we believe that the use of this oil only as an emergency food is related to the small size of the fruits of *S. comosa*, which makes its preparation difficult. Thus, it is disregarded in the face of *C. nucifera* oil, which is easier to prepare due to the size of the fruits.

Other species of the Arecaceae family that have been identified as emergency food are *Copernicia prunifera* (Mill.) H. E. Moore (syn. *Copernicia cerifera* mentioned by Castro 1946 – "macaubeira") and *Cocos mucronata* (mentioned by Castro 1946, "oricuri"). The stem of the first species is used in the form of heart of palm, while the same organ of the second species is grated and cooked to prepare a poor nutritional quality bread, known as "sinister bread," which only fill one's belly, but does not suppress hunger. This bread was first mentioned in the literary work *Os Sertões* by Euclides da Cunha, published in 1902. However, the author did not mention the scientific name of the species. Then, Castro (1946) gave it the name *C. mucronata*, as mentioned above. However, we did not find this species, nor the update of its scientific name in "Flora 2020," which causes uncertainty about the scientific name of this emergency food.

4.5 Bromeliaceae

The Bromeliaceae family is present throughout Brazil, occurring in all biomes. Some species of this family are referred to in the literature as emergency food. This is the case of *Bromelia laciniosa* Mart. ex Schult. & Schult.f. (Castro et al. 1947) and *Encholirium spectabile* Mart. ex Schult. & Schult.f. (Fig. 1c) commonly known as "macambira" (Castro 1946; Nascimento et al. 2012). Both species are abundant and endemic to the Caatinga region. The bulb is the organ consumed as an emergency food. The collection of these Bromeliaceae adds difficulty to their consumption since they have leaves with many thorns on the margin. For it to be served as food, the bulb is cooked and then sun-dried. Once dried, the organ is crushed to make some flour from which couscous is prepared. No toxic effect associated to the flours of these Bromeliaceae is known. From the nutritional point of view, both have considerable caloric value. *B. laciniosa* has about 118 kcal/100 g of product (Chaves et al. 2015), while *E. spectabile* has 124 kcal/100 g. In both cases, most of this caloric value is related to carbohydrates, since in both species, the values of lipids and proteins are low (Castro et al. 1947; Nascimento et al. 2012; Chaves et al. 2015). *B. laciniosa* also has thiamine, about 168 μg/100 g of flour (Castro et al. 1947), and is considered exceptionally rich in calcium (Castro 1946).

Another Bromeliaceae that is also mentioned as an emergency food is *Bromelia karatas* "caraguatá-acanga" (Castro 1966). However, the most recent botanical reviews "Flora 2020" consider that this species does not occur spontaneously in Brazil, which leads them to believe that some identification problem occurred at the time of the publication of Castro's work (1946).

4.6 Cactaceae

The Cactaceae family is among the most important ones when it comes to emergency food in the Caatinga, thanks to its ability to face long periods of drought. One of the species used for this purpose is *Pilosocereus gounellei* (F.A.C.Weber) Byles & Rowley "xiquexique" (Fig. 1d), a shrub cactus occurring only in Brazil (Zappi et al. 2015). It is a species used mainly for animal consumption (Brito Cavalcanti and Milanez 2007). However, humans consume it when drought worsens. In this situation, the cladodes are cut, and their abundant thorns are removed together with a water storage tissue typical of Cactaceae. The cuts are made until a woody chamber that protects a marrow is reached. The chamber is then cooked or roasted for a few minutes over fire and then removed with a knife, leaving only the marrow that can be immediately consumed or transformed into flour used to prepare the couscous. The preparation of this emergency food is difficult due to the presence of thorns; its nutritional yield is low, yielding only 27.5 kcal/100 g of product; and toxic properties related to its consumption are unknown (Nascimento et al. 2012).

Cereus setosus and *Pilocerius setosus* (also known as xiquexique) are other cacti mentioned by Castro (1946) and Castro et al. (1947), respectively, as emergency food. The useful parts, method of preparation, and nutritional composition are similar to those of *P. gounellei* described above. However, we cannot confirm the current identification of the species according to the most recent botanical reviews, since both were not mentioned in "Flora 2020." Thus, we do not know which species the publications mentioned above refer to.

Within the Cactaceae family, *Opuntia ficus-indica* (L.) Mill "palma forrageira" (Fig. 1e) needs to be addressed, a species that has its origin in Mexico but is currently considered naturalized in Brazil (Zappi et al. 2015). Despite being introduced in the country since the first years after the arrival of the Europeans, the plant had its cultivation encouraged in the northeast region in the beginning of the 1930s, right after a great drought that occurred in 1932. At the time, the purpose was to use it as food for farm animals (Duque 1980).

O. ficus-indica has been little addressed in ethnobotanical surveys focused on food plants, as it is not originally from Brazil. However, our experience in different communities in the Caatinga allows us to affirm that the plant is one of the first to be mentioned when talking about emergency foods these days. The organ consumed is the cladode known as rackets, from which the thorns are removed and then chopped and cooked for human consumption. Toxic effects related to their intake are unknown (personal observations of the authors). Cladodes are an important source of water, carbohydrates, fibers, proteins, lipids, and minerals such as calcium, as well as antioxidants (Stintzing and Carle 2005; Bensadón et al. 2010; Silva et al. 2015), and are considered a noble food in other countries. However, in the northeast region of Brazil, its human use is seriously prejudged, probably due to its widespread use as animal food (Cantwell 2001).

4.7 Dioscoreaceae

The tuber of the vine *Dioscorea coronata* Hauman "pinanga de coroa" (Fig. 1f) is an organ that can be used as an emergency food. The structure is collected and consumed after cooking, which helps not only to soften but also to reduce its bitter taste. No health problems attributed to its consumption were recorded, but the caloric value for each 100 g of product is as low as 73.6 g/100 g. It has less than 1% protein and approximately 17 g of carbohydrates (Nascimento et al. 2012).

4.8 *Euphorbiaceae*

Within this family, two species are used as emergency food, *Manihot dichotoma* Ule "maniçoba" and *Manihot glaziovii* Müll.Arg. "purnunça" (Fig. 1g) (Nascimento et al. 2012), known as wild cassava. The edible part of both species in times of food scarcity is the tuberous roots. To be consumed, the raw organ needs to be washed several times to eliminate its toxic properties that can cause a range of effects, from dizziness to death. After proper washing, the roots can be cooked or toasted to be used later in the preparation of flour, beijus, cakes, porridge, etc. *M. dichotoma* and *M. glaziovii* have 104 and 79 kcal/100 g of product, respectively, most of which are attributed to carbohydrates (24 and 18 g/100 g). Although lower than the caloric value of conventional cassava (*Manihot esculenta* Crantz) (Bezerra and Saldanha, 2002), we believe that if their toxic properties are properly eliminated, *M. dichotoma* and *M. glaziovii* can be important emergency foods, helping populations when food from conventional crops are temporarily unavailable.

4.9 *Fabaceae*

The Leguminosae are considered the most important source of emergency food for the population of the Caatinga, due to the number of species listed as useful for this purpose. Some plants, popularly called *Mucuna*, have very similar forms of preparation and problems caused by their ingestion. Some examples are *Dioclea grandiflora* Mart. ex Benth (Fig. 1h) (Nascimento et al. 2012), *Mucuna urens* (L.) Medik. (Castro 1946), and *Mucuna glabra* (Reinecke) Wilmot-Dear. The latter was cited by Castro et al. (1947) and Castro (1966); however this species currently has no resolved scientific name, which makes it difficult to know which species the author was referring to at the time of publication.

D. grandiflora and *M. urens* are native vines of Brazil, although only the first is considered endemic to the Caatinga. The organs consumed from *D. grandiflora* as an emergency food are the seeds, while the roots are consumed from *M. urens* in addition to the seeds. Both organs are used for the preparation of a flour that will later be transformed into couscous. The flour produced from the seeds of *D. grandiflora* has a high caloric value, around 367 kcal/100 g, due to its high levels of proteins (30.9 g/100 g) and carbohydrates (54.3 g/100) (Nascimento et al. 2012). The same is observed for *M. urens*, with 28.5% proteins and 54.6% carbohydrates in addition to calcium, iron, and vitamin B1 (Castro 1946). The nutritional value of these species is up to five times greater than that of *Vigna unguiculata* (L.) Walp. (Beans or *feijão de corda* as known locally) (Taco 2011), another Leguminosae traditionally consumed in northeast Brazil. These results were taken into account to categorize *D. grandiflora* and *M. urens* as two important emergency foods in the Caatinga region, the cultivation of which should be encouraged and the use applied to other times of the year.

However, it is worth noting that the consumption of both *D. grandiflora* and *M. urens* flours is usually associated with different stomach disorders and other problems such as the suspension of menstruation and swelling of the skin, and in more extreme cases, ingestion may lead to death. It is customary to wash the flour numerous times before consumption to alleviate such toxic effects, a treatment that according to users does not eliminate all the toxicity attributed to plants (Castro 1946; Nascimento et al. 2012). However, Castro (1946) considers such symptoms to be a myth, linked much more to the advanced state of malnutrition of the population than the toxicity of the plant. To justify this theory, Castro et al. (1947) tested the flour from the seeds of *M. glabra* and confirmed that it was an innocuous product. However, we must be careful with this statement because the species used in this experiment was precisely the one with unknown updated Latin name, and therefore it is impossible to extrapolate the results obtained for *D. grandiflora* and *M. urens* without further studies being undertaken.

Castro (1946) mentioned two more Leguminosae that are used in times of food shortage: *Tipuana tipu* (Benth.) Kuntze (updated name of *Tipuana especiosa*) "pau de mocó," a non-endemic native shrub distributed throughout Brazil (Lima 2015), and *Senna occidentalis* (L.) Link (updated name of *Cassia occidentalis*), a tree popularly known as "fedegoso" (Bortoluzzi et al. 2020). A flour used to make porridge is usually produced with the roots of *T. tipu*, while the seeds of *S. occidentalis* are roasted to make coffee. For both species, we did not obtain information about the nutritional value or any toxic effect related to their consumption.

Another species of this family consumed in times of famine is *Geoffroea spinosa* Jacq. (Tomchinsky and Ming 2019), "umarizeiro," a non-endemic native tree occurring in the Caatinga (Penington 2015). The plant has fruits that, when eaten fresh, cause stomach problems that are only relieved when they are cooked and subsequently macerated. The cream resulting from this preparation can be consumed with meat or fish, replacing bread (Piso and Marcgrave 1648; Pickel 2008; Medeiros and Albuquerque 2014).

4.10 *Menispermaceae*

Abutua platyphylla, parreira-brava, name not found in "Flora 2020," is a species whose fruits resemble grape clusters, although these are not edible. The part of this vine consumed as an emergency food is the root, from which a flour low in protein and vitamins can be produced (Castro et al. 1947). In its composition stands out thiamine content of 22.5 μg/100 g and high starch content of 50.3%/100 g. Information regarding the preparation of the flour or the presence of problems caused by ingestion of the species was not found (Castro et al. 1947; Castro 1966).

5 Sociocultural Aspects of the Use of "Famine Foods": A Focus on Food Taboos

Despite the context of scarcity of food resources for the populations living in the semiarid regions of Brazil, in times when there is a drastic decrease in the rainfall regime, there are species, such as those listed above, that can be used in such a situation. However, the fact that there is a resource available does not imply its use, since preferences or not for certain foods are rooted in social and cultural matters (MacBeth and Lawry 1997).

When social and cultural questions about the use of a plant are shared between members of a group, they can become food taboos, which act as social markers to show differences between individuals and groups, influence attitudes and behaviors, and facilitate the functioning of systems (Garine 1995). In relation to famine foods, food taboos are mainly linked to the scarcity of financial resources of human populations, leading to a social prejudice toward the use of these foods. These prejudice views often do not take into account the nutritional value of the species because, as demonstrated in the previous topic, many plants have a considerable nutritional value.

Josué de Castro in his work *The Geography of Hunger*, issued in 1946, pointed out that cultural issues and beliefs carried by our ancestors and, consequently, passed on to future generations influence the eating practices that we adopt throughout our trajectory. A classic example is the high consumption of farinaceous foods by populations of northeast Brazil, a practice encouraged by our colonizers, in which the consumption of more nutritious and available foods in the region gradually decreased (Castro 1946).

Another aspect that we consider relevant to determine the low use of emergency food in the Caatinga region, a real taboo, is the fact that many of these plants are used frequently for animal feeding. Thus, those who need to consume such food may feel equivalent to animals. This often leads them to deny the fact that they are consuming such food or even that they have already consumed them at some point in their lives (personal observations of the authors).

This trend of low consumption of native foods, emergency foods or not, is not seen in other regions of the world, especially in countries that have an arid climate, such as those on the African continent (Okigbo 1975; Ogle and Grivetti 1985; Campbell 1986). In this sense, we can mention some examples such as southeastern Nigeria, where leaves and fruits of different native species are used in times of famine (Okigbo 1975). In the region of Swaziland, Ogle and Grivetti (1985) reported the use of more than 200 species of wild plants that are commonly consumed. Wachiira (1987), in a study carried out in Kenya, found that native leaves and fruits contribute significantly to the diet and that many emergency plants are consumed mainly by children. In regions close to native forests in Zimbabwe, Campbell (1986) found that wild fruits are consumed only in periods of scarcity and when there is a decrease in agricultural work, a behavior that is comparable to that found in the populations of the Brazilian semiarid region.

Given this context of low utilization of wild food plants and particularly famine foods, consumers and policy makers may not be aware of the benefits that underutilized plants can offer, which results in little attention to their conservation, crop improvement, nutritional assessment, production technologies, and post-harvest handling and promotion or marketing (Hughes 2009). In relation to this trend of famine foods being accessed only in times of scarcity, the World Vegetable Center (WVC) has been working to recognize the potential that these foods have in alleviating poverty and malnutrition in developing countries, especially those located in Africa (Hughes 2009).

Another discussion related to the use of "famine foods" is linked to the palatability. In this sense, Crúz et al. (2014) reported that one of the factors that often interfere in the selection of a native species for food is precisely the taste, as well as the preparation time that is required to neutralize components that can interfere in the palatability resulting in bitter taste or undesired texture, for example. Another issue also described about the use of certain species is linked to their morphological characteristics. Often, there are plants with thorns that need a long period of preparation to be consumed. Chaves and Barros (2015), in their study on the consumption of Cactaceae, revealed that the most used parts of the species are cladodes, followed by fruits and flowers. A large part of the cladodes are used to make flour, which demands a fair amount of time for its preparation, making this practice very unusual nowadays. Therefore, considering the cost-benefit of the collection, treatment and cooking activities (nutritional value – energy), these species end up not being consumed in times when there are other resources that are easier to be found, prepared, and used.

Finally, the new model of social and economic development imposed by modern society may have a negative influence on the consumption of traditional foods, since there is an ease in acquiring products in the food markets that would take a long time to be prepared manually at home (Crúz et al. 2014). The purchase of these foods is often only possible, thanks to the financial support given to the populations by government agencies and international agencies to fight hunger (Minnis 1991; Nascimento et al. 2012), which can pose a problem if such support is deficient. Furthermore, we believe that this financial contribution can interrupt the flow of knowledge between generations, leaving information about famine foods restricted to older individuals in the localities (Nascimento et al. 2012), while younger people are not interested in having such knowledge because they prefer to buy the food.

6 Final Considerations

Considering the aspects brought up in this chapter, we understand the high importance that famine foods had at different periods for human populations and how valuable they can be to alleviate hunger in emergencies. However, the absence of studies that report this issue can lead to the discontinuity of the different practices related to the use of these foods, and, if in some time they are necessary, the populations need to relearn how to use them. Thus, it is important that the knowledge about these practices is not restricted to the elderly and that the improvement of the socioeconomic conditions of the populations does not interfere in the knowledge related to the practices of use famine foods. For this, it is important that the stigma of poverty is considered in the different local communities of the Caatinga, because the use of any species to supply a basic need cannot be a matter of shame for any population. However, there is a historical prejudice paradigm in this issue; thus actions to popularize the use of these resources may play an important role. This can help to de-mystify the practice that famine foods are animal food and that, if people feed on them, they will be equated to animals. Finally, we believe that studies on famine foods are important so that they can be used beyond local areas and commercialized, and thus, we must ensure that knowledge about these practices is maintained so that there is resilience in socioecological systems.

References

Albuquerque UP, Andrade LHC (2002) Conhecimento botânico tradicional e conservação em uma área de caatinga no estado de Pernambuco, Nordeste do Brasil. Act Bot Bras 16:273–285. https://doi.org/10.1590/S0102-33062002000300004

Amarante, CB, Müller RCS, Dantas, KGF et al (2010) Composição química e valor nutricional para grandes herbívoros das folhas e frutos de aninga (*Montrichardia linifera,* Araceae). A Amaz 40(4): 729–736

Bensadón S, Hervert-Hernández D, Sáyago-Ayerdi SG et al (2010) By-products of Opuntia ficus-indica as a source of antioxidant dietary fiber. P Food H Nut 65(3):210–216. https://doi.org/10.1007/s11130-010-0176-2

Bezerra VS, Saldanha VSB (2002) Valor Nutricional da Mandioca (*Manihot esculenta* Crantz) e Transformações Pós-Colheita. Macapá: Embrapa Amapá. 18p. ; 21 cm (Embrapa Amapá. Documentos, 36)

Bhansali RR (2011) Famine foods in the Rajasthan Desert. Ec Bot Desert Plants (28):73–81. https://doi.org/10.1007/BF02861383

Bortoluzzi RLC, Lima AG, Souza VC, Rosignoli-Oliveira LG, Conceição AS (2020) Senna in Flora do Brasil 2020. Jardim Botânico do Rio de Janeiro. http://floradobrasil.jbrj.gov.br/reflora/floradobrasil/FB23149. Accessed March 14, 2021.

Bravo Filho ES, Santana MC, Santos PAA et al (2018) Levantamento etnobotânico da família Cactaceae no estado de Sergipe. R Fitos. https://doi.org/10.5935/2446-4775.20180005

Brito Cavalcanti N, Milanez G (2007) Consumo de Xiquexique (*Pilocereus gounellei* (A. Weber ex K. Schum.) Bly. ex Rowl) por caprinos no semi-árido da Bahia. R Caatinga (20) 1

Campbell BM (1986) The importance of wild fruits for peasant households in Zimbabwe. F Nut 12(1):38–44

Cantwell M (2001) Manejo pós-colheita de frutas e verdura de palma forrageira. In: Barbera G, Inglese P (eds) Agroecologia, cultivos e usos da palma forrageira. SEBRAE/PB, Paraíba, pp 20–27

Castro J (1946) Geografia da fome. Editora Brasiliense, 9ª edição, 1996. São Paulo

Castro J, Pechnik E, Parahim O et al (1947) Os "Alimentos bárbaros" dos sertões do Nordeste. Arquivos Brasileiros de Nutrição 3(2):5–29

Castro J (1966) Death in the Northeast. Random House, New York

Chaves EMF, Barros RFM (2015) Cactáceas: recursos alimentar emergencial no semiárido, Nordeste do Brasil. G Scient 9(2):129–135

Chaves EMF, Silva JN, Lima A et al (2015) Potential of wild food plants from the semi-arid region of northeast Brasil: chemical approach ethnoguided. Revista Espacios 36(16)

Cox GW (1978) The ecology of famine: an overview†. E Food Nut 6(4):207–220. https://doi.org/10.1080/03670244.1978.9990500

Crúz MP, Medeiros PM, Comarizao IS et al (2014) "I eat the manofe so it is not forgotten": local perceptions and consumption of native wild edible plants from seasonal dry forests in Brazil. J Ethnobiol Ethnomed 10:45. https://doi.org/10.1186/1746-4269-10-45

Duque SG (1980) O Nordeste e as culturas xerófilas. Mossoró: 3. ed. Escola Superior de Agricultura de Mossoró/Fundação Guimarães Duque. ESAM (Coleção Mossoroense,143). 316p

Eaton SB, Konner M (1985) Paleolithic nutrition a consideration of its nature and current implications. T. N Engl J Med 31(5). https://doi.org/10.1056/nejm198501313120505

Etkin N, Ross PJ (1982) Food as medicine and medicine as food. An adaptive framework for the interpretation of plant utilization among the Hausa of northern. Nigeria S Sc and Med 16(17):1559–1573. https://doi.org/10.1016/0277-9536(82)90167-8

Flora do Brasil (2020) in construction. Jardim Botânico do Rio de Janeiro. Available from: http://floradobrasil.jbrj.gov.br/. Accessed July 23, 2020

Garine I (1995) The diet and nutrition of human populations. In: Ingolg T (ed) Companion Encyclopedia of anthropology. Routledge, London, pp 226–264

Guinand Y, Lemessa D (2000) Wild-food plants in Ethiopia: reflections on the role of wild foods and famine foods at a time of drought. In: Kenyatta C, Henderson A (eds) The potential of indigenous wild foods. USAID/OFDA, CRS/Southern Sudan, pp 31–46

Hughes J (2009) Just famine foods? What contributions can underutilized plants make to food security? Acta Hortic (806):39–48. https://doi.org/10.17660/ActaHortic.2009.806.2

Instituto Brasileiro de Geografia e Estatística. Mapa de biomas do Brasil [Internet]. IBGE. 2004 [cited 2020 Maio 20]. Available from: https://ww2.ibge.gov.br/home/presidencia/noticias/21052004biomashtml.shtm. Accessed June 30, 2020

Irvine FR (1952) Supplementary and emergency food plants of West Africa. Ec Bot 6:23–40. https://doi.org/10.1007/BF02859192

Jacob MCM, Albuquerque UP (2020) Biodiverse food plants: which gaps do we need to address to promote sustainable diets? Eth Conserv 9:9. https://doi.org/10.15451/ec2020-04-9.09-1-6

Johns T (1990) Human perception, cognition, and behavior in relation to plant chemicals. In: Johns T (ed) The origins of human diet and medicine. The University of Arizona Press, pp 160–194

Koch I, Rapini A, Simões AO et al (2015) Apocynaceae in Lista de Espécies da Flora do Brasil. Jardim Botânico do Rio de Janeiro. http://floradobrasil.jbrj.gov.br/jabot/floradobrasil/FB4676. Accessed June 15, 2020

Lima HC (2015) Tipuana in Lista de Espécies da Flora do Brasil. Jardim Botânico do Rio de Janeiro. http://floradobrasil.jbrj.gov.br/jabot/floradobrasil/FB23206. Accessed March 14, 2021.

Lins Neto EMF, Peroni N, Albuquerque UP (2010) Traditional knowledge and management of umbu (*Spondias tuberosa* Anacardiaceae): an endemic species from the semi-arid region of Northeastern Brazil. Ec Bot 64(1):11–21. https://doi.org/10.1007/s12231-009-9106-3

Lins Neto EMF, Almeida ALS, Peroni N et al (2013) Phenology of *Spondias tuberosa* Arruda (Anacardiaceae) under different landscape management regimes and a proposal for a rapid phenological diagnosis using local knowledge. J Ethnobio Ethnomed 9:10. https://doi.org/10.1186/1746-4269-9-10

MacBeth H, Lawry S (1997) Food preferences and taste: an introduction. In: Macbeth H (ed) Food preferences and taste: continuity and change, Anthropology of Food & Nutrition. Berghanhn Books, pp 1, 236–14

Mayo SJ, Andrade IM (2020) Montrichardia in Flora do Brasil 2020. Jardim Botânico do Rio de Janeiro. Available from:. http://floradobrasil.jbrj.gov.br/reflora/floradobrasil/FB5013. Accessed March 14, 2021.

Medeiros MFT, Albuquerque UP (2014) Food flora in 17th century northeast region of Brazil in Historia Naturalis Brasiliae. J Ethnobio Ethnomed 10:50

Minnis PE (1991) Famine foods of the Northern American Desert Borderlands in historical context. J Ethnobiol 11(2):231–257

Müller JG, Almedom A (2008) What is "Famine Food"? Distinguishing between traditional vegetables and special foods for times of hunger/scarcity (Boumba, Niger). H Ecol 36(4):599–607. https://doi.org/10.1007/s10745-008-9179-0

Nascimento VT, Vasconcelos MAS, Maciel MIS et al (2012) Famine foods of Brazil's seasonal dry forests: ethnobotanical and nutritional aspects. Ec Bot 66(1):22–34. https://doi.org/10.1007/s12231-012-9187-2

Nascimento VT, Lucena RFP, Maciel MIS et al (2013) Knowledge and use of wild food plants in areas of dry seasonal forests in Brazil. E Food Nut 52:1–26. https://doi.org/10.1080/03670244.2012.707434

Nimer E (1972) Climatologia da região Nordeste do Brasil: introdução à climatologia dinâmica. R Bras de Geo 34:3–51

Ogle BM, Grivetti LE (1985) Legacy of the chameleon. Edible wild plants in the Kingdom of Swaziland, southern Africa: a cultural, ecological and nutritional study. Eco Food Nut, in four parts 16(3):193-208, Parts II-IV: 17(11): 1-64. https://doi.org/10.1080/03670244.1985.9990881

Okigbo BN (1975) Neglected plants of horticultural and nutritional importance in traditional farming systems of tropical West Africa. Acta Hortic 53:131–150. https://doi.org/10.17660/ActaHortic.1977.53.18

Prado, DE (2003) As Caatingas da América do Sul. *In*: Ecologia e conservação da caatinga / editores Inara R. Leal, Marcelo Tabarelli, José Maria Cardoso da Silva.; prefácio de Marcos Luiz Barroso Barros. – Recife : Ed. Universitária da UFPE 822 p

Pennington T (2015) Geoffroea in Lista de Espécies da Flora do Brasil. Jardim Botânico do Rio de Janeiro. http://floradobrasil.jbrj.gov.br/jabot/floradobrasil/FB22792. Accessed June 20, 2020

Piso W, Marcgrave G (1648) In Historia Naturalis Brasiliae: in qua non tantum plantae et animalia, sed et indigenarum morbi, ingenia et mores describuntur et iconibus supra quingentas illustrantur. Edited by de Laet J. Amsterdam: Elsevier; 1648. http://biblio.etnolinguistica.org/marcgrave-1648-historia. Accessed June 22, 2020

Pickel BJ (2008) Flora do Nordeste do Brasil segundo Piso e Marcgrave : no século XVII / Argus Vasconcelos de Almeida (Editor). – Recife: EDUFRPE, 312 p

Santos MB, Cardoso RL, Fonseca AAO et al (2011) Caracterização e qualidade de frutos de umbu-cajá (*Spondias tuberosa* X.S. mombin) provenientes do Recôncavo Sul da Bahia. Rev Bras Frutic 32(4):1089–1097

Stintzing FC, Carle R (2005) Cactus stems (*Opuntia* spp.): a review on their chemistry, technology, and uses. M Nut Food Res 49(2):175–194. https://doi.org/10.1002/mnfr.200400071

Silva AP, Souza CCE, Ribeiro JES et al (2015) Características Físicas, Químicas e Bromatológicas de Palma Gigante (*Opuntia ficus-indica)* e miúda (*Nopalea cochenillifera*) Oriundas do Estado da Paraíba. Revista Brasileira de Tecnologia Agroindustrial 9(2)

Tomchinsky B, Ming LC (2019) As plantas comestíveis no Brasil dos séculos XVI e XVII segundo relatos de época. Rodriguésia doi:DOI:dx.doi.org/10.1590/2175-7860201970040

Universidade Estadual de Campinas. Tabela brasileira de composição de alimentos – TACO. UNICAMP/NEPA, 4ª ed., 2011. Campinas

Wachiira KK (1987) Women's use of off-farm and boundary lands: agroforestry potentials. Final report. ICRAF, Nairobi, p 139

Zappi D, Taylor N, Santos MR et al (2015). Cactaceae in Lista de Espécies da Flora do Brasil. Jardim Botânico do Rio de Janeiro http://floradobrasil.jbrj.gov.br/jabot/floradobrasil/FB1663. Accessed June 02, 2020

Part IV
Food Composition Data on Brazilian Edible Plants by Biome

Food Composition Data: Edible Plants in Cerrado

Elisa Flávia Luiz Cardoso Bailão, Matheus Gabriel de Oliveira, Luciane Madureira de Almeida, Vanessa Cristiane Santana Amaral, Lee Chen Chen, Samantha Salomão Caramori, Joelma Abadia Marciano de Paula, Wilson de Melo Cruvinel, and Leonardo Luiz Borges

Contents

E. F. L. C. Bailão · L. M. de Almeida · V. C. S. Amaral · S. S. Caramori · J. A. M. de Paula
Campus Central, Universidade Estadual de Goiás, Anápolis, GO, Brazil

M. G. de Oliveira
Faculdade de Farmácia, Universidade Federal de Goiás, Goiânia, GO, Brazil

L. C. Chen
Instituto de Ciências Biológicas I, Universidade Federal de Goiás, Goiânia, GO, Brazil

W. M. Cruvinel
Escola de Ciências Médicas, Farmacêuticas e Biomédicas, Pontifícia Universidade Católica de Goiás, Goiânia, GO, Brazil

L. L. Borges (✉)
Campus Central, Universidade Estadual de Goiás, Anápolis, GO, Brazil

Escola de Ciências Médicas, Farmacêuticas e Biomédicas, Pontifícia Universidade Católica de Goiás, Goiânia, GO, Brazil

M. C. M. Jacob, U. P. Albuquerque (eds.), *Local Food Plants of Brazil*, Ethnobiology, https://doi.org/10.1007/978-3-030-69139-4_10

1 Introduction

The use of plants is an ancient tradition in many places around world and in many cultures plays an important role in health, beauty, and nutrition. However, placing a new natural plant-derived product in the market is challenging (Box 1). Due to the difficulties, interest in natural products drug discovery declined in the 1990s. Yet despite this, a rapid increase in the number of scientific studies targeting plant-derived natural products has occurred in the twenty-first century. This reflects the fact that synthetic compound libraries are limited and represent only a small part of the great diversity of chemical compounds (Atanasov et al. 2015). Further, a

Box 1 Principal challenges for a new plant-derived natural product to reach the market

Botanical identification requires a specialist, which is increasingly rare nowadays.
The availability of natural products is low, and pharmacological testing consumes a lot of material.
Re-collecting wild species is difficult, since plant biomes can rapidly disappear under anthropic pressure.
A plant's chemical composition varies according to environmental factors, such as temperature, water availability, UV radiation, soil nutrients, altitude, and atmospheric composition, which limits the time window for recollection.
In plant-derived natural products, compound transformation and degradation can occur during extraction and isolation processes, making standardization of procedures difficult.
Due to the complexity of molecular mixtures, fewer chemical products or compounds are isolated.
The lack of knowledge concerning agricultural cultivation of certain wild species makes wildcrafting the only alternative, which makes the collection process very time-consuming and can result in ecological problems and the risk of species extinction.
Very stringent protective regulations in some countries make bio-prospecting of new plant-derived natural products difficult.
Development of methods for total synthesis or derivatization to obtain molecules similar to natural products is complex, since the chemical structures of natural products often contain numerous oxygen substituents and chiral centers.
Incompatibility of natural products with established high-throughput screening (HTS) platforms; test compounds should not decompose or precipitate and should not interfere with assay reagents nor show nonspecific effects; natural products often fail in fulfilling these requirements.
Determining the precise molecular mechanism of action of natural products is challenging.
Difficulties during clinical trials often occur since the therapeutic activity of plant extracts is usually based in the synergistic and simultaneous action of many chemicals.
The interest of pharmaceutical companies in natural products (that are not synthetically modified) is limited due to controversies concerning their patentability.

Sources: Corson and Crews (2007), Gobbo-Neto and Lopes (2007), Jones and Kinghorn (2012), Bucar et al. (2013), Atanasov et al. (2015), Li and Weng (2017), and Thomford et al. (2018)

negative correlation between the in vitro potential of many molecules and their ADME/T (absorption, distribution, metabolism, excretion/toxicity) profile is generally noted (Gleeson et al. 2011). Synthetic compounds present a significantly lower number of chiral centers and are both smaller and present greater flexibility, which results in weaker, less specific activity (Feher and Schmidt 2003). In contrast, plant-derived natural products present a diversity of bioactive compounds with differing chemical scaffolds (Atanasov et al. 2015). Plant-derived products possess properties, such as binding affinities for specific proteins, that are evolutionary optimized (Appendino et al. 2010; Hunter 2008). Such characteristics make plant-derived products more advantageous for conducting ADME/T tests (Atanasov et al. 2015). Another advantage in drug discovery using medicinal plants is the known good correlation between ethnopharmacological information (well documented) and the medical use of the plant. Around 80% of identified compounds present some ethnomedicinal use, whether identical or related to current usage of the active plant elements (Fabricant and Farnsworth 2001). Further, technological advances in systems biology, bioinformatics, plant specialized metabolism, and synthetic biology present opportunities through which the medicinal properties of plants are discovered, utilized, and expanded toward developing new herb-inspired medicines (Li and Weng 2017). Accordingly, plant-derived natural products still represent a valuable source for drug discovery, this, motivated by global public health challenges, such as COVID-19, HIV/AIDS, malaria, hypertension, diabetes, and cancer (Thomford et al. 2018). About a quarter of all Food and Drug Administration (FDA)- and/or European Medical Agency (EMA)-approved drugs are plant-derived (Patridge et al. 2016; Thomford et al. 2018).

In this context, Brazil possesses various biomes with a great portion of the world's biodiversity and approximately 20% of the plants and microorganisms on the planet (Calixto 2000). Among these biomes, the Cerrado is a neotropical savanna with rich flora (Klink and Machado 2005). Over 13,127 highly endemic plant species have been recorded within this biome (Myers et al. 2000; Overbeck et al. 2015; Silva and Bates 2002; Silveira et al. 2016). Yet, in regard to environmental conservation, only 41% of the original Cerrado still remains as native vegetation (Soares-Filho et al. 2014). This reduction is due to anthropogenic activities, such as monoculture, pasture expansions, mineral extraction, and the growth of urban areas (Faleiro et al. 2013). Considering only agricultural activity between 2002 and 2009, an estimated loss of 92,712 km^2 of natural Cerrado area has occurred (Overbeck et al. 2015). Preservation policies in Brazil need improvement, yet only 7% of the Cerrado is under legal protection (Soares-Filho et al. 2014).

Both the bio-prospecting potential of the Cerrado for medicinal species and knowledge concerning traditional medicines are disappearing faster than plant biodiversity (Appendino et al. 2010). Studies concerning plant-derived natural products found in the Cerrado could help valorization of the biome and call attention to the urgent need for conservation. Recent studies have highlighted the pharmacological importance of little explored Cerrado plants (Bailão et al. 2015; Dutra et al. 2016; Neri-Numa et al. 2018b; Schiassi et al. 2018). Unfortunately, a discrepancy between the number of scientific publications on plants by Brazilian scientists and the development of new phytotherapeutic agents exists and underlines the need of public policies focused on innovation (Dutra et al. 2016).

The native fruits of the Cerrado are increasingly valued in the world market as functional and nutraceutical foods because of their high nutritional value and attractive sensory characteristics (Arruda et al. 2016; Finco et al. 2012). Agriculturally, many Cerrado species grow spontaneously with amazing adaptability, surviving in soils with little natural fertility and which lack chemical additives (Finco et al. 2012; Haridasan 2008). Further, Cerrado fruits can be employed in multiple food applications (Araújo et al. 2018), and their residues from industrial processing are rich in bioactive compounds that possess many functional properties.

In summary, in view of their economic importance and relatively slight recognition, the aim of this chapter is to review the principal vegetal species of the Cerrado, focusing on their pharmaceutical, cosmetic, and foods potential.

2 Technological Methods to Obtain Products

It is important to highlight that quality control and standardization of all processing stages (Fig. 1) is fundamental and guarantees the reproducibility and safety of the plant-derived product. Initially, plant samples are collected, dried, and processed by comminution or pulverization (by mechanical forces) to achieve fragmentation of the plant material into small particles. Solvent selection for extracts and determination of particle size are key steps in the process of obtaining herbal products. Decoctions, infusions, macerates, and tinctures are used to obtain plant extracts. In general, the liquid extract obtained must be concentrated and standardized to the established control parameters, usually found in different pharmacopoeias (List and Schmidt 1990).

Extract concentration is a key step to obtaining intermediate products and preparing liquid material for the drying process. When the liquid presents alcohol, one approach to concentrate the extract consists in the use of a rotary evaporator system, equipment able to remove the most volatile solvents under reduced pressure. To optimize vaporization, water bath temperature, cylinder immersion depth, and degree of vacuum are the adjustable parameters. The method is widely employed in phyto-pharmaceutical processes (List and Schmidt 1990). Other techniques to concentrate extracts include tube vaporization with auto-circulation, tube vaporization with pumped circulation, falling films vaporization, thin layer vaporization, and centrifugal rotary vaporization. Once the concentrated extract is obtained, various quality parameters are important at this stage for standardization such as pH, solid content, density, chemical markers content, and viscosity (List and Schmidt 1990).

Applied to fruits and vegetables that require chemical and physical stability, drying is the most common preservation method. When water is removed, deterioration due to microbial activity is retarded, yet without causing significant physical, nutritional, or sensory loss. Further, for phyto-medicines, dry plant extracts allow preparation of solid tablets and capsules (Kho and Hadinoto 2011). To obtain plant-derived products, there are many drying methods. Among them, spray-drying

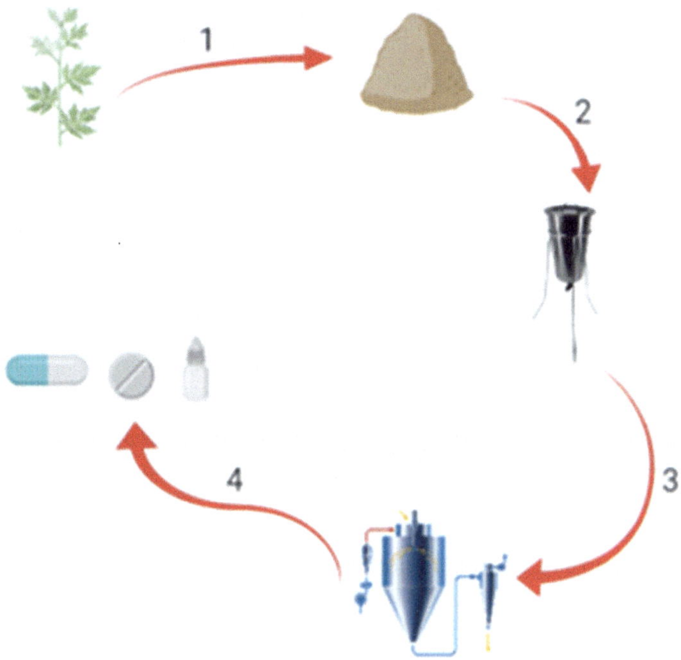

Fig. 1 Principal steps in plant processing. First, the plant is dried in an oven and pulverized (1). Upon obtaining the powder material, the extraction process is performed (2); once the liquid extract is obtained, the material is normally concentrated and submitted to quality control parameters before drying processes (3). When the dried powder is achieved, the material can be employed in final products for the pharmaceutical and food industries (4)

is a robust manufacturing process, often used in the food and pharmaceutical industries (Wang 2015). Spray-drying has various advantages, such as reproducibility, high capacity, fast drying capability, and short exposure time to high temperature processing (Al-Khattawi et al. 2018). Spray-dryer equipment rapidly and efficiently transforms liquids into powders (Schuck et al. 2009). The equipment also can be used to microencapsulate natural products. Microcapsule wall systems consist of polymers with chemical groups presenting hydrophilic or hydrophobic properties, which provide the emulsion characteristics of the system. The polymers must be chemically compatible and nonreactive with the encapsulated component. The approach is used for products with increased stability and flexibility (Maresca et al. 2016; Silva et al. 2020b).

Another drying technique that can be used for coarse particles is spouted bed technology. This technique provides a high solid circulation rate and increases contact between the drying agent and solid particles (Gao et al. 2017; Zhang et al. 2017). Other drying processes that can be used to obtain herbal products are centrifugal atomization, vacuum belt dryers, roller dryers, drying ovens, freeze dryers, and microwave dryers. In summary, dried materials are used to provide finalized products with applications in food, medicine, and cosmetics.

3 Species and Technological Products from the Cerrado

Numerous studies have demonstrated the technological potential of extracts from Cerrado plant species. In this section, we provide a brief review of the best-known and most-used fructiferous species of the Cerrado. Biological properties, intermediate products, characteristics, and potential technological applications for the species presented below are compiled in Table 1. Further, nutritional compositions of the Cerrado plant species highlighted in this chapter are compiled in Table 2.

4 *Anacardium occidentale* L. (*Anacardiaceae)*

Anacardium occidentale is a tree popularly known as the cashew tree or *cajueiro* (Fig. 2a). The pseudo-fruit, the large pulpy and juicy part commonly referred to as the "cashew apple," has a fine sweet flavor (Silva et al. 2007). In folk medicine, the cashew leaf and bark teas are used to treat diabetes, muscular debility, urinary disorders, and dyspepsia (Andrade Júnior et al. 2018).

In food applications, *A. occidentale* and *Psidium guajava* (guava) dried bagasse have been used to enrich cookies (Matias et al. 2005). Cashew is a rich source of vitamins, minerals, and other essential nutrients (Omorodion et al. 2017). Cookies respectively enriched with 10% and 5% cashew and guava fibers received a high rating for flavor acceptability (Matias et al. 2005).

Beverages based on *A. occidentale* aqueous extract from the fruits present an adequate concentration of viable probiotics and are lactose-free, which makes them both a functional food and good substitute for those who cannot consume milk (Firmo et al. 2020; Vergara et al. 2010). The *A. occidentale* juice is encapsulated using a spray-drying technique, thus stabilizing it for at least 14 days and protecting sensitive compounds present in the food matrix, like vitamin C (Bastos et al. 2012). By fermentation, a wine has been developed using *A. occidentale* juice and in sensory evaluations was rated as very acceptable as an alcoholic beverage. Significant differences (probably because of high tannin content in the cashew wine) in taste, aroma, flavor, and aftertaste exist between the *A. occidentale* wine and commercially sold grape wines (Mohanty et al. 2006).

In pharmaceutical applications, natural cashew gum has been used in conjunction with phthalocyanines to develop a layer-by-layer film for an electrochemical sensor to detect dopamine, with an acceptable detection limit for the pharmaceutical industry (Araújo et al. 2012). Cashew gum has also been used as a pharmaceutical excipient (a natural muco-adhesive polymer) in dental pastes containing aceclofenac (AC) for pain management in periodontitis (Hasnain et al. 2018); in bucal-adhesive curcumin tablets with the potential to improve the bioavailability of the active ingredient (Gowthamarajan et al. 2012); and in an Orabase formulation for periodontitis treatment (Souza Filho et al. 2018). The dental paste demonstrated sustained AC release over 6 h with good oral mucosa adhesion, resulting in effectual management

Table 1 Products obtained from Cerrado plant species highlighted in this chapter

Intermediate product	Part of the plant used	Yield	Technological methods	Favorable characteristics	Unfavorable characteristics	Possible applications	References
Anacardium occidentale L.							
Cookies	Fruit	Not specified	Drying in fixed bed convective dryer. Blend of cookie ingredients with 5%, 10%, and 15% dry residues of *A. occidentale*	Cookies with 10% dehydrated bagasse presented a high rate of acceptability	Cookies with 15% of the dehydrated bagasse were not accepted	Food application	Matias et al. (2005)
Drinks	Fruit	Not specified	Fermentation process	The symbiotic drink presented good physicochemical and microbiological characteristics	The acidity value and content of beneficial microorganisms in the probiotic drink were lower than recommended	Beverage application	Firmo et al. (2020)
Juice	Fruit	Not specified	Fermentation process	Contained prebiotic oligosaccharides	The yeast extract added as an external nitrogen source presented an unpleasant taste for human consumption	Beverage application	Vergara et al. (2010)
Juice	Fruit	40.7–55.3%	Microencapsulation	Physicochemical stability; high nutritional value	Not specified	Ingredient in the formulation of many food products	Bastos et al. (2012)
Wine	Fruit	Not specified	Fermentation process	Good nutritional and sensory properties and consumer acceptance	Lower acceptance than commercial grape wine	Beverage application	Mohanty et al. (2006)
Electrochemical sensor	Gum (tree exudate)	Not specified	Multilayer deposition	Stable smooth films with well-defined redox processes, punctuated by large globular features and exhibiting low roughness values	Not specified	Nano-biomedical devices	Araújo et al. (2012)

(continued)

Table 1 (continued)

Intermediate product	Part of the plant used	Yield	Technological methods	Favorable characteristics	Unfavorable characteristics	Possible applications	References
Dental paste	Gum (tree exudate)	19.22%	Conventional triturating method	Good tube extrusion and easily spreadable; sustained releasing of aceclofenac (the active ingredient) over 6 h; good adhesion to the oral mucosa	Not specified	Muco-adhesive dental pastes	Hasnain et al. (2018)
Buccal adhesive tablet	Gum (tree exudate)	70%	Development of a natural buccal adhesive polymer cashew nut tree gum (20% polymer concentration, 0.1% penetration enhancer, 40 mg backing layer, compressed at 2 tons/cm^2 for 10 s)	Unidirectional release; the rate of release of the drug substance (curcumin) as well as the bioadhesive bond strength of the formulation can be modulated; easy to formulate; inexpensive; provide easy application; with convenient removal from the mucosal surface	Not specified	Sustained release and muco-adhesive, tablet excipient for a variety of low-molecular-weight drug substances	Gowthamarajan et al. (2012)
Orabase formulation	Gum (tree exudate)	80%	Development of an Orabase gel with cashew gum polysaccharide	High yield; protein-free product; reduced alveolar bone loss and inflammation; nontoxic	Not specified	Potential adjuvant for periodontitis treatment and source of novel biotechnological discoveries	Souza Filho et al. (2018)
Gold nanoparticles	Leaves	Not specified	Green synthesized gold nanoparticle	Size between 10 and 60 nm and spherical in shape; antibacterial effect; permissible levels of cytotoxicity toward normal cells; high cytotoxicity toward MCF-7 cancer cells	Not specified	Antibacterial applications, such as food packing and wound dressing, and as an anthelmintic	Sunderam et al. (2019)

Silver nanoparticles	Testa (cashew nut husk)	Not specified	A simple economic and eco-friendly synthesis of silver nanoparticles	Average size of 25 nm and distorted spherical shape; high stability; good catalytic activity toward the reductive-degradation of azo dyes	Not specified	Catalyst for the reductive-degradation of carcinogenic azo dyes used in many industries such as food and pharmaceutics	Edison et al. (2016)
Annona crassiflora Mart.							
Dehydrated pulp	Fruit	~28.94%	Drying in forced air circulation	Good sensory properties and consumer acceptance	Not specified	Food application	Martineli et al. (2020)
Dehydrated pulp and carpel	Fruit and flower	Not specified	Freeze-drying and convective hot-air drying processes	Sources of alimentary fiber and derivatives from oleic and palmitic acids	Significant difference between fresh and dehydrated *marolo*, mainly in proteins	Food application	Corrêa et al. (2011)
Dehydrated pulp	Fruit	Not specified	Thin layer convective air drying using inulin as a drying aid agent	Lower moisture contents; satisfactory wettability results; lower hygroscopicity; better stability; prebiotic and sensory properties	Treatments without inulin produced powders with better solubility.	Food application	Botrel et al. (2016)
Bread	Fruit	Not specified	Drying in forced air circulation. Crushing. Blend of flour with bread ingredients	Good nutritional, sensory properties and consumer acceptance	Not specified	Food application	Villela et al. (2013)
Food bars	Fruit	Not specified	Drying in forced air circulation. Milling. Blend of flour with bars ingredients	Increase in antioxidant activity, vitamin C, and total carotenoids content. Good sensory properties and consumer acceptance	Not specified	Functional food	Silva et al. (2018b)
Liquor	Fruit	50%	Alcoholic infusion. Addition of syrup	Good sensory properties	Not specified	Food and beverage application	Oliveira et al. (2018a)

(continued)

Table 1 (continued)

Intermediate product	Part of the plant used	Yield	Technological methods	Favorable characteristics	Unfavorable characteristics	Possible applications	References
Caryocar brasiliense Camb.							
Cookies	Fruit	Not specified	Blend of wheat flour with 10% and 20% *C. brasiliense* flour	Good nutritional value	Cookies supplemented with flour (20%) presented higher than average moisture values	Food application	Silva et al. (2018a)
Pequi by-product flours	Peel	43.32 ± 0.72 g total dietary fiber/100 g pequi peel flour	Not applicable	Production of flours rich in antioxidant molecules and dietary fibers; total dietary fiber contents are in the same range of commonly employed fiber sources; total amount of extractable phenolics are significantly high in comparison to pequi pulp and almond and other fruit peels; high antioxidant capacity compared to extracts obtained from other tropical fruits	Not specified	Functional food	Leão et al. (2017)
Pectin	Peel	20.79 g/100 g	Microwave heating	Obtained pectins with high degree of esterification; shorter extraction times	Not specified	Functional food ingredient with use as a thickener, emulsifier, gelling agent, and stabilizer	Leão et al. (2018)
Prebiotic yogurt	Fruit	Not specified	Lyophilization	Lower pH, less syneresis, and higher water-holding capacity (characteristics favorable to conservation during storage); good sensory acceptance	Residual bitter taste	Functional prebiotic food	Souza (2015)

Pulp oil	Fruit	53.65 g oil/ 100 g pequi pulp	Supercritical fluid extraction	Attractive cost of manufacturing in the pilot and industrial scales; acceptable payback time (<1 year)	Not specified	Functional food	Johner et al. (2018b)
Pulp oil	Fruit	55 g oil/100 g pequi pulp	Supercritical fluid extraction assisted by pressing	Faster extraction rate and lower amounts of solvent consumption	Not specified	Functional food	Johner et al. (2018a)
Pulp oil random inter-esterified	Fruit	Not specified	Chemical inter-esterification	Increased thermal stability despite containing high amounts of oleic acid; total carotenoid content remained stable; increased range of applications; increased value along the entire pequi oil value chain	Not specified	Functional food fat,	Guedes et al. (2017)
Freeze-dried fruit pulp	Fruit	30 g/100 ml of extract	Freeze-drying after pretreatment with sucrose and ethanol	Crystalline structures with lower bulk porosity; particles showing uniformity and homogeneity, with a better distribution; good stability, with lower water sorption during storage	Low inter-particle interaction; low total carotenoid content	Food product	Alves et al. (2010)
Concentrated alcoholic and aqueous extracts	Fruit	Not specified	Nano-filtration	Concentration of polyphenols and carotenoids in aqueous extract	Nano-filtration of the alcoholic extract was not acceptable for concentrating carotencids and polyphenols	Functional food	Machado et al. (2013)
Pequi aqueous extract membrane-filtered	Fruit	Not specified	Ultrafiltration and nano-filtration	Concentration of polyphenols and carotenoids; removal of the sol-vent from the extract; process not submitted to high temperatures (energy saving); no change in the physical state of the solvent	Not specified	Products with functional properties	Machado et al. (2015)

(continued)

Table 1 (continued)

Intermediate product	Part of the plant used	Yield	Technological methods	Favorable characteristics	Unfavorable characteristics	Possible applications	References
Vacuum-dried fruit pulp	Fruit	71.78 ± 6.98 mg carotenoid/g pequi pulp	Vacuum drying at low temperatures (~40 °C)	Ascorbic acid and carotenoid contents stability; higher rehydration coefficient; minimal color change; less shrinkage; and reduced drying time	Osmotic dehydration substantially decreased moisture content, drying process time, and volumetric ratio of the dried product; yet also promoted leaching of bioactive constituents	Functional food	Mendonça et al. (2017)
Encapsulated carotenoid extract	Fruit	9.61 μg carotenoids/g pequi pulp	Encapsulation by emulsification (O/W) and foam-mat drying	Raise the bioavailability of carotenoid compounds; encapsulation alternative free of organic solvents, simple and inexpensive	Not specified	Functional food or natural dye	Pinto et al. (2018)
Biosensor for thiodicarb determination	Fruit	Not specified	Pequi polyphenol oxidase immobilized on chitosan cross-linked with cyanuric chloride	Construction simplicity; linear calibration range; low detection limit; good repeatability and reproducibility; long-term stability with the sensitivity of the inhibition-based polyphenol oxidase; the results were satisfactory when compared with those obtained using high-performance liquid chromatography	Not specified	Determination of thiodicarb in fresh fruits and vegetables	Lima et al. (2010a)
Methylic and ethylic biodiesels	Fruit	Methylic, 96%; ethylic, 80%	Alkaline trans-esterification via the methylic and ethylic routes	Thermally stable; in general, biodiesels emit lower levels of CO_2 and particulate material and originates from renewable source; it is easy to adapt biodiesel to diesel engines	Not specified	Substituent for conventional mineral diesel	Silva et al. (2014)

Chitosan coating	Fruit	Not specified	Orbital shaker extraction and dip coating application	The combination of *C. brasiliense* extract with chitosan was suitable as an effective edible coating, improving fresh weight retention and reducing color change in tomatoes together with retaining the benefits of chitosan in delaying fungal growth	Not specified	Preserving tomato quality	Breda et al. (2017)
Supercritical CO_2 extract	Leaves	Not specified	Supercritical CO_2 extraction	No cytotoxic and phototoxic hazards	Not specified	Development of cosmetic and/or pharmaceutical products	Amaral et al. (2014b)
Liquid soap and hand lotion	Leaves	Not specified	Supercritical CO_2 extraction	Antimicrobial and antioxidant activities	Not specified	Development of personal care products, especially antiseptic and anti-aging products	Amaral et al. (2014a)
Dipteryx alata Vog.							
Granola	Almonds	Not applicable	Not applicable	High levels of protein content, dietary fiber, and iron; lower moisture, water activity, sodium, lipids, and energy value; high stability, crunch texture, and consumer acceptability	Not specified	Health-promoting food preparations	Souza and Silva (2015)
Cupcake	Almond flour	Not specified	Blend of wheat flour with 30% *D. alata* almond flour	Good nutritional and sensory properties; formulation of a "light" product, with reduction of margarine and trans fatty acids; good consumer acceptance	Reduction in water absorption, stability, mixing tolerance index, and extensibility, increase in dough development time and resistance to extension	Reduced fat product. *D. alata* flour may be used in other food formulations	Paglarini et al. (2018)

(continued)

Table 1 (continued)

Intermediate product	Part of the plant used	Yield	Technological methods	Favorable characteristics	Unfavorable characteristics	Possible applications	References
Almond oil	Almonds	~22%	Supercritical CO_2 and/or ultrasound	High concentration of oleic and linoleic acids; same fatty acid profile when compared with oil extraction using organic solvents	Not specified	Fabrication of food and pharmaceutical products (cosmetics)	Santos et al. (2016)
Almond oil	Almonds	36.87% 32.62%	Compressed propane Supercritical CO_2 with ethanol as co-solvent	High concentration of oleic and linoleic acids; high levels of tocopherol; good antioxidant and antimicrobial activities; same fatty acid profile when compared with oil extraction using organic solvents; residues of the extraction presented high content of proteins (~32%)	Not specified	Food and pharmaceuticals applications	Fetzer et al. (2018)
Almond protein isolate	Proteinaceous cake remaining after almond oil extraction	884 ± 10 g/kg	Sequential extraction of almond defatted flour	High protein content; well-conserved protein arrangement; high in vitro digestibility; water and oil absorption capacity; emulsifying activity; foam formation; stability at mild and neutral pH	Loss of albumins and low-molecular-weight globulins during isolation process; low thermostability	Substitute ingredient in oily food formulations or in development of new products	Nunes et al. (2017)
Cookies	Partially defatted almond flour	53.7 g of cake/100 g of almonds	Continuous screw press; 25% substitution of wheat flour by *D. alata* almond flour	The flour presents high contents of proteins, calcium, and fibers; few content of carbohydrates; is rich in iron, zinc, and cooper; good content of total phenolics, total flavonoids, and condensed tannins; and present antioxidant activity	The flour presents high lipid content. The cookies were hard and brittle when 75% or more of the wheat flour was substituted by *D. alata* almond flour. The consumer acceptance of the cookies was not good even with only 25% *D. alata* almond flour wheat flour substitution	Positive enhancement of nutritional and antioxidant characteristics of cookies. *D. alata* almond flour may be used in other food formulations	Pineli et al. (2015)

Lamellar gel phase emulsion	Oil extracted from almonds	Not specified	10:10:80 *D. alata* oil/surfactants/purified water. Surfactants used: ceteareth-5 and steareth-2	Stable formulation with anisotropic structures (liquid-crystals); pH ~6.0; particle size ~12μm; structural similarity to *Stratum corneum* (SC) membranes; increase SC lipid fluidity	Not specified	Delivery system for drugs and cosmetic actives	Moraes et al. (2018)
Eugenia dysenterica DC.							
Wine	Fruit	Not specified	Batch fermentation	The acceptability of wines produced by immobilized cells was greater than 70% for color, flavor, and taste	Not specified	Beverage application	Oliveira et al. (2011)
Jelly	Fruit	Not specified	Blend of two types of pulp (filtered or unfiltered) with two amounts of pulp/sucrose/pectin (50:50:0.2 and 60:40:0.1)	All jelly formulations showed microbiologically safe and good acceptability. The formulation with pulp filtered/sucrose/pectin ratio 60:40:0.1 showed good nutritional value and stood out as a source of vitamin C	After 120 days of storage, the selected formulation presented significant reductions in all chemical parameters	Food application	Santos et al. (2012)
Raisins	Fruit	Not specified	Osmotic dehydration	Good nutritional and sensory properties and consumer acceptance	Not specified	Food application	Silva et al. (2015a)
Spray-dried extract	Leaves	34.64–63.92% (w/w)	Spray-drying	Low moisture contents and water activities; good recuperation ratios for total polyphenols, tannins, and flavonoids; dry product flow and compressibility varied over a range acceptable for pharmaceutical purposes	Not specified	Intermediate phyto-pharmaceutical products	Couto et al. (2013)

(continued)

Table 1 (continued)

Intermediate product	Part of the plant used	Yield	Technological methods	Favorable characteristics	Unfavorable characteristics	Possible applications	References
Spray-dried extract	Fruit	37–81.20%	Spray-drying	High process yield; high stability; low moisture content and water activity; high glass transition temperatures; low hygroscopicity; high solubility in water; an alternative to reduce transport costs, prevent deterioration or loss of phenolic compounds, and ensure the quality and availability of fruit	Processing temperatures and concentration of carrier agents influenced the physical properties of the powders	Functional food	Daza et al. (2016)
Encapsulated extract	Fruit	Not specified	Spray-drying	Powders were obtained with high total phenolic compounds retention and proanthocyanidins, antioxidant activity, inhibition of α-amylase and α-glucosidase, and antimicrobial activity against *S. aureus* and *L. monocytogenes*	Lower total phenolics and proanthocyanidins content than powder obtained by freeze-drying	Functional food	Daza et al. (2017)
Micro-emulsion	Leaves	Not specified	O/W and W/O micro-emulsions	High stability; good penetration into skin; antioxidant activity preservation; low irritability potential	Not specified	Dermatological treatments that require antioxidant action	Ferreira-Nunes et al. (2018)
Emulsion with chitosan microparticles	Leaves	31%	*E. dysenterica* aqueous extract captured in chitosan microparticles with further incorporation in a dermatologically acceptable formulation	Relative stability for 60 days at 6 °C; high skin penetration; angiogenic action	Not specified	Dermatological treatments	Silva et al. (2020a)

Hancornia speciosa Gomes							
Nectar	Fruit	Not specified	Formulations with different concentrations of pulps and sugar	Good sensory properties and consumer acceptance	Not specified	Food application	Assumpção et al. (2013)
Yogurt	Fruit	Not specified	Blend of yogurt ingredients with pulp and jam (0%, 5%, 6%, 7%, and 8%)	Good nutritional and sensory properties and consumer acceptance	Not specified	Functional food	Santos et al. (2017)
Jelly	Fruit	Not specified	Blend of pulp with jelly ingredients	High dietary fiber, total phenolic content, and antioxidant capacity; remained microbiologically stable	Not specified	Functional food	Souza et al. (2018a)
Bio-oil	Seeds	Not specified	Pyrolysis of mangaba seed to obtain bio-oil	Use of the mangaba industrial residues which contribute to environmental conservation	Not specified	Technological application	Santos et al. (2015)
Bio-membrane	Latex and silver nanoparticles	0.1–0.4% of AgNP	Dried at 40 °C	Biomaterial combines angiogenic, anti-inflammatory, and antibacterial properties and might be used to stimulate wound-healing	Not specified	Pharmaceuticals applications	Almeida et al. (2019)
Mauritia flexuosa L.f.							
Nectar	Fruits	Not specified	Crushing and sifting. formulations with different concentrations of pulp and sugar	Low-calorie product, source of vitamins, good sensory properties and consumer acceptance	Not specified	Functional food	Garcia et al. (2015)
Jelly	Fruit	Not specified	Blend of buriti pulp: water (1: 2) with jelly ingredients	Good nutritional and sensory properties and consumer acceptance	The diet pulp jelly requires formulation adjustment for better acceptability by tasters	Food application	Sousa et al. (2020)
Biscuits	Fruit	Not specified	Blend of biscuit ingredients with *M. flexuosa* flour (10%) and oats (4%)	The addition of oats resulted in cookies with smaller diameters and thicknesses and higher fiber content and consumer acceptability	Not specified	Food application	Santos et al. (2011)

(continued)

Table 1 (continued)

Intermediate product	Part of the plant used	Yield	Technological methods	Favorable characteristics	Unfavorable characteristics	Possible applications	References
Dehydrated and sprayed product	Fruits	Not specified	Drying in oven with forced air circulation. Crushing and sifting	The dehydrated and sprayed *M. flexuosa* retained good chemical and microbiological stability for at least 150 days of storage at temperatures of 4 °C and −12 °C	Not specified	Ingredient in formulated foods aimed at supplementation of pro-vitamin A	Aguiar and Souza (2017)
Flour	Fruit	Not specified	Lyophilization	Lowered triglyceride and total cholesterol levels and raised HDL levels in *Swiss* mice	Not specified	Nutraceutical	Pereira-Freire et al. (2019)
Flour	Fruit	2.53–11.84%	Crushing and bleaching. Separation of shell, endocarp, pulp, and seed was selected as a processing option	Peels and defatted pulp flours are highlighted as those with higher antioxidant potential. Carotenoids content was also higher in the peels and flours. All flours showed expressive amounts of total non-extractable proanthocyanidins	Blanching preserved only the extractable polyphenols, not carotenoids or total non-extractable proanthocyanidins	Source of dietary fiber and natural antioxidants in food	Resende et al. (2019)
Myrciaria cauliflora (Mart.) Berg							
Pomace powder	Fruit	Not specified	Freeze-drying, milling, and sieving	The co-product obtained during juice extraction was a source dietary fiber and polyphenols	Not specified	Functional food	Gurak et al. (2014)
Jelly extract	Not specified	~25%	Supercritical fluid extraction	Antioxidant capacity and low cost of manufacturing	Not specified	Food application	Cavalcanti et al. (2011)
Formulation of Bologna-type sausages containing *M. cauliflora* peel extract	Fruit	Not specified	Lyophilization. Maceration with stirring. Blend of the ingredients and extract (0.25%, 0.5%, 0.75%, and 1%)	The addition of extract protected the samples from color changes during storage, reduced the thiobarbituric acid reactive substance values, and prevented the decrease of sensory acceptance during storage	The extract had no positive effect on microbial stability during storage	Improving shelf life oxidative stability	Almeida et al. (2015)

Fermented dessert	Fruits	Not specified	Blend of fermented dairy dessert with hydroethanolic extract and syrup from *M. cauliflora* peel	The formulations were low in fat, presenting as acceptable for overall consumption, with attractive color and appreciable texture. Also presented antioxidant capacity	Not specified	Functional food	Almeida Neta et al. (2018)
Integral breads	Fruits	Not specified	Drying with forced air circulation. Crushing. Blend of bread ingredients with *M. cauliflora* flour (5%, 10%, and 15%)	The addition of flour in the loaves reduced carbohydrate contents, lipids, and caloric value. Fiber content increased by up to three times and phenolics by up to seven times, and consequently, antioxidant activity was higher in breads with *M. cauliflora* peel flour	In general, acceptance levels and acceptability indexes of the loaves lowered with increases in flour supplementation	Functional food	Ferreira (2017)
Juice	Fruit	Not specified	Maceration and lyophilization	Presence of chemical markers in fresh fruit extract which were not detected in commercial juice. The antioxidant capacity was higher in fresh fruit extract	During processing, anthocyanins and other polyphenols decreased significantly	Functional food	Wu et al. (2012)
Psidium guajava L.							
Powder	Fruits	Not specified	Hot-air drying and lyophilization	Vitamin C content and antioxidant activity, with the freeze-dried powder showing the highest values and also higher thermal stability. The powders had good nutritional and sensory properties and consumer acceptance	Powder morphologies of indicated possible separation of phases in the solids as a result of dehydration processes	Food with added value	Osorio et al. (2011)

(continued)

Table 1 (continued)

Intermediate product	Part of the plant used	Yield	Technological methods	Favorable characteristics	Unfavorable characteristics	Possible applications	References
Sheep meat nuggets	Fruit	94.89–95.61%	Incorporation of *P. guajava* powder (0.5% and 1%) in sheep meat nuggets	Total phenolics and total dietary fiber significantly increased and retarded lipid peroxidation of nuggets	Decreased emulsion stability, nuggets cooking yield; ash and moisture content of emulsion increased	Source of antioxidant dietary fiber in meat foods	Verma et al. (2013)
Treated seeds	Seeds	Not specified	Heat treatments (boiling or autoclaving)	The treatments reduced fat content significantly. Autoclaving at 121 °C for 15 min and germination for at least 7 days, have both been shown in lowering the phytic acid level to an extent that the treated seeds could be feasibly employed as a component of food or animal feed	Germination for 14 days caused a significant reduction in nutrient content in the range of 16–79%. Boiling did not reduce the phytic acid content substantially	Food application	Chang et al. (2014)
Guava cheese	Fruits	Not specified	Blend of softened guava and pectin (0.5%, 1%, and 1.5%). Cooking of the pulp	Addition of pectin to the guava cheese increased the hardness and phytochemical content, enhanced the shelf life stability and antioxidant potential	Not specified	Healthy fruit snack	Patel et al. (2016)
Meat	Not specified	Not specified	Substitution of corn with *P. guajava* (0%, 20%, 40%, and 60%) formulations in diets	Addition of 40% of *P. guajava* to the diet of lambs resulted in satisfying meat color, but less intense flavor and tenderness	Not specified	Alternative for the production of sheep meat	Costa et al. (2019)

Pterodon emarginatus Vogel							
Oil	Seeds	Not specified	Cold pressing	The supplementation of *P. emarginatus* oil (0.03%) in laying diets preserved the quality of the egg albums and also helped maintain the pH of yolks	The use of the oils in laying diets did not improve the internal quality of eggs stored under refrigeration	Preservation of egg albumen quality	Oliveira et al. (2018b)
Microencapsulated oil	Fruits	84.89–98.63%	Spray-drying	A 1:3:3.6 blend of essential oil/gum arabic/maltodextrin offered the best protection, with 98.63% of essential oil being retained and the same proportion of trapped b-caryophyllene	Not specified	Food and pharmaceutical applications	Alves et al. (2014)
Brosimum gaudichaudii							
Gingerbread	Seed flour	25–30%	Not specified	Improved sensory and nutritional characteristics	Not specified	Food application	Alves et al. (2017)
Byrsonima crassifolia							
Semidry table wine	Fruit	Not specified	Fermentation process of the pulp fruit was performed using *Saccharomyces cerevisiae*	Good nutritional, antioxidant and sensory properties	Not specified	Food application	Bizinoto (2017)
Campomanesia cambessedeana							
Dried fruit	Fruit	Not specified	Spray-drying	Efficient in increasing pulp shelf life	Changes compared to fresh pulp, such as browning and reduction of the total phenolic compounds content	Food application	Chung (2016)

Table 2 Nutritional composition of Cerrado plant species highlighted in this chapter

Scientific name	Common name	Plant part	Content	Amount (/100 g)
Anacardium occidentale	Cashew	Pulp	Total energy (kcal)	39
			Total energy (kJ)	167
			Humidity (g)	89.8
			Ash (g)	0.20
			Proteins (g)	0.50
			Total fats (g)	0.20
			Total carbohydrates (g)	9.30
Annona crassiflora	Araticum	Raw pulp	Total energy (kcal)	110
			Total energy (kJ)	459
			Humidity (g)	73.9
			Ash (g)	0.74
			Proteins (g)	1.37
			Total fats (g)	2.86
			Total carbohydrates (g)	20.43
Caryocar brasiliense	Pequi	Raw pulp	Total energy (kcal)	277
			Total energy (kJ)	1140
			Humidity (g)	54.6
			Ash (g)	0.80
			Proteins (g)	2.71
			Total fats (g)	25.50
			Total carbohydrates (g)	17.53
Dipteryx alata	Baru	Raw nut	Total energy (kcal)	504
			Total energy (kJ)	2095
			Humidity (g)	5.4
			Ash (g)	3.28
			Proteins (g)	24.21
			Total fats (g)	32.48
			Total carbohydrates (g)	33.72
Eugenia dysenterica	Cagaita	Raw pulp	Total energy (kcal)	36
			Total energy (kJ)	153
			Humidity (g)	90.4
			Ash (g)	0.29
			Proteins (g)	1.36
			Total fats (g)	0.51
			Total carbohydrates (g)	7.64

(continued)

Table 2 (continued)

Scientific name	Common name	Plant part	Content	Amount (/100 g)
Hancornia speciosa	Mangaba	Raw pulp	Total energy (kcal)	60
			Total energy (kJ)	252
			Humidity (g)	84.9
			Ash (g)	0.56
			Proteins (g)	0.88
			Total fats (g)	1.51
			Total carbohydrates (g)	11.56
Mauritia flexuosa	Buriti	Raw pulp	Total energy (kcal)	190
			Total energy (kJ)	786
			Humidity (g)	59.0
			Ash (g)	0.98
			Proteins (g)	2.19
			Total fats (g)	15.42
			Total carbohydrates (g)	25.80
Myrciaria cauliflora	Jabuticaba	Raw whole fruit	Total energy (kcal)	67
			Total energy (kJ)	281
			Humidity (g)	81.2
			Ash (g)	0.53
			Proteins (g)	0.17
			Total fats (g)	0.08
			Total carbohydrates (g)	18.04
Psidium guajava	Guava	Raw whole fruit	Vitamin A ERA (mcg)	68
			Vitamin A RE (mcg)	135
			Retinol (mcg)	0
			β-carotene (mcg)	810
			Lycopene (mcg)	5300

The dataset is available for public consultation on the Brazilian Biodiversity Information System – SiBBr (https://ferramentas.sibbr.gov.br/ficha/bin/view/FN)

of dental inflammation and pain through prolonged local delivery of AC in periodontitis treatment (Hasnain et al. 2018). The bucal-adhesive tablet bypasses first pass metabolism and provides sustained release of curcumin, a phenolic compound whose therapeutic effectiveness is often limited due to its poor absorption from the gastrointestinal tract (Gowthamarajan et al. 2012). The Orabase formulation with cashew gum polysaccharide decreased inflammation and bone loss (hallmarks in experimental periodontitis) (Souza Filho et al. 2018).

Fig. 2 Fructiferous species from the Cerrado highlighted in this chapter. (**a**) *Anacardium occidentale* (cashew); (**b**) *Mauritia flexuosa* (buriti); (**c**) *Hancornia speciosa* (mangaba); (**d**) *Psidium guajava* (goiaba); (**e**) *Caryocar brasiliense* (pequi); (**f**) *Myrciaria cauliflora* (jabuticaba); (**g**) *Pterodon emarginatus* (sucupira); (**h**) *Dipteryx alata* (baru); (**i**) *Annona crassiflora* (araticum); (**j**) *Eugenia dysenterica* (cagaita). All photographs were obtained from the Herbário da Universidade Estadual de Goiás (HUEG) and are available at https://www.gbif.org/pt/dataset/bbb1f181-3221-4a10-ad52-14f1da0dca26

A. occidentale was also used to produce nanoparticles in varied applications. Aqueous cashew leaf extract was used in gold nanoparticle preparation, presenting good antibacterial effect against *Escherichia coli* and *Bacillus subtilis* and demonstrating permissible levels of cytotoxicity toward normal cells and high cytotoxicity toward MCF-7 cancer cells (Sunderam et al. 2019). *A. occidentale* testa (the cashew nut husk) has been used in silver nanoparticles as a catalyst for reductive degradation of carcinogenic azo dyes (Edison et al. 2016).

5 *Annona crassiflora* Mart. (Annonaceae)

The species is popularly known as *araticum* or *marolo* (Fig. 2i). The fruits present unique sensory features such as attractive color, intense flavor, and exotic aroma, as well as a high nutrient content (Arruda et al. 2015). In folk medicine, the leaf infusions are used in oral administrations to treat inflammatory and painful ailments such as wounds, snakebites, diarrheas, malaria, and rheumatism (Formagio et al. 2015; Morzelle et al. 2011; Oliveira et al. 2019). Additionally, the fruits are used as a tonic and as an astringent, and its bark powder has antifungal and antirheumatic properties (Vilar et al. 2011).

The food potential of *A. crassiflora* is mainly associated with the pulps and carpels. The pulp and carpel when dehydrated retain their total carotenoid contents

(Martineli et al. 2020) and were shown to be a good source of alimentary fiber and derivatives such as oleic and palmitic acids (Corrêa et al. 2011). *A. crassiflora* pulp powder when produced by thin layer convective air drying using inulin as a drying aid presents lower moisture content, lower hygroscopicity, and better stability (Botrel et al. 2016).

When breads were enriched with *A. crassiflora* pulp flour, they presented excellent nutritional value, and the formulations were well accepted from a sensory point of view, a good alternative for using the fruit (Villela et al. 2013). Food bars enriched with *A. crassiflora* pulp flour present higher fiber content and a significant increase in antioxidant activity, vitamin C, and total carotenoids. In sensory evaluations, the modified food bar (adding 50% *A. crassiflora* flour) presented higher averages for all evaluated attributes and was the favorite of the evaluators (Silva et al. 2018b).

A. crassiflora pulp liqueur presents a translucent drink, dark yellow to light brown in color, with a strong flavor, and an aroma characteristic of the species. It can be consumed after meals as a digestive and can be used in preparations of drinks, sweets, chocolates, and various desserts (Oliveira et al. 2018a).

6 *Caryocar brasiliense* Camb. (Caryocaraceae)

The *Caryocar* genus presents 16 species, some of which possess economic potential since their fruits are often used as a source of edible oil (Ascari et al. 2013). Among *Caryocar* species, *Caryocar brasiliense* Camb., a native of the Brazilian Cerrado biome, is one of the most studied species. It is called *pequi*, and its fruit is widely consumed in Central Brazil (Fig. 2e). Local communities use *Caryocar* spp. to treat colds, bronchitis, liver disease, skin cancer, and ophthalmological problems, to regulate menstrual flow, and to cure hematomas and bruises as well (Ascari et al. 2013; Lopes et al. 2016). The hepatoprotective, gene-protective, and antioxidant effects of *C. brasiliense* fruits and oil have been scientifically verified (Colombo et al. 2015; Palmeira et al. 2016; Roesler et al. 2008; Torres et al. 2016). Moreover, *pequi* oil intake reduces hepatic triglycerides and improves cardiac function ex vivo in male rats (Oliveira et al. 2017). Toxicological investigations have indicated that oil extracts of *C. brasiliense* pulp present no toxicity, genotoxicity, maternal-embryo-toxic or teratogenic effects in rats (Traesel et al. 2016; Traesel et al. 2017a; Traesel et al. 2017b).

Although *C. brasiliense* has various applications, the principal use of *pequi* is as a food. Its flavor is used to aromatize many foods, especially in Central Brazil (Maia et al. 2008). Pequi is very nutritious and contains fibers, proteins, carbohydrates, and minerals (Ascari et al. 2013). Pequi flours are of industrial interest. Cookies made with a partial substitution of wheat flour with pequi pulp flour have been developed and present good nutritional value, revealing the great potential of pequi flour as an alternative for use in new food products (Silva et al. 2018a). The potential of pequi by-products (endocarp and mesocarp) as substrates to produce pequi flours rich in antioxidant molecules and dietary fibers has also been evaluated. Dietary fibers in *C. brasiliense* residue flour (exocarp and the mesocarp) ranged from 39.8

to 43.3 g/100 g (Leão et al. 2017). The pectin obtained from *C. brasiliense* peel flour is highly esterified (Leão et al. 2018) and can be employed in jams and preserves, as a stabilizer in acidic dairy products, or to promote viscosity in beverages (BeMiller 2019). *C. brasiliense* mesocarp flour has been added to yogurt, decreasing the pH and darkening the color of the final product. The high absorption rate of the flour (1.5%) favors less syneresis and increases water-holding capacity. The formulated prebiotic yogurt was suitable for consumption with viable lactic acid bacteria (Souza 2015). The antioxidant capacities of pequi by-product flours were higher than those of fruits and fruit by-products, such as blackberry and blueberry reported in the literature (Leão et al. 2017). Such results reveal a promising future for pequi peels as a potential source of dietary fiber and natural antioxidants (Leão et al. 2017).

Pequi has been widely used as a source of oil with nutritional and cosmetic value (Ascari et al. 2013), and its oil is high in monounsaturated (especially oleic) fatty acids and carotenoids (Oliveira et al. 2017). Many techniques can be used to extract pequi oil, such as supercritical fluid extraction (Johner et al. 2018b) and cold-press assisted supercritical fluid extraction by (Johner et al. 2018a). An unfavorable aspect of pequi oil in technological applications is that the native oil lacks the melting behavior and thermomechanical properties required for food functionality (Guedes et al. 2017). Thus, pequi oil can be modified through chemical inter-esterification, which increases tripalmitoylglycerol (a completely saturated triacylglycerol) content to ~6%. The inter-esterified oil does not fractionate and is thermally stable to 40 °C, and the total carotenoid content (~390 mg/kg) remains stable (Guedes et al. 2017).

Despite being highly coveted by the food and cosmetic industry (due to exposure to oxygen, light, heat, and the presence of enzymes, metals, and peroxides), carotenoids are highly susceptible to degradation during processing and storage (Boon et al. 2010). To preserve carotenoids, a freeze-dried pequi pulp has been developed (Alves et al. 2010). Membrane separation and vacuum drying to preserve the fruit's bioactive properties have also been tried. Nano-filtration processing has been shown to be efficient in concentrating bioactive compounds from pequi aqueous extracts (Machado et al. 2013, 2015). Low-temperature vacuum drying is recommended for preparing dried sliced pequi because the process retains higher ascorbic acid and carotenoid contents and is associated with a higher rehydration coefficient, minimal color change, less shrinkage, and reduced drying times (Mendonça et al. 2017). Another study proposed carotenoid extract encapsulation of pequi to obtain a powdered product using emulsification (5% emulsifier) with drying (at 60 °C) via the foam mat process (Pinto et al. 2018).

For biotechnological applications, a biosensor based on polyphenol oxidase obtained from *C. brasiliense* pulp was developed and used for thiodicarb pesticide determinations in fresh fruit and vegetable samples. The results were satisfactory (Lima et al. 2010a). *C. brasiliense* pulp has also been used to produce biodiesel. The obtained methylic and ethylic biodiesels by alkaline transesterification of pequi oil presented satisfactory thermal stability and were qualified as potential substituents of conventional mineral diesel (Silva et al. 2014).

The addition of *C. brasiliense* peel extract into a chitosan matrix reduced weight loss by 22% and total color variation in tomatoes by 50% (*Lycopersicon esculentum*

Mill.). Application of the extract alone was effective in controlling fungal growth on tomatoes, a 70% reduction when compared to uncoated fruits. The combination of the extract with chitosan was suitable as an effective edible coating, improving fresh weight retention and reducing color changes in tomatoes while retaining the benefits of chitosan in delaying fungal growth (Breda et al. 2017).

The antimicrobial and antioxidant activities of *C. brasiliense* leaf extract obtained by supercritical CO_2 extraction were also investigated. Supercritical *C. brasiliense* extractions presented antioxidant activity without cytotoxic or phototoxic hazards (Amaral et al. 2014a, b). Liquid soap and hand lotion containing this extract exhibited antimicrobial activity against *E. coli*, *Pseudomonas aeruginosa*, and *Staphylococcus aureus* (Amaral et al. 2014a). The data reveal that *C. brasiliense* leaf extract may be useful in the development of personal care products, especially for antiseptic skin products, as well as products that minimize damage caused by free radicals (Amaral et al. 2014a).

7 *Dipteryx alata* Vogel (Fabaceae)

Dipteryx alata is an oleaginous species native to the Cerrado known as *baru* that possesses an edible almond (Fig. 2h). Popularly, *D. alata* (*baru*) is used as an antirheumatic, tonic, and anti-ophidian and to regulate menstruation (Dey and De 2012; Sano et al. 2016). Baru almonds possess a high protein content and high in vitro digestibility relative to casein (Cruz et al. 2011). They also present a good essential amino acid composition, an appreciable amount of calcium, and high levels of iron, zinc, dietary fiber, and tannins (Fernandes et al. 2010; Martins et al. 2013). Baru almonds are rich in mono and sesquiterpenes, phytosterols, and tocopherol derivatives (Marques et al. 2015). The introduction of baru almonds into rat diets reduced triglycerides and VLDL and LDL cholesterol, increased HDL cholesterol, and yet did not interfere in weight gain, visceral fat, total cholesterol levels, or oxidative stress (Fiorini et al. 2017). A baru almond-enriched diet for 8 weeks in overweight and obese women reduced abdominal adiposity and increased HDL (Souza et al. 2018b). Some chemical safety assays indicate that *D. alata* bark extract is neither mutagenic (Ames test) nor toxic during pregnancy in rats, with no physical-neurobehavioral consequences on the development of the rats'offspring (Esteves-Pedro et al. 2012). Thus, besides its therapeutic importance, *D. alata* presents nutritional relevance while being chemically safe. This encourages its use, especially the almonds, as a raw material for food, pharmaceutical, and cosmetic industries (Moraes et al. 2018).

In food applications, baru almonds have already been used to produce a cereal supplement called granola (a mixture of cereals, whole grains, dried fruit, and nuts). Granola containing dried *caju-do-cerrado* (*Anacardium othonianum* Rizz) and baru almonds presents high protein levels, dietary fiber, and iron; lower moisture, water activity, sodium, and lipids; and high energy value, stability, crunchy texture, and consumer acceptability, potentially being a good food to bring to market (Souza and Silva 2015). Powdered baru almonds (baru flour) was used to produce reduced fat baru cupcakes. The flour was characterized as containing more protein, dietary fiber, minerals, and lipids than wheat flour. It was then used in cupcake formulations where 30% of the wheat flour was replaced with baru flour (Paglarini et al. 2018). A reduction in water absorption, stability, mixing tolerance index, and extensibility and an increase in dough development time and resistance to extension for the wheat-baru flour blend were observed. The baru flour mix allowed reduction in margarine use by 75% without reducing consumer acceptance, achieving a "reduced fat" claim (Paglarini et al. 2018).

Baru almonds are also a good source of polyunsaturated fatty acids, which are considered essential since the human metabolism is incapable of synthesizing them. To extract fatty acids from plant sources, organic solvents are normally used. However, severe legal restrictions have been proposed for reducing the use of organic solvents in industrial processes (Santos et al. 2016). Therefore, alternative techniques are being developed to obtain baru almond oil, such as supercritical CO_2 and compressed propane (Fetzer et al. 2018; Santos et al. 2016). The compressed propane technique presents the highest extraction yield (~37%) (Fetzer et al. 2018), and the oil product presents high concentrations of oleic and linoleic acids (Fetzer et al. 2018; Santos et al. 2016). Both supercritical fluid and compressed propane extractions avoid modifying the fatty acid profile of the oil (Fetzer et al. 2018; Santos et al. 2016). The proteinaceous cake remaining after baru almond oil extraction can be transformed into protein concentrates and isolates (adding value to the production chain), bringing new products to the food industry (Nunes et al. 2017). Partially defatted baru almond flour has already been used to produce cookies. Despite not having a good consumer acceptance, the cookies (25% baru almond flour) presented high fiber and bioactive compounds content, leading to higher antioxidant activity (Pineli et al. 2015). Baru almond flour can also be used in other food formulations.

For pharmaceutical and cosmetic applications, a lamellar gel phase emulsion using oil extracted from baru almonds was developed (Moraes et al. 2018). The system demonstrated stability during storage and was able to increase *stratum corneum* lipid fluidity, demonstrating its potential to act as a vehicle for drugs or in skin care (Moraes et al. 2018).

8 *Eugenia dysenterica* DC. (Myrtaceae)

Popularly known as *cagaita*, *Eugenia dysenterica* is a species native to the Cerrado with globular pale-yellow edible fruit highly appreciated by local populations (Fig. 2j). Different parts from this species have been employed in folk medicine to treat various disorders (Silva et al. 2015b; Souza et al. 2007). The leaves of *E. dysenterica* are used to treat diarrhea, cardiac diseases, diabetes, and jaundice and to reduce blood cholesterol. Cagaita fruits present laxative activity, and the flowers are employed to treat skin and bladder infections (Lima et al. 2010b; Lima et al. 2011; Silva et al. 2015b).

As to food applications, a fruit wine was elaborated from *E. dysenterica* pulp. The beverage presented good consumer acceptance for color, flavor, and taste (Oliveira et al. 2011). Jellies using two types of *E. dysenterica* pulp (filtered or unfiltered) and two different mixtures of pulp, sucrose, and pectin were obtained which were microbiologically safe and well accepted. The formulation with 60:40:0.1 (filtered pulp/sucrose/pectin) presented good nutritional value and was a good source of vitamin C (Santos et al. 2012). *E. dysenterica* raisins obtained using an osmotic dehydration process enjoyed good acceptance during sensory analyses (Silva et al. 2015a).

To prevent deterioration of the bioactive compounds in *E. dysenterica,* spray-drying methodology is used during dehydration of the leaves and fruit (Couto et al. 2013; Daza et al. 2016). The powdered *E. dysenterica* fruit extract can be used as a functional ingredient in food formulations or as an intermediate phyto-pharmaceutical product; it presents antioxidant activity, α-amylase and α-glucosidase inhibitory potential, and antimicrobial activity against *Staphylococcus aureus* and *Listeria monocytogenes* as well (Daza et al. 2017).

E. dysenterica leaf extract has been incorporated in micro-emulsions to develop stable, non-irritant topical formulations, improving cutaneous permeation without compromising its antioxidant activity (Ferreira-Nunes et al. 2018). The extract was also incorporated in chitosan microparticles with further incorporation in a dermatologically acceptable formulation. The emulsion presented relative stability for 60 days at 6 °C, good skin penetration, and angiogenic activity (Silva et al. 2020a). Thus, *E. dysenterica* leaf extract appears to be a promising alternative for dermatological treatments.

9 *Hancornia speciosa* Gomes (Apocynaceae)

Popularly known as mangabeira (Fig. 2c), *Hancornia speciosa* is a native Cerrado tree, and its fruit is called mangaba and is known by its exotic flavor and aroma (Cardoso et al. 2014). Economic interest in this fruit has been growing in recent years, because the pulp is rich in vitamins A, B_1, B_2, and C, as well as iron, phosphorus, and calcium (Narain et al. 2018). Mangaba can be consumed fresh or processed in the form of jelly, sweets, ice cream, juices, soft drinks, liquors, wine, and syrup (Almeida et al. 2016). Phytochemical studies have reported identification of differing classes among *H. speciosa* extract compounds. Flavonoids, catechins, proanthocyanidins, and tannins are present in the bark of the species. Steroids, triterpenes, and tannins are present in the leaves (Moraes et al. 2008), and chlorogenic acid, naringenin-7-O-glucoside, catechin, and proanthocyanidins were identified in the trunk latex (Neves et al. 2016).

With regard to its medicinal potential, the species has long been employed in folk medicine for treating wounds, rheumatism, hypertension, obesity, gastric lesions, and diabetes (Hirschmann and Arias 1990; Macedo and Ferreira 2004; Moraes et al. 2008). In addition to ethnobotanic surveys, certain pharmacological effects have also been associated with the plant extracts. For example, studies have shown that *H. speciosa* leaf extract presents endothelium-dependent vasodilatory effect (Ferreira et al. 2007), which is effective for blood pressure control (Silva et al.

2011). The leaf extract also has antidiabetic potential being able to reduce blood glucose concentration through inhibition of intestinal α-glucosidase and stimulation of glucose uptake by adipocytes (Pereira et al. 2015). Mangabeira bark extract possesses antioxidant activity and inhibits acetylcholinesterase production, which suggests that it can be used in the treatment of Alzheimer's, a disease linked with oxidative stress, acetylcholine deficiency in the brain, and inflammatory processes (Penido et al. 2017). The trunk latex possesses angiogenic (Almeida et al. 2014; D'Abadia et al. 2020), osteogenic (Floriano et al. 2016; Neves et al. 2016), anti-inflammatory (Marinho et al. 2011), and antioxidant (D'Abadia et al. 2020) activities, which may help in wound healing.

In food applications, a mixed nectar of *H. speciosa* and *E. dysenterica* has been developed achieving good acceptance by consumers due to an attractive sensory profile and good nutritional value (Assumpção et al. 2013). Yogurt made of goat's milk and *H. speciosa* pulp presented satisfactory sensory analyses, and the authors highlighted the product as a promising alternative for consumers who are allergic to cow's milk, as well as to those seeking a product with better digestibility (Santos et al. 2017).

A mixed jelly of *H. speciosa* and *Spondias tuberosa* Arr. C. ("umbu") presented high dietary fiber content, total phenolic content, and antioxidant capacity and remained microbiologically stable in moisture and total titratable acidity at 90 days of storage. The jelly presented sensory characteristics similar to commercialized jellies, but with more dietary fiber and functional properties as well (Souza et al. 2018a).

With regard to other technological applications, mangaba seeds can be used to produce bio-oil (Santos et al. 2015). The production of bio-oil from mangaba industrial processing residues can decrease wastes released into the environment. In another application, silver nanoparticles were added to *H. speciosa* latex biomembrane to develop a new biomaterial, combining angiogenic, anti-inflammatory, and antibacterial properties, which could potentially be used in wound healing (Almeida et al. 2019).

10 *Mauritia flexuosa* L. f. (Arecaceae)

Mauritia flexuosa is a palm that produces a fruit popularly known as *buriti* (Fig. 2b), which has an orange-colored thick mass and spongy endocarp that surrounds a very hard seed (Sampaio et al. 2008). The species presents antitumor (Siqueira et al. 2014), hypoglycemic (Bataglion et al. 2014), hypolipidemic (Aquino et al. 2012), antimicrobial (Koolen et al. 2013), and other healing activities (Batista et al. 2012). Due to its high β-carotene contents, buriti is extremely effective in treatment and prevention of xerophthalmia (Mariath et al. 1989); it also reduces the risk of developing cardiovascular disease (Neri-Numa et al. 2018a; Sandri et al. 2017). Although buriti possesses various ethnobotanical applications, the main use of buriti pulp is as a food, in the form of jams, jellies, juices, and ice creams (Gomes et al. 2011).

In food applications, the nectar obtained from *M. flexuosa* has low calories and presents iron, manganese, and fiber levels that meet daily requirements. It also presents good sensory properties and consumer acceptance (Garcia et al. 2015). *M. flexuosa* jellies in their conventional, light, and diet versions were elaborated (Sousa et al. 2020). The light and diet versions presented physical-chemical alterations in relation to the traditional jelly, yet without compromising quality in the storage periods studied. The conventional and light jelly formulations were well accepted. However, the diet pulp jelly required formulation adjustments for better taste acceptance (Sousa et al. 2020). *M. flexuosa* biscuit flour with addition of oats resulted in cookies with reduced diameters and thickness; higher moisture, lipid, fiber, and caloric contents; and lower protein and carbohydrate contents. Sensory characteristic testing revealed good consumer acceptability (Santos et al. 2011). *M. flexuosa* fruits processed by dehydration and spraying maintained good chemical and microbiological storage stability for 150 days, at temperatures of 4 °C and 12 °C (Aguiar and Souza 2017).

Freeze-dried flours from *M. flexuosa* were produced with its by-products which reduced triglyceride and total cholesterol levels and raised HDL levels in *Swiss* mice (Pereira-Freire et al. 2019). The flours can be used for preparing nutraceuticals with lipid-lowering characteristics and may be useful in support therapy for patients with dyslipidemia. *M. flexuosa* by-product flours present high dietary fiber contents and the presence of pectic polysaccharides, arabinoxylans, and xyloglucans. The peels and defatted pulp flours were highlighted for possessing higher antioxidant potentials as compared to the endocarp and to manually produced bran flours. Carotenoid content was also higher in the peel flours, and all of the flours produced presented expressive amounts of total non-extractable proanthocyanidins (Resende et al. 2019).

Although due to an elevated β-carotene content, *M. flexuosa* is considered a good source of carotenoids, and as a supplement of pro-vitamin A (Aguiar and Souza 2017), its fibers and polyphenols may bind to macromolecular compounds that are not dialyzable, or generate mineral complexes, decreasing their solubility and bioaccessibility (Bouayed et al. 2011).

11 *Myrciaria cauliflora* (Mart.) O. Berg. (Myrtaceae)

Myrciaria cauliflora is a native plant in southern and central South America. The fruit is known as *jabuticaba* and presents a white soft juicy pulp, sweet and acidic flavor, and a dark purple color (Fig. 2f). The pulp is rich in minerals and ascorbic acid (Fortes et al. 2011; Teixeira et al. 2011). The dark peel is rich in polyphenols and anthocyanins. Folk medicine uses jabuticaba to treat asthma, throat inflammation, and gastrointestinal and cardiovascular disturbances (Giraldi and Hanazaki 2010). Studies have also demonstrated jabuticaba peels significantly reduce blood cholesterol (Dragano et al. 2013), obesity-associated insulin resistance (Araújo et al. 2013), and promote vasorelaxant and hypotensive effects (Andrade et al. 2015).

With regard to food applications, *M. cauliflora* juice extraction yielded three powders obtained from the whole fruit (JWF), peel (JPE), and pomace (JPO). The pomace (JPO) was rich in total dietary fiber, insoluble dietary fiber, and phenolic compounds, especially monomeric anthocyanin, while the peel (JPE) presented a large amount of soluble dietary fiber (Gurak et al. 2014). Supercritical fluid extraction was used to prepare *M. cauliflora* jelly, which presented a good yield of antioxidant compounds (Cavalcanti et al. 2011). The addition of *M. cauliflora* peel extract to Bologna-type sausages decreased the pH at the start of processing and protected the samples from changes in color during storage. However, it had no positive effect on storage microbial stability (Almeida et al. 2015). A fermented dairy dessert containing *M. cauliflora* peel was characterized as a low-fat product, being considered acceptable for overall consumption, with an attractive color and appreciable texture, despite an acidic taste. These formulations are seen as a viable alternative for the use of *M. cauliflora* peel, as well as a potential functional food due to the concentration of lactobacilli, besides the presence of antioxidant phenolic compounds (Almeida Neta et al. 2018). Addition of *M. cauliflora* peel flour to loaves reduced carbohydrate content, lipids, and caloric value while increasing humidity. Fiber content increased up to three times and phenolics up to seven times, and consequently, antioxidant activity was higher in breads supplemented with *M. cauliflora* peel flour. Bread formulated with 5% *M. cauliflora* peel flour presented greater sensory acceptance in overall impression and flavor attributes (Ferreira 2017).

It is important to highlight that certain marker compounds present in the fresh *M. cauliflora* fruit extracts were not detected in commercial juice and jams, and thus, the antioxidant capacity of the fresh fruit extract was higher (Wu et al. 2012).

12 *Psidium guajava* L. (Myrtaceae)

Psidium guajava, popularly known as guava or goiaba (Fig. 2d), is traditionally consumed (fruit) due to its flavor and nutritional value (Satyal et al. 2015). The aqueous extract of guava leaves has been reported to be efficacious in the treatment of various types of gastrointestinal disturbances such as diarrhea, peristaltic reflex inhibition, and gastroenteritis (Gutiérrez et al. 2008). The whole plant is used as a skin tonic and is employed to treat female-related diseases like dysmenorrhea, miscarriage, uterine bleeding, and premature labor (Rishika and Sharma 2012). Various flavonoids such as guajaverin, guayfotavolic acid, guajadial, and guavanoic acid were identified in this plant (Wang et al. 2014).

P. guajava fruit powders, when obtained by lyophilization, present a higher residual content of vitamin C, antioxidant activity, and thermal stability, than when obtained by hot-air drying. Sensory analyses revealed similarities between the two dehydrated powders, but a pleasant guava jelly-resembling flavor was detected for both solids. Both powders enjoyed high consumer acceptance (Osorio et al. 2011).

Incorporation of *P. guajava* powder in sheep meat nuggets resulted in significant decrease in emulsion and nugget pH, emulsion stability, cooking yield, and nuggets' moisture content. Emulsion ash and moisture content increased. Total phenolics, total dietary fiber, and ash content significantly increased in the nuggets with powder added. The powder was also found to retard lipid peroxidation in cooked sheep meat nuggets, did not affect sensory characteristics of the products, and can be used as a source of antioxidant dietary fiber in meat food preparations (Verma et al. 2013).

Heat treatments (boiling or autoclaving) did not affect *P. guajava* seed total dietary fiber or ash contents. Boiling did not significantly reduce phytic acid content, but autoclaving caused a reduction of 91%, a level below the anti-nutritional threshold. Germination for 14 days caused significant reductions in nutrient contents of 16–79% and also reduced the phytic acid content by 90%. Thus, *P. guajava* seeds can be treated thermally or germinated to manipulate their composition and enable their use in both food and feed industries (Chang et al. 2014).

Addition of pectin (0.5%, 1%, and 1.5%) to *P. guajava* cheese, a semisolid concentrated fruit product, increased hardness, phytochemical content, and antioxidant potential while enhancing shelf life stability. The guava cheese formulated presented good nutrition, sensory properties, and consumer acceptance and may well be utilized as a healthy fruit snack when adding alternative sweeteners and functional ingredients to obtain further health benefits (Patel et al. 2016).

The sensory quality and physicochemical characteristics of meat from Santa Inês lambs fed with diets containing *P. guajava* agro-industry by-product were tested. As a result, inclusion of 40% guava in diets (in place of corn) produced meat with satisfactory consumer acceptance in terms of color, yet with a less intense flavor, and tenderness. It was also observed that the animals suffered weight reduction (Costa et al. 2019).

13 *Pterodon emarginatus* Vogel (Fabaceae)

P. emarginatus is a tree popularly known as *sucupira* (Fig. 2g). Its fruits are *cryptosamaras*, which in their central structure present an alveolus filled with bitter oil (Lorenzi 2008). The seeds are commercialized in popular markets, especially for their pharmacological properties. The population makes use of the seeds in hydroalcoholic macerations to treat laryngological diseases, as antirheumatic, anti-inflammatory for spinal problems, depurative, and fortifier (Hansen et al. 2010). Seeds of *P. emarginatus* are also used as raw material to obtain an amber-colored viscous oil that is considered a rich source of vouacapane diterpenes (Santos et al. 2010).

The effects of dietary supplementation in laying hens of *P. emarginatus* (0.03%) and *Copaifera langsdorffii* Desf (0.03% and 0.06%) oils on the quality of their fresh eggs – stored under refrigeration and at room temperature – were evaluated. Supplementation with these oils helped to preserve the quality of the egg albums (kept at room temperature for up to 14 days), similar to using refrigeration. It also helped to maintain the pH of the yolks in eggs stored at room temperature for up to 30 days (Oliveira et al. 2018b).

In pharmaceutical applications, microencapsulation of the *P. emarginatus* fruit essential oil (spray-drying) may contribute to the development of an herbal medicine, since 98.63% of the essential oil was retained and the same proportion of β-caryophyllene was captured (Alves et al. 2014).

14 Other Species with Technological Potential

Brosimum gaudichaudii Trécul (Moraceae) is popularly known as mama-cadela. Its roots are traditionally used for the treatment of patients with vitiligo. The activity is attributed mainly to the *furanocoumarins* psoralen and bergapten (Jacomassi et al. 2007; Morais et al. 2018). *B. gaudichaudii* seed flour in gingerbread improved the sensory characteristics and may be useful in partial replacement of wheat flour (Alves et al. 2017).

Byrsonima crassifolia (L.) Kunth (Malpiguiaceae) is popularly known as murici. Its bright yellow fruit is edible with a sweet taste and a slightly bitter aftertaste (Pereira and Freitas 2002). Its fruits and leaves are used to treat gastrointestinal problems, such as ulcers and diarrhea. Its roots are used to heal wounds and infections of the mouth and throat such as gingivitis, tonsillitis, and pharyngitis (Gellen and Silva 2016). An alcoholic fermented drink from *B. crassifolia* was obtained with physicochemical characteristics in accordance with the Brazilian legislation for grape wines, except for NaCl chlorides, ashes, and total titratable acidity parameters (Bizinoto 2017). The fermented alcohol of *B. crassifolia* was classified as a semidry table wine. Fruit-based beverage formulations with *B. crassifolia* and Spondias mombin L. (cajá) were well-accepted and presented antioxidant properties (Souza et al. 2020).

Campomanesia cambessedeana Berg (Myrtaceae) is popularly known as gabiroba or guavira. Its fruits are appreciated for their sensory characteristics and nutritional value. Its pulp is succulent, acidic, and slightly sweet (Malta et al. 2012). The leaves and fruits have anti-inflammatory, antidiarrheal, and antiseptic properties in the urinary tract and are also used to prevent rheumatism and liver disorders (Campos et al. 2012). Atomization processing of *C. cambessedeana* fruit pulp was efficient in increasing pulp shelf life. However, the powder underwent changes such as browning and reduction of the total phenolic compounds content when compared to the fresh pulp (Chung 2016).

15 Conclusion

Cerrado species have great technological potential as functional f*oods, and for nutraceutical, cosmetic, and pharmacological* applications *as well*. Yet consumption is normally limited to local communities. This may be due to scarce information regarding their potential use, the processing necessary, or the quality of products that can be obtained from these species. In the food industry, the potential of many Cerrado species was demonstrated for cookies, food bars, cupcakes, juices, jellies, liquors, wines, yogurt, and other formulations, which presented good nutritional values and high acceptability in sensory tests. Moreover, Cerrado species can be prospected to produce drugs to alleviate different human diseases such as inflammations, infections, and gastrointestinal disturbances or to produce cosmetics with antiaging or healing properties. Such applications enable commercialization, generate employment and income for the local populations, favor industrial expansion, minimize wastes, and promote the generation of various co-products.

References

Aguiar JPL, Souza FCA (2017) Dehydration and spraying of buriti pulp (*Mauritia flexuosa* L.): shelf-live evaluation. Rev Bras Frutic 39

Al-Khattawi A, Bayly A, Phillips A, Wilson D (2018) The design and scale-up of spray dried particle delivery systems. Expert Opin Drug Deliv 15(1):47–63. https://doi.org/10.1080/17425247.2017.1321634

Almeida Neta MC, Queiroga APR, Almeida RLJ, Soares AC, Gonçalves JM, Fernandes SS, Sousa MC, Santos KMO, Buriti FCA, Florentino ER (2018) Fermented dessert with whey, ingredients from the peel of jabuticaba (*Myrciaria cauliflora*) and an indigenous culture of *Lactobacillus plantarum*: composition, microbial viability, antioxidant capacity and sensory features. Nutrients 10(9):1214

Almeida LM, Floriano JF, Ribeiro TP, Magno LN, Mota LSLS, Peixoto N, Mrué F, Melo-Reis P, Lino-Júnior RS, Graeff CFO, Gonçalves PJ (2014) *Hancornia speciosa* latex for biomedical applications: physical and chemical properties, biocompatibility assessment and angiogenic activity. J Mater Sci Mater Med 25(9):2153–2162. https://doi.org/10.1007/s10856-014-5255-8

Almeida PL, Lima SN, Costa LL, Oliveira CC, Damasceno KA, Santos BA, Campagnol PCB (2015) Effect of jabuticaba peel extract on lipid oxidation, microbial stability and sensory properties of Bologna-type sausages during refrigerated storage. Meat Sci 110:9–14. https://doi.org/10.1016/j.meatsci.2015.06.012

Almeida LM, Nogueira CA, Borges PP, Prado ADL, Gonçalves PJ (2016) State of the art of scientific literature on *Hancornia speciosa*: trends and gaps. Rev Bras Frutic 38(4)

Almeida LM, Magno LN, Pereira AC, Guidelli EJ, Filho OB, Kinoshita A, Gonçalves PJ (2019) Toxicity of silver nanoparticles released by *Hancornia speciosa* (Mangabeira) biomembrane. Spectrochim Acta A Mol Biomol Spectrosc 210:329–334. https://doi.org/10.1016/j.saa.2018.11.050

Alves CCO, Resende JV, Prado MET, Cruvinel RSR (2010) The effects of added sugars and alcohols on the induction of crystallization and the stability of the freeze-dried peki (*Caryocar brasiliense* Camb.) fruit pulps. LWT Food Sci Technol 43(6):934–941. https://doi.org/10.1016/j.lwt.2010.01.029

Alves SF, Borges LL, dos Santos TO, de Paula JR, Conceição EC, Bara MTF (2014) Microencapsulation of essential oil from fruits of pterodon emarginatus using gum arabic and maltodextrin as wall materials: composition and stability. Dry Technol 32(1):96–105. https://doi.org/10.1080/07373937.2013.816315

Alves VM, Abreu DJM, Moraes IA, Lessa TRC, Silva NF, Amorim KA, Silva EP, Damiani C (2017) Avaliação da substituição parcial de farinha da semente da mama-cadela na elaboração de pão de mel. Work presented at III Simpósio de Alimentos e Nutrição

Amaral LF, Moriel P, Foglio MA, Mazzola PG (2014a) *Caryocar brasiliense* supercritical CO2 extract possesses antimicrobial and antioxidant properties useful for personal care products. BMC Complement Altern Med 14:73. https://doi.org/10.1186/1472-6882-14-73

Amaral LF, Moriel P, Foglio MA, Mazzola PG (2014b) Evaluation of the cytotoxicity and phototoxicity of *Caryocar brasiliense* supercritical carbon dioxide extract. BMC Complement Altern Med 14:450. https://doi.org/10.1186/1472-6882-14-450

Andrade Júnior FP, Alves TWB, Romano TKF, Medeiros FD (2018) Antibacterial and antifungal potential of extracts of *Anacardium occidentale*. Tchê Química 15(30):313–321

Andrade DML, Reis CF, Castro PFS, Borges LL, Amaral NO, Torres IMS, Rezende SG, Gil ES, Conceição EC, Pedrino GR, Rocha ML (2015) Vasorelaxant and hypotensive effects of jaboticaba fruit (*Myrciaria cauliflora*) extract in rats. Evid Based Complement Alternat Med 2015:696135. https://doi.org/10.1155/2015/696135

Appendino G, Fontana G, Pollastro F (2010) 3.08—Natural products drug discovery. In: Liu H-W, Mander L (eds) Comprehensive natural products II. Elsevier, Oxford, pp 205–236

Aquino JS, Pessoa DCNP, Araújo KLGV, Epaminondas PS, Schuler ARP, Souza AG, Stamford TLM (2012) Refining of buriti oil (*Mauritia flexuosa*) originated from the Brazilian The Cerrado : physicochemical, thermal-oxidative and nutritional implications. J Braz Chem Soc 23:212–219

Araújo IMS, Zampa MF, Moura JB, dos Santos JR, Eaton P, Zucolotto V, Veras LMC, de Paula RCM, Feitosa JPA, Leite JRSA, Eiras C (2012) Contribution of the cashew gum (*Anacardium occidentale* L.) for development of layer-by-layer films with potential application in nanobiomedical devices. Mater Sci Eng C 32(6):1588–1593. https://doi.org/10.1016/j.msec.2012.04.048

Araújo CRR, Esteves EA, Dessimoni-Pinto NAV, Batista AG (2013) *Myrciaria cauliflora* peel flour had a hypolipidemic effect in rats fed a moderately high-fat diet. J Med Food 17(2):262–267. https://doi.org/10.1089/jmf.2012.0256

Araújo ACMA, Menezes EGT, Terra AWC, Dias BO, Oliveira ER, Queiroz F (2018) Bioactive compounds and chemical composition of Brazilian The Cerrado fruits wastes: pequi almonds, murici, and sweet passionfruit seeds. Food Sci Technol 38:203–214

Arruda HS, Fernandes RVB, Botrel DA, Almeida MEF (2015) Frutos do The Cerrado: conhecimento e aceitação de *Annona crassiflora* Mart. (araticum) e *Eugenia dysenterica* Mart. (cagaita) por crianças utilizando o paladar e a visão. J Health Biol Sci 3(4):224. https://doi.org/10.12662/2317-3076jhbs.v3i4.168.p224-230.2015

Arruda HS, Botrel DA, Fernandes RVB, Almeida MEF (2016) Development and sensory evaluation of products containing the Brazilian Savannah fruits araticum (*Annona crassiflora* Mart.) and cagaita (*Eugenia dysenterica* Mart.). Braz J Food Technol 19

Ascari J, Takahashi JA, Boaventura MAD (2013) The phytochemistry and biological aspects of Caryocaraceae family. Rev Bras Plantas Med 15:293–308

Assumpção CF, Bachiega P, Santana ATMC, Morzelle MC, Vilas Boas BM, Souza EC (2013) Néctar misto de mangaba (*Hancoria speciosa* Gomes) e cagaita (*Eugenia dysenterica*) – perfil sensorial e características físico-químicas. Rev Bras Produtos Agroindustriais 15(3):219–224

Atanasov AG, Waltenberger B, Pferschy-Wenzig EM, Linder T, Wawrosch C, Uhrin P, Temml V, Wang L, Schwaiger S, Heiss EH, Rollinger JM, Schuster D, Breuss JM, Bochkov V, Mihovilovic MD, Kopp B, Bauer R, Dirsch VM, Stuppner H (2015) Discovery and resupply of pharmacologically active plant-derived natural products: a review. Biotechnol Adv 33(8):1582–1614. https://doi.org/10.1016/j.biotechadv.2015.08.001

Bailão EFLC, Devilla IA, Conceição EC, Borges LL (2015) Bioactive compounds found in Brazilian The Cerrado fruits. Int J Mol Sci 16(10):23760–23783. https://doi.org/10.3390/ijms161023760

Bastos DS, Gonçalves MP, Andrade CT, Araújo KGL, Rocha Leão MHM (2012) Microencapsulation of cashew apple (*Anacardium occidentale*, L.) juice using a new chitosan–commercial bovine whey protein isolate system in spray drying. Food Bioprod Process 90(4):683–692. https://doi.org/10.1016/j.fbp.2012.04.005

Bataglion GA, Silva FMA, Eberlin MN, Koolen HHF (2014) Simultaneous quantification of phenolic compounds in buriti fruit (*Mauritia flexuosa* L.f.) by ultra-high performance liquid chromatography coupled to tandem mass spectrometry. Food Res Int 66:396–400. https://doi.org/10.1016/j.foodres.2014.09.035

Batista JS, Olinda RG, Medeiros VB, Rodrigues CMF, Oliveira AF, Paiva ES, Freitas CIA, Medeiros AC (2012) Atividade antibacteriana e cicatrizante do óleo de buriti *Mauritia flexuosa* L. Ciênc Rural 42:136–141

BeMiller JN (2019) Pectins. In: BeMiller JN (ed) Carbohydrate chemistry for food scientists, 3rd edn. AACC International Press, St. Paul, pp 303–312. https://doi.org/10.1016/B978-0-12-812069-9.00015-7

Bizinoto CS (2017) Desenvolvimento do fermentado alcoólico de murici (*Byrsonima crassifolia* (L.) Kunth) – Malpighiaceae. Instituto Federal de Educação, Ciência e Teconologia do Triângulo Mineiro, Uberaba

Boon CS, McClements DJ, Weiss J, Decker EA (2010) Factors influencing the chemical stability of carotenoids in foods. Crit Rev Food Sci Nutr 50(6):515–532. https://doi.org/10.1080/10408390802565889

Botrel DA, Rodrigues ICB, Souza HJB, Fernandes RVB (2016) Application of inulin in thin-layer drying process of araticum (*Annona crassiflora*) pulp. LWT Food Sci Technol 69:32–39. https://doi.org/10.1016/j.lwt.2016.01.018

Bouayed J, Hoffmann L, Bohn T (2011) Total phenolics, flavonoids, anthocyanins and antioxidant activity following simulated gastro-intestinal digestion and dialysis of apple varieties: bioaccessibility and potential uptake. Food Chem 128(1):14–21. https://doi.org/10.1016/j.foodchem.2011.02.052

Breda CA, Morgado DL, Assis OBG, Duarte MCT (2017) Effect of chitosan coating enriched with pequi (*Caryocar brasiliense* Camb.) peel extract on quality and safety of tomatoes (*Lycopersicon esculentum* Mill.) during storage. J Food Process Preserv 41(6):e13268. https://doi.org/10.1111/jfpp.13268

Bucar F, Wube A, Schmid M (2013) Natural product isolation–how to get from biological material to pure compounds. Nat Prod Rep 30(4):525–545. https://doi.org/10.1039/c3np20106f

Calixto JB (2000) Efficacy, safety, quality control, marketing and regulatory guidelines for herbal medicines (phytotherapeutic agents). Braz J Med Biol Res 33:179–189

Campos RP, Hiane PA, Ramos MIL, Ramos Filho MM, Macedo MLR (2012) Conservação pós-colheita de guavira (*Campomanesia* sp.). Rev Bras Frutic 34:41–49

Cardoso LM, Reis BL, Oliveira DS, Pinheiro-Sant'Ana HM (2014) Mangaba (*Hancornia speciosa* Gomes) from the Brazilian The Cerrado : nutritional value, carotenoids and antioxidant vitamins. Fruits 69(2):89–99. https://doi.org/10.1051/fruits/2013105

Cavalcanti RN, Veggi PC, Meireles MAA (2011) Supercritical fluid extraction with a modifier of antioxidant compounds from jabuticaba (*Myrciaria cauliflora*) byproducts: economic viability. Proc Food Sci 1:1672–1678. https://doi.org/10.1016/j.profoo.2011.09.247

Chang YP, Tan MP, Lok WL, Pakianathan S, Supramaniam Y (2014) Making use of guava seed (*Psidium guajava* L): the effects of pre-treatments on its chemical composition. Plant Foods Hum Nutr 69(1):43–49. https://doi.org/10.1007/s11130-013-0396-3

Chung MMS (2016) Polpa de guavira (*Campomanesia cambessedeana* Berg) desidratada em spray dryer: efeitos das condições de processo e composição da alimentação nas propriedades físico químicas e atividade antioxidante. Universidade de São Paulo, Pirassununga

Colombo NBR, Rangel MP, Martins V, Hage M, Gelain DP, Barbeiro DF, Grisolia CK, Parra ER, Capelozzi VL (2015) *Caryocar brasiliense* Camb. protects against genomic and oxidative damage in urethane-induced lung carcinogenesis. Braz J Med Biol Res 48:852–862

Corrêa SC, Clerici MTPS, Garcia JS, Ferreira EB, Eberlin MN, Azevedo L (2011) Evaluation of dehydrated marolo (*Annona crassiflora*) flour and carpels by freeze-drying and convective hot-air drying. Food Res Int 44(7):2385–2390. https://doi.org/10.1016/j.foodres.2011.02.052

Corson TW, Crews CM (2007) Molecular understanding and modern application of traditional medicines: triumphs and trials. Cell 130(5):769–774. https://doi.org/10.1016/j.cell.2007.08.021

Costa RG, Cavalcanti MCA, Nobre PT, Queiroga RCRE, Medeiros GR, Silva NV, Batista ASM, Araújo Filho JT (2019) Sensory quality of meat from Santa Inês lambs fed with guava (*Psidium guajava* L.) agroindustrial by-product. Food Sci Technol 40(3):653–658

Couto RO, Martins FS, Chaul LT, Conceição EC, Freitas LAP, Bara MTF, Paula JR (2013) Spray drying of *Eugenia dysenterica* extract: effects of in-process parameters on product quality. Rev Bras 23(1):115–123. https://doi.org/10.1590/S0102-695X2012005000109

Cruz KS, Silva MA, Freitas O, Neves VA (2011) Partial characterization of proteins from baru (*Dipteryx alata* Vog) seeds. J Sci Food Agric 91(11):2006–2012. https://doi.org/10.1002/jsfa.4410

D'Abadia PL, Bailão EFLC, Lino-Júnior RS, Oliveira MG, Silva VB, Oliveira LAR, Conceição EC, Melo-Reis PR, Borges LL, Gonçalves PJ, Almeida LM (2020) *Hancornia speciosa* serum fraction latex stimulates the angiogenesis and extracellular matrix remodeling processes. An Acad Bras Cienc 92:e20190107

Daza LD, Fujita A, Fávaro-Trindade CS, Rodrigues-Ract JN, Granato D, Genovese MI (2016) Effect of spray drying conditions on the physical properties of cagaita (*Eugenia dysenterica* DC.) fruit extracts. Food Bioprod Process 97:20–29. https://doi.org/10.1016/j.fbp.2015.10.001

Daza LD, Fujita A, Granato D, Fávaro-Trindade CS, Genovese MI (2017) Functional properties of encapsulated Cagaita (*Eugenia dysenterica* DC.) fruit extract. Food Biosci 18:15–21. https://doi.org/10.1016/j.fbio.2017.03.003

Dey A, De JN (2012) Phytopharmacology of antiophidian botanicals: a review. Int J Pharm 8:62–79

Dragano NRV, Marques AC, Cintra DEC, Solon C, Morari J, Leite-Legatti AV, Velloso LA, Maróstica-Júnior MR (2013) Freeze-dried Jaboticaba peel powder improves insulin sensitivity in high-fat-fed mice. Br J Nutr 110(3):447–455. https://doi.org/10.1017/S0007114512005090

Dutra RC, Campos MM, Santos AR, Calixto JB (2016) Medicinal plants in Brazil: pharmacological studies, drug discovery, challenges and perspectives. Pharmacol Res 112:4–29. https://doi.org/10.1016/j.phrs.2016.01.021

Edison TNJI, Atchudan R, Sethuraman MG, Lee YR (2016) Reductive-degradation of carcinogenic azo dyes using *Anacardium occidentale* testa derived silver nanoparticles. J Photochem Photobiol B Biol 162:604–610. https://doi.org/10.1016/j.jphotobiol.2016.07.040

Esteves-Pedro NM, Borim T, Nazato VS, Silva MG, Lopes PS, dos Santos MG, Dal Belo CA, Cardoso CRP, Varanda EA, Groppo FC, Gerenutti M, Oshima-Franco Y (2012) *In vitro* and *in vivo* safety evaluation of *Dipteryx alata* Vogel extract. BMC Complement Altern Med 12:9. https://doi.org/10.1186/1472-6882-12-9

Fabricant DS, Farnsworth NR (2001) The value of plants used in traditional medicine for drug discovery. Environ Health Perspect 109(Suppl 1):69–75

Faleiro FV, Machado RB, Loyola RD (2013) Defining spatial conservation priorities in the face of land-use and climate change. Biol Conserv 158:248–257. https://doi.org/10.1016/j.biocon.2012.09.020

Feher M, Schmidt JM (2003) Property distributions: differences between drugs, natural products, and molecules from combinatorial chemistry. J Chem Inf Comput Sci 43(1):218–227. https://doi.org/10.1021/ci0200467

Fernandes DC, Freitas JB, Czeder LP, Naves MM (2010) Nutritional composition and protein value of the baru (*Dipteryx alata* Vog.) almond from the Brazilian savanna. J Sci Food Agric 90(10):1650–1655. https://doi.org/10.1002/jsfa.3997

Ferreira SPL (2017) Aproveitamento da casca de jabuticaba para o enriquecimento nutricional de pães de forma integrais. Instituto Federal de Educação, Ciência e Tecnologia do Triângulo Mineiro, Uberaba

Ferreira HC, Serra CP, Endringer DC, Lemos VS, Braga FC, Cortes SF (2007) Endothelium-dependent vasodilation induced by *Hancornia speciosa* in rat superior mesenteric artery. Phytomedicine 14(7):473–478. https://doi.org/10.1016/j.phymed.2006.11.008

Ferreira-Nunes R, Silva SMM, Souza PEN, Magalhães PO, Cunha-Filho M, Gratieri T, Gelfuso GM (2018) Incorporation of *Eugenia dysenterica* extract in microemulsions preserves stability, antioxidant effect and provides enhanced cutaneous permeation. J Mol Liq 265:408–415. https://doi.org/10.1016/j.molliq.2018.06.023

Fetzer DL, Cruz PN, Hamerski F, Corazza ML (2018) Extraction of baru (*Dipteryx alata* vogel) seed oil using compressed solvents technology. J Supercrit Fluids 137:23–33. https://doi.org/10.1016/j.supflu.2018.03.004

Finco A, Silva FDB, Oliveira IG, Botelho R (2012) Physicochemical characteristics and antioxidant activity of three native fruits from Brazilian Savannah (the Cerrado). Braz J Food Nutr 23:179

Fiorini AMR, Barbalho SM, Guiguer EL, Oshiiwa M, Mendes CG, Vieites RL, Chies AB, Oliveira PB, Souza MSS, Nicolau CCT (2017) *Dipteryx alata* Vogel may improve lipid profile and atherogenic indices in Wistar rats *Dipteryx alata* and atherogenic indices. J Med Food 20(11):1121–1126. https://doi.org/10.1089/jmf.2017.0052

Firmo AQ, Sousa MM, Cavalcanti MS (2020) Desenvolvimento e caracterização de bebidas produzidas à base de castanha de caju (*Anacardium occidentale* L.). Res Soc Dev 9(1):84911645. https://doi.org/10.33448/rsd-v9i1.1645

Floriano JF, Neto FC, Mota LSLS, Furtado EL, Ferreira RS, Barraviera B, Gonçalves PJ, Almeida LM, Borges FA, Herculano RD, Graeff CFO (2016) Comparative study of bone tissue accelerated regeneration by latex membranes from *Hevea brasiliensis* and *Hancornia speciosa*. Biomed Phys Eng Express 2(4):045007. https://doi.org/10.1088/2057-1976/2/4/045007

Formagio ASN, Vieira MC, Volobuff CRF, Silva MS, Matos AI, Cardoso CAL, Foglio MA, Carvalho JE (2015) *In vitro* biological screening of the anticholinesterase and antiproliferative activities of medicinal plants belonging to Annonaceae. Braz J Med Biol Res 48:308–315

Fortes GAC, Naves SS, Godoi FFF, Duarte AR, Ferri PH, Santos SC (2011) Assessment of a maturity index in jabuticaba fruit by the evaluation of phenolic compounds, essential oil components, sugar content and total acidity. Am J Food Technol 6:974–984

Gao H, Gong X, Hu G (2017) Statistical and frequency analysis of pressure fluctuation in an annular spouted bed of coarse particles. Powder Technol 317:216–223. https://doi.org/10.1016/j.powtec.2017.05.007

Garcia LGC, Becker FS, Damiani C (2015) Néctar de buriti (*Mauritia flexuosa*): a bebida funcional do the Cerrado. Revista Verde de Agroecologia e Desenvolvimento Sustentável 10(1):263–268

Gellen L, Silva E (2016) Atividade antimicrobiana de extratos de raízes de *Byrsonima crassifolia*. J Bioenergy Food Sci 3:63–71. https://doi.org/10.18067/jbfs.v3i2.88

Giraldi M, Hanazaki N (2010) Uso e conhecimento tradicional de plantas medicinais no Sertão do Ribeirão, Florianópolis, SC, Brasil. Acta Botanica Brasilica 24:395–406

Gleeson MP, Hersey A, Montanari D, Overington J (2011) Probing the links between in vitro potency, ADMET and physicochemical parameters. Nat Rev Drug Discov 10(3):197–208. https://doi.org/10.1038/nrd3367

Gobbo-Neto L, Lopes NP (2007) Medicinal plants: factors of influence on the content of secondary metabolites. Quim Nova 30(2):374–381

Gomes LRP, Lopes MTG, Bentes JLS, Barros WS, Costa-Neto PQ, Contim LAS (2011) Genetic diversity in natural populations of buriti (*Mauritia flexuosa* L. f.). Crop Breeding Appl Biotechnol 11:216–223

Gowthamarajan K, Jawahar N, Wake P, Jain K, Sood S (2012) Development of buccal tablets for curcumin using *Anacardium occidentale* gum. Carbohydr Polym 88(4):1177–1183. https://doi.org/10.1016/j.carbpol.2012.01.072

Guedes AMM, Antoniassi R, Galdeano MC, Grimaldi R, Carvalho MG, Wilhelm AE, Marangoni AG (2017) Length-scale specific crystalline structural changes induced by molecular randomization of pequi oil. J Oleo Sci 66(5):469–478. https://doi.org/10.5650/jos.ess16192

Gurak PD, de Bona GS, Tessaro IC, Marczak LDF (2014) Jaboticaba pomace powder obtained as a co-product of juice extraction: a comparative study of powder obtained from peel and whole fruit. Food Res Int 62:786–792. https://doi.org/10.1016/j.foodres.2014.04.042

Gutiérrez RMP, Mitchell S, Solis RV (2008) *Psidium guajava*: a review of its traditional uses, phytochemistry and pharmacology. J Ethnopharmacol 117(1):1–27. https://doi.org/10.1016/j.jep.2008.01.025

Hansen D, Haraguchi M, Alonso A (2010) Pharmaceutical properties of 'sucupira' (*Pterodon* spp.). Braz J Pharm Sci 46:607–616

Haridasan M (2008) Nutritional adaptations of native plants of The Cerrado biome in acid soils. Braz J Plant Physiol 20:183–195

Hasnain MS, Rishishwar P, Rishishwar S, Ali S, Nayak AK (2018) Extraction and characterization of cashew tree (*Anacardium occidentale*) gum; use in aceclofenac dental pastes. Int J Biol Macromol 116:1074–1081. https://doi.org/10.1016/j.ijbiomac.2018.05.133

Hirschmann GS, Arias AR (1990) A survey of medicinal plants of Minas Gerais, Brazil. J Ethnopharmacol 29(2):159–172. https://doi.org/10.1016/0378-8741(90)90052-U

Hunter P (2008) Harnessing Nature's wisdom. Turning to Nature for inspiration and avoiding her follies. EMBO Rep 9(9):838–840. https://doi.org/10.1038/embor.2008.160

Jacomassi E, Moscheta IS, Machado SR (2007) Morfoanatomia e histoquímica de *Brosimum gaudichaudii* Trécul (Moraceae). Acta Botanica Brasilica 21:575–597

Johner JCF, Hatami T, Meireles MAA (2018a) Developing a supercritical fluid extraction method assisted by cold pressing: a novel extraction technique with promising performance applied to pequi (*Caryocar brasiliense*). J Supercrit Fluids 137:34–39. https://doi.org/10.1016/j.supflu.2018.03.005

Johner JCF, Hatami T, Zabot GL, Meireles MAA (2018b) Kinetic behavior and economic evaluation of supercritical fluid extraction of oil from pequi (*Caryocar brasiliense*) for various grinding times and solvent flow rates. J Supercrit Fluids 140:188–195. https://doi.org/10.1016/j.supflu.2018.06.016

Jones WP, Kinghorn AD (2012) Extraction of plant secondary metabolites. Methods Mol Biol 864:341–366. https://doi.org/10.1007/978-1-61779-624-1_13

Kho K, Hadinoto K (2011) Optimizing aerosolization efficiency of dry-powder aggregates of thermally-sensitive polymeric nanoparticles produced by spray-freeze-drying. Powder Technol 214(1):169–176. https://doi.org/10.1016/j.powtec.2011.08.010

Klink CA, Machado RB (2005) A conservação do the Cerrado brasileiro. Megadiversidade 1(1):147–155

Koolen HHF, Silva FMA, Gozzo FC, Souza AQL, Souza ADL (2013) Antioxidant, antimicrobial activities and characterization of phenolic compounds from buriti (*Mauritia flexuosa* L. f.) by UPLC–ESI-MS/MS. Food Res Int 51(2):467–473. https://doi.org/10.1016/j.foodres.2013.01.039

Leão DP, Franca AS, Oliveira LS, Bastos R, Coimbra MA (2017) Physicochemical characterization, antioxidant capacity, total phenolic and proanthocyanidin content of flours prepared from pequi (*Caryocar brasilense* Camb.) fruit by-products. Food Chem 225:146–153. https://doi.org/10.1016/j.foodchem.2017.01.027

Leão DP, Botelho BG, Oliveira LS, Franca AS (2018) Potential of pequi (*Caryocar brasiliense* Camb.) peels as sources of highly esterified pectins obtained by microwave assisted extraction. LWT Food Sci Technol 87:575–580. https://doi.org/10.1016/j.lwt.2017.09.037

Li FS, Weng JK (2017) Demystifying traditional herbal medicine with modern approach. Nat Plants 3:17109. https://doi.org/10.1038/nplants.2017.109

Lima F, Lucca BG, Barbosa AMJ, Ferreira VS, Moccelini SK, Franzoi AC, Vieira IC (2010a) Biosensor based on pequi polyphenol oxidase immobilized on chitosan crosslinked with cyanuric chloride for thiodicarb determination. Enzym Microb Technol 47(4):153–158. https://doi.org/10.1016/j.enzmictec.2010.05.006

Lima TB, Silva ON, Oliveira JTA, Vasconcelos IM, Scalabrin FB, Rocha TL, Grossi-de-Sá MF, Silva LP, Guadagnin RV, Quirino BF, Castro CFS, Leonardecz E, Franco OL (2010b) Identification of *E. dysenterica* laxative peptide: a novel strategy in the treatment of chronic constipation and irritable bowel syndrome. Peptides 31(8):1426–1433. https://doi.org/10.1016/j.peptides.2010.05.003

Lima TB, Silva ON, Silva LP, Rocha TL, Grossi-de-Sa MF, Franco OL, Leonardecz E (2011) In vivo effects of cagaita (*Eugenia dysenterica*, DC.) leaf extracts on diarrhea treatment. Evid Based Complement Alternat Med 2011:1–10. https://doi.org/10.1155/2011/309390

List PH, Schmidt PC (1990) Phytopharmaceutical technology. CRC Press, Boca Raton

Lopes PSN, Pereira AV, Pereira EBC, Martins ER, Fernandes RC (2016) *Caryocar brasiliense* – Pequi. In: Vieira RF, Camillo J, Corandin L (eds) Espécies nativas da flora brasileira de valor econômico atual ou potencial: plantas para o futuro: Região Centro-Oeste, vol 44. Série Biodiversidade. MMA, Brasília, DF,

Lorenzi H (2008) Árvores brasileiras: manual de identificação e cultivo de plantas arbóreas nativas do Brasil, vol 1. 5 ed. edn. Instituto Plantarum de Estudos da Flora, Nova Odessa

Macedo M, Ferreira AR (2004) Plantas medicinais usadas para tratamentos dermatológicos, em comunidades da Bacia do Alto Paraguai, Mato Grosso. Rev Bras 14:40–44

Machado MTC, Mello BCBS, Hubinger MD (2013) Study of alcoholic and aqueous extraction of pequi (*Caryocar brasiliense* Camb.) natural antioxidants and extracts concentration by nanofiltration. J Food Eng 117(4):450–457. https://doi.org/10.1016/j.jfoodeng.2012.12.007

Machado MTC, Mello BCBS, Hubinger MD (2015) Evaluation of pequi (*Caryocar brasiliense* Camb.) aqueous extract quality processed by membranes. Food Bioprod Process 95:304–312. https://doi.org/10.1016/j.fbp.2014.10.013

Maia JGS, Andrade EHA, Silva MHL (2008) Aroma volatiles of pequi fruit (*Caryocar brasiliense* Camb.). J Food Compos Anal 21(7):574–576. https://doi.org/10.1016/j.jfca.2008.05.006

Malta LG, Ghiraldini FG, Reis R, Oliveira MV, Silva LB, Pastore GM (2012) *In vivo* analysis of antigenotoxic and antimutagenic properties of two Brazilian The Cerrado fruits and the identification of phenolic phytochemicals. Food Res Int 49(1):604–611. https://doi.org/10.1016/j.foodres.2012.07.055

Maresca D, Prisco A, La Storia A, Cirillo T, Esposito F, Mauriello G (2016) Microencapsulation of nisin in alginate beads by vibrating technology: preliminary investigation. LWT Food Sci Technol 66:436–443. https://doi.org/10.1016/j.lwt.2015.10.062

Mariath JGR, Lima MCC, Santos LMP (1989) Vitamin A activity of buriti (*Mauritia vinifera* Mart) and its effectiveness in the treatment and prevention of xerophthalmia. Am J Clin Nutr 49(5):849–853. https://doi.org/10.1093/ajcn/49.5.849

Marinho DG, Alviano DS, Matheus ME, Alviano CS, Fernandes PD (2011) The latex obtained from *Hancornia speciosa* Gomes possesses anti-inflammatory activity. J Ethnopharmacol 135(2):530–537. https://doi.org/10.1016/j.jep.2011.03.059

Marques FG, Neto JRO, Cunha LC, Paula JR, Bara MTF (2015) Identification of terpenes and phytosterols in *Dipteryx alata* (baru) oil seeds obtained through pressing. Rev Bras 25(5):522–525. https://doi.org/10.1016/j.bjp.2015.07.019

Martineli M, Magalhães PAC, Castricini A, Rocha LAC, Rivas JC (2020) Quality and consumer acceptance of dehydrated araticum pulp. Rev Sítio Novo 4(1):276–286

Martins FS, Borges LL, Paula JR, Conceição EC (2013) Impact of different extraction methods on the quality of *Dipteryx alata* extracts. Rev Bras 23:521–526

Matias MFO, Oliveira EL, Gertrudes E, Magalhães MMA (2005) Use of fibres obtained from the cashew (A*nacardium ocidentale*, L) and guava (P*sidium guayava*) fruits for enrichment of food products. Braz Arch Biol Technol 48:143–150

Mendonça KS, Corrêa JLG, Junqueira JRJ, Cirillo MA, Figueira FV, Carvalho EEN (2017) Influences of convective and vacuum drying on the quality attributes of osmo-dried pequi (*Caryocar brasiliense* Camb.) slices. Food Chem 224:212–218. https://doi.org/10.1016/j.foodchem.2016.12.051

Mohanty S, Ray P, Swain MR, Ray RC (2006) Fermentation of cashew (*Anacardium occidentale* L.) "apple" into wine. J Food Process Preserv 30(3):314–322. https://doi.org/10.1111/j.1745-4549.2006.00067.x

Moraes TM, Rodrigues CM, Kushima H, Bauab TM, Villegas W, Pellizzon CH, Brito ARMS, Hiruma-Lima CA (2008) *Hancornia speciosa*: indications of gastroprotective, healing and anti-*Helicobacter pylori* actions. J Ethnopharmacol 120(2):161–168. https://doi.org/10.1016/j.jep.2008.08.001

Moraes C, Anjos JLV, Maruno M, Alonso A, Rocha-Filho P (2018) Development of lamellar gel phase emulsion containing baru oil (*Dipteryx alata* Vog.) as a prospective delivery system for cutaneous application. Asian J Pharm Sci 13(2):183–190. https://doi.org/10.1016/j.ajps.2017.09.003

Morais MC, Almeida PHG, Ferreira NLO, Arruda RL, Borges LL, Freitas O, Conceição EC (2018) Validation of a photostability indicating method for quantification of furanocoumarins from *Brosimum gaudichaudii* soft extract. Rev Bras 28:118–123

Morzelle MC, Souza EC, Assumpção CF, Vilas Boas BM (2011) Development and sensory evaluation of mixed nectar of passion fruit (*Passiflora edulis* Sims) and araticum fruit (*Annona crassiflora*). Rev Bras Produtos Agroindustriais 13(2):131–135

Myers N, Mittermeier RA, Mittermeier CG, da Fonseca GA, Kent J (2000) Biodiversity hotspots for conservation priorities. Nature 403(6772):853–858. https://doi.org/10.1038/35002501

Narain N, França FRM, Neta MTSL (2018) Mangaba—*Hancornia speciosa*. In: Rodrigues S, de Oliveira SE, de Brito ES (eds) Exotic fruits. Academic Press, pp 305–318. https://doi.org/10.1016/B978-0-12-803138-4.00040-X

Neri-Numa IA, Sancho RAS, Pereira APA, Pastore GM (2018a) Small Brazilian wild fruits: nutrients, bioactive compounds, health-promotion properties and commercial interest. Food Res Int 103:345–360. https://doi.org/10.1016/j.foodres.2017.10.053

Neri-Numa IA, Soriano Sancho RA, Pereira APA, Pastore GM (2018b) Small Brazilian wild fruits: nutrients, bioactive compounds, health-promotion properties and commercial interest. Food Res Int 103:345–360. https://doi.org/10.1016/j.foodres.2017.10.053

Neves JS, Franchin M, Rosalen PL, Omar NF, Santos MA, Paschoal JAR, Novaes PD (2016) Evaluation of the osteogenic potential of *Hancornia speciosa* latex in rat calvaria and its phytochemical profile. J Ethnopharmacol 183:151–158. https://doi.org/10.1016/j.jep.2016.02.041

Nunes AA, Favaro SP, Miranda CHB, Neves VA (2017) Preparation and characterization of baru (*Dipteryx alata* Vog) nut protein isolate and comparison of its physico-chemical properties with commercial animal and plant protein isolates. J Sci Food Agric 97(1):151–157. https://doi.org/10.1002/jsfa.7702

Oliveira MES, Pantoja L, Duarte WF, Collela CF, Valarelli LT, Schwan RF, Dias DR (2011) Fruit wine produced from cagaita (*Eugenia dysenterica* DC) by both free and immobilised yeast cell fermentation. Food Res Int 44(7):2391–2400. https://doi.org/10.1016/j.foodres.2011.02.028

Oliveira LG, Moreno LG, Melo DS, Costa-Pereira LV, Carvalho MMF, Silva PHE, Alves AM, Magalhães FC, Dias-Peixoto MF, Esteves EA (2017) *Caryocar brasiliense* oil improves cardiac function by increasing Serca2a/PLB ratio despite no significant changes in cardiovascular risk factors in rats. Lipids Health Dis 16:37. https://doi.org/10.1186/s12944-017-0422-9

Oliveira ECT, Gonçalves CAA, Costa LL, Mesquita MS, Rigo GMO, Silva MR, Jardim FBB, Oliveira Filho JHO, Santos ENF (2018a) Licor de araticum (Annona crassiflora Mart.). Boletim Técnico IFTM 4(2):6–13

Oliveira GR, Lima CB, Ribeiro LMCS, Café MB, Moreira JS, Oliveira EM, Racanicci AMC (2018b) Adição de óleo de copaíba (*Copaifera langsdorffii*) e sucupira (*Pterodon emarginatus*) na alimentação de poedeiras: qualidade física de ovos armazenados em diferentes temperaturas. Ciência Animal Brasileira 19:1–12

Oliveira CC, Matos NA, Veloso CC, Lage GA, Pimenta LPS, Duarte IDG, Romero TRL, Klein A, Perez AC (2019) Anti-inflammatory and antinociceptive properties of the hydroalcoholic fractions from the leaves of *Annona crassiflora* Mart. in mice. Inflammopharmacology 27(2):397–408. https://doi.org/10.1007/s10787-017-0426-0

Omorodion NT, Iyare GI, Erameh TO, Achukwu PU, Ogochukwu AG (2017) The effects of *Anacardium occidentale* leave extract on histology of selected organs of Wistar rats. MOJ Biol Med 2(2):216–221. https://doi.org/10.15406/mojbm.2017.02.00046

Osorio C, Carriazo JG, Barbosa H (2011) Thermal and structural study of guava (*Psidium guajava* L) powders obtained by two dehydration methods. Química Nova 34:636–640

Overbeck GE, Vélez-Martin E, Scarano FR, Lewinsohn TM, Fonseca CR, Meyer ST, Müller SC, Ceotto P, Dadalt L, Durigan G, Ganade G et al (2015) Conservation in Brazil needs to include non-forest ecosystems. Divers Distrib 21(12):1455–1460. https://doi.org/10.1111/ddi.12380

Paglarini CS, Queirós MS, Tuyama SS, Moreira ACV, Chang YK, Steel CJ (2018) Characterization of baru nut (*Dipteryx alata* Vog) flour and its application in reduced-fat cupcakes. J Food Sci Technol 55(1):164–172. https://doi.org/10.1007/s13197-017-2876-1

Palmeira SM, Silva PRP, Ferrão JSP, Ladd AABL, Dagli MLZ, Grisolia CK, Hernandez-Blazquez FJ (2016) Chemopreventive effects of pequi oil (*Caryocar brasiliense* Camb.) on preneoplastic lesions in a mouse model of hepatocarcinogenesis. Eur J Cancer Prev 25(4):299–305

Patel P, Ellis K, Sunkara R, Shackelford L, Ogutu S, Walker LT, Herring J, Verghese M (2016) Development of a functional food product using guavas. Food Nutr Sci 7:927–937

Patridge E, Gareiss P, Kinch MS, Hoyer D (2016) An analysis of FDA-approved drugs: natural products and their derivatives. Drug Discov Today 21(2):204–207. https://doi.org/10.1016/j.drudis.2015.01.009

Penido AB, Morais SM, Ribeiro AB, Alves DR, Rodrigues ALM, Santos LH, Menezes JESA (2017) Medicinal plants from Northeastern Brazil against Alzheimer's disease. Evid Based Complement Alternat Med 2017:1753673. https://doi.org/10.1155/2017/1753673

Pereira J, Freitas B (2002) Estudo da biologia floral e requerimentos de polinização do muricizeiro (*Byrsonima crassifolia* L.). Rev Ciênc Agron 33:5–12

Pereira AC, Pereira ABD, Moreira CCL, Botion LM, Lemos VS, Braga FC, Cortes SF (2015) *Hancornia speciosa* Gomes (Apocynaceae) as a potential anti-diabetic drug. J Ethnopharmacol 161:30–35. https://doi.org/10.1016/j.jep.2014.11.050

Pereira-Freire JA, Amorim VR, Ribeiro FC, Medeiros SR, Silva JD, Ferreira PM (2019) Compostos bioativos e potencial nutracêutico do fruto de buriti (*Mauritia flexuosa* L) na terapia coadjuvante em portadores de dislipidemia. In: Cardoso NA, Rocha RR, Laurindo MV (eds) As Ciências Biológicas e da Saúde na contemporaneidade, vol 2. Atena Editora, Ponta Grossa, pp 103–115

Pineli LLO, Carvalho MV, Aguiar LA, Oliveira GT, Celestino SMC, Botelho RBA, Chiarello MD (2015) Use of baru (Brazilian almond) waste from physical extraction of oil to produce flour and cookies. LWT Food Sci Technol 60(1):50–55. https://doi.org/10.1016/j.lwt.2014.09.035

Pinto MRMR, Paula DA, Alves AI, Rodrigues MZ, Vieira ÉNR, Fontes EAF, Ramos AM (2018) Encapsulation of carotenoid extracts from pequi (*Caryocar brasiliense* Camb) by emulsification (O/W) and foam-mat drying. Powder Technol 339:939–946. https://doi.org/10.1016/j.powtec.2018.08.076

Resende LM, Franca AS, Oliveira LS (2019) Buriti (*Mauritia flexuosa* L. f.) fruit by-products flours: evaluation as source of dietary fibers and natural antioxidants. Food Chem 270:53–60. https://doi.org/10.1016/j.foodchem.2018.07.079

Rishika D, Sharma R (2012) An update of pharmacological activity of *Psidium guajava* in the management of various disorders. Int J Pharm Sci Res 3(10):3577–3584

Roesler R, Catharino RR, Malta LG, Eberlin MN, Pastore G (2008) Antioxidant activity of *Caryocar brasiliense* (pequi) and characterization of components by electrospray ionization mass spectrometry. Food Chem 110(3):711–717. https://doi.org/10.1016/j.foodchem.2008.02.048

Sampaio MB, Schmidt IB, Figueiredo IB (2008) Harvesting effects and population ecology of the buriti palm (*Mauritia flexuosa* L. f., Arecaceae) in the Jalapão Region, Central Brazil. Econ Bot 62(2):171. https://doi.org/10.1007/s12231-008-9017-8

Sandri DO, Xisto ALRP, Rodrigues EC, Morais EC, Barros WM (2017) Antioxidant activity and physicochemical characteristics of buriti pulp (*Mauritia flexuosa*) collected in the city of Diamantino – MT. Rev Bras Frutic 39(3)

Sano SM, Brito MA, Ribeiro JF (2016) *Dipteryx alata*: baru. In: Vieira RF, Camillo J, Corandin L (eds) Espécies nativas da flora brasileira de valor econômico atual ou potencial: plantas para o futuro: Região Centro-Oeste, vol 44. Série Biodiversidade. MMA, Brasília

Santos AP, Zatta DT, Moraes WF, Bara MTF, Ferri PH, Silva MRR, Paula JR (2010) Composição química, atividade antimicrobiana do óleo essencial e ocorrência de esteróides nas folhas de *Pterodon emarginatus* Vogel, Fabaceae. Rev Bras 20:891–896

Santos CA, Ribeiro RC, Silva EVC, Silva NS, Silva BA, Silva GF, Barros BCV (2011) Elaboration of biscuit flour buriti (*Mauritia flexuosa* L. f) without added oat (*Avena sativa* L.). Rev Bras Tecnologia Agroindustrial 5(1):262–273

Santos PRG, Cardoso LM, Bedetti SF, Hamacek FR, Moreira AVB, Martino HSD, Pinheiro-Sant'Ana HM (2012) Geleia de cagaita (*Eugenia dysenterica* DC.): desenvolvimento, caracterização microbiológica, sensorial, química e estudo da estabilidade. Rev Instituto Adolfo Lutz (Impresso) 71:281–290

Santos RM, Santos AO, Sussuchi EM, Nascimento JS, Lima AS, Freitas LS (2015) Pyrolysis of mangaba seed: production and characterization of bio-oil. Bioresour Technol 196:43–48. https://doi.org/10.1016/j.biortech.2015.07.060

Santos P, Aguiar AC, Viganó J, Boeing JS, Visentainer JV, Martínez J (2016) Supercritical CO2 extraction of cumbaru oil (*Dipteryx alata* Vogel) assisted by ultrasound: global yield, kinetics and fatty acid composition. J Supercrit Fluids 107:75–83. https://doi.org/10.1016/j.supflu.2015.08.018

Santos EA, Bessa VR, Bonnas DS (2017) Development of a goat milk yogurt flavored with the mangaba fruit (*Hancornia speciosa* Gomes) Iniciação Científica. CESUMAR 19(1):35–45

Satyal P, Paudel P, Lamichhane B, Setzer WN (2015) Leaf essential oil composition and bioactivity of *Psidium guajava* from Kathmandu, Nepal. Am J Essent Oils Nat Prod 3(2):11–14

Schiassi MCEV, Souza VR, Lago AMT, Campos LG, Queiroz F (2018) Fruits from the Brazilian The Cerrado region: physico-chemical characterization, bioactive compounds, antioxidant activities, and sensory evaluation. Food Chem 245:305–311. https://doi.org/10.1016/j.foodchem.2017.10.104

Schuck P, Dolivet A, Méjean S, Zhu P, Blanchard E, Jeantet R (2009) Drying by desorption: a tool to determine spray drying parameters. J Food Eng 94(2):199–204. https://doi.org/10.1016/j.jfoodeng.2008.08.014

Silva JMC, Bates JM (2002) Biogeographic patterns and conservation in the south American the cerrado: a tropical savanna hotspotthe cerrado, which includes both forest and savanna habitats, is the second largest South American biome, and among the most threatened on the continent. Bioscience 52(3):225–234. https://doi.org/10.1641/0006-3568(2002)052[0225:BPACIT]2.0.CO;2

Silva JG, Souza IA, Higino JS, Siqueira-Júnior JP, Pereira JV, Pereira MSV (2007) Atividade antimicrobiana do extrato de *Anacardium occidentale* Linn. em amostras multiresistentes de *Staphylococcus aureus*. Rev Bras 17:572–577

Silva GC, Braga FC, Lima MP, Pesquero JL, Lemos VS, Cortes SF (2011) *Hancornia speciosa* Gomes induces hypotensive effect through inhibition of ACE and increase on NO. J Ethnopharmacol 137(1):709–713. https://doi.org/10.1016/j.jep.2011.06.031

Silva TA, Assunção RMN, Vieira AT, Oliveira MF, Batista ACF (2014) Methylic and ethylic biodiesels from pequi oil (*Caryocar brasiliense* Camb.): production and thermogravimetric studies. Fuel 136:10–18. https://doi.org/10.1016/j.fuel.2014.07.035

Silva CDM, Pires CRF, Lima JP, Pereira AS, Silva CA (2015a) Osmotic dehydration to obtain cagaita raisins. J Bioenergy Food Sci 2(4):226–233

Silva SMM, Silva CAG, Fonseca-Bazzo YM, Magalhães PO, Silveira D (2015b) *Eugenia dysenterica* Mart. Ex DC. (cagaita): planta brasileira com potencial terapêutico. Infarma – Ciências Farmacêuticas 27(1):49–95

Silva CLM, Santos TC, Oliveira MLP, Silva LMSF, Araújo CIA, Jesus CA, Vieira CR (2018a) Centesimal composition of biscuits made with flour *Caryocar brasiliense* Camb. (Caryocaraceae). Caderno de Ciências Agrárias 10(1). https://doi.org/10.35699/2447-6218.2018.3024

Silva JS, Damiani C, Silva EP, Ruffi CRG, Asquieri ER, Silva TLL, Vilas Boas EVB (2018b) Effect of marolo (*Annona crassiflora* Mart.) pulp flour addition in food bars. J Food Qual 2018:8639525. https://doi.org/10.1155/2018/8639525

Silva SMM, Ferreira-Nunes R, Amore CR, Martins DHN, Pic-Taylor A, Fonseca-Bazzo YM, Silveira D, Gelfuso GM, Magalhães PO (2020a) Emulsion incorporating *Eugenia dysenterica* aqueous extract captured in chitosan microparticles as a novel topical treatment of cutaneous infections. J Drug Delivery Sci Technol 55:101372. https://doi.org/10.1016/j.jddst.2019.101372

Silva TM, Borges LL, Alves SF, Vaz PDS, Silva LP, Paula JR, Bara MTF, Conceição EC, Caramori SS (2020b) New spray-dried microcapsule based on Brazilian cashew polysaccharide (*Anacardium othonianum* Rizz.) and maltodextrin as wall material. J Polym Environ 28(3):854–862. https://doi.org/10.1007/s10924-019-01643-4

Silveira FAO, Negreiros D, Barbosa NPU, Buisson E, Carmo FF, Carstensen DW, Conceição AA, Cornelissen TG, Echternacht L, Fernandes GH, Garcia QS, Guerra TJ, Jacobi CM, Lemos-Filho JP, Stradic S, Morellato LPC, Neves FS, Oliveira RS, Schaefer CE, Viana PL, Lambers H (2016) Ecology and evolution of plant diversity in the endangered campo rupestre: a neglected conservation priority. Plant Soil 403(1–2):129–152

Siqueira EP, Andrade AA, Souza-Fagundes EM, Ramos JP, Kohlhoff M, Nunes YRF, Veloso MDM, Campos FF, Johann S, Alves TMA, Zani CL, Cota BB (2014) *In vitro* antibacterial action on methicillin-susceptible (MSSA) and methicillin-resistant (MRSA) *Staphylococcus aureus* and antitumor potential of *Mauritia flexuosa* L. f. J Med Plants Res 8(48):1408–1417

Soares-Filho B, Rajao R, Macedo M, Carneiro A, Costa W, Coe M, Rodrigues H, Alencar A (2014) Land use. Cracking Brazil's forest code. Science 344(6182):363–364. https://doi.org/10.1126/science.1246663

Sousa PB, Machado MRG, Moura Filho JM, Feitosa ISC, Fialho Filho A, Sousa AB, Rocha NG, Chaves CC, Silva MJM, Reis DCC, Silva J, Pinto LIF (2020) Conventional, light and diet buriti jelly: development, physical-chemical, microbiological and sensory characterization. Braz J Dev 6(4):21272–21293

Souza JLF (2015) Mesocarp pequi flour (*Caryocar brasiliense* Cambess): drying kinetics, nutritional, and functional properties and yogurt's enrichment. Dissertação, Instituto Federal Goiano, Rio Verde, Goiás, Brazil

Souza Filho MD, Medeiros JVR, Vasconcelos DFP, Silva DA, Leódido ACM, Fernandes HF, Silva FRP, França LFC, Lenardo D, Pinto GR (2018) Orabase formulation with cashew gum polysaccharide decreases inflammatory and bone loss hallmarks in experimental periodontitis. Int J Biol Macromol 107:1093–1101. https://doi.org/10.1016/j.ijbiomac.2017.09.087

Souza PL, Silva MR (2015) Quality of granola prepared with dried caju-do-the Cerrado (*Anacardium othonianum* Rizz) and baru almonds (*Dipteryx alata* Vog). J Food Sci Technol 52(3):1712–1717. https://doi.org/10.1007/s13197-013-1134-4

Souza ERB, Carneiro IF, Naves RV, Borges JD, Leandro WM, Chaves LJ (2007) Emergência e crescimento de cagaita (*Eugenia dysenterica* DC.) em função do tipo e do volume de substratos. Pesquisa Agropecuária Tropical 31(2):89–95

Souza HRS, Carvalho MG, Santos AM, Ferreira IM, Silva AMO (2018a) Compostos bioativos e estabilidade de geleia mista de umbu (*Spondias tuberosa* Arr. C.) e mangaba (*Hancornia speciosa* G.). Rev Bras Higiene e Sanidade Animal 12(2):236–248

Souza RGM, Gomes AC, Castro IA, Mota JF (2018b) A baru almond-enriched diet reduces abdominal adiposity and improves HDL concentrations: a randomized, placebo-controlled trial. Nutrition 55-56:154–160. https://doi.org/10.1016/j.nut.2018.06.001

Souza VR, Aniceto A, Abreu JP, Montenegro J, Boquimpani B, Jesuz VA, Campos MBE, Marcellini PS, Freitas-Silva O, Cadena R, Teodoro AJ (2020) Fruit-based drink sensory, physicochemical, and antioxidant properties in the Amazon region: murici (*Byrsonima crassifolia* (L.) Kunth and verbascifolia (L.) DC) and tapereba (*Spondia mombin*). Food Sci Nutr 8(5):2341–2347. https://doi.org/10.1002/fsn3.1520

Sunderam V, Thiyagarajan D, Lawrence AV, Mohammed SSS, Selvaraj A (2019) *In-vitro* antimicrobial and anticancer properties of green synthesized gold nanoparticles using *Anacardium occidentale* leaves extract. Saudi J Biol Sci 26(3):455–459. https://doi.org/10.1016/j.sjbs.2018.12.001

Teixeira GHA, Teixeira GHA, Durigan JF, Santos LO, Hojo ETD, Cunha Júnior LC (2011) Changes in the quality of jaboticaba fruit (*Myriciaria jaboticaba* (Vell) Berg. cv. Sabará) stored under different oxygen concentrations. J Sci Food Agric 91(15):2844–2849. https://doi.org/10.1002/jsfa.4530

Thomford NE, Senthebane DA, Rowe A, Munro D, Seele P, Maroyi A, Dzobo K (2018) Natural products for drug discovery in the 21st century: innovations for novel drug discovery. Int J Mol Sci 19(6):1578. https://doi.org/10.3390/ijms19061578

Torres LRO, Santana FC, Torres-Leal FL, Melo ILP, Yoshime LT, Matos-Neto EM, Seelaender MCL, Araújo CMM, Cogliati B, Mancini-Filho J (2016) Pequi (*Caryocar brasiliense* Camb.) almond oil attenuates carbon tetrachloride-induced acute hepatic injury in rats: antioxidant and anti-inflammatory effects. Food Chem Toxicol 97:205–216. https://doi.org/10.1016/j.fct.2016.09.009

Traesel GK, Menegati SELT, Santos AC, Souza RIC, Boas GRV, Justi PN, Kassuya CAL, Argandoña EJS, Oesterreich SA (2016) Oral acute and subchronic toxicity studies of the oil extracted from pequi (*Caryocar brasiliense*, Camb.) pulp in rats. Food Chem Toxicol 97:224–231. https://doi.org/10.1016/j.fct.2016.09.018

Traesel GK, Araújo FHS, Castro LHA, Lima FF, Menegati SELT, Justi PN, Kassuya CAL, Cardoso CAL, Argandoña EJS, Oesterreich SA (2017a) Safety assessment of oil from Pequi (*Caryocar brasiliense* Camb.): evaluation of the potential genotoxic and clastogenic effects. J Med Food 20(8):804–811. https://doi.org/10.1089/jmf.2017.0021

Traesel GK, Lima FF, Santos AC, Souza RIC, Cantadori DT, Kretschmer CR, Navarini VJ, Oesterreich SA (2017b) Evaluation of embryotoxic and teratogenic effects of the oil extracted from *Caryocar brasiliense* Cambess pulp in rats. Food Chem Toxicol 110:74–82. https://doi.org/10.1016/j.fct.2017.10.018

Vergara CMAC, Honorato TL, Maia GA, Rodrigues S (2010) Prebiotic effect of fermented cashew apple (*Anacardium occidentale* L) juice. LWT Food Sci Technol 43(1):141–145. https://doi.org/10.1016/j.lwt.2009.06.009

Verma AK, Rajkumar V, Banerjee R, Biswas S, Das AK (2013) Guava (*Psidium guajava* L.) powder as an antioxidant dietary fibre in sheep meat nuggets. Asian-Australas J Anim Sci 26(6):886–895. https://doi.org/10.5713/ajas.2012.12671

Vilar JB, Ferri PH, Chen-Chen L (2011) Genotoxicity investigation of araticum(*Annona crassiflora* Mart., 1841, Annonaceae) using SOS-Inductest and Ames test. Braz J Biol 71:197–202

Villela P, Batista AG, Dessimoni-Pinto NAV (2013) Nutritional composition of *Annona crassiflora* pulp and acceptability of bakery products prepared with its flour. Food Sci Technol 33:417–423

Wang B (2015) Development of spray drying technology for microencapsulation of bioactive materials. Auburn University, Auburn

Wang F, Chen YH, Zhang YJ, Deng GF, Zou ZF, Li AN, Xu DP, Li HB (2014) Chemical components and bioactivities of *Psidium guajava*. Int J Food Nutr Saf 5(2):98–114

Wu SB, Dastmalchi K, Long C, Kennelly EJ (2012) Metabolite profiling of jaboticaba (*Myrciaria cauliflora*) and other dark-colored fruit juices. J Agric Food Chem 60(30):7513–7525. https://doi.org/10.1021/jf301888y

Zhang H, Liu M, Li T, Huang Z, Sun X, Bo H, Dong Y (2017) Experimental investigation on gas-solid hydrodynamics of coarse particles in a two-dimensional spouted bed. Powder Technol 307:175–183. https://doi.org/10.1016/j.powtec.2016.11.024

Food Plants in the Caatinga

Daline Fernandes de Souza Araújo, Maria Elieidy Gomes de Oliveira, Paloma Oliveira Antonino Assis de Carvalho, Emanuella de Aragão Tavares, Gerlane Coelho Bernardo Guerra, Rita de Cássia Ramos do Egypto Queiroga, Silvana Maria Zucolotto Langassner, Fabrícia França Bezerril, Ana Cristina Silveira Martins, Gracy Kelly Vieira de Vasconcelos Medeiros, Tamires Alcântara Dourado Gomes Machado, and Nadja Maria da Costa Melo

Contents

D. F. S. Araújo (✉)
Health Sciences College of Trairi, Federal University of Rio Grande do Norte, Santa Cruz, Rio Grande do Norte, Brazil

M. E. G. de Oliveira · P. O. A. A. de Carvalho · R. C. R. E. Queiroga · F. F. Bezerril
G. K. V. V. Medeiros · T. A. D. G. Machado
Departament of Nutrition, Federal University of Paraíba, João Pessoa, Paraíba, Brazil

E. d. A. Tavares · S. M. Z. Langassner
Departament of Phamacy, Federal University of Rio Grande do Norte, Natal, Rio Grande do Norte, Brazil

G. C. B. Guerra · N. M. C. Melo
Department of Biophysics and Pharmacology, Federal University of Rio Grande do Norte, Natal, Rio Grande do Norte, Brazil

A. C. S. Martins
Food of Engineering, Federal University of Paraíba, João Pessoa, Paraíba, Brazil

M. C. M. Jacob, U. P. Albuquerque (eds.), *Local Food Plants of Brazil*, Ethnobiology, https://doi.org/10.1007/978-3-030-69139-4_11

1 Introduction

"Famine foods," "alternative food plants," "wild edible plants," "unconventional vegetables," "traditional vegetables," and "unconventional food plants" are names given to a group of underutilized plants which have received these designations in reaction to their extinction due to the expansion of monocultures (Bhandari 1974; Brasil 2010; Kinupp 2004; Leal et al. 2018; Uprety et al. 2012). These native and exotic plants are cultivated spontaneously in nature; however, it appears that they have been little explored. They are often wiped out to make room for the production of other foods that boost the economy, favoring the reduction of an area's biodiversity (Leal et al. 2018). An example of this destruction is cacti that represent a large number and variety of species, found throughout the world due to their great adaptive capacity. Their ecological advantages can be attributed to crassulacean acid metabolism, which allows absorption of CO_2 during the night, reducing the loss of water during the process of photosynthesis (Guevara-Figueroa et al. 2010; Mancuso 2019).

The Cactaceae family is widely distributed, encountered from Canada, across the USA, Mexico, and Central and South America. In some countries, such as Mexico and Colombia, several species of Cactaceae family are used in folk medicine. The family is subdivided into three subfamilies Cactoideae, Opuntioidea, Pereskioideae (Barthlott and Hunt 1993) and a new subfamily Maihunioideae qualified to Maihuenia (Weber) Schumann (Anderson 2001).

In Brazil, species like *Cereus jamacaru* DC., *Harrisia adscendens* (Gürke) Britton & Rose, *Opuntia ficus-indica* (L.) Mill, and *Pilosocereus gounellei* are popularly used in various diseases as analgesics, antibiotics, diuretics and for coughs and heart disease and to cure certain types of ulcers (Andrade et al. 2006; Lucena et al. 2012). In this chapter, the nutritional composition, biological properties, and use of some of the Caatinga food plants will be presented.

2 *Cereus jamacaru* DC. (Mandacaru)

2.1 Species Characteristics

The *Cereus jamacaru* is a typical species of the Caatinga biome, located in the semiarid region in northeastern Brazil, including almost all its states (Ceará, Rio Grande do Norte, Paraíba, Pernambuco, Alagoas, Sergipe, Bahia), and extending to the southeast region, in the north of Minas Gerais. This species is also known as mandacaru, cardeiro, jamacaru, thistle, cardon, caxabú, and wild fig (Andrade et al. 2006; Lucena et al. 2015; Santos et al. 2017; Silva et al. 2019). Other species of *Cereus* are found in other regions of the country. Microsatellites have identified a total of 31 alleles in mandacaru plants, with those from the northeast region showing the lowest polymorphism when compared to their relatives from the south, southeast, and midwest regions (Fernandes et al. 2016).

C. jamacaru is usually found on stony soils, alongside other species of cacti. It is formed by porous pulp cladodes and full of thorns, which bear fruit in the rainiest months, between February and May, and can reach 3–7 m in height (Rocha and Agra 2002). Its fruits have an ovoid shape, with approximately 12 cm in length, white pulp with numerous black and very small seeds, reminiscent of the pitaya fruit, and are very juicy (Fig. 1). The flowers open only at night, between January

Fig. 1 Pictures of *Cereus jamacaru* DC. (mandacaru): (**A**) fruit; (**B**) cladode (Font: The authors)

and August, being visited by moths and bats (Rosado and Rosado 1960). Due to their specific morphological and physiological characteristics, they can withstand high temperatures and a long period of drought or low water availability.

2.2 *Popular Use and Possible Application*

Because they are spongy, cacti can store water for long periods, serving as food for animals in times of drought (Sales et al. 2014). Cacti have become popular as ornamental plants in addition to their potential as a source of substances for medicinal, cosmetic, and food use (Biavatti et al. 2007; Park et al. 2001). Human consumption of cactus is reported to have begun in Brazil in the 1980s by groups considered vulnerable, suffering from famine and drought (Santos et al. 2001). Currently, there has been increased interest in studying its nutrients, health effects, and preparation as food products.

Medicinal properties have been attributed to the mandacaru, for example, as infusion for treatment of various diseases, including renal, hepatic, respiratory, stomach, and sinus problems (Agra et al. 2007; Albuquerque et al. 2007; Saraiva et al. 2015; Silva et al. 2014). Other studies have reported the cytotoxic activity of an aqueous extract (non-protein) from the stem (Silva 2015); antibacterial activity of ethanolic extract from the stem (Davet et al. 2009); as well as the purification and characterization of two proteins from *C. jamacaru* seeds (Aragão et al. 2000; da Costa et al. 2001).

In addition to its medicinal properties, mandacaru is used as fodder in times of drought. Its branches are used in civil construction, to make wooden spoon, doors, windows, boards, and slats, in addition to its wide use as an ornamental plant (Agra et al. 2008; Lucena et al. 2013; Rodrigues and Elesb 2009). The mandacaru has also been cited as a bio-indicator of natural phenomena, its flowering being the sign of a good winter (Lucena et al. 2012).

2.3 *Physical-Chemical Characteristics and Nutritional Composition*

The mandacaru fruit, specifically its pulp, presents in its physical-chemical composition significant values of proteins (1.8–2.35 g/100 g), lipids (1.08–1.98 g/100 g), carbohydrates (9.76–9.86 g/100 g), minerals (0.20–1.30 g/100 g), total soluble solids (TSS; 10.3–14.93 g/100 g), pH (3.73–4.93), titratable acidity (TA; 0.26–0.32 g/100 g), TSS/TA (32.65–50.15), moisture (82.75–90.58 g/100 g), and total phenolic compounds (28.35–326.78 mg EGA/100 g) (Almeida et al. 2009; Bahia et al. 2010; Melo et al. 2017; Moreira et al. 2018; Nascimento et al. 2011; Santos 2018; Santos Neto et al. 2019). These differences may be related to the stage of maturation, location, cultivation attributes (such as scarcity of water and soil nutrients), and climatic changes in the year in which the samples used in the aforementioned studies were collected.

A study aiming at the physical-chemical characterization, carried out by our research group with the cladode and the fruit of the mandacaru collected in different regions of the state of Paraíba, Brazil, evaluated the following parameters: pH, molar acidity, moisture, ash, proteins, lipids, insoluble and soluble fiber contents by enzymatic-gravimetric method (AOAC 2016), and total phenolic and flavonoid contents by procedures described by Liu et al. (2002), Sousa et al. (2011), and Zhishen et al. (1999). In vitro antioxidant activity of cladode and fruit of the mandacaru was assessed by an iron reduction method (ferric reducing antioxidant power-FRAP) (Rockenbach et al. 2011) and the ABTS method (Sariburun et al. 2010). Values of the measured physicochemical parameters of cladode and fruit of the mandacaru are presented in Table 1.

Lima (2016), Oliveira et al. (2004), and Sousa et al. (2014) highlight that the pulp of the mandacaru fruit is semi-acidic and low in vitamin C (~10 mg/100 g), carotenoids (~0.06 mg/100 g), and anthocyanin (0.23–1.83 mg/100 g), having high levels of calcium (~585 mg/100 g), magnesium (~238 mg/100 g), and potassium (~136 mg/100 g). The small amount of carotenoids and anthocyanins may be related to the presence of other pigments, such as betalains (Lima 2016). Bahia et al. (2010), in a study of the physicochemical characteristics of the mandacaru fruit, identified other carbohydrates in the fruit pulp, such as soluble fibers (pectin 4.36%) and

Table 1 Physical-chemical characterization of *Cereus jamacaru* DC (mandacaru)

	Parts of mandacaru	
Variables	Cladode	Fruit
pH	4.88 ± 0.31	5.03 ± 0.29
Molar acidity (g/100 g)	2.61 ± 0.50	0.38 ± 0.25
Moisture (g/100 g)	91.86 ± 0.88	89.70 ± 0.64
Ash (g/100 g)	1.53 ± 0.93	0.40 ± 0.01
Proteins (g/100 g)	0.82 ± 0.25	1.60 ± 0.17
Lipids (g/100 g)	0.74 ± 0.39	0.60 ± 0.17
Total fibers (g/100 g)	7.24 ± 1.20	4.13 ± 0.83
Soluble fibers (g/100 g)	4.57 ± 0.59	1.99 ± 0.19
Insoluble fibers (g/100 g)	2.67 ± 0.61	2.14 ± 0.64
Total phenolic (mg EGA/100 g)[a]	14.47 ± 3.10	14.79 ± 5.54
Antioxidant activity – FRAP (μmol TEAC/g)[b]	0.28 ± 0.22	0.39 ± 0.04
Antioxidant activity – $ABTS^{\bullet+}$ (μmol TEAC/g)[b, c]	1.84 ± 0.56	3.43 ± 0.57

Font: The authors

[a]The results are expressed in milligram equivalents of gallic acid (EGA) per hundred grams of sample (mg EGA/100 g)

[b]The results are expressed as micromoles of Trolox equivalent antioxidant capacity (TEAC) per hundred grams of sample (μmol TEAC/100 g)

[c]$ABTS^{\bullet+}$ cation - 2,2-azino-bis (3-etilbenzo-tiazoline)-6-sulfonic acid

insoluble fibers (total fibers 0.88%), which are important for human health, as these balance the absorption of blood fats, sugar, and cholesterol.

The literature also shows that with regard to antioxidant activities ($ABTS^+$ and DPPH), the peel of mandacaru fruit *in natura* presents greater activities (11.62 ± 1.34μmol Trolox/g and 8.46 ± 0.90 g/g of DPPH, respectively) when compared to the pulp (9.71 ± 0.52μmol Trolox/g and 6.93 ± 0.86 g/g of DPPH, respectively). Pasteurization processes decrease these activities (Santos et al. 2020). Some studies have revealed the presence of tannins and flavonoids in the cladode and fruit of *C. jamacaru* (Davet et al. 2009; Dutra et al. 2019; Nascimento et al. 2011), which could be associated with these antioxidant properties in addition to its anti-inflammatory, antifungal, and anticancer activities, as well as others.

According to Coelho et al. (2004), the seeds of the mandacaru fruit constitute a significant source of carbohydrates (~66 g/100 g), proteins (~17 g/100 g), and lipids (~5 g/100 g), with nutritionally interesting levels for food and feed. Mayworm and Salatino (1996) in characterizing the oil extracted from the seeds of *C. jamacaru* reported that the mandacaru seed oils are rich in unsaturated fatty acids, mainly oleic acid (30.2%) and linoleic acid (43.4%), but saturated, palmitic (14.6%) and stearic (3.7%) oils were also found. Further, according to these authors, the composition obtained from mandacaru seeds is similar to that found in soybean oil, which is why this author suggests both species may have common uses.

Silva (2017) evaluating the phytochemical profile and cytotoxic activity of hydroalcoholic extract of *C. jamacaru* DC. by HPLC, FTIR, and UV-VIS observed the absorbance of several metabolites with important therapeutic properties, such as gallic acid, ferrulic acid, caffeine, quercitin, and rutin. In that study, toxicity assessment was performed using *Artemia salina* cysts, resulting in an LC50 of 1509.17μg/mL, considered non-toxic, as its LC50 was greater than 1000μg/mL. In addition, cytotoxic activity was analyzed in the cell lines NCI-H292 (pulmonary mucoepidermoid carcinoma), HEp-2 (laryngeal squamous cell carcinoma), MCF-7 (human breast adenocarcinoma), and HL-60 (promyelocytic leukaemia cell). Thus, the extract showed a higher percentage of cell growth inhibition given NCI-H292 cell lines of 24.1%. Hemolytic activity showed a percentage of hemolysis of 3.33%, considered low.

2.4 Biological Properties

Mandacaru is a cactaceous symbol of Caatinga vegetation in the Brazilian Northeast region; however, there are few studies about the biological properties of this species. Much of the research with mandacaru is focused on ethnobotanical studies, showing that this cactus is widely used in traditional medicine in the form of teas prepared from the root and used to treat diseases such as rheumatism; wounds; urinary infections; kidney (Albuquerque et al. 2007; Lucena et al. 2013), liver, and respiratory problems; flu; cough; bronchitis; constipation; nausea; vomiting; hypertension (Albuquerque et al. 2007); ovarian cysts; and menstrual regulation (Saraiva et al.

2015) and to combat scurvy (Paulino et al. 2011; Scheinvar 1985). The mandacaru is also used as a food alternative for animals and humans, especially in the dry season (Lucena et al. 2013; Peron 2011; Rodrigues and Elesb 2009).

Other research has shown that the mandacaru has benefits in the treatment of obesity and has anti-cytotoxic, antitumor, and antioxidant properties, which makes it a matrix with great potential for the development of new drugs (Dutra et al. 2018; Dutra et al. 2019; Medeiros et al. 2019).

Mota et al. (2019) were the first to report the prospecting of proteins of biotechnological importance in different parts of the *C. jamacaru* (stem, roots, fruit peel, and seed) by enzymatic activities and protease inhibitory activity. Results of this study identified the presence of protease, peroxidase, chitinase, β-1,3-glucanase, and protease inhibitors, mainly for protein extract of the root. This extract presents significant antifungal activity against the *Colletotrichum gloeosporioides*, which causes anthracnose in the fruit from several plant species. The inhibitory activity in the vegetative development of the phytopathogen results from morphological alterations in the cell surface, increased permeability in the membrane, and induction of reactive oxygen species, events that culminate in cell death of the fungus. In addition, the research demonstrated that the *C. jamacaru* root extract was able to inhibit the germination of *C. gloeosporioides* spores.

Davet et al. (2009) verified the antibacterial effect of the crude ethanolic extract of wood (EBLE) and the cortex (EBCO) of *C. jamacaru* against pure colonies of eight pathogenic microorganisms. Further, they showed that EBCO had more pronounced antimicrobial activity than EBCE and that the microorganisms whose growth was most influenced by EBCO were *Streptococcus epidermidis*, *Staphylococcus aureus*, *Pseudomonas aeruginosa*, and *Escherichia coli*.

It is known that vegetables, including the Cactaceae family, have an autochthonous microbiota and that certain bacteria may have a probiotic effect. The probiotic effect and its benefits on the intestinal health and on the general health status of an individual have been increasingly studied. Allied to this, the demand for foods with functional potential and for non-dairy matrix probiotics has increased over time. In view of the above, our research group has been conducting tests with mandacaru for the following purposes: (*i*) to isolate and identify native bacteria from the fruit and mandacaru cladode, (*ii*) to evaluate the probiotic potential of isolated bacteria, (*iii*) to evaluate the prebiotic and protective potential of the lyophilized cladode on these isolated strains, and (*iv*) to ferment the mandacaru fruit with the native microorganisms and to evaluate in vitro bioactive activities of the fermented fruit. For this experiment, 12 collections have been carried out to date, totaling 377 isolates. Of these, 58 isolates passed the gram stain test, assessment for catalase activity, and remained viable for the next stages of the study. Studies of this magnitude are scarce on the subject in question, and the research findings are expected to be of importance. The research that has been carried out with this cactus, especially by our research group, is expected to contribute to the valorization of mandacaru as a source of bioactive compounds with a positive attraction for consumer health. This in turn will stimulate the agroindustrial sector to valorize this matrix as an ingredient with added value for the functional food industry.

Due to the fiber and phenolic content and, consequently, the antioxidant capacity, mandacaru stands out as a matrix of interest for the development of nutraceutical products that will add functional characteristics and thus increase its market potential. Martins (2018) studied the development of a prebiotic goat yogurt using gelatinous substances from mandacaru and passion fruit (*Passiflora edulis* Sims) to evaluate their physicochemical characteristics. This gel was found to be in compliance with Brazilian Resolution n° 12 of 1978 (Brasil 1978), regarding the relative values for moisture (10.04 ± 0.09 g/100 g) and total soluble solids (65.5 ± 0.70 g/100 g). In addition, parameters such as pH (3.28 ± 0.04), molar acidity (33.28 ± 1.31 g/100 g), ash (0.74 ± 0.03 g/100 g), and proteins (0.77 ± 0.00 g/100 g) showed the potential of the gel being added to other food matrices, such as fermented milk, thus improving and/or adding nutritional and sensory value to these types of food products.

Since 2002, the World Health Organization has stimulated the recuperation of data from plants used in ancient medical practices, as these are considered potentially useful in the development of new drugs (World Health Organization 2002). The American plants were widely used long before the arrival of the Europeans to the continent in the fifteenth century. Besides being one of the richest countries in biodiversity, Brazil is also one of the most diverse in terms of Amerindian culture (Forzza et al. 2012; Neves 2006).

Studies on the mandacaru fruit indicate its potential for exploitation in various technological applications and industrial processes, such as the production of fermented drinks and preparation of dehydrated fruit powder, sweets, jellies, and ice cream (Almeida et al. 2006; Almeida et al. 2011; Moreira et al. 2018; Oliveira et al. 2015; Santos Neto et al. 2019).

Despite this, research on the economic potential of use of mandacaru fruit for food and beverages or its manufacture by agribusiness is scarce. Moreover, there is a need for scientific studies on post-harvest usability, mainly related to enzymatic browning, a limiting factor for valorization in the productive chain of this fruit. This suggests that further studies should be encouraged on the sustainable use of this fruit species in view of its cultural valorization of the Northeast and its contribution of extra resources to the population living in the region, especially in times of drought.

3 *Opuntia ficus-indica* (L.) Mill and *Nopalea cochenillifera* (L.) Salm-Dyck (Palm Species)

3.1 Species Characteristics

The *Opuntia fícus-indica* (L.) Mill (prickly pear) and *Nopalea cochenillifera* (L.) Salm-Dyck (cochineal cactus) palm species (Fig. 2) with origins from Mexico are widely cultivated in Brazil as well as in different parts of the world (Nobel 2001).

Fig. 2 Pictures of *Opuntia ficus-indica* (L) Mill (prickly pear) and *Nopalea cochenillifera* (L) Salm-Dyck (cochineal cactus). (**A**) Flower; (**B**) cladode; (**C**) fruit (Font: The authors)

They are a species well adapted to the northeastern semiarid region, which is characterized by having shallow, stony, or sandy soils and little organic matter, although the soil is rich in soluble minerals (Oliveira et al. 2011). It is thought that this species was introduced into Brazil because of its property of natural carmine dye, called carminic acid. This dye is biosynthesized by a small insect, the mealybug (*Dactylopius* sp.), which lives in the cladodes of the plant (Alves et al. 2008; Pessoa 1967).

Regarding botanical classification, forage palms belong to the Cactaceae family, genus *Opuntia* and *Nopalea* (Zappi and Taylor 2020). The two most common types of palm that grow in the northeastern region of Brazil differ from each other by their morphological features: the giant or big palm and the small or sweet palm (Lyra et al. 2015).

The giant palm is a variation of that which belongs to the species *Opuntia ficus-indica*. Its cladode[1] is green-matte and oval-elliptical or sub-oval, can be up to 50 cm long, and weighs about 1 kg. The flowers are hermaphroditic, medium sized, and bright yellow, and their corolla remains open at maturation. The fruit is an ovoid berry, large, and yellow or purple. The small or sweet palm is a small plant and has a highly branched stem. Its cladode weighs about 350 g and is almost 25 cm long, obovate in shape, and bright green. The flowers are red, and their corolla remain half closed during the cycle. The fruit is a purple colored berry. The small or sweet palm

[1] Green, flat branch with leaf-like shape and function present in some plants, mainly in cacti.

is considered more palatable and easier to manage because it does not contain spines in its structure, although it has less resistance to drought (da Silva and Santos 2006).

3.2 Popular Use and Possible Application

The use as feed for animals in times of drought has been the main use for palm plantations in Brazil (Menezes et al. 2005). In some regions of Brazil and in several other countries in the world, palms are also used for human consumption, where people consume their fruits and young cladodes "nopalitos" as vegetables. In Mexico and in southern USA, palm cultivation is focused on fruit production, although young cladodes are also consumed. The palm fruits are also largely marketed in Europe, being produced and imported from countries around the Mediterranean sea (Moussa-Ayoub et al. 2011; Saenz 2000).

Due to agricultural problems related to the increase in arid zones and the scarcity of water resources, some cacti have gained importance as an effective food source for humans (Shetty et al. 2012; Stintzing and Carle 2005). Therefore, investigations on the chemical components and nutritional values of cactus have become the subject of research in diverse scientific fields (Fernández-López et al. 2010).

3.3 Physical-Chemical Characteristics and Nutritional Composition

The chemical composition of forage palm varies according to the species, the cultivation area, and the age of the cladode. The main characteristics of the forage palm are high water, minerals, soluble carbohydrates, and vitamin contents. Palm cladodes are also considered a source of mucilage. Usually, they can be consumed fresh or cooked, and studies have demonstrated the feasibility of processing the cladodes to obtain juice, jellies, gels, liquid sweeteners, pickles, jams, sauces, and other foods (Moussa-Ayoub et al. 2011; Saenz 2000). Data on the composition of *O. ficus-indica* and *N. cochenillifera* are shown in Table 2.

Regarding the constitution of soluble solids, a content of 6.60% was found for the giant palm and 5.60% for the small palm. The acidity of the giant palm and small palm was significantly different, with citric acid values of 0.20% and 0.07%, respectively. The pH of the giant palm was 4.40 and of the small palm was 4.70, a significant difference between the two species analyzed (Silva et al. 2015a, b).

The palm cladodes are covered by a cuticle that controls evaporation and allows water to be stored up to 90–93% of the plant volume (da Silva and Santos 2006). The average values of humidity found in an experiment with the species of giant and small palm were 91.00% and 89.67%, respectively. Santos et al. (2006) also found high humidity values for the giant palm (89.80%) and small palm (84.60%).

Table 2 Physical-chemical composition of *Opuntia fícus-indica* (giant palm) and *Nopalea cochenillifera* (small palm)

Variables	*Opuntia fícus-indica*	*Nopalea cochenillifera*
Soluble solids (%)	6.60	5.60
Titratable acidity	0.20	0.07
pH	4.40	4.70
Humidity (%)	91.00	89.60
Dry matter (%)	9.00	10.33
Ash (%)	1.19	1.17
Calcium (%)	6.20	7.20
Phosphorus (%)	0.13	0.10
Total protein (%)	0.86	0.86
Crude fiber (%)	1.65	1.37
Reducing sugars (%)	1.69	1.95
Lipids (%)	0.40	0.27

Font: (Silva et al. 2015a, b)

Similar to the moisture content, the dry matter contents of the two species also presented different values. Dry matter content was higher for the smaller species, although the ash contents showed no significant difference (Batista et al. 2003). Independent of the genus, palm species have a considerable amount of mineral matter, although these values vary according to the species. Differences might be due to the age of the cladodes, the geographical area, and the time of year when the material was collected. Regarding the levels of calcium, phosphorus, protein, crude fiber, reducing sugars, and lipids, there is no significant difference between species (Santos et al. 2006; Viana et al. 2014).

The palms are also important because of their high mucilage content. Mucilage is mainly composed of galactose, mannose, xylose, and other sugars. Therefore, it has a high capacity to retain water, like pectins and some algae polysaccharides. Due to this high water absorption capacity, mucilage can be used in food, cosmetics, and pharmaceutical products where it dissolves, disperses, and forms colloids (Del-Valle et al. 2005). The density obtained for the mucilage of the *N. cochenillifera* is comparable to that reported for the same concentrations of arabic gum (Tahir et al. 2007). Indeed, several cultures have traditionally used the cladodes of *O. ficus-indica* and *N. cochenillifera* as an important ingredient for cooking. Its characteristics of density, viscosity, pH, and conductivity fit the recommendations for use as an additive in the food and medicine formulations. Therefore, the mucilage from the palm can be considered a good alternative emulsifier and stabilizer.

No less important than the cladode, the fruit of the palm is a berry with many pleasant-tasting seeds, which facilitates its inclusion in human diets. Moreover, it presents readily absorbable sugars, high vitamin C content (12.7 mg/100 g), β-carotene (12.9μg/100 g), minerals, polyphenols, and amino acids (Stintzing et al. 2001). The palm fruit, considered a source of nutrients and vitamins in some countries, is processed together with some products such as sweets, alcoholic drinks, and additives (Karababa et al. 2014; Kinupp and Lorenzi 2014; Sáenz-Hernández 2001).

Palms can be considered an effective alternative to fight hunger and sub nutrition in the northeastern semiarid region, as they are rich in vitamins A, B, and C and in minerals, such as calcium, magnesium, sodium, and potassium, and have 17 types of amino acids. Comparatively, palm has greater nutritional value than foods such as cabbage, beets, and bananas, in addition to their economic advantage. The palm is commonly consumed in juices and salads and in stewed and cooked foods (Flores-Valdez 2001; Kinupp and Lorenzi 2014; Sáenz-Hernández 2001). In some regions, the high resistance to the consumption of palm as a food for humans is due to prejudice, since the traditional use of the plant is to feed animals (Nunes 2011).

Furthermore, there is a growing interest in the use of palms other than for food. Traditional medicine has recognized some benefits attributed to the use of these species, as they have traditionally been used to treat urinary problems and hypertension (Lans 2006). There are reports about the consumption of palm for treating diabetes (Andrade-Cetto and Heinrich 2005; Lans 2006). Some people *use* thin slices of cladodes on burned skin or on swelling, similar to the use of other species such as aloe vera (*Aloe* sp.) (Hoffmann 2001).

3.4 Biological Properties

Some studies have also demonstrated pharmacological uses of these species. The gastroprotective activity of mucilages and pectins extracted from *O. ficus-indica* cladodes was tested by three different in vivo ulcer models induced by ethanol in rats (Galati et al. 2002a, 2007; Vázquez-Ramírez et al. 2006). The in vivo study by Hwang et al. (2017) suggests that the aqueous extract of *O. ficus-indica* can be used to prevent and/or control blood glucose levels; therefore, it is a potential dietary supplement. Another in vitro and in vivo study also showed that the *O. ficus-indica* extract improved hyperglycemia, hyperinsulinemia, and glucose tolerance due to augmented pancreatic function caused by the increase in β-cell mass (Leem et al. 2016). In the same context, a pilot clinical trial was conducted with Mexican patients in order to verify the effects of a drink based on *N. cochenillifera* (Fabela-Illescas et al. 2015). The studies by Trombetta et al. (2006) evaluated the healing potential of two lyophilized polysaccharide extracts obtained from *O. ficus-indica* cladodes in induced wounds in rats. The authors concluded that the hygroscopic, rheological, and viscoelastic properties of these polysaccharides may be essential to promote healing. Mouhaddach et al. (2017) reported on the analgesic activity of *O. ficus-indica* extracts obtained by decoction carried out in hot plate and tail movement of in vivo models.

Potential diuretic (Galati et al. 2002b), antioxidant, anti-inflammatory (Ammar et al. 2018; Benayad et al. 2014; Matias et al. 2014; Necchi et al. 2011), antimicrobial, and antiviral activities (Bargougui et al. 2019; Gomez-Flor et al. 2006) were also attributed to these species in nonclinical studies.

The pharmacological potential of the palms is generally related to the presence of components already identified in these species. These are primarily phenolic compounds such as quinic acid, gallic acid, protocatechuic acid, syringic acid, derivatives of hydroxycinnamic acid, and flavonoids as isorhamnetin, quercetin, naringenin, luteolin, apigenin, kaempferol, cirsiliol, nicotiflorin, and rutin (Belhadj Slimen et al. 2017; De Leo et al. 2010; Guevara-Figueroa et al. 2010; Moussa-Ayoub et al. 2011).

Taken together, most studies conclude that the species *O. ficus-indica* and *N. cochenillifera* exhibit nutritional, economic, and pharmacological potential. Nonclinical and clinical trials have been carried out in an attempt to explore, identify, and test new bioactive and pharmacological activities of these species.

4 *Pilosocereus gounellei* (Xique-xique)

4.1 Species Characteristics

Pilosocereus gounellei (A. Weber ex K. Schum. Bly. ex Rowl), popularly known as the xique-xique (Fig. 1a), is a species of the Cactaceae family, belonging to the subfamily Cactoideae, of the genus *Pilosocereus* Byles & Rowley (Dias et al. 2015), exclusively found in Brazilian Caatinga (Oliveira et al. 2018). Its occurrence has been confirmed in the northeast part of Brazil, in Alagoas, Bahia, Ceará, Maranhão, Paraíba, Pernambuco, Piauí, Rio Grande do Norte, and Sergipe (Zappi et al. 2015).

Fig. 3 *Pilosocereus gounellei* (xique-xique). (**A**) Xique-xique. (**B**) Fruit. (**C**) Cladode (Font: The authors)

This cactaceous grows in dry areas of the semiarid northeast region and shallow soils and on rocks and multiplies regularly, covering large areas of Caatinga. It has an erect trunk with wide lateral branches, gently describing a wide curve upward, its branches composed of large thorns, skin an opaque green color (Fig. 3a), white tubular flowers (Cavalcanti and Resende 2007), and fruits reddish, rounded berries with small seeds (Fig. 3b) (Almeida et al. 2007). The cladodes of xique-xique can be divided into vascular cylinder and central stem (Fig. 3c).

4.2 *Popular Use and Possible Application*

The xique-xique is a natural resource that makes an important contribution to the livelihood of local populations in the semiarid region of northeastern Brazil (Monteiro et al. 2015). It has been used as animal food, human food, human medicine, and veterinary medicine, besides serving for construction, as ornament, in personal hygiene, as bio-indicator of rain when blooming, and in technology, among other uses (Silva 2015) (Table 3).

The use of xique-xique in food has been reported in a number of studies. In addition to the consumption of its *in natura* fruit (Lucena et al. 2015), the pulp extracted from the stem of xique-xique has been used in the elaboration of different products, such as *cocada*, other candies, flour, and couscous (Almeida et al. 2007; Lucena et al. 2013). The dried and powdered pulp can be incorporated into wheat for the preparation of bakery products (Almeida et al. 2007). The flour to prepare the couscous can be produced from the cladode and can be consumed cooked or roasted (Nascimento et al. 2012).

In addition to the nutritional potential of xique-xique, popular knowledge about this species involves its medicinal use. Xique-xique parts such as stems, roots, and flowers are popularly used to treat constipation (Lucena et al. 2015), gastritis (Lucena et al. 2012), as well as urethra inflammation (Roque et al. 2010), prostate inflammation, hypoglycemia, injuries (Agra et al. 2008; Albuquerque et al. 2007), and jaundice (Albuquerque et al. 2007).

4.3 *Physical-Chemical Characteristics and Nutritional Composition*

When evaluating the physical and chemical characteristics of the xique-xique, it was observed that the vascular cylinder and the central stem of the xique-xique branches presented, respectively, soluble solids (1.50 and 1.50 °Brix), pH (4.66 and 5.18), ash (1.38% and 1.34%), ascorbic acid (0.33% and 0.25%), total solids (5.83% and 13.96%), and insoluble solids (3.61% and 2.98%). This leads to the conclusion that the central stem is more suitable for the production of flour, and the vascular

Table 3 Description of popular use and possible application of *Pilosocereus gounellei* (xique-xique)

Purposes	Part used	Mode of use	Medicinal use	References
Human feed	Fruit	*In natura*	–	Lucena et al. (2015) Machado et al. (2018)
	Pulp	Roasted or baked (cookies, cakes, *cocada*, sweets, flour, couscous)	–	Almeida et al. (2007) Nascimento et al. (2012) Lucena et al. (2013) Machado et al. (2018)
	Pulp	Dried and converted into powder (incorporated into wheat for use in baking)	–	Almeida et al. (2007)
Animal feed	Fruit	*In natura*	–	Bravo Filho et al. (2018) Lucena et al. (2015) Machado et al. (2018)
	Branch	Cut or burned		
	Entire plant			
Human medicine	Mucilage (aquifer parenchyma)	–	To remove thorns from the skin	Lucena et al. (2015)
	Pulp			
	Fruit	*In natura*	Constipation	Lucena et al. (2015)
	Pulp	Soak and drink the water	Gastritis	Lucena et al. (2012)
	Root	Macerate	Urethra inflammation	Roque et al. (2010)
	Root	Decoction	Prostate inflammation	Agra et al. (2008)
	Stem, root, and flowers	–	Prostate inflammation, jaundice, hypoglycemia, and injuries	Albuquerque et al. (2007)
Veterinary medicine	Mucilage	–	Used on animal wounds	Lucena et al. (2013, 2015)

(continued)

Table 3 (continued)

Purposes	Part used	Mode of use	Medicinal use	References
Ornament	Entire plant	Gardens and backyards	–	Lucena et al. (2013, 2015) Machado et al. (2018) Bravo Filho et al. (2018)
Personal hygiene	Pulp	Shampoo	–	Lucena et al. (2013)
Bio-indicator	Flower	Rain indicator	–	Lucena et al. (2015)
Technology	Thorns	"needles" (Making lace)	–	Lucena et al. (2012)
Shadow	Entire plant	–	–	Machado et al. (2018)
Construction	Entire plant	Hedge	–	Lucena et al. (2015) Bravo Filho et al. (2018)
Construction	Mucilage	Surface treatment	–	Nobréga (2019)

Table 4 Physical-chemical characterization of *Pilosocereus gounellei* (xique-xique)

	Xique-xique cladodes	
Variables	Vascular cylinder	Central stem
Moisture (g/100 g)	94.16 ± 0.99	87.13 ± 1.64
Acidity (g/100 g)	0.07 ± 0.01	0.09 ± 0.02
pH	4.77 ± 0.21	4.87 ± 0.21
Ash (g/100 g)	1.99 ± 0.04	1.59 ± 0.00
Proteins (g/100 g)	0.49 ± 0.00	0.76 ± 0.01
Lipids (g/100 g)	0.28 ± 0.00	0.77 ± 0.04
Total fiber (g/100 g)	2.70 ± 0.01	6.54 ± 0.01
Insoluble fiber (g/100 g)	1.83 ± 0.00	5.18 ± 0.05
Soluble fiber (g/100 g)	0.87 ± 0.01	1.37 ± 0.04
TSS (°Brix)	2.00 ± 0.42	2.71 ± 0.23
Fructose (g/100 g)	0.31 ± 0.02	1.07 ± 0.02
Glucose (g/100 g)	0.14 ± 0.02	0.54 ± 0.01
Xylose (g/100 g)	0.06 ± 0.01	0.37 ± 0.02
Arabinose (g/100 g)	0.06 ± 0.01	0.20 ± 0.01
Potassium (mg/100 g)	308.4 ± 3.65	101.6 ± 1.65
Magnesium (mg/100 g)	182.40 ± 2.10	167.10 ± 4.11

Font: (Bezerril 2017)

cylinder is more appropriate for the production of products with a high water content (Almeida et al. 2007).

Preliminary studies on the xique-xique cladodes have reported high water content, soluble and insoluble fibers, and high values of minerals, especially potassium, magnesium, and calcium in the vascular cylinder. The presence of fructose, glucose, xylose, and arabinose was also found as the main constituents (Bezerril 2017) (Table 4).

The stem of *P. gounellei* contains an anionic trypsin inhibitor stable to 50 °C and active on medically important bacteria; the yeast, *Candida krusei*, may lessen the use of antibiotics and thus contribute to reduce the problem of antibiotic resistance. Outcomes reported have revealed new insights into the biochemistry of *P. gounellei*, increasing the biotechnological value of this plant (Rocha Filho et al. 2019).

Flours (Fig. 4a) developed from the central stem of xique-xique have a high content of soluble and insoluble fibers, resistant starch, and minerals, mainly calcium, iron, potassium, magnesium, and manganese. In addition, the plant has high viscosity and gel texture, indicating a potential for use as a thickening or gelling agent in food. In addition, the flours had compounds with antioxidant activity and few antinutritional factors. Cookies (Fig. 4b) made from the flour of this plant also proved to be rich in fibers and minerals. In sensory terms, they were tasty and well received by consumers (Machado 2019).

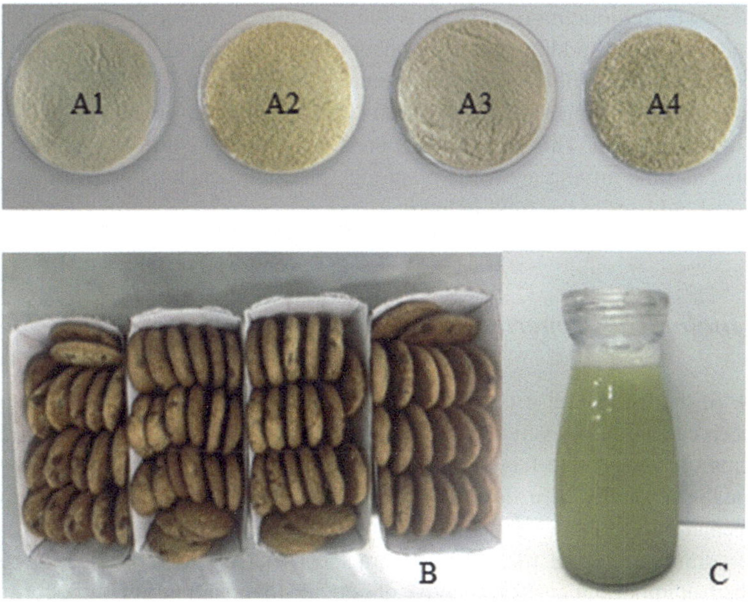

Fig. 4 Products made from *Pilosocereus gounellei* (xique-xique). (Font: The authors) Legend: (**A**1) Fresh xique-xique flour (100 mesh); (**A**2) natural xique-xique flour (28 mesh); (**A**3) autoclaved xique-xique flour (100 mesh); (**A**4) xique-xique autoclaved flour (28 mesh); (**B**) xique-xique cookie (100 mesh); (**C**) xique-xique Juice

Several genera of the Cactaceae family have bioactive compounds in their composition (Agostini-Costa 2020). The xique-xique contained triterpenes and phenolic compounds (Almeida et al. 2005), flavonoids (Nascimento et al. 2012), and catechin and epigallocatechin gallate, in addition to other compounds belonging to the group of phenolic acids and procyanidins (Bezerril 2017).

Xique-xique fruits contain various functional compounds, such as phenolic compounds and betalains, which provided high antioxidant activity, with potential for adding value (da Silva et al. 2018). A study isolated and identified ten compounds of *Pilosocereus gounellei*: pinostrobin, β-sitosterol, a mixture of β-sitosterol/stigmasterol, 13^2-hydroxyphaeophytin a, phaeophytin a, sitosterol 3-O-β-D-glucopyranoside/stigmasterol 3-O-β-D-glucopyranoside, kaempferol, quercetin, 7′-ethoxy-*trans*-feruloyltyramine, and *trans*-feruloyltyramine. In addition, same authors have demonstrated that the fruit ethanol extract possesses excellent antioxidant activity, mainly because of the presence of phenolic compounds reported in the genus and the Cactaceae family (Maciel et al. 2015).

The juice (Fig. 4c) made from the xique-xique cladode showed high fiber content, mainly soluble, as well as a variety of phenolic compounds, such as catechin, epicatechin, epicatechin gallate, epigallocatechin gallate, quercetin 3-glucoside, rutin, kaempferol 3-glucoside, gallic acid, caffeic acid, syringic acid, chlorogenic acid, naringenin, and hesperidin (Assis et al. 2019). In addition, the xique-xique juice induced a selective and intense fermentable activity to different probiotic Lactobacillus isolates, similar to the effects caused by fructooligosaccharide (FOS), a well-known prebiotic ingredient. This suggests that xique-xique juice could have prebiotic properties, adding value to an unconventional and still little exploited plant food as a source of bioactive compounds (Ribeiro et al. 2020).

Despite few existing studies, research has revealed the great potential of the whole of *P. gounellei*. This provides an enormous advantage due to the low residue generated and the infinite areas the plant may be used, which range from the pharmaceutical industry to the food industry.

4.4 Biological Properties

In view of the traditional use of xique-xique by the population, biological studies have been carried out to validate its effects on health. The crude ethanol extract of the stems of xique-xique has been reported to have low toxicity, exhibiting anti-inflammatory activity at a dose of 25 mg/kg over 4 hours as evaluated by the carrageenan-induced paw edema model in rats (Dias et al. 2015).

The administration of the ethanol extracts of root and cladodes of the xique-xique had an important gastroprotective effect, due to the inhibition of the formation of gastric lesions in animal models, without promoting toxic effects (Sousa et al. 2018).

The saline extract from the stem of xique-xique containing flavonoids and reducing sugars demonstrated antinociceptive activity, without showing toxicity or altering motor coordination in mice (Oliveira et al. 2018). In addition, the saline extract from the stem of *P. gounellei* did not present significant toxic effects over 28 consecutive days and demonstrated antipyretic activity, together with hypoglycemic and hypolipidemic effects (Oliveira et al. 2019). Still, when administered orally at the tested doses, the extract is genotoxically safe, when used with caution in doses above 1.000 mg/kg, and has a protective effect against CPA-induced DNA damage (Oliveira et al. 2020).

The ingestion of xique-xique cladode juice had a protective effect on intestinal inflammation and showed a decrease in pro-inflammatory markers and oxidative stress in an animal model of inflammatory bowel disease. These effects were attributed to the phenolic compounds and fibers present in the juice (Assis et al. 2019).

5 Final Considerations

This chapter has shown that the uses of cacti have gained prominence in several areas, especially as a food for human consumption and for pharmacological use, making cacti of interest in reducing hunger and fighting diseases. We have highlighted the diversity of nutrients in cladodes and in fruits, used for fresh consumption, in elaborated preparations, or even in extracts such as herbal medicines. The need for further experimental and clinical research on these plants is highlighted in order to elucidate their toxicity related to health considerations. The use of cacti is important for the restoration and conservation of the Caatinga's biodiversity, the development of local culture, and the promotion of food and nutritional security, especially in the northeast part of Brazil.

References

Agostini-Costa TS (2020) Bioactive compounds and health benefits of Pereskioideae and Cactoideae: a review. Food Chem 327:126961. https://doi.org/10.1016/j.foodchem.2020.126961

Agra M, Baracho G, Nurit K et al (2007) Medicinal and poisonous diversity of the flora of "Cariri Paraibano", Brazil. J Ethnopharmacol 111:383–395. https://doi.org/10.1016/j.jep.2006.12.007

Agra MF, Silva KN, Lima J et al (2008) Survey of medicinal plants used in the region Northeast of Brazil. Braz J Pharmacogn 18:472–508

Albuquerque UP, de Medeiros PM, de Almeida ALS et al (2007) Medicinal plants of the Caatinga (semi-arid) vegetation of NE Brazil: a quantitative approach. J Ethnopharmacol 114:325–354. https://doi.org/10.1016/j.jep.2007.08.017

Almeida C, Lima e Silva T, Amorim E et al (2005) Life strategy and chemical composition as predictors of the selection of medicinal plants from the Caatinga (Northeast Brazil). J Arid Environ 62:127–142. https://doi.org/10.1016/j.jaridenv.2004.09.020

Almeida MM, Passos D, De Almeida TS et al (2006) Cinética da produção do fermentado do fruto do mandacaru. Rev Bras Prod Agroindustriais 8:35–42

Almeida C, Figueirêdo R, Queiroz A, Oliveira F (2007) Physical and chemical characteristics of xiquexique pulps. Rev Cien Agron 38:440–443

Almeida M, Silva F, Conrado L et al (2009) Physical and physical-chemical characterization of mandacaru fruits. Rev Bras Prod Agro 11:15–20

Almeida MM, Luiz F, Da Silva H et al (2011) Revista verde de agroecologia e desenvolvimento sustentável Grupo Verde de Agricultura Alternativa (GVAA) estudo cinético e caracterização da bebida fermentada do Cereus jamacaru P. DC. Rev Verde Agroecol Desenvolv Sustent 6:176–183

Alves A, De Souza M, Carla A et al (2008) Fruto de palma [Opuntia fícus-indica (L) Miller, Cactaceae]: morfologia, compocicão química, fisiologia, índices de colheita e fisiologia pós-colheita. Rev Iberoam Tecnol Postcosecha 9:16–25

Ammar I, Ben Salem M, Harrabi B et al (2018) Anti-inflammatory activity and phenolic composition of prickly pear (Opuntia ficus-indica) flowers. Ind Crop Prod 112:313–319. https://doi.org/10.1016/j.indcrop.2017.12.028

Anderson E (2001) The cactus family. Timber Press, Portland

Andrade C, Marques JGW, Zappi DC (2006) Utilização de cactáceas por sertanejos baianos. Tipos conexivos para definir categorias utilitárias. Sitientibus 6:3–12

Andrade-Cetto A, Heinrich M (2005) Mexican plants with hypoglycaemic effect used in the treatment of diabetes. J Ethnopharmacol 99:325–348. https://doi.org/10.1016/j.jep.2005.04.019

Aragão T, Souza P, Uchôa A et al (2000) Characterization of a methionine-rich protein from the seeds of Cereus jamacaru Mill. (Cactaceae). Braz J Med Biol Res 33:897–903. https://doi.org/10.1590/S0100-879X2000000800005

Assis POA, Guerra GCB, Araújo DF et al (2019) Intestinal anti-inflammatory activity of xique–xique (Pilosocereus gounellei A. Weber ex K. Schum. Bly. Ex Rowl) juice on acetic acid-induced colitis in rats. Food Funct 10:7275–7290. https://doi.org/10.1039/C9FO00920E

Association Official Analytical Chemists AOAC (2016) Methods of analysis of AOAC international, 20th edn. AOAC International, Rockville

Bahia E, Morais L, Silva M, et al (2010) Estudo das Características físico-químicas do fruto do mandacaru (Cereus jamacaru P.DC.) cultivado no sertão pernambucano. In: VIII SEMEALI

Bargougui A, Maatoug Tag H, Bouaziz M, Triki S (2019) Antimicrobial, antioxidant, total phenols and flavonoids content of four cactus (Opuntia ficus-indica) cultivars. Biomed Pharmacol J 12:1353–1368. https://doi.org/10.13005/bpj/1764

Barthlott W, Hunt DR (1993) Cactaceae. Springer Berlin Heidelberg, Berlin, Heidelberg

Batista AM, Mustafa AF, McAllister T et al (2003) Effects of variety on chemical composition, in situ nutrient disappearance and in vitro gas production of spineless cacti. J Sci Food Agric 83:440–445. https://doi.org/10.1002/jsfa.1393

Belhadj Slimen I, Mabrouk M, Hanene C, Najar TT (2017) LC-MS analysis of phenolic acids, flavonoids and Betanin from spineless Opuntia ficus-indica fruits. Cell Biol 5:17. https://doi.org/10.11648/j.cb.20170502.12

Benayad Z, Martinez-Villaluenga C, Frias J et al (2014) Phenolic composition, antioxidant and anti-inflammatory activities of extracts from Moroccan Opuntia ficus-indica flowers obtained by different extraction methods. Ind Crop Prod 62:412–420. https://doi.org/10.1016/j.indcrop.2014.08.046

Bezerril FF (2017) Caracterização nutricional e de compostos bioativos do Xique-Xique (Pilosocereus gounellei (A. Weber ex. K. Schum.) Bly. ex Rowl.). Dissertation, Federal University of the Paraíba

Bhandari MM (1974) Famine foods in the Rajasthan desert. Econ Bot 28:73–81. https://doi.org/10.1007/BF02861383

Biavatti MW, Marensi V, Leite SN, Reis A (2007) Ethnopharmacognostic survey on botanical compendia for potential cosmeceutic species from Atlantic Forest. Rev Bras Farmacogn 17:640–653. https://doi.org/10.1590/S0102-695X2007000400025

Brasil (1978) Resolução Normativa n°. 15 de 4 de maio de 1978. Ministério da Saúde. Secretaria de Vigilância Sanitária.

Brasil (2010) Manual de Hortaliças Não convencionais. Ministério da Agricultura. Pecuária e Abastecimento, Brasília, p 92

Bravo Filho ES, de Santana MC, Santos PAA, Ribeiro AS (2018) Levantamento etnobotânico da família Cactaceae no estado de Sergipe. Rev Fitos 12:41–53. https://doi.org/10.5935/2446-4775.20180005

Cavalcanti N, Resende G (2007) Consumo de xiquexique (Pilocereus gounellei (A. Weber ex K. Schum.) Bly. ex Rowl) por caprinos no Semi-Árido da Bahia. Rev Caatinga 20:22–27

Coelho R, Ferreira Neto C, Figuerêdo R (2004) Avaliação química de sementes de mandacaru (Cereus jamacaru P. DC.). In: XXIX Congresso Brasileiro de Ciência e Tecnologia de Alimentos. Recife, Brazil

da Costa IR, de Souza PAS, Bloch C Jr et al (2001) Isolation and characterisation of a reserve protein from the seeds of Cereus jamacaru (Cactaceae). Braz Arch Biol Technol 44:331–335. https://doi.org/10.1590/S1516-89132001000400002

da Silva CCF, Santos LC (2006) Palma Forrageira (Opuntia Fícus- Indica Mill) como alternativa na alimentação de ruminantes (Forage Palm (Opuntia Fícus- Indica Mill) as alternative in ruminant feeding). Redvet 7:1–11

da Silva AA, Dantas RL, Véras CDS et al (2018) Physical and physicochemical characteristics, bioactive compounds and antioxidant activity of xiquexique (Pilosocereus gounellei (A. Weber ex K. Schum.)) fruit. Semin Ciências Agrárias 39:1969. https://doi.org/10.5433/1679-0359.2018v39n5p1969

Davet A, Virtuoso S, Dias JFG et al (2009) Atividade antibacteriana de Cereus jamacaru DC, Cactaceae. Rev Bras Farmacogn 19:561–564. https://doi.org/10.1590/S0102-695X2009000400009

De Leo M, De Abreu MB, Pawlowska AM et al (2010) Profiling the chemical content of Opuntia ficus-indica flowers by HPLC–PDA-ESI-MS and GC/EIMS analyses. Phytochem Lett 3:48–52. https://doi.org/10.1016/j.phytol.2009.11.004

Del-Valle V, Hernández-Muñoz P, Guarda A, Galotto MJ (2005) Development of a cactus-mucilage edible coating (Opuntia ficus indica) and its application to extend strawberry (Fragaria ananassa) shelf-life. Food Chem 91:751–756. https://doi.org/10.1016/j.foodchem.2004.07.002

Dias G, Guedes E, Maciel J et al (2015) Acute oral toxicity and anti-inflammatory activity of ethanolic extract from Pilosocereus gounellei (Fac Weber) in rats. Int J Pharmacogn Phytochem Res 7:1–6

Dutra JV, Moisés Ferreira J, Costalonga Pereira P et al (2018) Cereus jamacaru D.C. Hydroalcoholic extract promotes anti-cytotoxic and antitumor activity. Pharmaceuticals 11:1–18. https://doi.org/10.3390/ph11040130

Dutra JV, de Oliveira J-H, dos Santos V et al (2019) Fruiting increases total content of flavonoids and antiproliferative effects of Cereus jamacaru D.C. cladodes in sarcoma 180 cells in vitro. Asian Pac J Trop Biomed 9:66–72. https://doi.org/10.4103/2221-1691.250857

Fabela-Illescas HE, Ávila-Domínguez R, Hernández-Pacheco A et al (2015) Efecto de una bebida a base de nopal (Nopalea cochenillifera (l) Salm-dyck) en pacientes de una población rural de Hidalgo, México; ensayo clínico piloto. Nutr Hosp 32:2710–2714. https://doi.org/10.3305/nh.2015.32.6.9762

Fernandes VN, das Neves AF, Martin PG et al (2016) Genetic structure and molecular divergence among samples of mandacaru (Cereus spp.; Cactaceae) as revealed by microsatellite markers. Biochem Syst Ecol 64:38–45. https://doi.org/10.1016/j.bse.2015.11.003

Fernández-López JA, Almela L, Obón JM, Castellar R (2010) Determination of antioxidant constituents in cactus pear fruits. Plant Foods Hum Nutr 65:253–259. https://doi.org/10.1007/s11130-010-0189-x

Flores-Valdez C (2001) Produção, industrialização e comercialização de verdura de palma forrageira. In: Barbera G, Inglese P (eds) Agroecologia, cultivos e usos da palma forrageira. Sebrae Paraiba, Brazil, pp 94–102

Forzza RC, Baumgratz JFA, Bicudo CEM et al (2012) New Brazilian floristic list highlights conservation challenges. Bioscience 62:39–45. https://doi.org/10.1525/bio.2012.62.1.8

Galati E, Pergolizzi S, Miceli N et al (2002a) Study on the increment of the production of gastric mucus in rats treated with Opuntia ficus indica (L.) Mill. cladodes. J Ethnopharmacol 83:229–233. https://doi.org/10.1016/S0378-8741(02)00243-X

Galati E, Tripodo M, Trovato A et al (2002b) Biological effect of Opuntia ficus indica (L.) Mill. (Cactaceae) waste matter. J Ethnopharmacol 79:17–21. https://doi.org/10.1016/S0378-8741(01)00337-3
Galati EM, Monforte MT, Miceli N et al (2007) Opuntia ficus indica (L.) Mill. mucilages show cytoprotective effect on gastric mucosa in rat. Phyther Res 21:344–346. https://doi.org/10.1002/ptr.2075
Gomez-Flor R, Tamez-Guer P, Tamez-Guer R et al (2006) In vitro antibacterial and antifungal activities of nopalea cochenillifera pad extracts. Am J Infect Dis 2:1–8. https://doi.org/10.3844/ajidsp.2006.1.8
Guevara-Figueroa T, Jiménez-Islas H, Reyes-Escogido ML et al (2010) Proximate composition, phenolic acids, and flavonoids characterization of commercial and wild nopal (Opuntia spp.). J Food Compos Anal 23:525–532. https://doi.org/10.1016/j.jfca.2009.12.003
Hoffmann W (2001) Etnobotânica. In: Barbera G, Inglese P (eds) Agroecologia, cultivos e usos da palma forrageira. Sebrae Paraiba, Brazil, pp 12–19
Hwang SH, Kang I-J, Lim SS (2017) Antidiabetic effect of fresh nopal (*Opuntia ficus-indica*) in low-dose streptozotocin-induced diabetic rats fed a high-fat diet. Evid Based Complement Altern Med 2017:1–8. https://doi.org/10.1155/2017/4380721
Karababa E, Coşkuner Y, Aksay S (2014) Some physical fruit properties of cactus pear (Opuntia spp.) that grow wild in the Eastern Mediterranean region of Turkey. J Prof Assoc Cactus Dev 2014:1–8
Kinupp V (2004) Barros I, Levantamento de dados e divulgação do potencial das plantas alimentícias alternativas no Brasil. 22:1–4
Kinupp V, Lorenzi H (2014) Plantas alimentícias não convencionais (PANC) no Brasil. Instituto Plantarum de Estudos da Flora, São Paulo
Lans CA (2006) Ethnomedicines used in Trinidad and Tobago for urinary problems and diabetes mellitus. J Ethnobiol Ethnomed 2:45. https://doi.org/10.1186/1746-4269-2-45
Leal ML, Alves RP, Hanazaki N (2018) Knowledge, use, and disuse of unconventional food plants. J Ethnobiol Ethnomed 14:1–9. https://doi.org/10.1186/s13002-018-0209-8
Leem K-H, Kim M-G, Hahm Y-T, Kim H (2016) Hypoglycemic effect of Opuntia ficus-indica var. saboten is due to enhanced peripheral glucose uptake through activation of AMPK/p38 MAPK pathway. Nutrients 8:1–15. https://doi.org/10.3390/nu8120800
Lima R (2016) Caracterização e potencial antioxidante do fruto da palma (Tacinga inamoena) e do mandacaru (Cereus jamacaru). Dissertation, Federal Rural University of the Semi-arid Region
Liu M, Li XQ, Weber C et al (2002) Antioxidant and antiproliferative activities of raspberries. J Agric Food Chem 50:2926–2930. https://doi.org/10.1021/jf0111209
Lucena CM, da Costa GM, de Sousa RF et al (2012) Conhecimento local sobre cactáceas em comunidades rurais na mesorregião do sertão da Paraíba (Nordeste, Brasil). Biotemas 25:281–291. https://doi.org/10.5007/2175-7925.2012v25n3p281
Lucena CM, Lucena RF, Costa GM et al (2013) Use and knowledge of Cactaceae in Northeastern Brazil. J Ethnobiol Ethnomed 9:1–11. https://doi.org/10.1186/1746-4269-9-62
Lucena C, Carvalho T, Ribeiro JES et al (2015) Conhecimento botânico tradicional sobre cactáceas no semiárido do Brasil. Gaia Sci 9:77–90
Lyra M, Silva M, Mergulhão A et al (2015) Molecular studies of forage prickly-pear cactus from the semiarid of Pernambuco State-Brazil. J Appl Biol Biotechnol 3:1–5. https://doi.org/10.7324/jabb.2015.3201
Machado T (2019) Desenvolvimento de farinha de xiquexique (Pilosocereus gounellei): caracterização nutricional, propriedades funcionais e sua utilização na elaboração de cookies. Thesis, Federal University of the Paraíba
Machado J, Lucena C, Santos S et al (2018) Conhecimento botânico local sobre cactáceas: um estudo de caso no município de Boqueirão, Paraíba, Nordeste, Brasil. Flovet 1:1–21
Maciel J, Chaves O, Brito Filho S et al (2015) New alcamide and anti-oxidant activity of Pilosocereus gounellei A. Weber ex K. Schum. Bly. ex Rowl. (Cactaceae). Molecules 21:1–13. https://doi.org/10.3390/molecules21010011

Mancuso S (2019) Revolução das plantas: um novo modelo para o future. Ebu, São Paulo
Martins A (2018) Elaboração de iogurte caprino com potencial funcional adicionado de geleia de frutos de mandacaru (Cereus jamacaru) e de maracujazeiro amarelo (Passiflora edulis Sims): Caracterização e avaliação do efeito protetor da matriz alimentar. Dissertation, Federal University of Campina Grande
Matias A, Nunes SL, Poejo J et al (2014) Antioxidant and anti-inflammatory activity of a flavonoid-rich concentrate recovered from Opuntia ficus-indica juice. Food Funct 5:3269–3280. https://doi.org/10.1039/C4FO00071D
Mayworm M, Salatino A (1996) Teores de óleo e composição de ácidos graxos de sementes de Cereus jamacaru DC. (Cactaceae), Zizyphus joazeiro Mart. (Rhamnaceae) e Anadenanthera colubrina 90 (Benth) Brenan var cebil (Griseb.) Von Altschul. (Mimosaceae). Sitientibus 15:201–209
Medeiros IU, Medeiros RA, Bortolin RH et al (2019) Genotoxicity and pharmacokinetic characterization of Cereus jamacaru ethanolic extract in rats. Biosci Rep 39:1–11. https://doi.org/10.1042/BSR20180672
Melo RDS, Silva SDM, De Sousa ASB et al (2017) Maturação e qualidade de frutos de mandacaru (Cereus jamacaru P.DC.) de diferentes bioclimas do estado da Paraíba. Agropecuária Técnica 38:160–168. https://doi.org/10.25066/agrotec.v38i3.33818
Menezes R, Simões D, Sampaio E (2005) A palma no Nordeste do Brasil – Conhecimento atual e novas perspectivas de uso. Rural Federal University of Pernambuco, Recife
Monteiro ER, Strioto DK, Meirelles ACS et al (2015) Genetic structure of Pilosocereus gounellei (Cactaceae) as revealed by AFLP marker to guide proposals for improvement and restoration of degraded areas in Caatinga biome. Genet Mol Res 14:16966–16974. https://doi.org/10.4238/2015.December.15.2
Moreira IS, Silva LM, Castro DS et al (2018) Fruit of mandacaru: kinetics of drying and physical-chemical characterization. J Agric Sci 10:461–470. https://doi.org/10.5539/jas.v10n11p461
Mota TR, Linhares HVS, Araújo-Filho JH et al (2019) Protein extract from Cereus jamacaru (DC.) inhibits Colletotrichum gloeosporioides growth by stimulating ROS generation and promoting severe cell membrane damage. Microb Pathog 130:71–80. https://doi.org/10.1016/j.micpath.2019.02.033
Mouhaddach A, El-hadi A, Taghzouti K et al (2017) Assessment of Opuntia ficus-indica in vivo following ethnobotanical survey: confirmation of its analgesic activity. Phytothérapie. https://doi.org/10.1007/s10298-017-1109-4
Moussa-Ayoub TE, El-Samahy SK, Kroh LW, Rohn S (2011) Identification and quantification of flavonol aglycons in cactus pear (Opuntia ficus indica) fruit using a commercial pectinase and cellulase preparation. Food Chem 124:1177–1184. https://doi.org/10.1016/j.foodchem.2010.07.032
Nascimento VT, de Moura NP, da Silva Vasconcelos MA et al (2011) Chemical characterization of native wild plants of dry seasonal forests of the semi-arid region of northeastern Brazil. Food Res Int 44:2112–2119. https://doi.org/10.1016/j.foodres.2010.12.024
Nascimento VT, Vasconcelos MA, Maciel MIS, Albuquerque UP (2012) Famine foods of Brazil's seasonal dry forests: ethnobotanical and nutritional aspects. Econ Bot 66:22–34. https://doi.org/10.1007/s12231-012-9187-2
Necchi RMM, Maki TDT, do Canto GS et al (2011) Antiinflammatory activity and biochemical parameters of the ethanol extract of Nopalea cochenillifera (L.) Salm-Dyck (Cactaceae). Lat Am J Pharm 30:786–789
Neves E (2006) Arqueologia da Amazônia. Rio de Janeiro
Nobel P (2001) Biologia ambiental. In: Agroecologia, cultivo e usos da palma forrageira. Sebrae Paraiba, João Pessoa, pp 36–48
Nobréga A (2019) Tratamento superficial para Blocos de Terra Comprimida (BTC): avaliação do potencial da mucilagem de xique-xique (Pilosocereus gounellei). Dissertation, Federal University of Paraíba

Nunes CS (2011) Usos e aplicações da palma forrageira como uma grande uses and applications palm forage as a major source of economy for the semiarid. Rev verde Agroecol e Desenvolv sustentável 6:58–66

Oliveira F, Alexandre H, Figueirêdo R (2004) Características físico-químicas da polpa e casca do fruto do mandacaru. In: Congresso Brasileiro de Ciência e Tecnologia de Alimentos. Recife, Brazil

Oliveira ASC, Cavalcante Filho FN, Rangel AHDN, Lopes KBDP (2011) A palma forrageira: alternativa para O semi-árido the cactus pear: alternative to the semi-arid. Rev Verde Agroecol e Desenvolv Sustentável 6:49–58

Oliveira AS, Figueirêdo RM, Queiroz AJ, Brito JG (2015) Estabilidade da polpa do Cereus jamacaru em pó durante o armazenamento. Rev Bras Eng Agrícola e Ambient 19:147–153. https://doi.org/10.1590/1807-1929/agriambi.v19n2p147-153

Oliveira AM, Luna Freire MO, Silva WAV et al (2018) Saline extract of Pilosocereus gounellei stem has antinociceptive effect in mice without showing acute toxicity and altering motor coordination. Regul Toxicol Pharmacol 95:289–297. https://doi.org/10.1016/j.yrtph.2018.04.004

Oliveira AM, Silva WAV, Ferreira MRA et al (2019) Assessment of 28-day oral toxicity and antipyretic activity of the saline extract from Pilosocereus gounellei (Cactaceae) stem in mice. J Ethnopharmacol 234:96–105. https://doi.org/10.1016/j.jep.2019.01.036

Oliveira AM, Freitas AFS, Paiva PMG, Napoleão TH (2020) Genotoxicity assessment of saline extract from Pilosocereus gounellei (Cactaceae) and its chemopreventive effect against cyclophosphamide-induced DNA damage. Heliyon 6:e03811. https://doi.org/10.1016/j.heliyon.2020.e03811

Park E-H, Kahng J-H, Lee SH, Shin K-H (2001) An anti-inflammatory principle from cactus. Fitoterapia 72:288–290. https://doi.org/10.1016/S0367-326X(00)00287-2

Paulino R, Henriques G, Coelho M, Araújo V (2011) Riqueza e importância das plantas medicinais do Rio Grande do Norte. Rev Biol e Ciências da Terra 11:157–168

Peron A (2011) Esterases para a análise de variabilidade genética em mandacaru (Cactaceae). State University of Maringá

Pessoa A (1967) Cultura da palma-forrageira. Sudene, Recife

Ribeiro TS, Sampaio KB, Menezes FNDD et al (2020) In vitro evaluation of potential prebiotic effects of a freeze-dried juice from *Pilosocereus gounellei* (A. Weber ex K. Schum. Bly. Ex Rowl) cladodes, an unconventional edible plant from Caatinga biome. 3 Biotech 10:1–9. https://doi.org/10.1007/s13205-020-02442-8

Rocha EA, Agra MF (2002) Flora do Pico do Jabre, Paraíba, Brasil: Cactaceae juss. Acta Bot Brasilica 16:15–21. https://doi.org/10.1590/S0102-33062002000100004

Rocha Filho CAA, Karla Amorim P, de Albuquerque Lima T et al (2019) PgTI, the first bioactive protein isolated from the Cactus Pilosocereus gounellei, is a trypsin inhibitor with antimicrobial activity. Adv Res:1–11. https://doi.org/10.9734/air/2019/v20i330160

Rockenbach II, Rodrigues E, Gonzaga LV et al (2011) Phenolic compounds content and antioxidant activity in pomace from selected red grapes (Vitis vinifera L. and Vitis labrusca L.) widely produced in Brazil. Food Chem 127:174–179. https://doi.org/10.1016/j.foodchem.2010.12.137

Rodrigues L, Elesb R (2009) Avaliação da composição físico-química de frutos de mandacaru (Cereus jamacaru P.). Acta Agronómica 58:245–250

Roque A, Rocha R, Loiola MI (2010) Uso e diversidade de plantas medicinais da Caatinga na comunidade rural de Laginhas, município de Caicó, Rio Grande do Norte (Nordeste do Brasil). Rev Bras Plantas Med 12:31–42. https://doi.org/10.1590/S1516-05722010000100006

Rosado V, Rosado A (1960) Plantas do Nordeste especialmente do Ceará, 2nd edn. Imprensa Oficial, Fortaleza

Saenz C (2000) Processing technologies: an alternative for cactus pear (Opuntia spp.) fruits and cladodes. J Arid Environ 46:209–225. https://doi.org/10.1006/jare.2000.0676

Sáenz-Hernández D (2001) Fabricação de alimentos e obtenção de subprodutos. In: Barbera G, Inglese P (eds) Agroecologia, cultivos e usos da palma forrageira. Sebrae Paraiba, Brazil, pp 140–146

Sales M, Martins L, Souza I et al (2014) Cereus jamacarude Candolle (Cactaceae), o mandacaru do nordeste brasileiro. Publ UEPG Ciencias Biol e da Saude 20:135–142. https://doi.org/10.5212/Publ.Biologicas.v.20i2.0006

Santos I (2018) Compostos fenólicos totais e atividade antioxidante da casca do fruto do mandacaru (Cereus jamacaru) em pó obtido em secador de leito fixo. Dissertation, Federal University of Campina Grande

Santos Neto JP, Silva VDN, Silva PA et al (2019) Características Físico-Químicas de Frutos de Mandacaru (Cereus Jamacaru P. Dc.) Cultivados no Sertão Alagoano. Rev Craibeiras Agroecol 4:e7741

Santos LA, Lima AMP, Passos IV et al (2001) Uso e percepções da alimentação alternativa no estado da Bahia: um estudo preliminar. Rev Nutr 14:35–40. https://doi.org/10.1590/S1415-52732001000400006

Santos D, Farias I, Lira M et al (2006) Manejo e utilização da palma forrageira (Opuntia e Nopalea) em Pernambuco. IPA, Recife

Santos CAG, Dantas MK, Medeiros MFT (2017) Reports of knowledge and the uses of cactaceae species in the historia naturalis brasiliae (1648): comparative past-present analyses. Ethnos 2:1–19. https://doi.org/10.22276/ethnoscientia.v2i1.43

Santos NC, Almeida RLJ, da Silva LRI et al (2020) Pasteurização da polpa e da casca do fruto do mandacaru (Cereus jamacaru). Res Soc Dev 9:e403974027. https://doi.org/10.33448/rsd-v9i7.4027

Saraiva ME, Ulisses AVR, Ribeiro DA, et al (2015) Plant species as a therapeutic resource in areas of the savanna in the state of Pernambuco, Northeast Brazil. J Ethnopharmacol 171:141–153. https://doi.org/10.1016/j.jep.2015.05.034

Sariburun E, Şahin S, Demir C et al (2010) Phenolic content and antioxidant activity of raspberry and blackberry cultivars. J Food Sci 75:328–335. https://doi.org/10.1111/j.1750-3841.2010.01571.x

Scheinvar L (1985) Cactáceas. Flora Ilustrada Catarinense, Itajaí

Shetty AA, Rana MK, Preetham SP (2012) Cactus: a medicinal food. J Food Sci Technol 49:530–536. https://doi.org/10.1007/s13197-011-0462-5

Silva V (2015) Diversidade de uso das cactáceas no Nordeste do Brasil: uma revisão. Gaia Sci 9:175–182. https://doi.org/10.21707/gs.v9i2.26476

Silva G (2017) Cereus jamacaru DC.: Perfil fitoquímico e avaliação citotóxica. Dissertation, Federal University of Pernambuco

Silva F, Albuquerque UP, Costa Júnior LM et al (2014) An ethnopharmacological assessment of the use of plants against parasitic diseases in humans and animals. J Ethnopharmacol 155:1332–1341. https://doi.org/10.1016/j.jep.2014.07.036

Silva APG, Souza CCE, Ribeiro JES et al (2015a) Características físicas, químicas e bromatológicas de palma gigante (Opuntia ficus-indica)e miúda (Nopalea cochenillifera) oriundas do estado da Paraíba. Rev Bras Tecnol Agroindustrial 9:1810–1820. https://doi.org/10.3895/rbta.v9n2.1616

Silva DM, Carvalho FR, Moura AG et al (2015b) Cytotoxic action of the stem aqueous extract of the stem of Cereus jamacaru DC. (mandacaru) Acción citotóxica del extracto acuoso del tallo de Cereus jamacaru DC. (mandacaru). Rev Cuba Plantas Med 19:226–234

Silva LFCR, Valle L, Nascimento ARC, Medeiros MFT (2019) Cereus jamacaru DC. (Cactaceae): from 17th century naturalists to modern day scientific and technological prospecting. Acta Bot Brasilica 33:191–197. https://doi.org/10.1590/0102-33062018abb0352

Sousa MSB, Vieira LM, de Lima A (2011) Fenólicos totais e capacidade antioxidante in vitro de resíduos de polpas de frutas tropicais. Braz J Food Technol 14:202–210. https://doi.org/10.4260/BJFT2011140300024

Sousa A, Lima L, Calou I (2014) Teor de antocianinas e flavonoides amarelos presentes na casca e polpa do fruto do mandacaru (Cereus jamacaru). In: VII Simpósio Iberoamericano de Plantas Medicinais and II Simpósio Iberoamericano de Investigação em Câncer

Sousa GA, Oliveira IS, Silva-Freitas FV et al (2018) Gastroprotective effect of ethanol extracts of cladodes and roots of Pilosocereus gounellei (A. Weber ex K. Schum.) Bly. Ex Rowl

(Cactaceae) on experimental ulcer models. J Ethnopharmacol 218:100–108. https://doi.org/10.1016/j.jep.2018.02.009

Stintzing FC, Carle R (2005) Cactus stems (Opuntia spp.): a review on their chemistry, technology, and uses. Mol Nutr Food Res 49:175–194. https://doi.org/10.1002/mnfr.200400071

Stintzing FC, Schieber A, Carle R (2001) Phytochemical and nutritional significance of cactus pear. Eur Food Res Technol 212:396–407. https://doi.org/10.1007/s002170000219

Tahir A, Elkheir M, Yagoub A (2007) Effect of tree and nodule age on some physicochemical properties of gum from Acacia senegal (L.) Wild., sudan. Res J Agric Biol Sci 3:866–870

Trombetta D, Puglia C, Perri D et al (2006) Effect of polysaccharides from Opuntia ficus-indica (L.) cladodes on the healing of dermal wounds in the rat. Phytomedicine 13:352–358. https://doi.org/10.1016/j.phymed.2005.06.006

Uprety Y, Poudel RC, Shrestha KK et al (2012) Diversity of use and local knowledge of wild edible plant resources in Nepal. J Ethnobiol Ethnomed 8:16. https://doi.org/10.1186/1746-4269-8-16

Vázquez-Ramírez R, Olguín-Martínez M, Kubli-Garfias C, Hernández-Muñoz R (2006) Reversing gastric mucosal alterations during ethanol-induced chronic gastritis in rats by oral administration of *Opuntia ficus-indica* mucilage. World J Gastroenterol 12:4318. https://doi.org/10.3748/wjg.v12.i27.4318

Viana AD, Corrêa JLG, Justus A (2014) Optimisation of the pulsed vacuum osmotic dehydration of cladodes of fodder palm. Int J Food Sci Technol 49:726–732. https://doi.org/10.1111/ijfs.12357

World Health Organization WHO (2002) Traditional Medicine Strategy 2002–2005

Zappi D, Taylor N (2020) Cactaceae. In: Flora do Brasil 2020 em construção. Botanical Gardel of Rio de Janeiro

Zappi D, Taylor N, Santos M, Larocca J (2015) Cactaceae. In: Lista de Espécies da Flora do Brasil. Botanical Gardel of Rio de Janeiro

Zhishen J, Mengcheng T, Jianming W (1999) The determination of flavonoid contents in mulberry and their scavenging effects on superoxide radicals. Food Chem 64:555–559. https://doi.org/10.1016/S0308-8146(98)00102-2

Food Composition Data: Edible Plants in the Pampa

Adriana Carla Dias Trevisan, Vanuska Lima da Silva, Josué Schneider Martins, Gabriela Coelho-de-Souza, Stefany Areva Severo, and Mariana Oliveira Ramos

Contents

1 Introduction

Biomes with non-forest vegetation systems have been particularly neglected by public policies for conservation and sustainable development. This is precisely the case of the Pampa biome. Also known as southern Brazilian grasslands, the Pampa is a mosaic of different types of vegetation with a landscape that comprises savannah, steppe, steppic savannah, coast, transition areas, and semi-deciduous and deciduous forests (Chomenko and Bencke 2016) with a particular flora and fauna and great biodiversity (Quadros et al. 2008). In Brazil, it is only found in Rio Grande do Sul (RS) state, where it occupies nearly 63% of the territory (Suertegaray and Silva 2009). It has a unique cultural, genetic, and natural heritage with regional, national, and global importance. Despite this, from an official conservation perspective, only 0.4% of the Pampa is legally protected in conservation units, which is far from the global average of 17% per biome (Sosinski et al. 2019). Figure 1 presents an idea of the landscape of this biome, in which fields and forest islands coexist.

The conservation and sustained management of agrobiodiversity, through actions to support consumption of edible native species, are essential strategies for food and

A. C. D. Trevisan (✉) · S. A. Severo
Universidade do Estado do Rio Grande do Sul/Ecos do Pampa, Santana do Livramento, RS, Brazil
e-mail: adriana-trevisan@uergs.edu.br

V. L. da Silva · J. S. Martins · G. Coelho-de-Souza · M. O. Ramos
Universidade Federal do Rio Grande do Sul, Porto Alegre, RS, Brazil

M. C. M. Jacob, U. P. Albuquerque (eds.), *Local Food Plants of Brazil*, Ethnobiology, https://doi.org/10.1007/978-3-030-69139-4_12

Fig. 1 Mosaic of ecosystems in the Pampa, Santana do Livramento, RS. (Photo: Trevisan ACD/ Echoes of the Pampa)

nutritional sovereignty and security. Knowledge about the species' nutritional composition has an essential role in advancing these strategies. Some of the species have higher nutrient concentrations of carotenoids, fibers, anthocyanins, vitamins, and minerals than foods commonly consumed (Martins et al. 2019; Rockett et al. 2020; Schmidt 2018b). From a nutritional and phytochemical perspective, the edible native species of the Pampa biome have strong potential to be included in the diet of the population.

In this chapter, we present nine native food plants from the Pampa with different common uses among local people: six are eaten, one used as a spice, and two taken as tea for medicinal purposes. In addition, the three species used as spices and medicine have an important role in the ecological landscape because they provide ecosystem services to food systems. *Schinus molle* L. offers an abundance of food to avifauna that are predators of many undesirable insects. The intense fluorescence of *Aloysia gratissima* (Gillies & Hook) Tronc helps to maintain pollinators, while *Elionurus muticus* (Spreng.) Kuntze is a companion plant to *Butia yatay* (Mart.) Becc. and marks the territory of family farms because this medicinal plant is found in the butiazal understory where there is animal grazing (Curicaca et al. 2018; Lemos et al. 2014). This chapter is composed of three sections, in addition to the introduction. The first explains how the species were chosen, considering their ecology and nutritional composition. The final considerations reflect on gaps in existing knowledge.

2 The Pampa and the Selection of its Edible Species

The long history of environmental degradation and contamination caused by industrial husbandry, associated with landownership concentrated in large properties, has devalued the use of species from the Pampa. Cultural practices forged by pressure from the dominant economic model have led to neglect of the Pampa's natural heritage, which has been forgotten by inhabitants of the territory. Local species with great food and economic potential are not only neglected but are often also suppressed, given that they are thought to "pollute the fields." Nevertheless, studies are underway in the territory that demonstrate opportunities for sustainable use of the Pampa's natural heritage, serving to generate income and stimulate social development and reproduction of local communities.

From the perspective of sustainable use of edible species from biodiversity, it is necessary to understand the ecology of each species, to be able to guide practices linked to handling of natural populations and/or domestication of the species. To manage the natural population, it is understood that the strategies used should be specific to each species, correlating with their ecological plasticity. When planning domestication, it is important to emphasize that given the large genetic variation within populations, it is always necessary to define the objectives of selection, that is, to determine if the intention is production of fruits, juices, spices, teas, essential oils, fibers, crafts, or phytotherapeutic products. Studies demonstrate that various native vegetable species of Brazilian flora have significant quantities and varieties of nutrients such as food fibers, minerals, and vitamins (Rockett et al. 2020; Schmidt 2018b; Seraglio et al. 2018; SiBBr 2020), and they have revealed antioxidant (Dalla Nora et al. 2014a; Rockett et al. 2020), antihyperglycemic (Vinholes et al. 2017), and anti-inflammatory properties. It is also important to highlight that scientific studies have related the consumption of anthocyanins and carotenoids to protection of the human organism and to reducing risk of developing non-transmissible chronic diseases (Sampaio and Almeida 2009; Strack and Souza 2012) including Alzheimer's (Silveira et al. 2011). Each one of the fruits has structural peculiarities, such as strong or fragile seeds, smooth or fibrous bark, fluid, or dense pulp, so that their processing becomes an important challenge that must be studied and understood.

Thus, in addition to the need to revive local knowledge, the creation of environments of innovation that offer support by linking the knowledge between local peoples and research institutions is of strategic importance today. One of the environments of innovation being developed in the Pampa is found at the Universidade Estadual do Rio Grande do Sul (UERGS), in the research and extension group "Ecology of Knowledge in agroecosystems of the Pampa biome – Echoes of the Pampa" that has been collecting results with the species presented in this chapter. In addition to "Echoes of the Pampa," the choice and description of the species was assisted by research groups at the Universidade Federal do Rio Grande do Sul (UFRGS) that work with food and nutritional sovereignty while emphasizing the ecology of knowledge and sustainable use of biodiversity anchored in species important to local communities. The species presented here (Table 1) were selected

Table 1 Selected food and medicinal species of the Pampa biome

	Scientific name	Popular names	Family	Habit
1	*Butia odorata* (Barb. Rodr.) Noblick)	Butiá, coquinho-azedo, butiazeiro	Arecaceae	Arboreal
2	*Butia yatay* (Mart.) Becc.	Butiá-jataí, jataí, yatai,yatay	Arecaceae	Arboreal
3	*Eugenia involucrata* D.C.	Cerejeira-do-rio-grande, cerejeira-do-mato, cereja do-rio-grande, cereja-do-mato	Myrtaceae	Arboreal
4	*Myrcianthes pungens* (O.Berg.) D. Legrand	Guabijú, guabijuzeiro	Myrtaceae	Arboreal
5	*Guettarda uruguensis* Cham. & Schltd	Veludinha, veludinho	Rubiaceae	Arboreal
6	*Acanthosyris spinescens* (Mart. & Eichler) Griseb	Sombra de touro, quebracho-rojo	Santalaceae	Arboreal
7	*Schinus molle* L.	Anacauita, piriquiteira, aroeira-sala	Anacardiaceae	Arboreal
8	*Aloysia gratíssima* (Gillies & Hook) Tronc.	Garupá, alfazema-do-Brasil, mimo do Brasil, cedrón-del-monte	Verbenaceae	Shrub
9	*Elionurus muticus* (Spreng.) Kuntze	Capim-corona, Capim-limão	Poaceae	Herbaceous

with a focus on three principal criteria: (1) having popular use, (2) proven multifunctionality, and (3) being linked to agroecological productive systems for management of pasture and or agroforestry systems.

The species presented in this chapter represent only a sample of the potential that the botanic biodiversity of the Pampa offers to local communities in terms of food and nutritional security as well as the potential to generate income. The information is based on the results of studies being conducted by UERGS/Echoes of the Pampa and at the Universidad de la Republica del Uruguay in the Pampa territory. Specifically, to select the priority species we present here, we prepared an analysis matrix with data from floristic surveys in home gardens, interviews with farmers, and discourse analysis of participants in a radio program that each week discusses a plant species from the Pampa. The data matrix allowed identifying the most common species in home gardens and those best known to Pampa residents. This led to the selection of thirty-five species from which the authors of this chapter selected nine species based on the criteria presented in the previous paragraph.

Due to the scarcity of scientific publications, only three species (*B. odorata*, *M. pungens*, and *E. involucrata*) will have their nutritional and phytochemical composition presented. For the other species, articles were found that examine their biological properties, such as fungicidal, bactericidal, and insecticidal or as food preservatives, but not their use as food. It is essential to establish links between local production systems and the diffusion of knowledge about biodiversity to expand reflections about cultural habits and preferences and to intensify the sustainable use and conservation of the biome. The dominant productive model in the territory, which gives little value to local knowledge, combined with regional asymmetries of development and gaps in scientific studies about the Pampa, is an impediment to increasing value and stimulating innovation with sociobiodiversity. Figure 2 presents images of some of the species presented, to introduce readers not only to the colors, textures, and shapes described but to the diversity found in the Pampa.

Sociobiodiversity is understood to be the relation between biological diversity, traditional agricultural systems, and the use and management of these resources combined with the knowledge and culture of traditional communities and family farmers (BRASIL 2000, 2009). This concept comes from the Brazilian discussion around the sustainability of extractivism while recognizing the diversity and importance of traditional peoples and family farmers. Experiences in the construction of sociobiodiversity product chains have demonstrated the capacity for inclusion of family farmers, extractivists, traditional peoples, and communities that manage and conserve native biodiversity. This reality has contributed to the supply of healthy,

Fig. 2 Fruits and flowers of sociobiodiversity of the Pampa biome. (**a**) Anacauíta (*S. molle*), (**b**) garupá (*A. gratissima*), (**c**) guabijú (*M. pungens*), (**d**) cerejeira-do-rio-grande (*E. involucrata*), (**e**) butiá (*B. odorata*), (**f**) veludinha (*G. uruguensis*) (Photos from Echoes of the Pampa: Lika Furtado (**a**–**e**) and Franciely Pando (**f**))

fresh food or with little processing, with high nutritional values, which is increasingly necessary considering the poor diets, malnutrition, and overweight condition of most of the Brazilian population (Ramos et al. 2017; Tutwiler et al. 2017). In the case of the Pampa biome, it is necessary to overcome the negligence and underutilization of many native species that have multiple potentials for food, medicinal, aromatic, and cosmetic uses. As we present in this chapter, we believe that overcoming the existing gaps in knowledge and giving value to the use of these species should take place by supporting the ecology of knowledge, guaranteeing the qualified participation of local communities. The governance processes used in this production of knowledge are essential to protect traditional knowledge, guarantee the sharing of benefits, and avoid that knowledge is appropriated by distant actors with greater economic and political power. Moreover, to overcome the current situation, it is necessary to redesign the productive units in the Pampa to promote resilience based on biodiverse productive systems from an agroecological perspective.

3 Edible and Medical Species of the Pampa: Ecology and Nutrition

The species selected are potential resources for the development of sustainable productive systems in the Pampa. The ecological aspects presented are the result of a review of the literature and primary data from extension and research projects being executed by the Echoes of the Pampa at UERGS. The databases researched were Portal de Periódicos Capes, SciELO, ScienceDirect, and Redalyc (from 1959 to present). The search descriptors were the scientific names of the nine species presented here, and the inclusion criteria were articles and theses that address species use, sociobiodiversity, or species ecology. With the same search descriptors, the data about nutritional composition are also presented from a review conducted by researchers at UFRGS in the following databases: Web of Science, Bireme, SciELO, Scopus, ScienceDirect, and CAPES (without limiting the time period), in addition to research in The Brazilian Biodiversity Information System (SiBBr) of the Ministry of Science, Technology, Innovation and Communications (MCTIC), with technical support from the United Nations Environment Program (UNEP) and financial support from the Global Environmental Fund (GEF). The inclusion criteria were articles and theses with results that address phytochemical and nutritional qualities.

The Butiás

The butiá or Brazilian palms are composed of the *Butia* genus in the Arecaceae family and are found from Bahia and Goiás to Rio Grande do Sul and neighboring countries (Rodrigues et al. 2008). All the butiás originate from mother plants whose stem grows below ground level and form a natural gregarious population with large production of edible fruits (Meerow et al. 2009). The *butiazais*, as butiá palm groves are known, have a strategic role in the conservation and restoration of the Pampa biodiversity, because by forming islands in the landscape of open fields, they increase ecological resilience by promoting connectivity. The Brazilian palm groves are extremely threatened by the absence of regeneration due to their elimination by intensive cattle grazing and agriculture and expansion of urban areas (Fonseca 2014). Given this situation, the Rota dos Butiazais, a project that attracts tourists to family farms, is an example of a strategy for environmental conservation and sustainable use of biodiversity associated with the *butiazais* that has been developed since 2016, promoting integration between Brazil, Uruguay, and Argentina (Embrapa 2017; Sosinski et al. 2019).

In the Pampa biome, six species stand out, and the most characteristic are *B. odorata*, *B. yatay*, and *B. lallemantii*. *B. yatay* and *B. lallemantii* are endemic to the Pampa (Leitman et al. 2015; Soares et al. 2014). The large intra-populational genetic variation as well as the small number of individuals within these natural populations is a challenge to the conservation and use of the butiazais. Beyond the direct use of the fruits as food, there is also unexplored potential of the compounds in the pulp and seeds of the butiá fruit. Due to the large number of publications about them and because they are more commonly explored commercially, and are even among the species processed and sold by the Cadeia Solidária das Frutas Nativas no RS (Ramos 2019), we are able to present data for *B. odorata* and *B. yatay*.

(a) Butiá

Butia odorata

The species *B. odorata* was initially described in 1826 as *Cocos capitata* and until 2010 was denominated as *B. capitata* (Noblick 2011). It is known locally as bitter small coconut, butiazeiro, or butiá and occurs in ecosystems of the Atlantic Forest and Pampa biomes. It produces fruit once a year, flowers from October to December, and bears fruit from January to April (Büttow et al. 2009). Ecologically it occurs in savannah formations, as palm or butiá groves in poor soil with low water retention capacity. Its fruits are dispersed between the summer and early fall in arrangements of dormant seeds that require the conditions found in the islands of *butiazais* to be able to germinate and thus naturally regenerate (Schlindwein 2012). Thus, like the inflorescences, the fruits of lemon yellow to fire red coloring vary considerably in size (Mistura et al. 2012). The seed bunch can be considered ripe and ready for picking when at least half of the fruits have uniform color.

Among the edible uses of the butiá we can cite manufacturing of oil from the seed endosperm, collection of the sap that flows from the apical meristem after cutting the plant, extraction of hearts of palm, and raw consumption and processing of the fruits in the form of jellies, liquors, ice cream, and sweets (Büttow et al. 2009; Rivas and Barbieri 2017). Moreover, numerous crafts are made from them, and they are elegantly used in landscaping (Büttow et al. 2009; Geymonat and Rocha 2009; Sganzerla 2010). Important projects in the agroecological movement have given value to the fruits of the species and their associated ecosystems by production of butiá juice and frozen pulp (Ramos 2019). The yield of the pulp production varies between 50 and 60% of total fruit weight, without addition of water, and the pulp can be maintained frozen for up to 1 year, without significant sensory or nutritional changes.

Both the pulp and kernels have important nutritional properties. Although they are normally discarded after pulping, the butiá seeds have a kernel of high nutritional value. The *B. odorata* seed was found to have a high protein content (from 12 to 15 g/100 g), fat (10 g/100 g), as well as minerals such as phosphorus, magnesium, and potassium (Faria et al. 2008). Its pulp is generally acidic (pH lower than 4,5) and has high fiber (1 to 4 g/100 g) and vitamin (vitamins A and C) content as well as various minerals (including iron, zinc, and magnesium) (Table 2).

Another important factor is the presence of bioactive compounds and chemical elements with antioxidant properties, among which it is worth noting carotenoids and phenolic compounds. The maximum and minimum values available in the literature for total carotenoids in *B. odorata* are proportional to other recognized nutritional species such as cajá (3,86 mg/100 g), araticum (1249 mg/100 g), and another butiá species, *Butia eriospatha* (1,73 mg/100 g).

(b) Butiá-jataí

Butia yatay

B. yatay has many botanical similarities with *B. witeckii*, *B. quaraimana*, and *B. paraguayensis*, which generates controversies among researchers, including hypotheses that the natural populations are the result of plantings. *B. yatay* has its etymology linked to its small hard fruit and is known as butiá-jataí, jataí, or yatay in Spanish. It is widely distributed in Argentina and Uruguay but in Brazil occurs only in two specific regions of Rio Grande do Sul, one in a typical Pampa region and

Table 2 Nutritional composition of butiá (*Butia odorata*) per 100 g of fruit pulp with natural moisture content[a]

Nutrient	Min.	Max.	Med.	%RDI
Carbohydrates (g/100 g)[b]	6,58	14,84	10,71	8,24
Proteins (g/100 g)[d]	0,60	5,28	2,94	5,25
Fats (g/100 g)[c]	0,11	2,41	1,26	2,29
Food fiber (g/100 g)[b]	0,84	4,89	2,87	20,46
Carotenoids (mg/100 g)	2,80	5,50	4,15	–
Anthocyanins (C3Gmeq/100 g)	1,05	25,13	13,09	–
Vitamin C (mg)[b]	23,00	63,84	43,42	48,24
Vitamin K (mg)[e]	165	462	313,50	9,22
Mg (mg)[b]	4,0	12,0	8,00	1,90

Caption: Min, minimum amounts; max, maximum amounts; med, medium of the minimum and maximum amounts found in literature; % RDI, percent of recommended daily intake

[a]Conversion was realized, when necessary, from a dry base (dehydrated food) to food with its natural moisture content

[b]Institute of Medicine (2006), to adult men

[c]RDC 269/2005 Brasil (2005); g, gram; μg, micrograms; mg, milligrams

[d]Calculated to a man of 70 kg

[e]Institute of Medicine (2019)

Sources: Hoffman et al. (2014), Ferrão et al. (2013), and Fonseca (2012)

another in Giruá, a municipality in the Atlantic Forest (Marchiori et al. 1995; Patterer et al. 2019).

In the Pampa, it is found in the well-known Palmar de Coatepe palm grove, in the municipality of Quaraí, which occupies 60 km^2 and is a disjunction from its center of distribution in Argentina, the "Palmar Grande" (Marchiori et al. 1995; Marchiori and Alves 2011). This natural population with 4–8-m-tall solitary stems that are 20–30 cm in diameter has low densities of individuals due to cattle grazing, which compromises natural regeneration and population viability (Marchiori et al. 1995; Marchiori and Alves 2011). The flowers may be yellow, green, or purplish and the fruits yellow and purplish when ripe (Soares et al. 2014).

Considering its phytochemical composition, four classes of phenolic compounds stand out, coumarins, flavonoids, saponins, and tannins (Matos and Muller 2019), with potential for multiple uses. From a nutritional perspective, Martins and collaborators (2019) identified a significant presence of food fiber (9,0 g/100 g in the fruit and 4,7 g/100 g in the pulp) and vitamin C (72,8 mg/100 g in the pulp in 100 grams of butiá-yatai. These amounts allow categorizing this food as rich in food fibers and vitamin C, to the degree that consumption of 100 g of the fruit represents more than 15% and 100%, respectively, of RDI of these nutrients by the Ministry of Health (Martins et al. 2019).

Cerejeira-do-rio-grande
Eugenia involucrata

The Myrtaceae family is a broad family with 27 genera and 1026 species in Brazil, with 73% of these endemic (Sobral et al. 2015; Vasconcelos et al. 2017). In the Pampa biome, 8 genera and 17 species are described (Flora do Brasil 2020a) and, as in the other biomes where it is found, have important value in the floristic component of the arboreal stratum with a large vector of diversity because of their endemism. One characteristic of the seeds of this family is a low tolerance to drying; thus, the production of seedlings must be conducted soon after pulp removal or collection. The Myrtacea have aromatic flowers and many glands, normally trichomes, which store essential oils (Oliveira 2018). Among the genera in the family, *Eugenia* stands out as the most representative in Brazil, with 392 species, 303 of which are endemic (Flora do Brasil 2020b), with strong food potential due to their fruits with a marked flavor.

One of the important species of this genus is *Eugenia involucrata*, whose center of diversity is the Atlantic Forest with dispersal to the Pampa. It is known as cerejeira-do-mato, cerejeira-do-rio-grande, cerejeira-da-terra, pitanga-preta, cereza-del-monte (espanhol), and îwîrá-yepiró (guarani) and occurs at a range of altitudes and latitudes (Carvalho 2008; Martinez-Crovetto 1968; Praderi 1959). The average trees are from 10 to 15 meters tall, and it is considered a rare species, with low densities per area (Pase do Prado 2009). It is commonly found in backyards on farms in the Pampa. Its fruit production is intense and for this reason is highly sought by animals, which are its natural dispersers, and generates an important bank of plantules for regeneration. In addition to consumption of the raw fruit, the tea from an infusion of the leaves has popular local use for gastric problems, diarrhea, ulcers, and inflammations. It has proven antioxidant, anti-inflammatory, and antinociceptive effects (Barzotto 2019; Marin et al. 2008; Nicácio et al. 2017).

E. involucrata is among the Myrtacea species found in the Pampa with strong potential for production. Data from research and extension projects point to fruit production; however, there is a strong trend toward a multiple use of the species that includes the production of essential oils, natural extracts, and isolation of active phytochemical compounds. It is a soft, sweet, and slightly acidic fruit with yellowish-red to dark red color and is found in various sizes. It has up to four stiff and large seeds, with relatively sensitive integument. For this reason, the processing of this fruit is a bit more complex than that of butiá and guabiju, because breaking the seeds releases compounds into the pulp that are sensorially undesirable. However, after extraction of the pulp, various products can be prepared such as juices, sweets, and cakes.

Table 3 presents the nutritional data for the cerejeira-do-rio-grande. As is common in fruits, the cereja has a low energy density due to its high moisture (of about 90%) and low fat and protein content. In general, the data indicate the presence of significant quantities of iron, copper, and magnesium. The carotenoid content in the pulp of the cereja-do-rio-grande is significant and is comparable to other fruits rich

Table 3 Nutritional composition of the pulp of the cerejeira-do-rio-grande (*E. involucrata*) with natural moisture content[a]

Nutrient	Min.	Max.	Med.	%RDI
Carbohydrates (g/100 g)[b]	3,46	8,27	5,87	4,51
Proteins (g/100 g)[d]	0,62	0,94	0,78	1,39
Fats (g/100 g)[c]	0,21	0,37	0,29	0,53
Food fiber (g/100 g)[b]	1,30	1,70	1,50	10,71
Fe (mg/100 g)[b]	0,41	2,43	1,42	17,75
Zn (mg/100 g)[b]	0,09	0,14	0,12	1,05
K (mg/100 g)[e]	154,80	204,68	179,74	5,29
Na (mg/100 g)[e]	0,63	1,29	0,96	0,06
Ca (mg/100 g)[b]	11,20	13,74	12,47	1,25
P (mg/100 g)[c]	5,73	6,24	5,99	0,83
Mg (mg/100 g)[b]	6,96	12,26	9,61	2,29
Mn (mg/100 g)[b]	0,09	0,16	0,13	5,43
Cu (mg/100 g)[b]	0,06	0,09	0,08	8,33
Total carotenoids (mg/100 g)	8,92	20,21	14,57	–
Vitamin A (μg.EAR/100 g)[b]	26,00	899,38	462,69	51,41

Caption: Min, minimum amounts; max, maximum amounts; med, medium; % RDI, percent of recommended daily intake

[a]Conversion was realized, when necessary, from a dry base (dehydrated food) to food with its natural moisture content

[b]Institute of Medicine (2006), to adult men

[c]RDC 269/2005 Brasil (2005); g, gram; μg, micrograms; mg, milligrams. EAR: Retinol equivalent

[d]Calculated to a man of 70 kg

[e]Institute of Medicine (2019)

Source: Camlofski (2008), Schmidt (2018b), SiBBr (2020), and Schmidt (2018a)

in this nutrient such as guabiroba (24,58 mg/100), pequi (15,28 mg/100 g), and tucumã (6,26 mg/100 g) according to the data available from the Brazilian Biodiversity Information System (SiBBr 2020).

The quantities of bioactive compounds such as phenols, vitamins, and pigments presented in this table may vary with processing factors such as light incidence, contact with oxygen, and exposure to heat, as well as agronomic factors such as production location, climate conditions, and the degree of ripeness of fruits. The levels of carotenoids and vitamin A present in the butiá pulp should be emphasized, given that they are very important elements in a healthy diet.

Guabiju

Myrcianthes pungens

Also, in the Myrtacea family, *M. pungens*, known as guabijú or guabijuzeiro, grow as trees up to 30 m tall and are found in the Atlantic Forest and Pampa (Lorenzi 2008; Sobral et al. 2015; Venzke 2012). There are registers of blossoming from September to January and fruit bearing from January to March (Fior et al. 2010; Guollo 2019). It is a species that ecologically occupies an advanced successional state and for this reason requires specific conditions of shade and soil quality to develop. The species is important to the promotion of local fauna because of its attractive and succulent fruits. Thus, among 181 species described as priority for recovery of typical Pampa ecosystems, *M. pungens* is indicated for forest and savannah formations (Guarino et al. 2018). Its wood is considered commercially important and can be used in civil construction and fine cabinet making and in the production of handles for tools and agricultural equipment. It is a heavy, elastic, strong, and dense wood (dos Santos and Marchiori 2014).

The guabiju stands out for its sweet flavor, and for this reason, together with *E. involucrata*, over time, it was selected for inclusion in domestic backyards in the Pampa. It is a small fruit with a quite fragile structure, with smooth skin that is easily separable from the pulp. Table 4 presents scientific data about the nutritional and phytochemical properties of *M. pungens*.

This data indicates strong nutritional potential, considering that guabiju has significant amounts of carotenoids and anthocyanins. These two compounds are responsible for the more orange-yellow internal and dark purple external color of the fruit. Fruit processing in general has significant impact on the degradation of bioactive compounds such as anthocyanins and total carotenoids in guabiju. Dalla Nora et al. (2014a) report that while the anthocyanin content in 100 g of fresh fruit is 231,2 mg, it is reduced to 13,53 mg in 100 g of dry fruit due to the sensibility of this compound to exposure to heat.

Veludinha

G. uruguensis

The Rubiaceae family is the fourth largest family among the angiosperms. Among the Rubiaceae, the *Guettarda* stands out with 21 species identified in Brazil in various types of vegetation (Pereira and Kinoshita 2013). *G. uruguensis* is the only species that occurs in Rio Grande do Sul and is also found in Uruguay and Argentina (Barbosa et al. 2015; Piaggio and Delfino 2009; Takáts and Toselli 2015).

Table 4 Nutritional composition of guabiju pulp (*M. pungens*) with natural moisture content[a]

Nutrient	Min.	Max.	Méd.	%RDI
Carbohydrates (g/100 g)[b]	12,39	12,41	12,40	9,54
Proteins (g/100 g)[d]	0,55	0,56	0,55	0,99
Fats (g/100 g)[c]	0,05	0,06	0,06	0,10
Food fiber (g/100 g)[b]	6,21	6,23	6,22	44,43
Carotenoids (mg/100 g)	0,96	0,96	0,96	–
Anthocyanins (meq C3G/100 g)	13,52	295,41	154,47	–
Na (mg/100 g)[e]	16,86	52,10	34,48	2,30

Caption: Min, minimum amounts; max, maximum amounts; med, medium; % RDI, percent of recommended daily intake
[a]Conversion was realized, when necessary, from a dry base (Dehydrated food) to food with its natural moisture content
[b]Institute of Medicine (2006), to adult men
[c]RDC 269/2005 Brasil (2005) g, gram; µg, micrograms; mg, milligrams
[d]Calculated to a man of 70 kg
[e]Institute of Medicine (2019)
Source: Andrade et al. (2011), Dalla Nora et al. (2014a, b), and Seraglio et al. (2018)

This species is known as *veludinha*, due to its furry leaves and branches, or palo-cruz, in Spanish, due to the stipules that look like thorns in the form of a cross (Newton 1986).

It is described as a small tree, 3 to 10 m tall, with opposite leaves, pedunculated flowers, and drupaceous fruits with 2–9 seeds (Duarte et al. 2014; Pereira 2007). It has a caducous characteristic, losing up to 90% of its leaves in winter, blossoming peaks in November–December and fruit bearing in January–February (Athayde et al. 2009). To highlight its edible potential, it begins to fruit 2–3 years after planting (Teixeira et al. 2019). Biological studies indicate that the veludinha has antimicrobial and antioxidant activities (Duarte et al. 2014). Moreover, its phytochemical compounds have potential use as allelopathic products (Duarte et al. 2017).

It is an ornamental species that has aromatic flowers and is highly sought by bees, being indicated for apiculture (Duarte et al. 2017). It has strong potential in landscaping and urban tree planting, due to its fast growth, ease of pruning, the aesthetic of its architecture, white flowers, and purplish fruits. Landscapers often use the veludinha in clumps in large areas with harmonious results.

Sombra-de-touro

Acanthosyris spinescens

The Santalaceae family has 39 genera with 13 of them found in Brazil (SiBBr 2020). Among the genera occurring in Brazil stands out *Acanthosyris*, with only two Brazilian species and only one that occurs naturally in the Pampa biome: *A. spinescens* (Dettke and Caires 2015). It is known as sombra-de-touro, quebracho-rojo, or quebrachillo, in Spanish, and îwá-he'e, in Guarani (Martinez-Crovetto 1968).

The species flowers between October and November, and the fruits mature between January and May and are naturally dispersed by animals (Kinupp and De Barros 2007). From the perspective of use with livestock in the Pampa, native groves with *A. spinescens* offer feed for cattle (Guarino et al. 2018) from fruits fallen on the ground, and their dense canopies offer shade in the hottest periods of the year, highlighting potential in silvopasture systems. In addition to its use as forage, its fruit with sweet pulp has high potential as a food source and its seeds for extraction of vegetable oil (Kinupp and De Barros 2007; Teixeira et al. 2019). The popular use indicates that its seeds can also be consumed raw or roasted. In the region, mainly in Uruguay, they are commonly used for pies and are crunchy, like Brazil nuts and almonds.

Since its natural populations have low rates of value of importance (0,98%) and the groves along rivers in the Pampa have been increasingly devastated, the species is now considered threatened in the biome (Ramos 2008). In the 1960s, studies found high concentration of fatty acids with high molecular weight, not common in essential oils of other local species (Powell and Smith Jr. 1966); however, it is still a neglected species in terms of scientific research. The vegetable oil of *A. spinescens* is registered as a major component oleic acid, which has high cicatrizant power (Dauber 2016). Popular knowledge also highlights that when consumed raw in a large quantity, still warm from the sun (picked from the tree), it can have laxative power.

Anacauita

Schinus molle

The Anarcadiaceae family has important representatives in Brazilian food culture, such as mango (*Mangifera indica*) and caju (*Anacardium occidentale*). One of the characteristics of this family is the presence of an anatomic structure that secretes terpenes and resins, which confer a special aroma to its representatives. In Brazil, there are 13 genera, 57 described species (17 endemic), highlighted by the genus *Schinus*, with 10 Brazilian species, and among them, 9 are found in the Pampa (Silva-Luz et al. 2020).

In this context, *S. molle* stands out, known as *anacauita*, *piriquiteira*, *aroeira-salsa*, or anacahuita in Spanish and aguará-ybá in Guarani and is quite a representative of the biome (Gomes et al. 2013). It is a perennial tree, 4 to 10 m; it has a dense, rounded, and open canopy (Gomes et al. 2013). It flowers from August to November and bears fruit from December to March. It is important to honey production, and its fruits are highly appreciated by local avifauna.

Studies demonstrate that its leaves, stems, and fruits have medicinal properties. They are composed of volatile oils with a strong odor and various biological activities (Díaz et al. 2008; Giuffrida et al. 2020). The mature berries are used as an alternative to pepper in foods. The essential oil of the anacauita is registered to have important antibacterial, antifungal, and insecticidal activity, which supports the traditional use of the species for infections (Fenner et al. 2006; Santos et al. 2010; Tomazoni et al. 2017). Due to its tannin content, it is also used to cure leather. In addition to the potential of its resins and oils, its use as food is related to its aromatic potential, and it establishes its potential for extractivism or commercial planting, as is realized with *S. terebinthifolius* (Pich 2019) for the production of red pepper, an export product.

Garupá

Aloysia gratissima

In Brazil, the Verbenaceae family includes 15 genera, 286 species with 63% of these endemic, and many with strong aromatic characteristics (Flora do Brasil 2020c). The *Aloysia* genera stands out and includes ten Brazilian species and three varieties (Moroni and O'Leary 2020), while Argentina is a possible center of dispersion (Silva et al. 2006). Among them, *A. gratissima* is found in Mexico to northeast Argentina, concentrated in the southern region of Brazil (Santos 2007). The botanic identification of the species is complex, and some authors affirm that there are two varieties: *A. gratissima* var. *gratissima* and *A. gratissima* var. *sellowi (*Moroni et al. 2016*). A. gratissima* is a bush that can reach 3 meters in height, with an irregular growth pattern. It flowers intensely through spring and summer (Ross et al. 2018). It is called garupá, alfazema-do-Brasil, mimo-do-Brasil, or cedrón-del-monte in Spanish and is well-known as a highly aromatic wild species.

It is locally used in popular medicine and is especially effective for bronchial, pulmonary, and bladder infections (Santos 2007). Moreover, the aqueous extracts of *A. gratissima* are efficient in the treatment of neurological, depressive, and inflammatory disorders, and its essential oils are effective in fighting bacteria that cause pneumonia and are used as insecticides against mosquito that carries dengue (Coelho and Coêlho 2006; Santos et al. 2016; Souza et al. 2020; Zeni et al. 2013).

Its direct use is highlighted by ornamental and landscaping applications for planting in groups and by its therapeutic potential due to its intense aromatic flowering. It is also ecologically important for honey production, and in traditional Gaucho cooking, garupá leaves are used to flavor meat dishes and are included in the popular infusion with erva-mate (*Ilex paraguariensis*), a species native to the Atlantic Forest originally used by the Indigenous Guarani who live in southern Brazil. In contemporary cooking, it is used as flavoring for other savory dishes and for juices and sweets.

Capim-corona

Elionurus muticus

Elionurus is a small genus with three species in Brazil that is part of the large Poaceae family (Flora do Brasil 2020d). Botanic classification of the genera is controversial, and the characteristic that separates the species has been the presence or not of a citric odor (Welker and Longhi-Wagner 2007). *E. muticus*, known as capim-limão or capim-corona in Brazil and pasto-limón or colia-peluda in Uruguay, is an herbaceous plant found in sandy soils (Hess et al. 2007; Marchi and Barbieri 2015). It flowers from October to December in erect, pilose, solitary stalks 8–10 cm, which are silvery white in color with tones of pink and aromatic (Marchi and Barbieri 2015).

It is a native Gramineae that, since dominant in sandy soils of the Pampa, is not desired in cattle pasture. However, if well managed, in periods when it is more palatable to animals, it has potential use as forage. With rapid growth, its composition contains proteins, fibers, and minerals that can be combined with supplements to compose beef cattle and sheep diets (Bernardis 1998; O'Connor et al. 2011), especially the young plants, as a strategy for facing the Pampa winter. Another contemporary use of the species is related to the production of essential oils whose products can be found for sale by companies in this sector. The volatile oils of *E. muticus* has as their principal component citral, a compound of strong interest in the food, perfume, and cosmetics industries (Fuller 2013). Beyond these uses, local communities use capim-corona to treat various diseases. Their roots are chewed for tooth and stomach pain, and their aerial portions are sudoriparous and control fever (Chagonda and Fungirayi 2016). Studies have found *E. muticus* to have strong antibacterial and antifungal activity (Hess et al. 2007; Puppin et al. 2018).

4 Final Considerations

The consumption of native edible plants is an important strategy for supporting sustainable food systems. These foods have an important role in the promotion of a suitable and healthy diet, given that they can contribute to the conservation of environments and cultures while simultaneously expanding the range of nutrients accessed by the individuals who consume them. In the case of the Pampa, the expansion in the offer and consumption of edible, aromatic, and medicinal plants from the biome involves innovation processes that, although underway, are still a weak element in research and extension activities in the fields of agronomy and nutrition.

The supply of these plants now available for use as food, in large part from extractivism, and research underway, creates an opportunity to diversify diets and reflect on food choices of the local population. These plants, especially the fruits presented here, represent sources of fibers, minerals, vitamins, and antioxidant compounds whose deficiency is related to countless health problems that afflict the Brazilian population. The fruits are joined by medicinal and aromatic plants that have a proven quantity of bioactive compounds that promote good health. Thus, improving knowledge and the sustainable use of this nutritionally healthy food biodiversity, including its regular consumption in daily diets, is an action that exemplifies the win-win paradigm that is so necessary in times of crisis of human and planetary health. The Pampa territory comports innovative productive designs that can promote both environmental health and conservation of ecosystems and territorial development based on sustainability.

References

Andrade JMM, Aboy AL, Apel MA et al (2011) Phenolic composition in different genotypes of guabiju fruits (*Myrcianthes pungens*) and their potential as antioxidant and antichemotactic agents. J Food Sci 76. https://doi.org/10.1111/j.1750-3841.2011.02375.x

Athayde EA, Giehl ELH, Budke JC et al (2009) Fenologia de espécies arbóreas em uma floresta ribeirinha em Santa Maria, sul do Brasil. Rev Bras Biociências 7:43–51

Barbosa MR, Sobral M, Proença C, et al (2015) *Guettarda* in Lista de Espécies da Flora do Brasil. In: Jard. Botânico do Rio Janeiro. http://floradobrasil.jbrj.gov.br/jabot/floradobrasil/FB171. Accessed 15 Sept 2014
Barzotto ILM (2019) Otimização de extratos de *Eugenia involucrata* DC. Universidade do Vale do paraíba
Bernardis AC (1998) Nutritive value of the grass *Elionurus muticus* (Spreng) Kuntze. Inf Tecnológica 9:227–230
BRASIL (2000) Programa Nacional de Conservação da Biodiversidade. A Convenção sobre Diversidade Biológica – CDB. In: Decreto Legis. no. 2. http://www.terrabrasilis.org.br/ecotecadigital/pdf/serie-biodiversidade%2D%2D02-convencao-sobre-diversidade-biologica.pdf%3E. Accessed 7 Oct 2020
Brasil (2005) RDC 269 de 22 de setembro de 2005: Regulamento Técnico sobre a ingestão diária recomendada (IDR) de proteína, vitaminas e minerais
BRASIL (2009) PORTARIA INTERMINISTERIAL N° 239 de 21 DE JULHO de 2009
Büttow MV, Barbieri RL, Neitzke RS, Heiden G (2009) Conhecimento tradicional associado ao uso de butiás (*Butia* spp., arecaceae) no sul do Brasil. Rev Bras Frutic 31:1069–1075. https://doi.org/10.1590/s0100-29452009000400021
Camlofski AM (2008) Caracterização do fruto de cerejeira (*Eugenia involucrata* DC) visando seu aproveitamento tecnológico. Universidade Estadual de Ponta Grossa
Carvalho PE (2008) Espécies arbóreas brasileiras. Embrapa Florestas, Colombo
Chagonda LS, Fungirayi B (2016) Antifungal activity of the essential oil of *Elionurus muticus* (Spreng) Kunth from Zimbabwe against *Candida albicans*, *C. krusei* and *Cryptococcus neoformans*. J Multidiscip Eng Sci Technol 3:5331–5335
Chomenko L, Bencke GA (2016) Nosso Pampa desconhecido. Fundação Z, Porto Alegre
Coelho AAM, Coêlho AAA (2006) Análise inseticida de extratos de plantas do bioma Cerrado sobre triatomíneos e larvas de *Aedes aegypti*. Universidade de Brasília
Curicaca, FLD, et al (2018) Conservação dos butiazais e salvaguarda da pecuária familiar em campo nativo na região de Quaraí, Rio grande do Sul
Dalla Nora C, Jablonski A, Rios AO et al (2014a) The characterisation and profile of the bioactive compounds in red guava (*Psidium cattleyanum* Sabine) and guabiju (Myrcianthes pungens (O. Berg) D. Legrand). Int J Food Sci Technol 49:1842–1849. https://doi.org/10.1111/ijfs.12493
Dalla Nora CD, Müller CDR, de Bona GS et al (2014b) Effect of processing on the stability of bioactive compounds from red guava (*Psidium cattleyanum* Sabine) and guabiju (Myrcianthes pungens). J Food Compos Anal 34:18–25. https://doi.org/10.1016/j.jfca.2014.01.006
Dauber B (2016) Elucidação dos componentes e avaliação da atividade antiquimiotáxica do óleo sa semente de *Acanthosyris spinencens*. Universidade Federal do Rio Grande do Sul
Dettke GA, Caires CS (2015) *Santalaceae* in Lista de Espécies da Flora do Brasil. In: Jard. Botânico do Rio Janeiro
Díaz C, Quesada S, Brenes O et al (2008) Chemical composition of *Schinus molle* essential oil and its cytotoxic activity on tumour cell lines. Nat Prod Res 22:1521–1534. https://doi.org/10.1080/14786410701848154
Santos SR dos, Marchiori JNC (2014) Anatomia da madeira de *Myrcianthes pungens* (O. Berg) D. Legrand (Myrtaceae). Balduinia 0:25–30. https://doi.org/10.5902/2358198014089
Duarte A, Hirota B, Campos R (2014) Avaliação da atividade antioxidante e antimicrobiana do extrato etanólico bruto e frações orgânicas obtidas a partir da casca do caule da espécie *Guettarda uruguensis* Cham & Scthdl. (Rubiaceae). Rev Ciênc Farm Básica Apl 35:607–614
Duarte AFS, Krause MS, Martins ALC, Miguel OG (2017) Atividade fitotóxixa de *Guettarda uruguensis* (Rubiaceae) sobre a germinação e crescimento de sementes de Lactuca sativa. Visão acadêmica 18:5–13
Embrapa (2017) Rota dos Butiazais. 6
Faria JP, Almeida F, Silva LCR et al (2008) Caracterização da polpa do coquinho-azedo (*Butia capitata* var *capitata*). Rev Bras Frutic 30:827–829. https://doi.org/10.1590/s0100-29452008000300045

Fenner R, Betti AH, Mentz LA, Rates SMK (2006) Plantas utilizadas na medicina popular brasileira com potencial atividade antifúngica. Rev Bras Ciencias Farm J Pharm Sci 42:369–394. https://doi.org/10.1590/S1516-93322006000300007

Ferrão TS, Ferreira DF, Flores DW et al (2013) Evaliation of composition and quality parameters of jelly palm (butia odorata) fruits from different regions of Southern Brazil. Food Research International 54:57–62. https://doi.org/10.1016/j.foodres.2013.06.002

Fior CS, Rodrigues LR, Calil AC et al (2010) Qualidade fisiológica de sementes de guabijuzeiro (*Myrcianthes pungens* (Berg) Legrand - Myrtaceae) em armazenamento. Rev Arvore 34:435–442. https://doi.org/10.1590/s0100-67622010000300007

Flora do Brasil (2020a) Myrtacea. In: em construção. http://floradobrasil.jbrj.gov.br/reflora/floradobrasil/FB171. Accessed 2 Jul 2020

Flora do Brasil (2020b) *Eugenia*. In: em construção. http://floradobrasil.jbrj.gov.br/reflora/floradobrasil/FB10338. Accessed 2 Jul 2020

Flora do Brasil (2020c) *Verbenaceae*. In: em construção. http://floradobrasil.jbrj.gov.br/reflora/floradobrasil/FB246. Accessed 20 Oct 2020

Flora do Brasil (2020d) *Poaceae*. In: em construção. http://floradobrasil.jbrj.gov.br/reflora/floradobrasil/FB19. Accessed 2 Jul 2020

Fonseca LX (2012) Caracterização de frutos de butiazeiro (Butia odorata Barb. Rodr.) Noblick & Lorenzi e estabilidade de seus compostos bioativos na elaboração e armazenamento de geleias. Universidade Federal de Pelotas

Fonseca MM (2014) Biologia reprodutiva de *Butia odorata* (Barb. Rodr.) Noblick. Universidade Federal de Pelotas

Fuller TN (2013) Caracterização genética e química e atividade biológica do óleo essencial de populações naturais de *Elionurus muticus* humb. & bompl ex willd. Universidade Federal do Rio grande do Sul

Geymonat G, Rocha N (2009) M'botiá. Ecosistema unico en el mundo. Castillos: Casa Ambiental, 1st edn. Casa Ambiental, Montevideo

Giuffrida D, Martínez N, Arrieta-Garay Y et al (2020) Valorisation of *Schinus molle* fruit as a source of volatile compounds in foods as flavours and fragrances. Food Res Int 133:109103. https://doi.org/10.1016/j.foodres.2020.109103

Gomes GC, Cardoso JH, Ferrer RS et al (2013) Árvores da Serra dos Tapes: Guia de Identificação com Informações Ecológicas, Econômicas e Culturais. Embrapa Clima Temperado, Pelotas

Guarino E, Overbeck GE, Boldrini II et al (2018) Espécies de Plantas Prioritárias para Projetos de Restauração Ecológica em Diferentes Formações Vegetais no Bioma Pampa: Primeira Aproximação. Doc/Embrapa Clima Temperado 457:79

Guollo K (2019) Biologia floral e reprodutiva de guabijuzeiro, sete-capoteiro e ubajaizeiro. Universidade Tecnológica Federal do paraná

Hess SC, Peres MTLP, Batista AL et al (2007) Evaluation os seasonal changes in chemical composition and antibacterial activity of *Elionurus muticus* (Sprengel) O. Kuntze (Gramineae). Quim Nov 30:370–373

Hoffmann JF, Barbieri RL, Rombaldi CV, Chaves FC (2014) Butia spp. (Arecaceae): An overview. Scientia Horticulturae 179:122–131. https://doi.org/10.1016/j.scienta.2014.08.011

Institute of Medicine (2006) Dietary DRI reference intakes. National Academy of Sciences, Washington, D.C.

Institute of Medicine (2019) Dietary reference intakes for sodium and potassium. The National Academies Press, Washington, D.C.

Kinupp VF, De Barros IBI (2007) Observações Biológicas, Estudo Bromatológico e Potencial Econômico de Sombra-de-Touro. Rev Bras Biociências 5:66–68

Leitman P, Soares K, Henderson A, et al (2015) *Arecaceae* in Lista de Espécies da Flora do Brasil. In: Jard. Botânico do Rio Janeiro. http://floradobrasil.jbrj.gov.br/jabot/floradobrasil/FB34028. Accessed 15 Mar 2013

Lemos RPM, D'Oliveira CB, Rodrigues CR et al (2014) Modeling distribution of *Schinus molle* L. in the Brazilian Pampa: insights on vegetation dynamics and conservation of the biome. Ann For Res 57:1. https://doi.org/10.15287/afr.2014.272

Lorenzi H (2008) Árvores brasileiras: Manual de identificação e cultivo de plantas arbóreas nativas do Brasil, 5th edn. Editora Plantarum, Nova Odessa – SP
Marchi MM, Barbieri RL (2015) Cores e Forma no Bioma Pampa: Gramíneas e ornamentais nativas, 1st edn. Embrapa, Brasilia
Marchiori JNC, Alves FS (2011) Palmar de Coatepe (Quaraí, RS): enfoque fitogeográfico. Balduina 25:21–26
Marchiori INC, Elesbão LEG, Alvarez Filho A (1995) O palmar de Coatepe. Ciência Ambient 1:93–104
Marin R, Apel MA, Limberger RP et al (2008) Volatile components and antioxidant activity from some myrtaceous fruits cultivated in southern Brazil. Lat Am J Pharm:27
Martinez-Crovetto R (1968) Etnobiologica. Etnobiologica 4:1–24
Martins JS, Melo EM, Fallavena LP, Hertz PF (2019) Avaliação nutricional de Butiá (*Butia yatai*) processado. Segurança Aliment e Nutr 26:1–7
Matos TM, Muller NTG, M. D, et al (2019) Characterization of foliar secondary metabolites in *Butia yatay* (Mart.) Becc. Rev Tecnológica 28:31–38. https://doi.org/10.4025/revtecnol.v28i1.50962
Meerow AW, Noblick L, Borrone JW et al (2009) Phylogenetic analysis of seven WRKY genes across the palm subtribe attaleinae (Areceaceae) identifies *Syagrus* as sister group of the coconut. PLoS One 4:10–13. https://doi.org/10.1371/journal.pone.0007353
Mistura CC, Barbieri RL, Castro CM et al (2012) Caracterização de uma população natural de *Butia odorata* (Barb.Rodr.) Noblick&lorenzi:cor e tamanho dos frutos. In: Encontro sobre pequenas frutas e frutas nativas do Mercosul. Embrapa Clima Temperado, Pelotas, p 80
Moroni P, O'Leary N (2020) *Aloysia* in Flora do Brasil 2020 em construção. In: Jard. Botânico do Rio Janeiro. http://floradobrasil.jbrj.gov.br/reflora/floradobrasil/FB15122. Accessed 2 Jul 2020
Moroni P, O'Leary N, Filloy J (2016) Species delimitation in the *Aloysia gratissima* complex (Verbenaceae) following the phylogenetic species concept. Bot J Linn Soc 180:193–212. https://doi.org/10.1111/boj.12369
Newton J (1986) Estudo anatômico da madeira de veludinho, *Guettarda uruguensis* Cham. et Schlecht. Ciência e Nat 1:115–123
Nicácio AE, Rotta EM, Boeing JS et al (2017) Antioxidant activity and determination of phenolic compounds from *Eugenia involucrata* DC. Fruits by UHPLC-MS/MS. Food Anal Methods 10:2718–2728. https://doi.org/10.1007/s12161-017-0840-3
Noblick LR (2011) Validation of the name *Butia odorata*. Palms 55:48–49
O'Connor TG, Martindale G, Morris CD et al (2011) Influence of grazing management on plant diversity of Highland Sourveld grassland, Kwazulu-Natal, South Africa. Rangel Ecol Manag 64:196–207. https://doi.org/10.2111/REM-D-10-00062.1
Oliveira AG (2018) Sistemática de *Eugenia* (Myrtaceae, Myrteae): evolução da flor e da inflorescência e implicações taxonômicas. Universidade de São Paulo
Pase do Prado A (2009) Aspectos autocológicos e silviculturais de *Eugenia involucrata* DC. Universidade Federal de Santa Maria
Patterer NI, Zucol AF, Fernández Pepi MG, Brea M (2019) Plant and soil phytolith analysis of a relictual community of *Butia yatay* in temperate-humid savannas (Entre Ríos, Argentina). Quat Int 525:126–139. https://doi.org/10.1016/j.quaint.2019.07.019
Pereira ZV (2007) Rubiaceae Juss. do Parque Estadual das Várzeas do Rio Ivinhema, Mato Grosso do Sul: florística, sistema reprodutivo, distribuição espacial e relações alométricas de espécies distílicas. Universidade Estadual de Campinas
Pereira ZV, Kinoshita LS (2013) Rubiaceae Juss. do Parque Estadual das Várzeas do Rio Ivinhema, MS, Brasil. Hoehnea 40:205–251
Piaggio M, Delfino L (2009) Florística y fitosociología de un bosque fluvial en Minas de Corrales, Rivera, Uruguay. Iheringia - Ser Bot 64:45–51
Pich RC (2019) Obtenção e caracterização do produto seco da pimenta rosa (*Schinus terebinthifolius* R.) por cast tape drying. Universidade Federal de Uberlândia

Powell RG, Smith CR Jr (1966) New acetylenic fat acids from *Acanthosyris spinescens* seed oil. Biochemistry 5:625–631

Praderi R (1959) Notas preliminares acerca de la vegetación del río Uruguay medio. Boletín la Soc Taguató 1:33–45

Puppin DGPB, Barbosa JP, Oliveira TR et al (2018) Anti-Candida activity of the essential oil from *Elionurus muticus*: a preliminary study. Aust Dent Sci 3:2–4

Quadros FLF, Trindade JPP, Borba M (2008) A abordagem funcional da ecologia campestre como instrumento de pesquisa e apropriação do conhecimento pelos produtores rurais. In: Campos Sulinos. Ministério do Meio Ambiente, Brasilia, pp 206–213

Ramos G (2008) Florística y Fitosociología Leñosa de Rincón de Franquía, Bella Unión, Uruguay. Tesina, Universidad de la República del Uruguay

Ramos MO (2019) Cadeias agroalimentares sob o enfoque da soberania e segurança alimentar e nutricional: uma construção a partir do estudo de cadeias de frutas nativas da Mata Atlântica

Ramos MO, Cruz FT, Souza GC, Kubo RR (2017) Cadeias de Produtos da Sociobiodiversidade no Sul do Brasil: Valorização de Frutas Nativas da Mata Atlântica no Contexto do Trabalho com Agroecologia. Amaz – Rev Antropol 9:98. https://doi.org/10.18542/amazonica.v9i1.5485

Rivas M, Barbieri RL (2017) Butiá, *Butia odorata*. Instituto Interamericano de Cooperación para la Agricultura (IICA)

Rockett FC, Schmidt H, Pagno CH et al (2020) Native fruits from southern Brazil: Physico-chemical characterization, centesimal composition, and mineral content. J Food Process Preserv 44:1–11. https://doi.org/10.1111/jfpp.14582

Rodrigues PRF, Barbieri RL, Costa JC, Morselli TBG (2008) Ocorrência de *Butia lallemantii* no município de Alegrete (RS) e ameaças às populações naturais. In: Encontro de Pós-graduação da UFPEL. Pelotas, RS, pp 0–3

Ross P, Muñoz J, Cracco P (2018) Flora Indígena del Uruguay: árboles y arbustos ornamentales. Agropecuaria Hemisferio Sur, Montevideo

Sampaio L, Almeida C (2009) Vitaminas antioxidantes na prevenção do câncer do colo uterino. Rev Bras Cancerol 55:289–296

Santos FM (2007) Aspectos ecofisiológicos de *Aloysia gratissma* (Gillies et Hook) Troncoso [Verbenaceae] associados à composição do óleo essencial e sua ação microbiana. Universidade Federal de Lavras

Santos ACA, Rossato M, Serafini LA et al (2010) Efeito fungicida dos óleos essenciais de *Schinus molle* L. e *Schinus terebinthifolius* Raddi, Anacardiaceae, do Rio Grande do Sul. Rev Bras Farmacogn 20:154–159

Santos FM, Pinto JEBP, Bertolucci SKV et al (2016) Chemical composition and antimicrobial activity of the essential oil from the leaves and flowers of *Aloysia gratissima*. Acta Hortic 1125:215–221. https://doi.org/10.17660/ActaHortic.2016.1125.27

Schlindwein G (2012) Implicações ecológicas da dormência de sementes em *Butia odorata* (Arecaceae). Universidade Federal do Rio Grande do Sul

Schmidt H (2018a) Caracterização físico-química, nutricional e de compostos ativos de sete espécies da família Myrtaceae nativas da região sul do Brasil. Universidade Federal do Rio Grande do Sul

Schmidt S (2018b) Caracterização físico-química, nutricional e de compostos bioativos de sete espécies da família Myrtaceae nativas da região sul do Brasil. Universidade Federal do Rio Grande do Sul

Seraglio SKT, Schulz M, Nehring P et al (2018) Nutritional and bioactive potential of Myrtaceae fruits during ripening. Food Chem 239:649–656. https://doi.org/10.1016/j.foodchem.2017.06.118

Sganzerla M (2010) Caracterização físico-química e capacidade antioxidante do butiá. universidade Federal de Pelotas

SiBBr (2020) Sistema de informação sobre a Biodiversidade Brasileira. https://ala-bie.sibbr.gov.br/ala-bie/species/364658#classification. Accessed 1 Jul 2020

Silva CC, Vandresen F, de Oliveira CMA et al (2006) Chemical composition of *Aloysia gratissima* (Gill. et Hook) Tronc. (Verbenaceae). Biochem Syst Ecol 34:593–595. https://doi.org/10.1016/j.bse.2006.01.001
Silva-Luz CL, Mitchell JD, Pirani JR, Pell SK (2020) Anacardiaceae in Flora do Brasil 2020 em construção. In: Jard. Botânico do Rio Janeiro. http://floradobrasil.jbrj.gov.br/reflora/floradobrasil/FB1199. Accessed 20 Oct 2020
Silveira S, De Lucena EV, Pereira TF et al (2011) Atividade anticolinesterásica dos frutos de *Myrcianthes pungens* (O. Berg) D.Legrand (Myrtaceae). Arq Ciência da Saúde da Unipar 15:127–133. https://doi.org/10.25110/arqsaude.v15i2.2011.3707
Soares KP, Longhi SJ, Neto LW, De Assis LC (2014) Palmeiras (Arecaceae) no Rio Grande do Sul, Brasil. Rodriguesia 65:113–139
Sobral M, Proença C, Souza M, et al (2015) Lista de Espécies da Flora do Brasil. In: Myrtaceae List. Espécies da Flora do Bras
Sosinski ÊE, Urruth LM, Barbieri RL et al (2019) On the ecological recognition of Butia palm groves as integral ecosystems: why do we need to widen the legal protection and the in situ/on-farm conservation approaches? Land Use Policy 81:124–130. https://doi.org/10.1016/j.landusepol.2018.10.041
Souza RC, Baldisserotto B, Melo JFB et al (2020) Dietary *Aloysia triphylla* essential oil on growth performance and biochemical and haematological variables in Nile tilapia. Aquaculture 519:734913. https://doi.org/10.1016/j.aquaculture.2019.734913
Strack MH, Souza CG (2012) Antocianinas, catequinas e quercetina: evidências na prevenção e no tratamento das doenças cardiovasculares. Rev Bras Nutr Clin 27:43–50
Suertegaray DMA, Silva LAP (2009) Tchê Pampa: histórias da natureza gaúcha. In: Patta Pillar V et al (eds) Campos Sulinos: conservação e uso sustentável da biodiversidade, 1st edn. Ministério do Meio Ambiente, Brasília, p 403
Takáts A, Toselli C (2015) Estudio ambiental y turístico de la Reserva Natural Las Piedras, Gualeguaychú, Entre Ríos, Argentina. Tiempo de gestión 21:40
Teixeira N, Melo JCS, Batista LF et al (2019) Edible fruits from Brazilian biodiversity: a review on their sensorial characteristics versus bioactivity as tool to select research. Food Res Int 119:325–348. https://doi.org/10.1016/j.foodres.2019.01.058
Tomazoni EZ, Ribeiro RTS, Schwambach J (2017) Potencial fungitóxico dos óleos essenciais de *Schinus molle* L. e *Schinus terebinthifolius* Raddi contra fungos patogênicos do tomateiro. Rev Bras Agroecol 12:7
Tutwiler A, Padulosi S, Hunter D (2017) Securing sustainable and nutritious food systems through mainstreaming agricultural biodiversity: an interdisciplinary study. Lancet 389:S22. https://doi.org/10.1016/s0140-6736(17)31134-0
Vasconcelos TNC, Proença CEB, Ahmad B et al (2017) Myrteae phylogeny, calibration, biogeography and diversification patterns: increased understanding in the most species rich tribe of Myrtaceae. Mol Phylogenet Evol 109:113–137. https://doi.org/10.1016/j.ympev.2017.01.002
Venzke TS (2012) Florística de comunidades arbóreas no Município de Pelotas, Rio Grande do Sul. Rodriguésia 63:571–578
Vinholes J, Lemos G, Lia Barbieri R et al (2017) In vitro assessment of the antihyperglycemic and antioxidant properties of araçá, butiá and pitanga. Food Biosci 19:92–100. https://doi.org/10.1016/j.fbio.2017.06.005
Welker CAD, Longhi-Wagner HM (2007) A família Poaceae no Morro Santana, Rio Grande do Sul, Brasil. Rev Bras Biociências 5:53–92
Zeni ALB, Zomkowski ADE, Maraschin M et al (2013) Evidence of the involvement of the monoaminergic systems in the antidepressant-like effect of *Aloysia gratissima*. J Ethnopharmacol 148:914–920. https://doi.org/10.1016/j.jep.2013.05.042

Food Composition Data: Edible Plants from the Amazon

Bernardo Tomchinsky, Gabriela G. Gonçalves, and Almecina B. Ferreira

Contents

1 Edible Amazon

The Amazon rainforest comprises a biome with predominantly tropical moist broadleaf vegetation, which occupies the lowlands of the equatorial and humid region of South America. Its 7 million square kilometers are spread over the hydrographic basins of the Amazon, Orinoco, and Araguaia-Tocantins rivers in nine South American countries (Bolivia, Brazil, Colombia, Ecuador, Guyana, French Guiana, Peru, Suriname, Venezuela), of which Brazil holds 60% of the entire territory (IBGE 2016).

Despite this apparent uniformity, the Amazon rainforest is composed of a mosaic of different types of terrain, vegetation, and ecosystems whose distribution has been altered throughout history according to the Earth's geological and climate changes

B. Tomchinsky (✉)
Universidade Federal do Sul e Sudeste do Pará, Instituto de Estudos em Saúde e Biologia, Marabá, Pará, Brazil
e-mail: btomchinsky@unifesspa.edu.br

G. G. Gonçalves
Colégio da Villa, Jaguariúna, São Paulo, Brazil

A. B. Ferreira
Universidade Federal do Acre, Centro de Ciências Biológicas e da Natureza, Rio Branco, Acre, Brazil
e-mail: almecina.ferreira@ufac.br

M. C. M. Jacob, U. P. Albuquerque (eds.), *Local Food Plants of Brazil*, Ethnobiology, https://doi.org/10.1007/978-3-030-69139-4_13

(Prance 1973; Ab'saber 1977; Haffer and Prance 2002). Some of these Amazon landscapes are mountain ranges and mountains, tepuis; aquatic ecosystems; blackwater, whitewater, and transparent rivers of different origins and composition; beaches; islands; archipelagos; flood plains; metalliferous savannas; igapó/várzea (flooded) forests; tropical forests; terra firme forests; capoeira (secondary) forests; Amazonian caatingas; cerrado enclaves; babassu forests; restingas; taquara bamboo forests; and other formations with densely concentrated species (Nelson and Oliveira 2001).

We still know little about the vegetation of the Amazon. Recent studies on the entire Amazonian territory, including Brazil and neighboring countries, list between 11,000 and 14,000 species of known tree plants (Ter Steege et al. 2013; Cardoso et al. 2017), a number close to the 12,000 plants described in the Brazilian Amazon (Flora do Brasil 2020). Hopkins (2017) calculates that we know only 70% of the 16,000 species that might exist in the biome. Following the current pace of new discoveries of plant species in the biome, it would take at least another 300 years to get to know the entire Amazonian flora (Ter Steege et al. 2016). Just between 2014 and 2015, 216 new plant species were described for science in the biome (WWF 2017). These data are alarming when we consider the current rate of deforestation in the biome and the threats to the survival of traditional peoples and communities, those who hold knowledge about this biodiversity.

Recently, we have better understood of what consists the human (anthropogenic) influence in the formation of the Amazon rainforest. Some works of historical ecology relate the origin of various types of vegetation in the Amazon or the presence and distribution of certain species to the handling of the landscape by indigenous peoples, such as Brazil nut forests, babassu forests, Amazonian dark earth, bamboo forests, and liana shrublands (Balée 1989; Shepard Jr and Ramirez 2011; Clement et al. 2015; McMichael et al. 2013). These studies suggest that between 12% and 80% of the entire Amazonian forest could have anthropogenic origin (Balée 1989, 1994; Levis et al. 2017).

From 11,000 years ago, when the oldest known archaeological sites in the Amazon date back to (Roosevelt et al. 1996), the original population spread throughout the Amazon basin forming groups with different cultural practices. This socio-biocultural diversity is expressed in the more than 180 indigenous ethnicities that inhabit the Brazilian Amazon (Heck et al. 2005), without mentioning those that were exterminated in the process of European colonization (Nimuendaju 2017). The reduction of this indigenous population represented the loss of a part of the biodiversity that was in process of domestication, including at least 138 species in different stages of domestication (Clement 1999).

With the beginning of the European colonization, from the sixteenth century on, there was an abrupt rupture in the cultural, social, and environmental dynamics existing in the Amazon with the introduction of exotic species (Crosby 1972) and new cultural practices that were incorporated into the local diet (Cascudo 2011). At the same time, other native species started to have new uses, including as food (Tomchinsky and Ming 2019).

In this process, populations from different parts of the world and from Brazil contributed to the formation of different Amazonian identities, including influences from European, African, and Asian populations, as well as other traditional peoples and communities who settled in the region, such as fishermen, riverside dwellers, rubber tappers, Brazil nut pickers, babassu coconut breakers, extractivists, and small farmers among others. Given this complex sociocultural and environmental diversity, it is impossible to identify a single Amazonian dietary culture.

The isolated study of edible plants is not enough to construct an overview on the Amazonian dietary culture, as it is related to other foods of animal and mineral extraction; objects used for farming, hunting, and preparing food; and mythologies, in addition to cultural and social practices. The traditional agricultural system of the Rio Negro, declared a cultural heritage of Brazil, demonstrates how complex these dietary practices are and how they involve several social, cultural, and environmental factors (Emperaire 2010).

For a long time, Amazonian biodiversity has been exploited with an extractive approach, as in the economic cycles of the sertão drugs, or the Brazil nut and rubber trees. These Amazônia was known through a fetishized or exotic lens by the rest of the world through the eyes of the foreign travellers and naturalists. In addition to the domesticated and cultivated Amazon species that have been brought to the rest of the world, many edible Amazon species are used only by local populations and remain unknown to the rest of the world.

Open markets in small and large Amazonian cities are an important showcase to know this edible biodiversity and constitute an important center for agricultural diversity and exchange of germplasm among the local population (Emperaire and Eloy 2008). Some of the most noteworthy markets are the Ver-o-Peso and the market of the 25 de Setembro street in Belém, Pará; the Elias Mansur Market in Rio Branco, Acre; the Adolpho Lisboa Market; the fruit stands near the airport; and the Laranjeiras Market in Manaus, Amazonas, or the weekend fairs in São Gabriel da Cachoeira or Tabatinga, in the state of Amazonas.

The appreciation of traditional food plants or "unconventional food plants" (plantas alimentícias não convencionais, PANC) by the fine cuisine circuit and by movements aimed at valorize sociobiodiversity, such as Slow Food, brought a new stimulus to the study of Amazonian food plants.

However, there is yet much to be studied, considering that between 10% and 20% of the world flora has current or potential nutritional use to humans (Kinupp 2007; Kinupp and Lorenzi 2014), which means that, just in the Amazon, from 1.6 to 3.2 thousand plants are edible. Most of the studies carried out in the Amazon region, so far, have been restricted to ethnobotanical research on a specific species, region, or ethnic group, without pursuing research on the nutritional properties and botanical or agronomic aspects. This occur partly due to the bureaucracy for carrying out these researches within the Brazilian legislation (Tomchinsky et al. 2013) but also due to the lack of human and financial resources in research institutions in the northern region of Brazil, in comparison with other regions of the country, or due to the limitations of postgraduate research inside institutional deadlines.

The objective of this chapter is to give an overview on the knowledge about Amazonian edible plants, traditional or unconventional, and general aspects related to them. The species addressed here were separated into groups according to the type of dietary use and the part used including fruit species, roots and tubers, grains, pulses and nuts, vegetables, and condiment plants.

2 Amazonian Fruit Plants

The Amazon rainforest has a great diversity of edible fruit plants, Cavalcante (1972) lists 163 species and Rabelo (2012) lists at least 250 edible species, many of which known only by the local population, while a few have achieved national or international projection, such as açaí (*Euterpe oleracea* Mart.), cocoa (*Theobroma cacao* L.), and cupuassu (*Theobroma grandiflorum* (Willd. Ex Spreng.) K. Schum.).

Fruit-bearing trees are an important part of the Amazonian nutritional routine (Table 1); several ethnobotanical surveys show that they are by far the most representative group for the local population's diet. In indigenous communities from the Lower Uaupés River region, fruit species represent 78% of the total food plants cited (Gonçalves 2017). The percentage of fruit species is also large in ethnobotanical listings made with the Yanomami, Sateré-Mawé, and Baniwa indigenous groups who inhabit the Amazon region (Albert and Milliken 2009; Bustamante 2009; Fernandes 2012) (Fig. 1).

Among the fruit trees, some produce small fruits, such as buiuiu or buxixu, which belong to the family Melastomataceae, and are compared to berries. These small fruits do not have a large amount of pulp or high energy value and are consumed as an "appetizer" or as "snacks." They are shrub plants found in capoeiras, on the river banks or streams (Amazonian igarapés), in anthropic areas, or in places with high incidence of light and commonly consumed on the way to the farms or during fishing and hunting, especially by children. Some representatives are buxixu-azul (*Clidemia japurensis* DC.), buxixu (*Clidemia hirta* (L.) D. Don), and buiuiu-rosa (*Clidemia rubra* (Aubl.) Mart.) (Kinupp and Lorenzi 2014; Gonçalves 2017). From the same family, goiaba-de-anta (*Bellucia dichotoma* Cogn. and *Bellucia grossularioides* (L.) Triana) is a high tree, appreciated for its fruits by humans and game animals.

In the Western Amazon, from the Rivers Solimões and Negro, it is possible to find in the markets the cubiu or maná-cubiu (*Solanum sessiliflorum* Dunal.), a cultivated Solanaceae, with fruits that are used in salads, jams, or soft drinks or as a seasoning for dishes, substituting lemon. The fruit has a high content of vitamin C, and its medicinal use in the treatment of diabetes and cholesterol has become widespread (Lorenzi and Matos 2008; Yuyama et al. 2002). Several local ethnovarieties of cubiu are known (Silva Filho et al. 1993). This diversity was studied by Corrado (2018) in the Upper Rio Negro, who reported a total of 22 varieties with specific shapes, sizes, colors, flavors, and uses.

Another interesting Solanaceae is the camapum or gooseberry (*Physalis pubescens* L.), a spontaneous herbaceous plant that is grown in capoeiras and backyards, with a high content of vitamins and nutrients, but without commercial farming in

Table 1 Nutritional value of wild food plants from Amazon

Taxon	Local name	Moisture (g/100 g)	Carbohydrate (g/100 g) total	Lipids total (g/100 g)	Protein (g/100 g)	Fiber (g/100 g)	Total caloric value (kcal/100 g)
Anacardiaceae *Spondias mombin* L.	*cajá*	89.4	8.92	0.26	0.82	0.8	39
Araceae *Xanthosoma sagittifolium* (L.) Schott	*taioba*	89.2	5.5	0.9	2.9	4.5	33
Arecaceae *Astrocaryum aculeatum* G. Mey	*tucumã*	44.9	22.23	28.18	3.44	9.6	337
Attalea speciosa Mart.	*babaçu*	15.8	79.2	0.2	1.4	17.9	288
Bactris gasipaes Kunth	*pupunha*	54.5	29.5	12.8	2.5	4.3	235
Euterpe oleracea Mart.	*açaí-do-pará*	87.7	1.08	4.87	0.79	5.3	62
Euterpe precatoria Mart.	*açai-do-amazonas*	85.3	3.1	4.27	0.78	6.3	67
Mauritia flexuosa L.f.	*buriti*	59	25.8	12.02	2.19	15	190
Oenocarpus bataua Mart.	*patauá*	34.5	45.91	14.4	4.07	31.1	267
Oenocarpus distichus Mart.	*bacaba-de-leque*	86.1	4.83	7.65	1.17	3.9	85
Oenocarpus mapora H. Karts.	*bacabi*	85.4	0.32	5.98	0.87	7.2	299
Oenocarpus minor Mart.	*bacabinha*	83.5	8.51	6.44	1.26	4.7	362
Clusiaceae *Platonia insignis* Mart.	*bacuri*	78.5	17.83	2.01	1.22	5.2	84
Dioscoreaceae *Dioscorea trifida* L.f.	*cara-roxo*	74.7	23.56		1.21	1.7	96
Fabaceae *Hymenaea courbaril* L.	*jatobá*	11.2	76.23	2.53	5.65	51	248

(continued)

Table 1 (continued)

Taxon	Local name	Moisture (g/100 g)	Carbohydrate (g/100 g) total	Lipids total (g/100 g)	Protein (g/100 g)	Fiber (g/100 g)	Total caloric value (kcal/100 g)
Lecythidaceae *Bertholletia excelsa* Bonpl.	*castanha-do-brasil*	3.5	15.1	63.5	14.5	7.9	674
Malpighiaceae *Byrsonima crassifolia* (L.) Kunth	*murici*	76.5	19.37	2.47	0.88	6.8	91
Malvaceae *Theobroma grandiflorum* (Willd. ex Spreng.) K. Schum.	*cupuaçu*	81.9	14.87	1.18	1.1	2.6	69
Myrtaceae *Eugenia stipitata* McVaugh	*araça-boi*	90.3	5.5	0.3	0.51	3.2	32
Myrciaria dubia (Kunth) McVaugh	*camu-camu*	92	7.12	0.23	0.38	2.9	26
Rubiaceae *Genipa americana* L.	*jenipapo*	79.7	17.7	0.7	0.84	4.6	71
Sapindaceae *Talisia esculenta* (A. St.-Hil) Radlk.	*pitomba*	83.2	14.89	0.19	1.15	2.4	61
Sapotaceae *Pouteria caimito* (Ruiz & Pav.) Radlk.	*abiu*	83.1	15	0.7	0.8	1.7	66
Solanaceae *Capsicum chinense* Jacq.	*pimenta-de-cheiro*	90.8	7.13	0.15	1.39	3.4	29
Capsicum chinense Jacq.	*pimenta-murupi*	88.3	8.75	0.26	1.86	4.6	36

Source: Sibbr (2020)

Fig. 1 Edible fruits collected by multiethnic indigenous communities from the Upper Black River. (Source: Gabriela Gonçalves)

Brazil. Despite this, camapum is frequently imported to supermarkets in southeastern Brazil from Colombia, where it is commercially grown (Muniz et al. 2015).

The Myrtaceae family has 266 species in the Brazilian Amazon and produces many small fruits, some of which are edible and nutritious, mainly those from the genera *Eugenia*, *Myrcia*, and *Psidium* (Souza 2015; Flora do Brasil 2020). Some of the species appreciated and cultivated in the Amazon are araçá-boi (*Eugenia stipitata* McVaugh), goiaba-amarela (*Psidium guineense* Sw.), araça (*Psidium araca* Raddi.), pitanga or Brazilian cherry (*Eugenia uniflora* L.), and guava (*Psidium guajava* L.).

The species *Myrciaria dubia* (Kunth) McVaugh, known as camu-camu, azedinha, or caçari, is grown in the Amazon rivers in areas with rocks and is frequently harvested from wild plants but in recent years started to be cultivated and more studied. Due to its high content of ascorbic acid (vitamin C) that reaches up to 3000 mg per 100 g of fruit, an amount higher than any other known fruit species, industries have given attention to this plant (Yuyama et al. 2002).

One of the main native species of the Amazon and cultivated worldwide is the pineapple (*Ananas comosus* (L.) Merr.), Bromeliaceae (Clement et al. 2010). Pineapple is an interesting plant because, in addition to its adaptability to Amazonian ecological conditions, it grows in roças (slash and burn farms), backyards, and capoeira shrublands, serving as an important nutritional resource throughout the year. Local varieties of pineapple are poorly studied, but most of them reach a high sugar content, and some are heavier than 15 kg each, such as those grown in the city of Tarauacá, in Acre State. In the Lower Uaupés River region are grown many varieties with smooth or serrated edge leaves, with distinct leaf colorings and large fruits with high sugar content, such as the abacaxi-aracu (Gonçalves 2017).

Palm trees (Arecaceae) are abundant in a number of species and frequently in the Amazon region, with multiple registered uses, mainly as food (Henderson et al. 1995; Lorenzi et al. 2004; Clement et al. 2005). This wide presence, associated with the high energy value of its fruits, the absence of anti-nutritional factors, the abundant production of fruit throughout the year, and its palatability, makes palm trees a key group in the diet of the Amazonian population.

The açaí (*Euterpe oleracea*, *Euterpe precatoria* Mart., and *Euterpe catinga* Wallace), the bacaba (*Oenocarpus bacaba* Mart. and *Oenocarpus distichus* Mart.), and patauá (*Oenocarpus bataua* Mart.) are consumed in the form of juice (wine) and are easily found in all markets in the region during the harvest period. A recent work points out that these species are among the hyperdominant plants in the Amazon (Ter Steege et al. 2013), which helps to understand their importance for the local population.

The consumption of açaí is relevant in all Amazon cities. Just the metropolitan region of Belém do Pará is responsible for a daily consumption of 360,000 liters of this fruit's pulp (Oliveira et al. 2002). The açaí pulp has great energy value, mainly from lipids and carbohydrates, and high levels of antioxidants (Santos et al. 2008; Yuyama et al. 2013). The consumption of açaí heart of palm (apical meristem) is more frequent in the Southeast region of Brazil than in the North region, as the local population prefers to consume fruits.

In the eastern Amazon, in a wide transition area into the cerrado and caatinga shrublands, palm forests are frequent, with some palm species common to the different biomes, including babassu-palm (*Attalea speciosa* Mart.), a species of great cultural importance for the local population and of multiple uses (Araujo et al. 2016). The extractive activity of babassu is dominated by women, currently organized in the Interstate Movement of the Babassu Coconut Gatherers (Movimento Interestadual das Quebradeiras de Coco-Babaçu, MIQCB), in what can be described as a traditional agricultural system. The babassu mesocarp flour and seed oil have a high nutritional value, as well as the coconut beetle larvae (*Pachymerus* sp., Coleoptera) found in fruits and stipes in decomposition process. Another species common to the palm forests is the inajá (*Attalea maripa* (Aubl.) Mart.), whose fruits are consumed raw when ripe and boiled in water and from them edible oil can be extracted (Lorenzi et al. 2004).

The tucumã-do-amazonas (*Astrocaryum aculeatum* G. Mey.) and the tucumã-do-pará (*Astrocaryum vulgare* Mart.), species present in dense or secondary forest, are consumed raw. In the State of Amazonas, the tucumã is the main ingredient of a typical sandwich (x-caboquinho), made with bread or tapioca (manioc flour), cheese, and fried banana. The pulp of *A. aculeatum* fruits has high concentrations of lipids, dietary fiber, potassium, selenium, and beta-carotene (Yuyama et al. 2013). Other species of the genus *Astrocaryum* also have nutritional use, such as murumuru (*Astrocaryum murumuru* Mart.), with its fruits consumed raw, and the jauari (*Astrocaryum jauari* Mart.) from igapó forests, whose fruits are consumed after cooking and its heart of palm is collected for local consumption.

The buriti palm (*Mauritia flexuosa* L.f.) is grown in clusters in marshy and wet terrain, swampy, or permanently flooded areas. Its fruits are consumed raw, or from its pulp, which is rich in carotenoids (Rodriguez-Amaya et al. 2008), a wine or regional sweet can be produced. From its stipe, an edible sap could be extracted, and, in addition, an edible tapioca flour is produced (Correa 1984; Lorenzi et al. 2004).

Among the Amazonian palm trees, one of the few cultivated and domesticated is the peach palm (*Bactris gasipaes* Kunth), whose pulp, rich in vitamin C and carotenoids, is appreciated by the population after being cooked or as flour in other dishes. Several ethnovarieties of peach palm are known with variation in shape, composition, and color of the fruits and seeds, plant size, and quantity of thorns in the trunks and leaves (Clement and Santos 2002). The farming of peach palm has expanded in Brazil for the extraction of its heart of palm, appreciated in the southeast of the country.

Fabaceae is a pantropical and relevant family in the Amazon, where it is estimated that there are about 2,000 species (Flora do Brasil 2020). The ingas (*Inga* spp.) are widely distributed in the Amazon, some of which are cultivated, as the ice cream bean (*Inga edulis* Mart.) and inga assu (*Inga cinnamomea* Benth.) or spontaneous such as *Inga ingoides* (Rich.) Willd., *Inga laurina* (Sw.) Willd, *Inga macrophylla* Humb. & Bonpl. ex Willd., and *Inga vera* Willd.. Out of the ingas, the white aril that surrounds the seeds is edible and appreciated by the population for its sweet flavor but has a low nutritional content (Gonçalves 2017).

The jatobá or jutaí (*Hymenaea courbaril* L., Fabaceae) is another native Fabaceae, which is appreciated for its farinaceous pulp of strong and peculiar odor and taste, rich in carbohydrates and proteins, consumed raw, or processed as flour for use in breads, cakes, and porridges (Rabelo 2012; Kinupp and Lorenzi 2014).

The *Theobroma* genus, Malvaceae, is composed of 22 species distributed from the Amazon basin to the south of Mexico, of which cocoa (*T. cacao*) and cupuassu (*T. grandiflorum*) have great economic relevance (Garcia et al. 2014). In this genus, almonds are widely used for the production of chocolate and cocoa butter, in addition to the mesocarp in sweets and soft drinks. Other species of the genus are restricted to local use but with great commercial potential such as the mocambo or baraturi (*Theobroma bicolor* Bonpl.), cupuí (*Theobroma subincanum* Mart.), cacauí (*Theobroma speciosum* Willd. Ex. Spreng.), cacau-cabeça-de-urubu (*Theobroma obovatum* Klotzsch ex Bernoulli), and cacao-cabeça-de-jacaré (*Theobroma mariae* (Mart.) K. Schum.), which are still poorly studied. Theobromine is the main alkaloid present in cocoa seeds, and among its medicinal effects in humans are the tonic sensation, endorphin and serotonin stimulation, diuretic effect, vasodilator, and the relief of respiratory problems (Peres et al. 2018).

Sapotaceae is an important family in the Amazon, the *Pouteria* genus being one of the most representative and well-known (Monteiro et al. 2007). The abiu (*Pouteria caimito* (Ruiz & Pav.) Radlk. is often grown and sold in Amazonian cities where it is appreciated, despite the latex present in some varieties and in unripe fruits, which coagulates in contact with air, gluing together the lips of whoever eats them (Falcão and Clement 1999).

Of the same genus, the ucuqui (*Pouteria ucuqui* Pires & RE Schult.) is a quite peculiar example of fruit consumed by indigenous communities in the Lower Uaupés

River. According to them, it is only possible to eat the mesocarp of at most two or three untreated fruits, without causing irritation to the oral mucosa; however, when cooked and prepared as porridge, it no longer causes irritation (Gonçalves 2017). The lesions done to the oral mucosa, caused by raw ucuqui, are associated with the presence of calcium oxalate crystals, common in some species of the Sapotaceae family (Monteiro et al. 2007) which is deactivated by cooking.

The genip tree (*Genipa americana* L.), Rubiaceae, grown in backyards and found in capoeira shrublands and secondary forests, is consumed raw when ripe or in alcoholic beverages. However, the greatest interest by the indigenous population is for its dyeing property, of the genipin pigment of blue-violet color, present in unripe fruits and activated in the presence of oxygen after fruit maceration, used in body painting or as a dye for basketry fibers (Prance 1975).

The biribá or wild sugar apple (*Annona mucosa* Jacq.) is the main native and cultivated representative of the Annonaceae family in the region. Its fruits weighing up to 1.5 kg are available in fairs sporadically. Until today, there is no extensive commercial fruit farms developed, despite there being several varieties selected, including some conceived in Florida/the USA (cv. Prolific.) (Lorenzi et al. 2015).

Brazil and Colombia are the diversity centers of the Passifloraceae family, which together concentrate the largest number of *Passiflora* species in the world (Cervi 1997). The Amazon biome has a large number of passion fruit species that still are little known about their nutritional composition. Despite being potentially edible, few species are actually grown or handled by the population so as to be found in fairs. Among the wild species, picked in capoeira shrublands or as urban weeds, the maracujá-de-cesto (*Passiflora foetida* L.) is common, and among the plants cultivated most frequently are *Passiflora riparia* Mart. ex Mast. and *Passiflora nitida* Kunth. (Gonçalves 2017). In addition to nutritional use, many species are also used in traditional medicine (Lorenzi and Matos 2008).

In the Eastern Amazon, the bacuri (*Platonia insignis* Mart.), Clusiaceae, is poorly cultivated, despite the high trade value in markets, and is found densely just in certain regions. The pulp of its fruit is very aromatic and used in sweets and beverages and from its seed is extracted a medicinal oil. The fruit has high levels of starch, potassium, phosphorus, and calcium (Carvalho and Nascimento 2018).

Within the different Amazonian ecosystems, the murici (*Byrsonima* spp.) Malpighiaceae adapts well in open environments and can flourish spontaneously or cultivated. Its small fruits, which vary in color from yellow to red and green, have a characteristic odor and can be stored for a long period in closed bottles filled with water, until consumption, as soft drinks, sweets, with alcoholic beverage, or raw (Cavalcante 1972; Lorenzi et al. 2015).

The cajá ou taperebá (*Spondias mombin* L.), Anacardiaceae, is a large tree present in the Amazon and the Atlantic Forests, whose fruits are appreciated raw, in juices, sweets, or ice creams. Despite seasonality, the pulp can be frozen for consumption at other times of the year. The huge Amazonian biodiversity can be illustrated by the scientific description of the cajarana (*Spondias testudinis* J.D. Mitch. & D.C. Daly) just in 1998, a plant that was known until then only by the local population of Acre (Mitchell and Daly 1998).

The uva-da-amazônia, cucura, or mapati (*Pourouma cecropiifolia* Mart.), Urticaceae, is found in the Western Amazon, generally associated with human presence, cultivated or in secondary vegetation. The tasty pulp of its fruits is appreciated for a short period of the year when it is available. Its ingestion can cause irritation in the mouth for some people, due to the rough skin of the epicarp (Lorenzi et al. 2015; Pedrosa et al. 2018).

An important characteristic of Amazonian fruits is the seasonality of each species when they are available in the markets, since most of them are climacteric and hard to store unprocessed. This fact helps to understand why, even in locations far from the largest urban centers, it is easier to find exotic and imported fruits such as orange, pear, or apple than the native ones produced locally.

Despite the immense diversity of edible fruits in the Amazon region, research on them is scarce, mainly related to botanical, agronomical, and biochemical studies. Only the most well-known species such as cocoa, cupuassu, Brazil nut, and açaí have been extensively studied. One of the difficulties regarding the widespread use and cultivation of these plants is that most of them are long-cycle trees and have not been domesticated or are not cultivated.

3 Roots and Tubers

Many plants that have edible tuberous organs (roots, tubers, bulbs, rhizomes) are traditionally cultivated by Amazonian peoples. These plants are the basis of the local diet.

Despite the absence of a single Amazonian dietary culture, it is possible to consider that cassava, manioc, or mandioca (*Manihot esculenta* Crantz) is the main edible species in the region. Mandioca was domesticated in the Madeira River region (Clement et al. 2010) and had its use broadened throughout the entire Amazon and adjacent regions. Cassava is classified into two major groups: mandioca-brava (with a high content of hydrogen cyanide acid) and the mandioca-mansa (with a low content of hydrogen cyanide acid), and in each group, there are hundreds of local ethnovarieties. Its varieties are used in the preparation of at least 14 forms of food and 13 different drinks (Kerr and Clement 1980).

The Upper Rio Negro region and the traditional communities of Acre concentrate a large number of local varieties, mainly mandioca-brava. Emperaire and Peroni (2007) analyzed the case of cassava in the Upper Rio Negro and considered that the agrobiodiversity stems from sociocultural criteria that involve selection, circulation, denomination, and production norms, leading the varieties of plants to be regarded not only as a phylogenetic resource but also as a cultural heritage. Products derived from "harsh" cassava in the Upper Rio Negro constitute a central part of the indigenous diet and culture. Transformation techniques are complex and involve a long operational chain. In this region, more than 70 varieties are grown (Emperaire and Eloy 2008), and the derivative products are very diverse: many types of flour, beiju (kind of pancake), porridge, and caxiri (beer), in addition to tapioca, tucupi, and maniçoba (leaves of maniva).

The plants of the Dioscoreaceae family are also of great importance in the diet of the Amazonian peoples. Among the edible and cultivated species of Dioscoreaceae, *Dioscorea altíssima* Lam., *Dioscorea alata* L., *Dioscorea bulbifera* L., and *Dioscorea trifida* Lf are the most used species and with the largest number of local types, with purple, white, or mixed coloring tubers (Castro et al. 2012). The purple color is due to the presence of anthocyanins with antioxidant action (Kinupp and Lorenzi 2014).

Along with *Dioscorea* and *Manihot*, sweet potatoes (*Ipomoea batatas* (L.) Lam.) are among the most important roots in the regional diet. In addition to the most common commercial cultivars, the Amazonian populations maintain local varieties. This cultivation seems to be particularly important among the Krahô indigenous people, who have an entire cosmology and cultural practices associated with this plant (Lima 2017).

The ariá (*Goeppertia allouia* (Aubl.) Borchs. & S. Suárez) is a tuberous species native to the Amazon and traditionally farmed by indigenous and riverside populations in Tropical America. In addition to food, its cultivation is also for the purposes of ornamentation and traditional medicine (Bueno and Weigel 1981). The main feature, which sets it aside from other tuberous species, is that even after cooking, it maintains its crispness quality, rare in other tuberous species.

The species *Casimirella rupestris* (Ducke) R.A.Howard and *Casimirella ampla* (Miers) Baehni, Icacinaceae, are endemic to the Amazon and naturally abundant in terra firme forest. Its popular name, mairá potato, derives from Nheengatu and means "liana variety, whose root is edible" (Stradelli 1929; Duno-de-Stefano and Amorim 2015). Its starchy tuberous roots can reach over 200 kilos. Ethnographic studies and traveller's reports highlight the importance of the mairá potato for some indigenous groups in the Amazon in the past. Spruce (1851) reports in his contact with the Tapuya that they consumed it in an identical way to mandioca-brava, extracting the starch and producing flour (Ribeiro 2018). Even with this high productivity, the mairá potato does not currently have significant cultivation among the Amazonian peoples, who show preference for cassava.

In the Lower Uaupés River, a variety of other tuberous species are cultivated besides cassava, such as white yams (*Dioscorea* spp.), macoari (*Heliconia hirsuta* Lf), achira or macoari-doce (*Canna edulis* Ker-Gawl.), sweet potatoes, and some species of Araceae (*Xanthosoma* spp.) and Marantaceae yai-tutu (*Maranta ruiziana* Korn), yai-comprido (*Maranta arundinacea* L.), and aria (*G. allouia*). The consumption of these roots is mostly after cooking, in porridge, or in the preparation of fermented drinks locally named caxiri (Gonçalves 2017) (Fig. 2).

For the Kaxinawá people in the state of Acre, tuberous species also play an important role in the local diet, especially the taioba (*Xanthosoma sagittifolium* (L.) Schott), with local types consumed, with different color and flavor: *Kixtuku* (dark orange), *Atsa Yubi* (white), *Roxo Yubi* (orange with purple edges), *Kapa Nawa* (light orange), *Kapa Yubi* (light orange with white edges), and *Maspa Yubi* (light orange with green edges). All tubers have, aside from color, different shapes.

Fig. 2 Edible roots and tubers cultivated by multiethnic indigenous communities from the Upper Black River. (Source: Gabriela Gonçalves)

The production of tuberous edible plants can be carried out throughout the year in the Amazon. Plants can be left intact in the field for a long period, and the roots are harvested gradually, as needed, unlike grains and pulses, which are farmed more commonly in temperate regions. This is an important strategy for traditional populations in the tropics, in response to the problems of sazonality and storing other foods in hot and humid climate (Martins 2005).

Unlike fruit species, which are grown in backyards or picked in the forest, family farms are the place where the cultivation of tuberous plants predominates. For many of the indigenous peoples of the Amazon, these farms and the plants grown in them are usually the domain of women, whereas forests and rivers are male domains.

4 Grains, Pulses, and Nuts

With regard to carbohydrate intake, some grains and seeds have taken on an important role regionally, along with tuberous roots, with emphasis on cultivated and domesticated species.

Maize (*Zea mays* L.), native to Central America, has been cultivated in the Amazon for more than 8000 years (Bush et al. 1989), a period before agriculture had even spread in the region (Clement et al. 2015). During this period, it was selected and domesticated again in the region (Kistler et al. 2018). Traditional corn grown by the indigenous peoples of the Amazon has farinaceous endosperm (soft corn) as oppose to hard and crystalline, like most commercial cultivars. There is a large number of cultivated maize ethnovarieties, each with specific uses by certain populations (Teixeira and Landau 2016). The cultural importance of corn is evident in the mythology of several indigenous peoples where it is present; in festive food preparations, such as caxiri; and in traditional festivals (Pereira 1974).

The Kaxinawa of Acre cultivate two important ethnovarieties of corn for the community's diet, classified as "Nawa Sheki" and "Sheki Kui," varying in orange, dark red, yellow, and whitish colors, which are used to make porridge and different types of cakes.

The peanut or groundnut (*Arachis hypogaea* L.) is another important source of carbohydrates and lipids often cultivated by various indigenous peoples of the Amazon. The Kawabi, from Xingu, have more than 14 ethnovarieties of this species, each with different characteristics and uses (Suassuna et al. 2016). Among the Kaxinawá, the peanut is known as tama, where there are six ethnovarieties of different colors.

Beans (*Phaseolus vulgaris* L., *Phaseolus lunatus* L., and *Vigna unguiculata* (L.) Walp.) are grown in floodplains or in terra firme and are among the most important plants for the local population. The "muffled" is a specific system for the culture of beans, where the capoeira shrub is cut down so that the branches may provide support for the growth of the climbing varieties (Sivieiro et al. 2016). The local varieties are quite diverse, such as the purple and butter beans, common in the state of Pará, or the red gorgutuba, red Peruvian, rosinha, quarantão, manteiga-roxo, and arigozinho in Acre.

Currently, rice (*Oryza sativa* L.) has become a species of regional importance, being grown in large areas in the process of pasture formation or by small farmers and indigenous peoples. Among the Krahô, 26 cultivated rice ethnovarieties were identified (Rangel and Dias 2016). Studies indicate that there were species of *Oryza* native to Brazil that was in the process of domestication by indigenous populations (Hilbert et al. 2019). In the Rio Negro, some riverside and indigenous populations still collect seeds of *Oryza latifolia* Desv. for their own consumption.

Economically, castanha-do-Brasil, Brazil nut, or Amazon nut *(Bertholletia excelsa* Bonpl.), Lecythidaceae, has had great national relevance since the nineteenth century, when Brazil nut pickers moved in the region during the harvest season. In addition to the caloric potential, studies show that Brazil nuts have high selenium content, an important element in disease prevention. Although Brazil has large areas of Brazil nut forests, associated with human activity in domesticating the landscape (Shepard Jr and Ramirez 2011) in regions such as Acre and Southeast Pará, deforestation has made Bolivia the world's largest exporter of Brazil nut (Homma 2000; Shanley et al. 2010). Considering the long reproductive cycle of the Brazil nut tree, studies have successfully enabled the commercial cultivation of selected plants and the use of techniques that reduce the time for fruiting juvenility in young plants.

The sapucaia (*Lecythis Pisonis* Cambess.), another species of Lecythidaceae, is consumed by the local population. Its dehiscent capsule fruit opens when riped and the seeds fallen on the ground, where the local fauna appreciated it (Mori and Prance 1981).

Guarana (*Paullinia cupana* Kunth., Sapindaceae) is a shrubby climbing plant, whose cultivation technique and selection and domestication process were carried out by the Mawé indigenous people, where it is present in mythology and festivities. Despite having its use more closely related to medicinal properties due to the high content of caffeine and other substances, currently its commercial cultivation is done on a large scale for the production of medicines and soft drinks for the industry (Tricaud et al. 2016).

In spite of the relatively low number of species that provide seeds and grains for the Amazonian population's diet, they have considerable regional importance, given the large volume consumed, the cultivated area, the nutritional aspects, and the existence of a large number of ethnovarieties and species handled or domesticated. This importance is also made evident by the presence of these plants in the mythology and culture of some indigenous peoples.

5 Leafy Vegetables

The habit of consuming leafy vegetables is not very common among the Amazonian population, especially among indigenous peoples (Katz et al. 2012). Alves da Silva (1962) observes in the Upper Rio Negro that "If *the indigenous diet lacks vegetables and legumes, many are the fruits they consume*." It is likely that the possible lack of vitamins and minerals, which could be caused by the low consumption of leafy vegetables, is supplemented by the large number of fruit species consumed by the local population (Katz et al. 2012).

Of the native vegetables, one of the foremost is the jambu (*Acmella oleracea* (L.) R.K. Jansen), which can also be classified as spice or medicinal. Jambu plant leaves are often used sautéed in traditional recipes such as duck in tucupi sauce and tacacá. Due to its local importance, there are many studies carried out with this species and even varieties improved available for the farmers (Cardoso 1997; Kinupp and Lorenzi 2014).

Some leafy vegetables available in the region are potentially toxic and need to be cooked for a long time as to lose this toxicity. Cassava leaves (*M. esculenta*), known as maniçoba, have a high level of hydrogen cyanide acid and are traditionally cooked for 7 days, before being used in festive dishes. Caruru leaves (*Phytolacca rivinoides* Kunth & C.D.Bouché), of more restricted use in the Upper Rio Negro, are consumed only after cooking along with fish or quarry broths (Gonçalves 2017).

The cipó-kupá (*Cissus gongylodes* (Baker) Burch. Ex Baker, Vitaceae) is one of the rare horticultural species that have been selected by indigenous populations and is cultivated to this day among Kayapós and Timbira groups. The plant has a climbing habit, and its succulent steam is used in broths or roasts, after the use of temperature to reduce the action of the calcium oxalate it has (Kerr et al. 1978; Dias et al. 2016).

Taioba (*Xanthosoma* spp.), Araceae, cultivated for its tubers roots, is another important native vegetable whose leaves are appreciated after cooking to reduce calcium oxalate (Cardoso 1997, Kinupp and Lorenzi 2014).

The contribution of migrant populations to the consumption of vegetables in the Amazon is remarkable. The habit of consuming okra (*Hibiscus esculentus* L.), cuxá (*Hibiscus sabdariffa* L.), caruru (*Talinum paniculatum* Jacq-Gaertn), and junça (*Cyperus esculentus* L.) in the eastern Amazon is influenced by African populations of the states of the Pará and Maranhão (Cardoso 1997). Asian immigrants were important for the development of modern horticulture in the Amazon, with techniques for the cultivation of conventional vegetables such as lettuce (*Lactuca sativa* L., Asteraceae) and black pepper (*Piper nigrum* L., Piperaceae) (Homma 2009). In addition to these plants, it is common to cultivate exotic vegetables in backyards and jiraus (suspended location) in the region, such as pumpkin (*Cucurbita* spp.), cowpea (*Vigna unguiculata* (L.) Walp.), and kale (*Brassica oleracea* var. Acephala DC.) (Cardoso 1997).

The attempt to grow tomatoes, bell peppers, or other conventional farm vegetables has not been successful in the region due to the lack of cultivars adapted to local climatic conditions, difficulty in producing seeds or propagules by the farmers themselves, or cultivars resistant to fungal diseases that increase with the high temperature and humidity of the region. Kinupp and Lorenzi (2014) address other native species that could be used as farm vegetables in the region, such as caapeba (*Piper peltatum* L.), picão (*Bidens* spp.), and other spontaneous ones, but which are neglected by or unknown to most of the population.

6 Amazonian Condiment Species

Spice plants are those that, when added in small quantities in food preparation, have the capacity to modify the perception we have of them (Tomchinsky 2017). They are species that, for the most part, have a high concentration of secondary compounds in their composition such as terpenes, alkaloids, and phenols (Lorenzi and Matos 2008; Simões et al. 2017).

Due to the high concentration of these secondary compounds, many of these plants have multiple uses, such as medicinal, entheogenic, or aphrodisiac (Schultes and Hofmann 2001; Lorenzi and Matos 2008; Ratsch 2011). However, for this same reason, the ingestion of high doses of some condiments may be potentially toxic to humans, such as the myristic acid in the nutmeg (Schultes and Hofmann 2001), coumarin in the cumaru (Lake 1999), piperine in the black pepper, and capsaicin in *Capsicum* peppers (Lorenzi and Matos 2008). The use of condiments is also important for food preservation, whether because of its antimicrobial action (Ethurk 2008; Liu et al. 2017) or in overlapping undesired traits in degraded foods.

The Amazonian culture, of indigenous origin, has adopted as its main condiment the hot peppers of the *Capsicum* genus. There are indications that *Capsicum chinense* Jacq. was domesticated in the Amazon region more than 4000 years ago

(Clement et al. 2010). Among the dozens of types of hot peppers grown in the region, the most common are locally known as murupi, pimenta-de-cheiro, pimenta-bode, cumari-do-pará and biquinho (*C. chinense*), tabasco (*Capsicum frutescens* L.), and dedo-de-moça and cumari (*Capsicum baccatum* L.) (Carvalho et al. 2006; Nascimento Filho et al. 2007; Ribeiro et al. 2008; Roman et al. 2000).

In the Upper Rio Negro, the consumption of peppers seems to be even more appreciated than in other regions by the local and predominantly indigenous population. The quinhapira, a fish stew typical of the region, is made with a large quantity of hot peppers. In this region, the production technique of the jiquitaia or the baniwa peppers is preserved, which mixes up to 78 ethnovarieties of peppers that are dried, smoked (roasted), grinded, and added to salt (Silva et al. 2016).

Salt was difficult to obtain in the pre-Columbian Amazon, when compared to other coastal locations, and was obtained just through trade with other distant regions or derived from vegetables. These vegetable salts have potassium cation (KCl), while those of marine origin have sodium cation (NaCl). They are obtained through the process of pyrolysis of certain species of plants, mainly from the families Arecaceae, Amaranthaceae, and Lecythidaceae. This process has been described among several Amazonian ethnic groups such as the Waimiri-Atroari, Witoto, Yanomami, and Tupinambá (Martius 1979; Bastos 1987; Albert and Milliken 2009; Tomchinsky 2017; Kermath et al. 2020).

The cassava (*M. esculenta*), a central species in the Amazon diet, is also a source of tucupi, an edible yellow broth, but potentially toxic due to its high concentration of cyanide, extracted during the production of the flour. After boiled and/or fermented and purified, the tucupi broth is used in the preparation of various Amazonian dishes such as tacacá and duck in tucupi sauce, topped on pepper sauces. When refined and reduced, it has the appearance of a smoked sweet molasses, black in color, known as black tucupi, which is used as a spice, with peppers and citrus-like ants (*Atta sexdens* rubropilosa Forel 1908) (Brazi 2013).

The first reports about annatto (*Bixa orellana* L.) describe its medicinal use or its use in body painting (Tomchinsky and Ming 2019). Currently, it is widely used as a red or yellow food dye in homemade and industrialized products (dye code E-160b) due to carotenoids bixin (fat-soluble) and norbixin (water-soluble) extracted from the aril of its seeds, which also have antioxidant action for food preservation (Garcia et al. 2012).

The jambu (*A. oleracea*), also cited as horticultural, can be considered to be a condiment for the anesthetic effect caused by its inflorescences, rich in spilanthol terpene, when consumed in broths, stews, or the infusion of alcoholic beverages (Cardoso 1997; Lorenzi and Matos 2008; Barbosa et al. 2016).

The chicória-do Pará (*Eryngium foetidum* L.) is a biannual herbaceous cultivated and registered as one of the first species native to Brazil for use in spices (Tomchinsky and Ming 2019). The eryngial terpene is predominant in the composition of its essential oil, which has aromatic properties similar to that of the coriander (*Coriandrum sativum* L.) (Paul et al. 2011; Kinupp and Lorenzi 2014).

Other exotic and more conventional species are frequent in Amazonian cuisines and grown in backyards or found in markets such as chives (*Allium shoenoprasum*

L.), turmeric (*Curcuma longa* L.), parsley (*Petrosilum crispum* (Mill.) Fuss), citrus plants (*Citrus* spp.), clove basil (*Ocimum gratissimum* L.), coriander (*Coriandrum sativum* L.), black pepper (*Piper nigrum* L.), cumin (*Cuminum cyminum* L.), onion (*Allium cepa* L.), and garlic (*Allium sativum* L.). Although exotic, these plants were incorporated centuries ago in the diet of the local population and in their pharmacopoeia and are of daily use by the entire Amazonian population.

Given the Amazonian biodiversity, the absence of more native species with widespread condiment use in the region is surprising. Some native plants have their use restricted to a certain region, such as japura (*Erisma japura* Spruce ex Warm.) in the Upper Rio Negro, whose fermented fruit is used to season fish (Ribeiro 1995; Gonçalves 2017); others have been studied as substitutes for the most frequently used spices or within a new proposal of fine cuisine that looks for "novelties" within the "exotic" Amazonian biodiversity.

The various species of the genus *Xylopia* (*X. aromatica* (Lam.) Mart., *X. benthamii* R.E.Fr., *X. frutescens* Aubl., and *X. sericea* A.St.-Hil.) are popularly known as pimenta-de-macaco or pindaíba, and its dried fruits have aromatic compounds used for food and in alcoholic infusions as "drink seasoning" and as medicine by the local population (Lorenzi and Matos 2008; Silva et al. 2015).

The canela-preciosa (*A. canelilla*), a large Lauraceae used as a medicine and for timber, has a characteristic odor that may be a substitute for the cinnamon plants (*Cinnamomum burmannii* (Nees & T. Nees) Blume, *C. verum* J. Presl, and *C. cassia* (L.) J.Presl.). Its essential oil is concentrated in vessels at the heartwood and in a lower concentration in the leaves, with a predominance of the compound 1-nitro-2-phenylethane responsible for its characteristic aroma (Lorenzi and Matos 2008; Kinupp and Lorenzi 2014).

The cumaru or tonka bean (*D. odorata*), a large Fabaceae, has seeds rich in coumarin with multiple uses (cosmetics and perfumery, tobacco fragrance, medicinal, rodenticide), as well as timber. The consumption of cumaru illustrates the discussion about the safe dosage of condiment species, since this plant had its use restricted in the USA and Europe due to the coumarin's carcinogenic and hepatotoxic potential (Carvalho 2009; Rego et al. 2017).

The indiscriminate exploitation of pau-cravo (*Dicypellium caryophyllaceum* (Mart.) Nees) (Lauraceae), a species with high content of eugenol, the same compound found in the clove (*Syzygium aromaticum*), for timber and medicinal use, almost brought this species to extinction, being currently found only in small areas and critically endangered (CR) (CNCFlora 2020). Pau-cravo's endangered case raises the question of how important it is to study more about the management of native non-cultivated species of the Amazon, before using them.

Several species of genus *Vanilla* (Orchidaceae) native to the Amazon (*V. appendiculata* Rolfe, *V. gardneri* Rolfe, *V. mexicana* Mill., *V. palmarum* (Salz. Ex Lincl.) Lindl., *V. planifolia* Jacks. Ex Andrews, *Vanilla pompona* Schiede) have a high content of vanillin in their seeds and beans and can be used in gastronomy safely. Considering the little knowledge about some of these species and their rarity in nature, several studies would be necessary to enable their sustainable exploitation or cultivation in agroforestry systems, since they are epiphytic plants with a long flowering and fruiting cycles.

The leaves of the garlic vine (*Mansoa alliaceae* (Lam.) A.H. Gentry) are occasionally used as a condiment to replace the common garlic, in addition to various registered medicinal uses (Lorenzi and Matos 2008). The sulfurous compounds alliin and allyl sulfides in its leaves give it its characteristic odor.

Recently, the priprioca or jointed flatsedge (*Cyperus articulatus* L.), an herb native to Africa and naturalized in the Amazon for medicinal and ritualistic uses, is being used as a dessert fragrance in fine cuisine restaurants (Trajano 2008).

Aiming to expand the definition of condiment adopted in this chapter, we can cite the various species that provide edible oil present in the region, with emphasis on the Arecaceae and Fabaceae families (Lorenzi et al. 2004; Tomchinsky 2017) or the numerous pulp fruits with high degrees brix that can be fermented and acetified to use as vinegar and acidifier.

Other native species of the Amazonian flora have great potential for use as spices, but they should be better studied regarding their phytochemical composition, safety of use, handling, and cultivation before their use becomes widespread. Of 923 plant species with records of condiment use present in Brazil, among native (501) and exotic (422), 343 are present in the Amazon region, including plants with records of use as aromatic, flavoring, coloring, source of vegetable salt, sweetener, edible oil, acidifier, edible gums and resins, meat tenderizer, or food wrapping; as a substitute for hop, coffee, cocoa, and vanilla; for aromatic infusions; or as "spice for alcoholic beverages" (Tomchinsky 2017).

7 Final Considerations

This chapter provides a brief overview on the knowledge about Amazonian edible plants, traditional or unconventional, and general aspects related to them, but does not take the topic to exhaustion. The Amazon has a rich food biodiversity, and among the different food groups, fruits, grains, and tubers are the most important in the diet of the Amazonian population. We can verify that among the cultivated plants, there is a great amount of ethnovarieties maintained by the local populations. It is worrisome that all this sociobiodiversity is threatened by the deforestation of the biome and by other threats to traditional peoples and communities that hold the knowledge on the use of these plants. Finally, most of this biodiversity has been poorly studied, mainly in relation to botanical, agronomic, and phytochemical aspects. This lack of knowledge is due, in part, to the few number of institutions and researchers dedicated to these studies in Amazon, in relation to its size.

References

Ab'Saber NA (1977) Espaços ocupados pela expansão dos climas secos na América do Sul, por ocasião dos períodos glaciais quaternários. Paléo 3:1–19
Albert B, Milliken W (2009) Urihi A: a terra-floresta Yanomami. Socioambiental/IRD, São Paulo/Paris
Alves da Silva ABA (1962) A civilização indigena do Uaupés. Missão Salesiana do Rio Negro, São Paulo
Araujo FR, Gonzalez-Perez SE, Lopes MA, Viegas IJM (2016) Ethnobotany of babassu palm (Attalea speciosa Mart.) in the Tucuruí Lake Protected Areas Mosaic – Eastern Amazon. Acta Bot Bras 30(2):193–204. https://doi.org/10.1590/0102-33062015abb0290
Balée W (1989) Cultura vegetal da Amazônia brasileira. In: Neves WA (ed) Biologia e ecologia Humana na Amazônia: avaliação e perspectivas. Museu Paraense Emílio Goeldi, Belem, pp 95–110
Balée W (1994) Footprints of the forest: Ka'apor ethnobotany: the historical ecology plant utilization by an amazon people. Columbia University Press, New York
Barbosa AF, Carvalho MG, Smit RE et al (2016) Spilanthol: occurrence, extraction, chemistry and biological activities. Rev Bras 26(1):128–113. https://doi.org/10.1016/j.bjp.2015.07.024
Bastos A (1987) A pantofagia ou as entranhas práticas alimentares na selva. Editora nacional, São Paulo
Brazi D (2013) Cozinha Tradicional Amazônica. Editora Bei, São Paulo
Bueno CR, Weigel P (1981) Brotação e desenvolvimento inicial de rizomas de Ariá (Calathea allouia (Aubl.) Lindl.). Acta Amazon 11(2):407–410. https://doi.org/10.1590/1809-43921981112408
Bush MB, Piperno DR, Colinvaux PA (1989) A 6000 year history of Amazonia maize cultivation. Nature 340:303–305. https://doi.org/10.1038/340303a0
Bustamante GGF (2009) Frutos, sementes e órgãos tuberosos na alimentação da etnia Sateré-Mawé dos Rios Marau e Urupadi (Maués-Amazonas). Dissertation, Faculdade de Ciências Agronômicas, Universidade Federal do Amazonas
Cardoso MO (1997) Hortaliças não convencionais da Amazônia. Embrapa, Brasília
Cardoso D, Sarkinem T, Alexander S et al (2017) Amazon plant diversity revealed by a taxonomically verified species list. PNAS 114(40):10695–10700. https://doi.org/10.1073/pnas.1706756114
Carvalho PER (2009) Cumaru-ferro Dipteryx odorata. Comunicado Técnico 225. Embrapa, Colombo
Carvalho JEU, Nascimento WMO (2018) BACURI: Platonia insignis. Instituto Interamericano de Cooperación para la Agricultura, Costa Rica
Carvalho SIC, Bianchetti LB, Ribeiro CSC, LopesS CA (2006) Pimentas do gênero Capsicum no Brasil. Documento 94. Embrapa hortaliças, Brasília
Cascudo LC (2011) História da alimentação no Brasil, 4th edn. Global, São Paulo
Castro AP, Fraze TJP, Pereira HS et al (2012) Etnobotânica das variedades locais do cará (Dioscorea spp.) cultivados em comunidades no município de Caapiranga, estado do Amazonas. Acta Bot Bras 26(3):658–667. https://doi.org/10.1590/S0102-33062012000300015
Cavalcante PB (1972) Frutas comestíveis da Amazônia. Museu Paraense Emília Goeldi, Belem
Cervi AC (1997) Passifloraceae do Brasil: estudo do gênero Passiflora L., subgênero Passiflora. Fontqueria, Madrid
Clement CR (1999) 1492 and the loss of Amazonian crop genetic resources. I. The relation between domestication and human population decline. Econ Bot 53(2):188–202. https://doi.org/10.1007/BF02866498
Clement CR, Santos LA (2002) Pupunha no mercado de Manaus: Preferências de consumidores e suas implicações. Rev Bras Frutic 24(3):778–779. https://doi.org/10.1590/S0100-29452002000300055
Clement CR, Lleras E, Leeuwen JV (2005) O potencial das palmeiras tropicais no Brasil: Acertos e fracassos das últimas décadas. Agrociencia 9(1–2):67–71

Clement CR, Cristo-Araújo M, Coppens GD et al (2010) Origin and domestication of native Amazonian crops. Diversity 2:72–106. https://doi.org/10.3390/d2010072
Clement CR, Denevan WM, Heckenberger MJ et al (2015) Domestication of Amazonia before European conquest. Proc R Soc B 282(20150813):1–9. https://doi.org/10.1098/rspb.2015.0813
CNCFLORA (2020) Dicypellium caryophyllaceum in Lista Vermelha da flora brasileira versão 2012.2 Centro Nacional de Conservação da Flora. Available in http://cncflora.jbrj.gov.br/portal/pt-br/profile/Dicypelliumcaryophyllaceum. Accessed 3 Sept 2020
Corrado AR (2018) Diversidade cultural e genética de solanum sessiliflorum Dunal na Amazônia Brasileira. Dissertation, Faculdade de Ciências Agronômicas, Universidade Estadual Paulista
Correa MP (1984) Dicionário das plantas uteis do Brasil e das exóticas cultivadas. MINAGRI, Rio de Janeiro
Crosby AW (1972) The Columbian exchange: biological and cultural consequences of 1492. Greenwood Press, Westpost
Dias T et al (2016) Cupá, uma cultura alimentar indígena pré-colombiana: relatos de reintrodução. In: Dias T, Edit JS, Udry C (eds) Diálogos de saberes – relatos da Embrapa. Coleção povos e comunidades tradicionais, vol 2. Embrapa, Brasília, pp 73–85
Duno-de-Stefano R, Amorim B (2015) Icacinaceae in Lista de Espécies da Flora do Brasil. Jardim Botânico do Rio de Janeiro. http://floradobrasil.jbrj.gov.br/jabot/FichaPublicaTaxonUC/FichaPublicaTaxonUC.do?id=FB84145. Accessed 03 Oct 2020
Emperaire L (2010) Dossiê de registro: o sistema agrícola tradicional do Rio Negro. Associação das comunidades indígenas do Médio Rio Negro, Brasília
Emperaire L, Eloy L (2008) A cidade, um foco de diversidade agrícola no Rio Negro (Amazonas, Brasil) ?. Boletim do Museu Paraense Emílio Goeldi. Ciências Humanas 3:195–211. https://doi.org/10.1590/S1981-81222008000200005
Emperaire L, Peroni N (2007) Traditional management of agrobiodiversity in Brazil: a case study of manioc. Hum Ecol 35(2):761–768. https://doi.org/10.1007/s10745-007-9121-x
Ethurk O (2008) Antibacterial and antifungal activity of ethanolic extracts from eleven spice plants. Biologia 61(3):275–278. https://doi.org/10.2478/s11756-006-0050-8
Falcão MA, Clement CR (1999) Fenologia e produtividade do abiu (Pouteria caimito) na Amazônia Central. Acta Amazon 29(1):3–1. https://doi.org/10.1590/1809-43921999291011
Fernandes RS (2012) Frutos, sementes e amêndoas silvestres alimentícias na comunidade indígena Tunuì-Cachoeira-AM. 2012. Dissertation, Faculdade de Ciências Agronômicas, Universidade Federal de Lavras
Flora do Brasil 2020 (2020) Jardim Botânico do Rio de Janeiro. Available in http://www.floradobrasil.jbrj.br. Accessed in 01 Oct 2020
Garcia CER et al (2012) Carotenoides bixina e norbixina extraídos do urucum (Bixa orellana L.) como antioxidantes em produtos cárneos. Ciência Rural 42(8):1510–1517. https://doi.org/10.1590/S0103-84782012000800029
Garcia TB et al (2014) Leaf anatomical features of three Theobroma species (Malvaceae) native to the Brazilian Amazon. Acta Amazon 44(3):291–300. https://doi.org/10.1590/1809-4392201300653
Gonçalves GG (2017) Etnobotânica de plantas alimentícias em comunidades indígenas multiétnicas do Baixo Rio Uaupés – Amazonas. Dissertation, Faculdade de Ciências Agronômicas, Universidade Estadual Paulista
Haffer J, Prance GT (2002) Impulsos climáticos da evolução na Amazônia durante o Cenozóico: sobre a teoria dos Refúgios da diferenciação biótica. Estudos avançados 16(46):175–206. https://doi.org/10.1590/S0103-40142002000300014
Heck ECK, Loebens F, Carvalho PD (2005) Amazônia indígena: conquistas e desafios. Estudos avançados 19(53):237–255. https://doi.org/10.1590/S0103-40142005000100015
Henderson A, Galeano G, Bernal R (1995) Field guide to the palms of the Americas. Pinceton University Press, Princeton
Hilbert L, Neves EG, Pugliese F et al (2019) Evidence for mid-Holocene rice domestication in the Americas. Nat Ecol Evol 1(11):1693–1698. https://doi.org/10.1038/s41559-017-0322-4

Homma AKO (2000) Cronologia da ocupação e destruição dos castanhais no Sudeste paraense. Embrapa Amazônia Oriental, Belém

Homma AKO (2009) Os japoneses na Amazônia e sua contribuição ao desenvolvimento agrícola. Somanlu 9(1):113–113

Hopkins MJ (2017) Modelling the known and unknown plant biodiversity of the Amazon Basin. J Biogeogr 34(8):1400–1411. https://doi.org/10.1111/j.1365-2699.2007.01737.x

IBGE (2016) Manual Técnico da Vegetação Brasileira. Manuais técnicos em Geociências. IBGE, Brasília

Katz E, Léopez CL, Fleury M et al (2012) No greens in the forest? Note on the limited consumption of greens in the amazon. Acta Soc Bot Pol 81:283–293. https://doi.org/10.5586/asbp.2012.048

Kermath BM, Bennet BC, Pulsipher LM (2020) Food plants in the Americas: a survey of the domesticated, cultivated and wild plants used for human food in North, Central and South America and the Caribbean. https://www.academia.edu/1139225/Food_Plants_in_the_Americas_A_Survey_of_the_Domesticated_Cultivated_and_Wild_Plants_Used_for_Human_Food_in_North_Central_and_South_America_and_the_Caribbean. Accessed 14 Sept 2020

Kerr WE, Clement CR (1980) Práticas agrícolas de consequências genéticas que possibilitaram aos índios da Amazônia uma melhor adaptação às condições ecológicas da região. Acta Amazon 10(2):251–226. https://doi.org/10.1590/1809-43921980102251

Kerr WE, Posey DA, Wolter Filho W (1978) Cupá, ou cipó-babão, alimento de alguns índios amazônicos. Acta Amazon 8(4):702–705. https://doi.org/10.1590/1809-43921978084702

Kinupp VF (2007) Riqueza de plantas alimentícias não-convencionais na região metropolitana de Porto Alegre. In: Plantas alimentícias não-convencionais da Região Metropolitana de Porto Alegre, RS. Porto Alegre. Dissertation, Faculdade Agronomia, Universidade Federal do Rio Grande do Sul

Kinupp VF, Lorenzi H (2014) Plantas alimentícias não convencionais (PANC) no Brasil: guia de identificação, aspectos nutricionais e receitas ilustradas. Instituto Plantarum, São Paulo

Kistler L, Maezumi SY, Souza JG (2018) Multiproxy evidence highlights a complex evolutionary legacy of maize in South America. Science 362(6420):1309–1313. https://doi.org/10.1126/science.aav0207

Lake BG (1999) Coumarin metabolism, toxicity and carcinogenicity: relevance for human risk assessment. Food Chem Toxicol 37(3):423–453. https://doi.org/10.1016/s0278-6915(99)00010-1

Levis C, Costa FRC, Peña-Claros M (2017) Persistent effect of pre-columbian plant domestication on Amazon. Science 355(6328):925–931. https://doi.org/10.1126/science.aal0157

Lima AGM (2017) A cultura da batata-doce: cultivo, parentesco e ritual entre os Krahô. Mana 23(2):455–490. https://doi.org/10.1590/1678-49442017v23n2p455

Liu Q, Meng X, Li Y et al (2017) Antibacterial and antifungal activities of spices. Int J Mol Sci 18(6):1283. https://doi.org/10.3390/ijms18061283

Lorenzi H, Matos FJA (2008) Plantas medicinais no Brasil: nativas e exóticas. Instituto Plantarum, Nova Odessa

Lorenzi H, Moreira de Souza H, Medeiros-Costa JD et al (2004) Palmeiras brasileiras e exóticas cultivadas. Instituto Plantarum, Nova Odessa

Lorenzi H, Lacerda MTC, Bacher LB (2015) Frutas no Brasil: nativas e exóticas – de consumo in natura. Instituto Plantarum de estudos da Flora, São Paulo

Martins OS (2005) Dinâmica evolutiva em roças de caboclos amazônicos. Diversidade Biológica e Cultura da Amazônia. Estudos avançados 19(53):209–220. https://doi.org/10.1590/S0103-40142005000100013

Martius KV (1979) Natureza, doenças, medicina e remédios dos indíos brasileiros. Brasiliana v 154. Editora Nacional, São Paulo

McMichael CH, Palace MW, Bush MB et al (2013) Predicting pre-Columbian anthropogenic soils in Amazonia. Proc R Soc B 281(20132475):1–10. https://doi.org/10.1098/rspb.2013.2475

Mitchell JD, Daly DC (1998) The “tortoise’s cajá”—a new species of Spondias (Anacardiaceae) from southwestern Amazonia. Brittonia 50:447–451. https://doi.org/10.2307/2807753

Monteiro MHDA, Andreata RHP, Neves LJ (2007) Estruturas secretoras em Sapotaceae. Pesquisas, Botânica 58:253–262
Mori AS, Prance GT (1981) The "sapucaya" group of Lecythis (Lecythidaceae). Brittonia 33:70–80
Muniz J, Molina AR, Muniz J (2015) Physalis: panorama produtivo e econômico no Brasil. Hortic Bras 33(2). https://doi.org/10.1590/S0102-053620150000200023
Nascimento Filho HR, Barbosa RI, Luz FJF (2007) Pimentas do gênero Capsicum cultivadas em Roraima, Amazônia brasileira. Hábitos e formas de uso. Acta Amazônica 37(4):561–568. https://doi.org/10.1590/S0044-59672007000400011
Nelson BN, Oliveira A (2001) Estado do conhecimento florístico da Amazônia. In: Capobianco JPR, Veríssimo A, Moreira A et al (ed) Biodiversidade na Amazônia Brasileira: avaliação e ações prioritárias para a conservação, uso sustentável e repartição de benefícios. São Paulo, pp 132–182
Nimuendaju C (2017) Mapa etno-histórico do Brasil e regiões adjacentes. IPHAN, IBGE, Brasília
Oliveira MSP, Carvalho JEU, Nascimento WMO et al (2002) Cultivo do Açaizeiro para Produção de Frutos. Circular Técnica. Embrapa Amazônia Oriental, Belém
Paul JHA, Seaforth CE, Tikasingh T (2011) Eryngium foetidum L. a review. Fitoterapia 82(3):302–308. https://doi.org/10.1016/j.fitote.2010.11.010
Pedrosa HC, Clement CR, Schietti J (2018) The domestication of the Amazon tree grape (Pourouma cecropiifolia) under an ecological lens. Front Plant Sci 9(203):1–14. https://doi.org/10.3389/fpls.2018.00203
Pereira N (1974) Panorama da alimentação indígena. comidas, bebidas e tóxicos na Amazônia Brasileira. Livraria São José, Rio de Janeiro
Peres LG, Brandão VB, Rezende AJ (2018) Teobromina, substância encontrada no cacau. Revista JRG de Estudos Acadêmicos 1(3):2595–1661
Prance GT (1973) Phytogeographic support for the theory of Pleistocene forest refuges in the Amazon basin, based on evidence from distribution patterns in Caryocaraceae, Chrysobalanaceae, Dichapetalaceae and Lecythidaceae. Acta Amazônica 3:5–28. https://doi.org/10.1590/1809-43921973033005
Prance GT (1975) Estudos sobre a vegetação das Campinas Amazônicas: Introdução a uma série de publicações sobre a vegetação das Campinas Amazônicas. Acta Amazon 5(3):207–209. https://doi.org/10.1590/1809-43921975053207
Rabelo A (2012) Frutos nativos da Amazônia: comercialização nas feiras de Manaus AM. INPA, Manaus
Rangel PHR, Dias T (2016) Reintrodução de variedades tradicionais de arroz para o resgate do sistema de produção diversificado e sustentável dos índios krahô. In: Dias T, Edit JS, Udry C (eds) Diálogos de saberes: coleção povos e comunidades tradicionais, vol 2. Embrapa, Brasília, pp 63–72
Ratsch C (2011) Las plantas del amor, los afrodisíacos en los mitos, la história ay el presente. FCE, México
Rego LJS, Silva ML, Silva LF et al (2017) Caracterização do consumo de amêndoa de cumaru na Amazônia Oriental. Biota Amazônia 7(3):23–27
Ribeiro B (1995) Os índios das águas pretas. Edusp/Companhia das Letras, São Paulo
Ribeiro RG (2018) Estudo etnobotânico e físico-químico da batata-mairá (Casimirella spp. – Icacinaceae). Dissertation, Ciências Biológicas, Instituto Nacional de Pesquisas da Amazônia
Ribeiro CSC, Lopes CA, Carvalho SIC et al (2008) Pimentas Capsicum. Embrapa, Brasília
Rodriguez-Amaya DB, Kimura M, Amaya-Farfan J (2008) Fontes brasileiras de carotenóides: tabela brasileira de composição de carotenóides em alimentos. MMA/SBF, Brasília
Roman ALC, Ming LC, Sablayrolles MGP (2000) Pimentas Capsicum L. no Brasil: notas sobre botânica, história, concepções indígenas e folclore. SBEE, Porto Alegre
Roosevelt AC, Costa ML, Machado CL et al (1996) Paleoindian Cave Dwellers in the Amazon: the peopling of the Americas. Science 272(5260):373–384. https://doi.org/10.1126/science.272.5260.373

Santos GM, Maia GA, Sousa PHM et al (2008) Correlação entre atividade antioxidantes e composição de polpas comerciais de ação (Euterpe oleracea Mart.). Acrhivos latino-americanos de nutricion 58(2):2008

Schultes RE, Hofmann A (2001) Plants of the gods: their sacred, healing and hallucinogenic powers. Inner Traditions

Shanley P, Serra M, Medina G (2010). Frutíferas e plantas úteis na vida amazônica. CIFOR

Shepard G Jr, Ramirez H (2011) "Made in Brazil": human dispersal of the Brazil Nut (Bertholletia excelsa, Lecythidaceae) in Ancient Amazonia. Econ Bot 65(1):44–65. https://doi.org/10.1007/s12231-011-9151-6

Sibbr (2020) Biodiversidade e nutrição. Sistema de Informação sobre a biodiversidade brasileira. Availabre in https://ferramentas.sibbr.gov.br/ficha/bin/view/FN. Accessed 25 Oct 2020

Silva Filho DF, Noda H, Clement CR (1993) Genetic variability of economic characters in 30 accessions of cubiu (Solanum sessiliflorum Dunal) evaluated in Central Amazonia. Revista Brasileira de Genética 16(2):409–417

Silva LE, Reis RA, Moura EA et al (2015) Plantas do Gênero Xylopia: composição química e potencial farmacológico. Revista Brasileira de Plantas Medicinais 17(4):814–826. https://doi.org/10.1590/1983-084X/14_076

Silva AL, Garcia A, Baniwa AF et al (2016) Pimenta jiquitaia Baniwa. Instituto Socioambiental/OIBI, São Paulo/São Gabriel da Cachoeira

Simões CMO, Schenkel OS, Mello JCP et al (2017) Farmagnosia do produto natural ao medicamento. Artmed, Porto Alegre

Sivieiro A, Mattar EPC, Silva VB et al (2016) Feijão-comum e feijão-caupi cultivados nas comunidades ribeirinhas e indígenas do Acre. P. In: Dias T, Edit JS, Udry C (eds) Diálogos de saberes – relatos da Embrapa. Coleção povos e comunidades tradicionais, vol 2. Embrapa, Brasília, pp 247–259

Souza MAD (2015) Estudos em Eugenia L. (Myrtaceae) na Amazônia Central: taxonomia com o uso de ferramentas morfoanatômica. Dissertation, Faculdade de Ciências Biológicas, Instituto Nacional de Pesquisas da Amazônia

Spruce R (1851) Journal of a Voyage up the Amazon and Rio Negro. In: Hooker W, Hooker's J (eds) Journal of Botany and Kew Garden Miscellany, Londres, pp 210–212

Stradelli E (1929) Vocabularios da lingua geral portuguez-nheengatú e nheengatú-portuguez, precedidos de um esboço de grammatica nheênga-umbuêsáua mirî e seguidos de contos em lingua geral nheengatú poranduua. Revista do Instituto Histórico e Geográfico Brasileiro 158(104):9–768

Suassuna TMF, Matos RG, Freita FO (2016) Caracterização de variedades de amendoim provenientes do Parque Nacional do Xingu. In: Dias T, Edit JS, Udry C (eds) Diálogos de saberes – relatos da Embrapa. Coleção povos e comunidades tradicionais, vol 2. Embrapa, Brasília, pp 145–157

Teixeira FF, Landau EC (2016). Identificação de acesso de milho de origem indígena em banco de germoplasma. In: Dias T, Edit JS, Udry C (eds) Diálogos de saberes – relatos da Embrapa. Coleção povos e comunidades tradicionais, vol 2. Embrapa, Brasília, pp 53–61

Ter Steege H, Vaessen RW, Cárdenas-López D et al (2013) The discovery of the Amazonian tree flora with an updated checklist of all known tree taxa. Nat Sci Rep 6(29549):1–15. https://doi.org/10.1038/srep29549

Ter Steege H, Vaessen RW, Cárdenas-López D et al (2016) A descoberta da flora arbórea da Amazônia com uma lista atualizada de todos os taxa arbóres conhecidos. Boletim Museu Paraense Emílio Goeldi Ciências Naturais 11(2):231–261. https://doi.org/10.1038/srep29549

Tomchinsky B (2017) Prospecção de plantas aromáticas e condimentares no Brasil. Dissertation, Faculdade de Ciências Agronômicas, Universidade Estadual Paulista

Tomchinsky B, Ming LC (2019) As plantas comestíveis no Brasil dos séculos XVI e XVII segundo relatos de época. Rodriguésia 70(e03792017):1–16. https://doi.org/10.1590/2175-7860201970040

Tomchinsky B, Ming LC, Hidalgo AF et al (2013) Impactos da legislação na pesquisa etnobotânica no brasil, com ênfase na região amazônica. Amazônica. Revista de Antropologia 5(3):734–761. https://doi.org/10.18542/amazonica.v5i3.1603

Trajano AAL (2008) Brasil a gosto. São Paulo, Melhoramentos

Tricaud S, Pinton F, Pereira JS (2016) Saberes e práticas locais do produtores de guaraná (Paullinia cupana Junth var. sorbilis) do médio Amazpnas: dias organizações locais frenta a inovação. Boletim do Museu Paranense Emílio Goeldi Ciências humanas 11(1):33–53. https://doi.org/10.1590/1981.81222016000100004

WWF (2017) Novas espécies de vertebrados e plantas na Amazônia: atualização e composição da lista: 2014–2015. WWF-Brasil /instituto de desenvolvimento sustentável Mamirauá, WWF-Brasil, Brasíli

Yuyama K, Aguiar JPL, Yuyama LKO (2002) Camu-camu: um fruto fantástico como fonte de vitamina C. Acta Amazon 32(1):169–174. https://doi.org/10.1590/1809-43922002321174

Yuyama LKO, Yuyama K, Aguiar JPL et al (2013) Fruteiras da Amazônia: potencial nutricional. INPA, Manaus

Food Composition Data: Edible Plants in Pantanal

Ieda Maria Bortolotto, Rita de Cássia Avellaneda Guimarães, Raquel Pires Campos, Mariana Rodrigues da Silva Lopes, Laleska Pâmela Rodrigues da Silva, Rosa Helena Silva, Geraldo Alves Damasceno-Junior, Arnildo Pott, and Priscila Aiko Hiane

Contents

1 Introduction

The Pantanal is one of the largest wetlands on Earth, the most part in Brazil, besides Bolivia and Paraguay (Adámoli 1982). The vegetation is represented by a mosaic of physiognomies that occur in floodable and flood-free areas, with influence from

I. M. Bortolotto (✉) · R. H. Silva · G. A. Damasceno-Junior · A. Pott
Programa de Pós-Graduação em Biologia Vegetal, Instituto de Biociências, Universidade Federal de Mato Grosso do Sul, Campo Grande, MS, Brazil

R. de Cássia Avellaneda Guimarães · M. R. da Silva Lopes · L. P. R. da Silva
Programa de Pós-Graduação em Saúde e Desenvolvimento na Região Centro-Oeste, Faculdade de Medicina, Universidade Federal de Mato Grosso do Sul, Campo Grande, MS, Brazil
e-mail: rita.guimaraes@ufms.br

R. P. Campos · P. A. Hiane
Faculdade de Ciências Farmacêuticas, Alimentos e Nutrição, Universidade Federal de Mato Grosso do Sul, Campo Grande, MS, Brazil
e-mail: raquel.campos@ufms.br; priscila.hiane@ufms.br

M. C. M. Jacob, U. P. Albuquerque (eds.), *Local Food Plants of Brazil*, Ethnobiology, https://doi.org/10.1007/978-3-030-69139-4_14

Cerrado, Chaco, and Amazon Forest (Prance and Schaller 1982), with over 2500 catalogued species of angiosperms (Pott and Pott (in press)). Hundreds of native species have a potential for alimentary use for both human and animals (Pott and Pott 1986, 1994; Pott et al. 2004; Bortolotto et al. 2018). Among them, circa 70 species were recorded as part of the diet and culture of the human populations living in indigenous and traditional communities in the western edge of the Pantanal (Bortolotto et al. 2015; Bortolotto et al. 2019). Among them, there are some species associated with rich traditional knowledge on their use in the diet, relevant in the past, were abandoned, or have low use value now.

Nuts and pulps of fruits, leaves, rhizomes, and other parts of food plants native to the Pantanal have relevant nutritional value with high levels of fibers, sugars, proteins, minerals, fatty acids, vitamins, and carotenoids (Hiane et al. 2006; Ramos et al. 2008; Prates et al. 2015; Arakaki et al. 2020). Besides, they have bioactive compounds with antioxidant action, and their inclusion in the diet protects the organism from several chronic diseases associated with oxidative stress (Pereira and Cardoso 2012).

The consumption of wild foods, free of agrochemicals, is also associated with sustainable actions, being beneficial for health and the environment. The inclusion of regional foods and native fruits in the diet of the population can be an economical and sustainable way of preventing diseases associated with malnutrition and represents an alternative for consumers, constituting a new source of foods, raw materials, new products, and wealth for the country (Marin 2006; Silva et al. 2010).

Several epidemiological studies indicate that the high ingestion of plant products is associated with a reduced risk of a variety of chronic diseases such as atherosclerosis and cancer, effects that have been particularly attributed to the compounds in antioxidant activity: vitamins C and E, phenolic compounds, especially flavonoids, and carotenoids (Silva et al. 2010). The antioxidants ingested in the human diet, such as vitamin C and phenolic compounds, can prevent carcinogenesis for scavenging the free radicals and impeding the linkage of the carcinogens to DNA (Valko et al. 2006).

Some native food species have edible parts with peculiar organoleptic characteristics, which make them distinct and can be included in new compounds potentially toxic to human beings (Pinela et al. 2017). Thus, the native food plants with traditional use deserve a distinct value since they offer the necessary security to indicate their alimentary use proven over centuries by the people consuming them frequently (Pardo de Santayana et al. 2012). Many species with high cultural value have an essential role in the human diet for being functional foods, capable of providing benefits to health, besides nutritional value, and being strategic for food and nutrition security and food sovereignty (Wittman 2012). The native food plants can also be useful as supplementary, seasonal, or subsistence sources in many cultural groups and play an essential role in fighting food shortage (Lulekal et al. 2011). Additionally, the fruits of native species from Brazil have nutraceutical properties for their characteristics being comparable to commercial drugs (dos Santos et al. 2017).

This chapter compiles nutritional information about 11 traditional native food species of the Pantanal abundant in natural plant physiognomies that have great potential for sustainable use associated with food and nutrition security and food sovereignty. The species were selected among those with the highest use value by two ethnobotanical studies (Bortolotto et al. 2015; Bortolotto et al. 2019) developed in the municipalities of Corumbá and Porto Murtinho, in the State of Mato Grosso do Sul, in the extreme west of the Brazilian Pantanal, besides one traditionally used in the past in this region. For each species, we searched published information on their nutritional composition, traditional uses, distribution, and phenological and abundance data around the communities.

This region has an altitudinal gradient varying between 80 m on the plain and 1064 m in areas of the Urucum-Amolar hills (Borges et al. 1997). Besides aquatic and wetland plants present in water bodies, riparian forests, and floodplain, there is high species richness in the vegetation of Cerrado and in seasonal deciduous and semideciduous forest (Prance and Schaller 1982; Damasceno-Junior et al. 2017) in more elevated areas. The municipality of Porto Murtinho, in the southernmost portion of the Pantanal, contains an area with Chaco vegetation, distributed in Argentina, Paraguay, Bolivia, and Brazil, in Mato Grosso do Sul (Prado et al. 1992). Porto Murtinho has 15,372 inhabitants, being 5313 habitants in the rural and 10,059 in the urban zone; the municipality of Corumbá has a population of 103,772 inhabitants, being 93,452 in the urban zone and 10,251 in the rural area (IBGE 2010). The study of Porto Murtinho includes persons who live in rural and urban area, while the study of Corumbá includes only persons of an indigenous community and three traditional in the rural area.

2 Characteristics, Main Uses, and Nutritional Composition of the Food Plants Native to the Pantanal

The 11 wild food species traditionally utilized by the human populations of the urban and rural area along the Paraguay River in the Pantanal (Brazil) selected for this work are presented in Table 1. We included three species of Arecaceae, two Fabaceae, and one representative of Rubiaceae, Sapotaceae, Poaceae, Lamiaceae, Malpighiaceae, and Myrtaceae. Except for *Byrsonima cydoniifolia* A. Juss., mentioned only for the municipality of Corumbá, *Oryza latifolia* Desv. (that was in disuse), and *Inga laurina* (Sw.) Willd only for Porto Murtinho, the other species were cited for both municipalities.

Arecaceae
***Acrocomia aculeata* (Jacq.) Lodd ex. Mart.**

Acrocomia aculeata is a palm locally known as *bocaiuva* or *macauba* (Table 1). It has a wide distribution in South and Central America, except Amazonia (Table 1). The drupaceous fruit is rounded with a diameter of 2.5–5.0 cm. The kernel is resistant and dark (black). The pulp color varies from yellow to orange (Fig. 1a, b) with a slightly sweetish taste. In the Pantanal, *A. aculeata* has several traditional uses in

Table 1 List of the 11 species of food plants of the Pantanal, Mato Grosso do Sul, Brazil, with botanical family, species, local name, edible part used and mode of consumption, fructification season, and occurrences

Family/species	Local name	Edible parts and traditional consumption mode in Pantanal	Fruiting season (maturation)	Occurrences
Arecaceae *Acrocomia aculeata* (Jacq.) Lodd.	Bocaiuva, macauba	Palm heart roasted or baked; fruit (pulp and almonds), ripe pulp (mesocarp), used to make bocaiuva flour, drink (liquor, fresh or fermented juice), almonds eaten raw or toasted; used to make oil (1, 2, 3)	September to December (4)	South and Central America, from Argentina to México (5); Brazil, except in the Amazon (6); Pantanal: An, De, Sd, Ce, Rf (1); Ms
Attalea phalerata Mart. ex Spreng.	Acuri, bacuri	Pulp unripe or ripe fruits eaten cooked, palm heart baked (1); ripe pulp (mesocarp) used to make acuri flour (3); almonds eaten raw; used to make oil (1, 3)	April to October (2)	South America, from Brazil to Peru (5); Brazil, Cerrado and Pantanal (6); Pantanal, An, De, Sd, Ce, Rf (1); Ms (2)
Copernicia alba Morong	Carandá	Heart palm and ripe fruit (in natura)	February to May (2)	South America, Brazil, Bolivia, Paraguai and Argentina (5); Brazil, Pantanal, in Rf; Ms
Fabaceae *Inga laurina* (Sw.) Will.	Ingá	Sarcotest seed consumed fresh (7)	November to March (8)	South and Central America, from Mexico to Paraguay and Argentina (5); Brazil, Amazônia, Caatinga, and Cerrado (6); Pantanal, Sd, Rf
Fabaceae *Prosopis ruscifolia* Griseb.	Algarrobo	Fruits to make algarrobo flour, to prepare breads, cakes, and an alcoholic drink (chicha) (7)	November to February (9)	South America, Argentina, Paraguay, Bolivia, and Brazil (5); Brazil, Pantanal and Caatinga (6); Pantanal, De, Ce, Ch (1)
Lamiaceae *Vitex cymosa* Bertero ex Spreng.	Tarumã	Ripe fruits (pulp) eaten fresh (1), used to make jams (3)	November to February (2)	South and Central America (5); Brazil, Amazônia, Caatinga, Cerrado, Mata Atlântica and Pantanal (6); Pantanal, Rf, Sd (1)

(continued)

Table 1 (continued)

Family/species	Local name	Edible parts and traditional consumption mode in Pantanal	Fruiting season (maturation)	Occurrences
Malpighiaceae *Byrsonima cydoniifolia* A. Juss.	Canjiquinha, canjiqueira	Ripe fruits (pulp) eaten fresh (1), juice (acid) (3)	September to March (3) and April (com. Pess.)	Bolivia and Brazil (5); Brazil, Caatinga, Cerrado and Pantanal (6); Pantanal, Rf, Fl (1); Ms
Myrtaceae *Plinia cauliflora* (DC.) Kausel	Jabuticaba	Ripe fruits eaten fresh, like a jam and vinegar (1), liquor	October to December (2)	South, Central, and North America, as well as records in Australia and South Africa (6); Brazil, Mata Atlântica (5); Pantanal, De, An
Poaceae *Oryza latifolia* Desv.	Arroz-do-pantanal, arroz-do-campo	Seeds: galinhada (rice and jerk chicken) (3)	May to June (2)	South and Central America, from Brazil to México (5); Brazil, Amazônia, Cerrado, Mata Atlântica, and Pantanal (6); Pantanal, Fl (1); Ms
Rubiaceae *Genipa americana* L.	Jenipapo	Ripe fruits (pulp) used to make jams and liquor (1) and juice (3)	October to December (2)	South and Central America, from Argentina to México (5); Brazil Amazônia, Caatinga, Cerrado, Mata Atlântica and Pantanal (6); Pantanal, An, Rf
Sapotaceae *Pouteria glomerata* (Miq.) Radlk.	Laranjinha-de-pacu	Ripe fruits (pulp) eaten fresh (1)	January to August (2)	South and Central America, from Brazil to México (5); Brazil, Amazônia, Cerrado, and Mata Atlântica (6); Pantanal, Rf

Sources: (1) Bortolotto et al. (2015), (2) Pott and Pott (1994), (3) Damasceno-Junior and Souza (2010), (4) Salis and Mattos (2009), (5) GBIF.org (2018), (6) Flora do Brasil (2020), (7) Bortolotto et al. (2019), (8) Pennington (1997), (9) de Matos Alves (2014)
De Deciduous forest, *Sd* semideciduous forest, *Ce* Cerrado, *Rf* riparian forest, *Fl* floodplains, *Ch* Chaco, *An* anthropogenic, and *Ms* monodominant stands

the diet, utilizing the pulp, heart of palm, and nut (Table 1). The fruit pulp is used for consumption in natura and in cakes, ice creams, and flour; the nut has several alimentary uses in natura, in preparing coconut sweet and milk mix, and in the production of edible oil (Bortolotto et al. 2017) and jams (da Silva et al. 2017). It has excellent yield (each bunch produces 6.32 kg of pulp and 1.36 kg of endosperm (nut) (Sanjinez-Argandoña and Chuba 2011)). In Corumbá, the flour obtained from the dehydrated pulp is commercialized (Dias and Galvani 2017).

The stipe and the roots contain starch (Peña 1976), but their uses were not observed in the Pantanal. The flour obtained from the processed amylaceous fibers

Fig. 1 Fruits and seeds of wild food plants from the Pantanal. (**a**) *A. aculeata* (fruits); (**b**) *A. aculeata* (pulp); (**c**) *A. phalerata* (fruits); (**d**) *A. phalerata* (seeds); (**e**) *C. alba* (fruits); (**f**) *I. laurina* (fruit); (**g**) *P. ruscifolia* (fruits); (**h**) *V. cymosa* (fruits); (**i**) *Byrsonima cydoniifolia* (immature fruits); (**j**) *Plinia cauliflora* (immature fruit); (**k**) *O. latifolia* (grains; (**l**) *G. americana* (fruit); (**m**) *P. glomerata* (fruits). (Pictures by Paulo Robson de Souza, except **f** (A. Pott) and **b**, **d**, and **m** (Ieda Maria Bortolotto))

of *A. aculeata* stipes was recorded for the Ayoreo people of the Paraguayan Chaco (Schmeda-Hirschmann 1994). Women of the Terena indigenous ethnicity in Mato Grosso do Sul consumed the heart of palm of *Acrocomia* spp. after childbirth to stimulate milk production (Oberg 1949). In general, the main alimentary uses are the fruits and heart of palm. It is the species with the highest use value in rural communities in the western edge on the Pantanal (Bortolotto et al. 2015).

Nutritional Composition Data

The edible portion composed of pulp and nut of *A. aculeata* represents circa 48% of the total fruit weight; the pulp is rich in total lipids, carbohydrates, fibers, and β-carotene, with considerable concentration of potassium, calcium, and phosphorous, showing to be a fruit with excellent yield and nutritive value for consumption in natura or in culinary (Prates et al. 2015). The pulp oil has an intense orange color characterized by the presence of carotenoids and high concentration of oleic acid (Hiane et al. 2005; Ramos et al. 2007). The nut oil is transparent, with a predominance of oleic and lauric fatty acids (Prates et al. 2015). The oils of both pulp and nut are also utilized for consumption or in pharmaceutical and cosmetic industries (Ciconini et al. 2013).

Regarding its nutritional composition, *A. aculeata* fruit pulp contains 52.99% moisture, 22.10% of carbohydrates, 8.14% of lipids, 1.50% of protein, 13.76% of fibers, and 167.67 calories in 100 g of pulp. The fresh pulp presents high concentrations of calcium 61.96 mg/100 g, phosphorous 36.70 mg/100 g, and potassium 766.37 mg/100 g and lower concentrations of sodium 3.74 μg/g, iron 7.81 μg/g, manganese 138 μg/g, zinc 6.02 μg/g, and copper 2.43 μg/g. This fruit is also rich in β-carotene (49.0 μg/g of pulp), contributing to the enrichment of the regional diet (Ramos et al. 2008).

Attalea phalerata Mart. ex Spreng.

Characteristics and Traditional Uses

Attalea phalerata is a palm locally known as *acuri* or *bacuri* (Table 1). It is widespread from Brazil to Peru, occurring in several physiognomies in the Pantanal, mostly in monodominant formations called *acurizal* (Table 1). The fruit is an ellipsoid drupe of approximately 5 cm long, green when immature and yellow when ripe (Fig. 1c). The fruits have fleshy mesocarp, yellowish when ripe (Fig. 1d).

The pulp (ripe or green), as well as the nut of the fruits, is edible; the oil obtained from the nut, the flour made from the mesocarp, and the heart of palm have traditional uses in the diet (Table 1). The coco water of immature fruits can also be consumed (Pott and Pott 1994). The edible oil extracted from the nut was not mentioned in recent ethnobotanical works in the Pantanal but has economic importance in Bolivia (Moraes et al. 1996). The use of fruits (nut and pulp) and heart of palm in the diet and the utilization of *chicha*, an alcoholic drink (used only in the past), is associated with the Guató and Bororo indigenous culture in the Pantanal (Schmidt 1942; Hartmann 1967; de Oliveira 1996).

Nutritional Composition Data

The potential of the nut protein (de Lima Mendes Ramos et al. 2017) and the pulp oil (de Lima et al. 2016, 2018) of *A. phalerata* fruit has been explored in experimental research. The ripe fruits of *A. phalerata* have a pulp rich in carotenoid pigments (pro-vitamin A), copper, and magnesium (Hiane et al. 2010). Additionally, the pulp lipidic fraction presents fatty acids with a predominance of oleic and palmitic acids, which can have considerable anti-inflammatory effects with potential nutraceutical properties (Freitas de Lima et al. 2017); it also has cytoprotective activity, probably for its capacity to inhibit the action of free radicals (de Lima et al. 2018). The fruit nut has high oil content and is rich in phosphorous and a source of iron (Hiane et al. 2010). The oleic and lauric fatty acids were the main compounds found in the nut oil (da Silva Baldivia et al. 2018). In Bolivia, the oil extracted from the nut (60–70% of dry weight) is rich in lauric and myristic fatty acids, comparable with other tropical oil crops (Moraes et al. 1996).

Regarding nutritional composition, the pulp of *A. phalerata* presents 56.90 g/100 g of moisture, 35.13 g/100 g carbohydrates, 5.97 g/100 g lipids, and low levels of ashes (1.25 g/100 g) and protein (0.75 g/100 g) (Siqueira et al. 2016).

***Copernicia alba* Morong**

Characteristics and Traditional Uses

Copernicia alba is a palm locally known as *carandá* (Table 1). It has occurrence in Brazil, Bolivia, Argentina, and Paraguay in South America (Table 1). In Brazil, it occurs only in the Pantanal, mostly in monodominant formations called *carandazal* (Table 1). The fruit is globous to ovoid, of approximately 1.2 cm in diameter, green when immature and black when ripe, with one seed (Moraes 2014). The fruits have a fleshy mesocarp, dark brown or almost black when ripe (Fig. 1e). The fruits ripen over an extended period, between April and November; some trees keep fruits until January (Silva 2018). The fruit production is considered high, more than 20 t/ha (Silva 2018).

The ripe pulp and the heart of palm have traditional uses in the diet in Porto Murtinho (Table 1). *C. alba* is gathered by the Ayoreo of the Paraguayan Chaco; the heart of palm is eaten raw or cooked in water or in ashes, and the palm ashes were used as a salt substitute (Schmeda-Hirschmann 1994; Peña-Chocarro et al. 2006), made from the burned spathe; the heart of palm is also consumed baked or ground as a flour (Peña-Chocarro et al. 2006). The Toba and Whichí of the Paraguayan Chaco consume the fresh fruits, and the Lengua-Maskoy ferment them to prepare *aloja* (Peña-Chocarro et al. 2006). The apex extracted from the local palm *C. alba* was mentioned for use by the Chorote Indians in Argentina although unexploited at present (Arenas and Scarpa 2007). A liquor was developed from the pulp of fruits harvested (Silva 2018).

Nutritional Composition Data

In the pulp of *C. alba* collected in Corumbá and Porto Murtinho, a moisture content of 54.12%; good levels of total carbohydrates (27.92%), lipids (6.03%), and proteins (3.39%); energetic value of 232.8 Kcal/100 g; and considerable content of ashes (3.71%), with considerable potential for food products (Silva 2018). The contents of lipids are around 48% in seeds of ripe fruit and 40% in the immature; in contrast, the peel of the ripe has a lower fat level (0.5%) than in the immature fruit (0.8%), and in the pulp, both values are very low (0.1%) (Silva 2018). The fruit contains bioactive phenolic and medicinal compounds and unsaturated fatty acids (Silva 2018). Furthermore, the fruit has high vitamin C content, of 20.5 mg/100 g, equivalent in the peel and the pulp (Silva 2018), thus, a valuable raw matter for functional foods as an antioxidant source.

In a province in Argentina, the values found in this fruit seem discrepant, consisting in 4.70% moisture, 82.80% total carbohydrates, 24.30% crude fiber, 1.90% lipids, 8.60% proteins, and 6.70% ashes. The fruit was rich in minerals such as sodium (31.13 mg/100 g), potassium (856.83 mg/100 g), magnesium (54.96 mg/100 g), calcium (98.73 mg/100 g), and expressive levels of iron (1.37 mg/100 g), manganese (0.72 mg/100 g), and zinc (0.29 mg/100 g) (Gorostegui et al. 2011). However, because the analyses in the Brazilian study were expressed on a fresh basis and the Argentinian one on a dry basis, the discrepant values of nutritional composition between the studies are justified and also can be further explained by differences between regions where the fruits were collected since factors such as soil and climate influence its composition.

Fabaceae
***Inga laurina* (Sw.) Willd.**
Characteristics and Traditional uses

Inga laurina, known as *ingá*, is a tree species widely distributed in South and Central America (Table 1). Its fruits are plane to convex pods (flat to convex) (Fig. 1f), margins slightly raised or not, glabrous, and yellowish to green-yellowish color (Pennington 1997), which varies with the degree of ripening (Martins et al. 2014). The edible part of the fruit is a white sarcotest, commercialized in El Salvador (Pennington 1997), and has sweetish taste. In the Pantanal, they are consumed in natura (Table 1). It has potential for various culinary uses (Kinupp and Lorenzzi 2014), as well as other species of the genus, such as *I. vera* Willd. (Bortolotto et al. 2017). *I. vera* is a species that also occurs in the Pantanal, even more common and more abundant than *I. laurina*, forming monodominant populations especially in the riparian forests of the Paraguay River and affluents, with fruits mainly in the flood season (Damasceno-Junior et al. 2005). However, nutritional data are yet lacking on *I. vera*.

Nutritional Composition Data
The fruit of *I. laurina* has the shape of a slightly curved pod; the peel is thin, with green-yellow color, which varies with the degree of ripening (Schulz et al. 2014). The fruits of *I. laurina* are constituted by pulp (41%), peel (39%), and seed (20%). For industrial use, higher pulp yields are obtained when optimizing the selection for more fresh mass and fruit length, once they are good indicators of the high association of these characteristics with pulp yield (da Silva Oliveira et al. 2019).

Regarding nutritional values, *I. laurina* fruit has high moisture content (85.39%); thus, it is more prone to deterioration. It has 81.91 Kcal/100 g of energy value; low contents of proteins (0.13 g/100 g), lipids (0.0007 g/100 g), and ashes (0.14 g/100 g); and expressive values of carbohydrates (13.52 g/100 g). For containing high levels of phenolic compounds (110.67 mg GA 100 g) and considerable quantities of vitamin C (1.60 mg AA 100 g), the fruit can be considered having a promising antioxidant potential (de Lima and Portari 2019).

When compared with the conventional apple (*Malus domestica* Borkh.), widely commercialized in Brazil, *I. laurina* contains high levels of phenolic content. Moreover, for its functional properties to health, it is advisable to stimulate the development of products by pharmaceutical and food industries and to promote the sustainable utilization of native fruits in areas with easier access, for both consumers and industries (de Siqueira et al. 2013).

Prosopis ruscifolia Griseb.
Characteristics and Traditional Uses

Prosopis ruscifolia, known as *algarrobo*, is a tree species, also belonging to the Fabaceae family. It has occurrence in Brazil, Bolivia, Argentina, and Paraguay in South America (Table 1). In Brazil, it occurs in the Caatinga and in the Pantanal (de Souza-Lima et al. 2017) (Table 1). The fruit is a drupaceous loment (Noguchi et al. 2009) (Fig. 1g) ripe in the dry months (Table 1). The sweetish fruits (Pott et al. 2004) have known traditional food uses for production of flour, preparation of bread and cakes, and an alcoholic drink (*chicha*) (Table 1). These were mentioned as food when going to the field, for example, and are used sporadically (Bortolotto et al. 2019). The seeds of *P. ruscifolia* are edible, roasted, and grinded (Boeri 2016). Other three species with edible fruits, *P. alba* Griseb, *P. nigra* Hieron., and *P.* Hassl., were also recorded for the Brazilian Chaco in the municipality of Porto Murtinho (Sartori et al. 2018). An *algarroba* beer prepared from fruits of *P. alba*, the "*aloja de algarroba*" (in local Spanish), is an ancient alcoholic drink of the Whichís (Argentina and Bolivia) (Cano et al. 2020) and other indigenous peoples from the Gran Chaco in South America (Arenas and Scarpa 2007).

Nutritional Composition Data

The fruit of *P. ruscifolia* is an excellent source of proteins (12.7 g/100 g) and has high content of carbohydrates (78.5 g/100 g), lipids, (4.32 g/100 g) and fibers (17.8 g/100 g) (Freyre et al. 2003) (Table 2). It has been shown (Freyre et al. 2003) (Table 2) that the pulp flour contains a nutritional complement of adequate amino acids; thus, it is a fruit with high biological value proteins as a food and also for enrichment of food products. For the digestibility nearly complete in the gastrointestinal tract, the flour has a low allergenic potential (Mamone et al. 2019).

Regarding micronutrients, the seeds have an excellent content of iron (4.57 mg/100 g), zinc (3.89 mg/100 g), considerable values of calcium (1.528 mg/100 g), phosphorous (4.719 mg/100 g), and potassium (5.887 mg/100 g) (Freyre et al. 2003). For its high iron content, the fruit can be consumed with other foods (lemon, orange, and guava) that have ascorbic or citric acid to increase the availability of iron, thus, helping in an adequate and healthy diet (Bernardi et al. 2004).

Lamiaceae
Vitex cymosa **Bertero ex Spreng**
Characteristics and Traditional Uses

Vitex cymosa, known as *tarumã*, is a tree belonging to the family Lamiaceae, with wide occurrence in Brazil and several countries of South and Central America (Table 1). The fruit is a globous drupe, with dark red or purple color at ripening (Fig. 1h). Its pulp is mucilaginous and juicy and has a sweetish taste. The fruits are abundant and cover the ground at maturity time (November to February) (Table 1). Besides traditional uses in the diet, for consumption in natura or as jam (Table 1), the ripe fruit has a potential to aromatize salty dishes, especially meats (Damasceno-Junior and Souza 2010). Although the fruit has a very strong sour-sweetish smell, unpleasant to some people, from it can be produced a syrup with economical potential for use in the cover of ice creams and cakes.

Nutritional Composition Data
Vitex cymosa fruit has a high moisture content (83.74g/100 g) that can cause fast deterioration since the proliferation of microorganisms is favored and consequently hinders fruit quality. In 100 g of whole fruits, low content of lipids (0.03%) and proteins (0.49%) were observed; however, it presents considerable values of total carbohydrates (9.34%) and fibers (4.66%). The fruit of *V. cymosa* is considered a food of low caloric value (36.6 kcal/100 g). However, the fruit is rich regarding micronutrients, potassium (287.8 mg/100 g), phosphorous (21.1 mg/100 g), and iron (0.43 mg/100 g), as well as vitamin C for children of 1–3 years of age and a source of fiber for children and adults (Caldeira et al. 2004). That was corroborated by results of fruits analyzed in Ecuador, showing that phosphorous and potassium were the main macro-minerals found in natura fruits (Guevara et al. 2020).

Malpighiaceae
Byrsonima cydoniifolia **A. Juss.**
Characteristics and Traditional Uses

Byrsonima cydoniifolia, locally known as *canjiqueira*, is shrub or treelet that belongs to the Malpighiaceae. It occurs in several countries of South and Central America and most of Brazil (Table 1). It occurs in monodominant formations in the Pantanal (Pott and Pott 1994) and the Araguaia wetland (Brazil) (Marimon and de Souza Lima 2001). The fruits are globose, drupaceous, and juicy and measure circa 2 cm diameter, with color varying from yellow to orange when ripe (Fig. 1i). Besides the traditional consumption of the fruit pulp in natura in the Pantanal (Table 1), it can be used to prepare liquor, jam, ice cream, and sweets.

Nutritional Composition Data
The fruit composition of *B. cydoniifolia* revealed 655.5 g/kg moisture, 47.42 g/kg carbohydrates, and 252.6 g/kg de lipids; furthermore, the fruit presented 45.9 g/kg carotenoids and 1.82 g/kg ascorbic acid, showing to be a fruit with high antioxidant

Table 2 Nutritional value of wild food plants of the Pantanal, Mato Grosso do Sul, Brazil

Taxon	Local name	Moisture (g/100 g)	Carbohydrate (g/100 g)	Lipids (g/100 g)	Protein (g/100 g)	Fiber (g/100 g)	Total caloric value (kcal/100 g)	Reference
Arecaceae *Acrocomia aculeata* (Jacq.) Lodd.	Bocaiuva	52.99	22.10	8.14	1.50	13.76	167.67	Ramos et al. (2008)
Attalea phalerata Mart. *ex* Spreng.	Acuri; bacuri	56.90	35.13	5.97	0.75	–	–	Siqueira et al. (2016)
Copernicia alba Morong	Carandá	54.12	27.92	6.03	3.39	–	–	Silva (2018)
Fabaceae *Inga laurina* (Sw.) Willd.	Ingá	85.39	13.52	0.0007	0.13	–	81.91	de Lima and Portari (2019)
Prosopis ruscifolia Griseb.	Algarrobo	–	78.5	4.32	12.7	17.8	–	Freyre et al. (2003)
Lamiaceae *Vitex cymosa* Bertero *ex* Spreng.	Tarumã	83.74	9.34	0.03	0.49	4.66	36.60	Caldeira et al. (2004)
Malpighiaceae *Byrsonima cydoniifolia* A. Juss.	Canjiqueira; Canjiquinha	65.55	4.74	25.26	–	–	–	Marcelino et al. (2018)
Myrtaceae *Plinia cauliflora* (DC.) Kausel	Jabuticaba	83.6	15.30	0.10	0.60	2.30	58.00	UNICAMP (2011)
Poaceae *Oryza latifolia* Desv.	Arroz-do-pantanal; arroz-do-campo	9.62	64.51	2.05	9.83	13.15	315.81	Barbosa et al. (2017)
Rubiaceae *Genipa americana* L.	Jenipapo	70.00* – 80.42**	14.57** – 22.10*	0.00* – 1.60**	0.5* – 1.59**	1.09** – 6.30*	79.04** – 90.4*	*Pacheco et al. (2014) ** Hamacek et al. (2013)
Sapotaceae *Pouteria glomerata* (Miq.) Radlk.	Laranjinha-de-pacu	85.22	10.93	0.32	0.69	–	49.36	Nogueira et al. (2018)

activity (Marcelino et al. 2018). These results show that knowing the nutritional properties of native fruits allows indicating their consumption in natura or as culinary ingredients; besides, the regular consumption of fruits with a considerable content of antioxidant compounds, such as *B. cydoniifolia*, prevents chronic non-communicable diseases (Gonzalez 2006).

Chemical analyses of the fruit composition of *B. cydoniifolia* showed derivatives of flavonoids and stilbenes, such as trans-piceatannol and resveratrol as the main secondary metabolites, demonstrating their anti-inflammatory and anti-hyperalgesic effect and sustaining its potential as a nutraceutical food (dos Santos et al. 2017). The fruits of *B. cydoniifolia* have potential as functional ingredients, and the oil has potential for edible uses (Marcelino et al. 2018). The pulp and the seed are significant sources of potassium and sodium, the seed having higher concentrations of calcium, copper, iron, manganese, magnesium, selenium, and zinc (Arakaki et al. 2020). Jam is an excellent alternative for the utilization of *B. cydoniifolia* fruits, due to the reduction of antinutritional factors and retention of bioactive compounds in processing this product (Prates et al. 2015).

Myrtaceae
***Plinia cauliflora* (DC.) Kausel**
Characteristics and Traditional Uses

Plinia cauliflora (*Myrciaria cauliflora* (Mart.) O. Berg), the *jabuticaba*, is a treelet circa 6 m tall, of the Myrtaceae family, and occurs in South, Central, and North America (Table 1), often grown in home gardens. Many Neotropical Myrtaceae have edible fruits. The fruit is a globose berry with approximately up to 5 cm in diameter (Fig. 1j), green when immature and purple when ripe. The pulp is whitish, soft, and sweet. The fruits have traditional consumption in natura and also as jams, vinegar, and liquor (Table 1). The fruits must be harvested fully ripe (soft), but for transport and storage, it is necessary to harvest them just before ripening (Damasceno-Junior and Souza 2010).

Nutritional Composition Data

For being widely consumed in Brazil, the nutritional composition of *P. cauliflora* is shown in the nutritional table of UNICAMP (2011); 100 g of fruit have contents of energy of 58 kcal, 0.6 g proteins, 0.1 g lipids, 15.3 total carbohydrates, and 2.3 g fibers. Regarding micronutrients, the fruit presents a high level of potassium (130 mg) and considerable levels of calcium (8 mg) and phosphorous (15 mg) (UNICAMP 2011).

Biazotto et al. (2019) evidenced in *P. cauliflora* high contents of carotenoids (326.70 μg/100 g) and total phenolics (109.65 mg GAE/100 g EF). Thus, their study demonstrated that this fruit represents an excellent resource with a technological and economic potential of bioactive compounds found in the Brazilian biodiversity of unexplored fruits, mainly in the nutritional and pharmaceutical areas.

An analysis of two cultivated varieties of *P. cauliflora* in Minas Gerais (Brazil) showed high contents of minerals, 2.75 g/100 g in Paulista and 3.82 g/100 g in

Sabará; in both, the component of insoluble fibers of 2.57 g/100 g and 3.30 g/100 g is higher than the soluble ones (de Lima et al. 2008).

Poaceae
***Oryza latifolia* Desv.**
Characteristics and Traditional Uses

Oryza latifolia is an emergent aquatic herb, approximately 2 m tall, perennial, belonging to Poaceae, the grass family (Pott and Pott 2000). This wild rice occurs in several countries of South and Central America (Table 1). In the Pantanal, it occurs in monodominant stands called *arrozal* in the floodplain of the Paraguay River (Bertazzoni and Damasceno-Júnior 2011). The grains are of the "agulhinha" type, small and reddish (Fig. 1k), with almond-like flavor. The rice ripens at the flood peak in the Pantanal (Table 1). Its cultural alimentary value is associated with indigenous people (Guató) from the Pantanal, but it has no longer been used in their diet (Bortolotto et al. 2015).

It is a species that can occupy extensive areas over the floodplain of the Paraguay River in years of more intense flooding. There are other four species of *Oryza* in the Pantanal: *O. alta* Swallen, *O. glumaepatula* Steud., and *O. grandiglumis* (Döll) Prod. (Filgueiras et al. 2015; Flora do Brasil 2020). Among these, *O. glumaepatula* presents high levels of total fractions of protein, albumin, and glutelin (protein, albumin, and glutelin fractions) (Santos et al. 2013).

Nutritional Composition Data

Its whole grain contains 9.62% moisture, 2.05% lipids, 9.83% proteins, 64.51% starch, 13.15% fibers, and 315.81 kcal of energy; these values are similar to *Oryza sativa* L., the commercial rice, a staple food in Brazil (Barbosa et al. 2017).

Rubiaceae
***Genipa americana* L.**
Characteristics and Traditional Uses

Genipa americana, *jenipapo*, is a tree belonging to the Rubiaceae family (Table 1). It has a wide distribution in Central and South America, including most of Brazil (Table 1). It is a dioecious species, i.e., only the female plant fructifies. The fruit is an oval berry (Fig. 1l) circa 8 cm, brownish when ripe. The pulp is spongy, juicy, and sweet, traditionally consumed in natura and as for sweets and juices (Table 1).

For being an easily perishable fruit, with peculiar sensorial characteristics that reduce its acceptability in natura, thus, aiming to avoid losses and increase the potential of its consumption, strategies that help its conservation and acceptance are searched, such as drying by osmotic dehydration using an osmotic agent, such as sugar, increasing the shelf life, palatability, and acceptance in comparison with the in natura fruit (Andrade et al. 2003).

Nutritional Composition Data

Regarding chemical composition, *G. americana* presents 70% moisture, 22.1% carbohydrates, 0.0% lipids, 0.5% proteins, and 6.3% alimentary fibers, besides

2.0 mg/100 g de vitamin C, 176.3 mg/100 phenolic compounds, and 70.2% of antioxidant capacity (Pacheco et al. 2014). Another nutritional study that also assessed the *G. americana* fruit found 80.42 g/100 g moisture, 14.57 g/100 g carbohydrates, 1.60 g/100 g lipids, 1.59 g/100 g protein, 1.09 g/100 g fibers, and total energetic value of 76.92 kcal/100 g (Hamacek et al. 2013). Thus, *G. americana* is a fruit with considerable nutritional values, besides high antioxidant capacity, and its inclusion in the regional diet is indicated. The fruit contains essential oil, tartaric acid, and glucose (Prance 1989).

Sapotaceae
***Pouteria glomerata* (Miq.) Radlk.**
Characteristics and Traditional Uses

Pouteria glomerata is a tree of the Sapotaceae family, locally known as *laranjinha-de-pacu* (Table 1), in allusion to its use as fishing bait. It occurs in several countries of South and Central America, frequently found in Pantanal riparian forests (Table 1). The fruits are fleshy, globose berries, with green peels when immature and yellow when ripe. It can present the shape of a squash pumpkin (undulated), and so it also receives the name of *moranguinha* (Fig. 1m). The edible part is the pulp, with a pleasant acid taste, traditionally consumed in natura (Table 1). For its acidity (presence of tartaric and malic acids) and pectin content, it forms a gel, being excellent to prepare jams. It has been utilized for the production and commercialization of frozen pulp for juices, popsicle, ice cream, and jams (Bortolotto et al. 2017) or sold in natura (Damasceno-Junior and Souza 2010).

Nutritional Composition Data
The pulp of this fruit contains 85.22% moisture, 10.93% carbohydrates, 0.32% lipids, 0.69% proteins, 0.64% ashes, 4.94% tilted acidity, and 3.19 pH (Nogueira et al. 2018). *P. glomerata*, besides vitamin C (34.87 mg/100 g), can be considered an excellent source of bioactive compounds, with antioxidant potential attributed to its phenolic compounds, anthocyanins (0.65 mg/100 g), carotenoids (0.93 mg/100 g), and flavonoids (9.63 mg/100 g), with beneficial effects on human health (Teixeira et al. 2020).

3 Discussion

The results show that the selected food plants in this work have rich nutritional values, wide geographical distribution, and abundance in the native physiognomies, with available nutrients spread over the year (Table 1). These species have been the target of projects that incentivize their utilization in communities of various municipalities. The combination of vast populations of the native species here presented with high production of fruits and high nutritional value offers support to nutritional security most of the year.

3.1 Nutritional Value

The three species of the Arecaceae family, *A. aculeata, A. phalerata,* and *C. alba*, have abundant fruits over a great part of the year and wide distribution in the Pantanal (Table 1), besides the heart of palm, little used. The endosperms of the *A. aculeata* and *A. phalerata* represent excellent sources of calories and edible oil for the local diet (Table 2), especially of the oleic and lauric fatty acids (Hiane et al. 2005; da Silva Baldivia et al. 2018). *A. aculeata* is rich in Mn, Cu, P, Mg (nut), Cu, and pro-vitamin A (pulp) (Hiane et al. 2010), what certainly justifies its broad use by the human populations, with the highest value of local use among the native food species, either in the present or in the past (Herberts 1998; Bortolotto et al. 2015).

Vitamin A is an essential component in the diet, and its deficiency still is a public health problem in several places in the world (Greiner 2013). Its availability to the rural human populations is a relevant health factor. A recent study evaluated the deficiency of vitamin A (DVA) in children and the associated factors, including all Brazilian regions, including the Central-West where the Pantanal is, in both rural and urban areas (Lima et al. 2018). Those authors found that DVA prevailed in the urban zone, even that the population resident in the rural zone presents a higher vulnerability to nutritional deficiencies at the world level. Considering that the precursors of vitamin A, mainly β-carotene, are found in the fruits of *A. aculeata* and *A. phalerata*, their continuity in the diet ought to be stimulated. Moreover, the fatty acids found in nuts of *A. phalerata* are important regulators of the metabolism and are frequently associated with a reduction of serum cholesterol and body fat and a lower risk of developing cardiovascular diseases (da Silva Baldivia et al. 2018).

In general, the fruits of *V. cymosa*, *G. americana*, *I. laurina*, and *P. glomerata* present nutritional compositions with reduced values of lipids and considerable quantities of ashes, moisture, and carbohydrates. *B. cydoniifolia* has slightly higher contents of lipids than those fruits. *O. latifolia* contains a higher protein level, and the palms *A. aculeata* and *A. phalerata* show significant values of proteins and lipids, plus other constituents. Besides these nutrients, all fruits exhibit phenolic compounds and vitamin C and carotenoids, especially the orangish fruits, which contain bioactive compounds with antioxidant properties. For the antioxidant action to occur, it needs to inhibit or impede the auto-oxidative process caused by free radicals, besides having stability about their intermediate compounds formed (Finco et al. 2012). *P. ruscifolia* fruit has a high alimentary and technological potential for the traditional communities of the Chaco since its nutritional composition is rich in several nutrients that help in the human body organic functions, besides the food security of communities more susceptible to nutritional deficiencies.

It is known that a diet containing essential nutrients and with the addition of substances of nutraceutical potentials, in a healthy lifestyle, can exert a fundamental role in prevention and or treatment of non-communicable diseases (Pereira and Cardoso 2012) and other illnesses, e.g. severe acute respiratory syndrome coronavirus 2 (SARS-CoV-2) caused by the coronavirus 2019 (COVID-19) (de Faria Coelho-Ravagnani et al. 2020).

The nutritional status of people infected by SARS-CoV-2 is indispensable to indicate a better prognostic and can influence the severity of the condition. The supplementation from a diet rich in probiotics, vitamins, and minerals does not act preventively neither in the treatment of the infection by COVID-19 but can improve the immune response in an auxiliary way. As such, the intake of vitamins C and D, zinc, and selenium is beneficial to individuals with or under risk of viral respiratory infections (de Faria Coelho-Ravagnani et al. 2020).

In Brazil, the most immediate consequence of the health crisis caused by COVID-19 pandemic is the worsening of food insecurity due to income restrictions limiting food access (da Silva Filho and Gomes Júnior 2020). Thus, incentives are needed to the consumption and valuation of native fruits available locally since most contain considerable quantities of vitamin C and other vital nutrients. With the social distance to slow the spread of COVID-19, the search for local products is being driven by economic reasons in evaluating the consumption of seasonal and regional foods, once these fruits are available to the local community.

3.2 Geographic Distribution and Abundance of WFP in Native Vegetation

Most food species of this chapter occur in several phytogeographic domains in South and Central America (Table 1) and are part of the food culture there, besides the Pantanal. That is especially evident in the neighboring countries Bolivia and Paraguay, with similar recorded uses to the mentioned for the Pantanal for species such as *A. aculeata* and *A. phalerata*, with production of oil, flours, and other products (Schmeda-Hirschmann 1994; Moraes et al. 1996). The fruits as the most important species culturally can be obtained over all months of the year (Table 1), what is strategic in terms of food and nutrition security and food sovereignty for the local populations.

Moreover, several species mentioned here as the most important culturally have high nutritional value and are abundant in the Pantanal. *B. cydoniifolia*, *A. phalerata*, *C. alba*, and *O. latifolia* grow as monodominant formations (Table 1), besides the occurrence of *Inga vera*. Monodominant formations are vegetation types characterized by the dominance of a single species, generally over 50% of the individuals or the cover (Hart 1990). One of the causes of monodominance in wetlands is the flood (Hart 1990). The names of these formations are associated with the local names of the species names: *canjiqueiral*, the formation of *B. cydoniifolia*, *acurizal* of *A. phalerata*, *carandazal* of *C. alba*, and *arrozal* of *O. latifolia* (Damasceno-Junior et al. (in press)). *B. cydoniifolia* is a pioneer species with high capacity to colonize open environments such as grasslands, a process called encroachment (Barbosa da Silva et al. 2016). Many ranchers control such species that take over native grasslands, e.g., *B. cydoniifolia* is cut before the flood or pulled down by tractors (Pott and Pott 1994). The occupation of *A. phalerata* and the abundance of *A. aculeata* are also associated with disturbed areas.

Indeed, all these species seem monocultures for the characteristics of their occurrence. Given their nutritional value, they constitute a great opportunity for utilization as a source of alternative income for rural properties and for local communities. Every year these species produce tons of products such as fruits and grains without an associated productive chain yet. As an aggravating factor, stands of valuable plants are eliminated in favor of grasslands. The use of these species by traditional communities and rural properties can configure an excellent opportunity for conservation of these ecosystems, as well as can offer unique products and could receive the denomination of origin, for example.

3.3 Products and Services Associated with Conservation (Sustainable Use) and Innovation

The use of wild foods, especially in low-income rural communities, which still have access to the biodiversity and detain knowledge on how to use it, is a strategy that can be very important for the food security under risks of climatic changes (Smith et al. 2019) and to improve health in the face of the COVID-19 pandemics, as mentioned earlier. Knowledge on the nutritional value of these fruits gives support to public policies that promote the sustainability for their subsistence, mainly in the aspect of food security of the Pantanal riverine communities.

Despite abandoning the use of traditional (*O. latifolia*) or neglected species (e.g., oil of palms), considering their abundance and wide distribution, they can still supply the daily dietary needs of the populations, especially those under food shortage. The situation of neglected food species and the need for initiatives to improve the acceptance of wild foods is under discussion in ethnobotanical studies (do Nascimento et al. 2013). In Brazil as well as in other South American countries, there are already advances aiming to conserve the culture and biodiversity associated with incentives to the use of native flood plants (May and da Vinha 2013; Depenthal and Meitzner Yoder 2017).

The Universidade Federal de Mato Grosso do Sul and several partners have projects in this line such as the program "Sabores do Cerrado & Pantanal" (Bortolotto et al. 2017) to strengthen technological practices to aggregate value in the income of traditional riverside communities, besides promoting knowledge in the academic area, qualification of the external community, symposia, and courses on the valuation of food plants of the Cerrado and Pantanal. Research and extension actions were developed in the last 15 years to incentive the consumption of native species abundant in the Pantanal, with cultural bound and with potential to improve the income of small communities (Damasceno-Junior and Souza 2010; Bortolotto et al. 2017). Thereby, the utilization commercialization (to obtain income) in small communities of the Pantanal was awakened and started a process of cultural rescue with incentives to the use in the local diet that has to be associated with the conservation of biodiversity and sustainable development (Candil et al. 2007).

Although neglected in the past century by the big market and without attention from public policies aiming their scientific study until the middle of the 1980 decade, the native species have gained importance over the last years, following a worldwide movement related to the use of wild food plants (Termote et al. 2011; Menendez-Baceta et al. 2012; Molina et al. 2012; Smith et al. 2019; Cano et al. 2020). However, they still need to be better explored in the nutritional aspect. Most are present in preparations of flours and oils or are tasty ingredients of ice creams, preserves, and other products. Thus, they deserve more attention as potential sources of nutrients, mainly in natura, and also as an income source for the population. The production of flours and oils is an essential strategy of the local populations to store foods for periods with low availability of the resources and for commercialization and to improve income.

The pulp flour *A. phalerata*, for example, is utilized for cakes, bread, and sauces; from the nut can be made coco sweet and extract edible oil. The pulp flour arises high interest for its nutritive value, consistency, taste, and flavor of excellent quality (Damasceno-Junior and Souza 2010). Recent studies showed that the pulp and the nut of *A. phalerata* are adequate ingredients for the formulation of müsli since they demonstrated improvement in the characteristics of color, texture, and nutritional and energetic value of the product compared with the analyzed commercial granola (Mendoza et al. 2016). *I. laurina* has functional properties for health, being important to stimulate the development of products in pharmaceutical and food industries, with sustainable use of the fruits (de Siqueira et al. 2013). Similarly, *P. ruscifolia* as well as other species of *Prosopis* have a high potential for food utilization (Boeri 2016). *C. alba* fruit has an excellent nutritional value with high vitamin C and represents a great potential for new food products (Silva 2018).

Despite this potential for traditional utilization in the Pantanal and potential for innovation, the products are still not conventional in the big markets and somehow neglected, called unconventional food plants (UFP) (Kinupp and Lorenzzi 2014). An UFP can be exemplified by *B. cydoniifolia*, with relevant nutritional and nutraceutical value, but has been neglected. With adoption of sustainable technologies, it will be possible to innovate and to valorize the use of traditional wild food plants of the Pantanal and to conserve both biodiversity and culture. The identification of challenges and solutions able to influence practices in the chains of local supplies can favor the sustainability of the global production ecosystem (Tomas et al. 2019). The collaboration of scientific research represents an important role to play, besides actors outside the academy encompassing local communities, indigenous groups, manager agencies, *non-government organizations* or NGOs, big transnational corporations, and governmental actors. The consumers can also be influent in the promotion of sustainability, guiding their purchases with sustainable thinking (Nyström et al. 2019).

The market trend for growth of sustainable production and consumption stands out, besides some emergent questions ending up to insert, influence, or press for transformations in the ways of relationship with the environment, the form to produce, transform, and consume the foods (da Veiga Dias et al. 2015). Some actions were suggested to strengthen the initiative "Plants for the Future" of the Ministry for Environment: the incentive to actions that stimulate people to consume products from the native biodiversity, creating demand for new products and diversifying the diet of the families; the increase and promotion of the use of these species, also in school diets; and stimulation to higher participation of components of the native biodiversity in strengthening the national gastronomy (MMA 2016).

Family agriculture is conceived as the guardian of the biological diversity, being responsible for the introduction of new species of products from the native biodiversity in the commercial circuits of the agro-alimentary system, as well as supplier for cosmetic and pharmaceutical industries (Mendes et al. 2014). From the utilization of fruits of the Pantanal, new rural family enterprises and the traditional and extractivist communities can stand out in food production and increase their income, with increased quality of life and the potential of regional sustainable development. The socioeconomic and cultural potential of native plant species of the Pantanal is valuable, and it can contribute to the supply of new products with nutritional and functional quality, bringing benefit to consumer health.

The combination of large populations of the species with the high production of fruits with high nutritional value offers nutritional security in most part of the year. Considering only these 11 species (Tables 1 and 2), it is noticeable that the availability of fruits over the year can provide essential nutrients in nearly all months.

4 Final Considerations

Until recently, most morphotypes of *bocaiuva* palms were identified as *A. aculeata*, while *A. totai* was considered a synonym. Later, both species became valid again, with occurrence in Central Brazil (including Pantanal) (Lorenzi 2010). However, a morphometric and molecular investigation showed the species occurring in the Pantanal is *A. totai*, besides the very rare and newly described *A. corumbaensis* (Vianna and Campos-Rocha 2020). The scientific works about nutritional data compiled in this chapter refer to *A. aculeata*, most times to an herbarium voucher, but that has not been updated in herbaria, to differentiate *A. aculeata* from *A. totai*.

The botanical collection of palms is not a simple task, demanding a specific training; however, there is a strong recommendation that a voucher shall be deposited in herbarium to allow tracking the data. Studies on food plants or with food potential included in genera such as *Oryza*, *Hymenaea*, *Campomanesia*, *Passiflora*, and *Arachis* (Bortolotto et al. 2018) represented in the Pantanal by various species ought to fulfill this requirement. *O. alta*, for example, is considered a synonym of *O. latifolia*, but it is still accepted as a distinct species. Such taxonomic controversies are frequent until they are settled among specialists. Since many native species have gained visibility recently, by a function of the results of studies on the nutritional value of wild food plants so far neglected, there is still a need for taxonomic revisions of species with doubtful identification. Also, care should be taken when searching a plant in the literature since it may have been reported with a name that became a synonym and so is not retrieved, e.g., *Scheelea phalerata*, now *Attalea phalerata*.

However, such situation is not restricted to palms either to our work. Many reports on nutritional value did not indicate a herbarium specimen, despite the recommendations. The use of wild plants for food and nutrition requires precise data, reliable and accessible about their composition as the data users must be sure about the reliability of the identification and naming of flood plants (Nesbitt et al. 2010). Those authors analyzed the identification and nomenclature of plants in 50 articles referring to 502 sampled species, each one associated with one or more nutritional data. They noted that from the 502 sampled plants, only 36 followed the best practices for plant identification and recommend that researchers should identify, name, and publish the species correctly.

Given the relevance of the Central-West native species destined to food and health, generally produced and commercialized by small farmers and local communities, we suggested to research funding organs to direct resources and incentives to the priority native species in the scope of the initiative "Plants for the Future," including public policies, as well as to maximize the application of resources in multidisciplinary and multi-institutional network efforts (MMA 2016). We point out that among the 11 species selected for this chapter, only *A. aculeata*, *P. cauliflora*, *G. americana*, and *O. latifolia* were included in the proposed "Plants for the Future," demonstrating the need to improve knowledge about species of the Pantanal, which also have widespread occurrence and high food potential.

Acknowledgments To the Universidade Federal de Mato Grosso do Sul for infrastructure and support, we thank the Brazilian Coordination for the Improvement of Higher-Level Education Personal (CAPES) for scholarships to M. R. da S. Lopes and L. P. R. da Silva, Finance Code 001, and the Graduate Program in Health and Development in the Central-West – Universidade Federal de Mato Grosso do Sul – and to the National Council for Scientific and Technological Development (CNPq) for the researcher grants to G. A. Damasceno-Junior, process number 307429/2019-9 and A. Pott, process number 303191/2017-1. Special thanks to Prof. Dr. Paulo Robson de Souza who kindly provided images of fruits to compose Fig. 1.

References

Adámoli J (1982) O Pantanal e suas relações fitogeográficas com os cerrados: discussão sobre o conceito de complexo do Pantanal. In: Anais do 32° Congresso nacional da Sociedade Botânica do Brasil. Universidade Federal do Piauí, Teresina, pp 109–119

Alves F de M (2014) Análise genética populacional de Prosopis rubriflora Hassl. ("espinheiro") e Prosopis ruscifolia Griseb. ("algaroba") (Leguminosae, Mimosoideae) em áreas de Chaco brasileiro. Universidade Estadual de Campinas

Andrade SAC, Metri JC, de Barros Neto B, Guerra NB (2003) Desidratação osmótica do jenipapo (Genipa americana L.). Ciência e Tecnol Aliment 23:276–281. https://doi.org/10.1590/s0101-20612003000200029

Arakaki DG, dos Santos VS, de Melo EP, Pereira H, Figueiredo PS, Cortês MR, Carollo CA, de Oliveira LCS, Tschinkel P, Reis F, Souza I, Rosa R, Sanches F, dos Santos EF, do Nascimento

VA (2020) Canjiqueira fruit: are we losing the best of it? Foods 9:1–15. https://doi.org/10.3390/foods9040521

Arenas P, Scarpa GF (2007) Edible wild plants of the Chorote Indians, Gran Chaco, Argentina. Bot J Linn Soc 153:73–85. https://doi.org/10.1111/j.1095-8339.2007.00576.x

Barbosa da Silva FH, Arieira J, Parolin P, Nunes da Cunha C, Junk WJ (2016) Shrub encroachment influences herbaceous communities in flooded grasslands of a neotropical savanna wetland. Appl Veg Sci 19:391–400. https://doi.org/10.1111/avsc.12230

Barbosa MM, Macedo MLR, Aiko Hiane P, Damasceno-Júnior GA, Braga Neto JA, Mendonça LABM, De Cássia Avellaneda Guimarães R, do Nascimento VA (2017) Proteins of Oryza latifolia from Brazilian Pantanal: impact on the use of aminoacids and weight gain of animals. Int J Dev Res 07:16885–16891

Bernardi C, Freyre M, Sambucetti ME, Pirovani ME (2004) Use of ascorbic and citric acids to increase dialyzable iron from vinal (Prosopis ruscifolia) pulp. Plant Foods Hum Nutr 59:175–179. https://doi.org/10.1007/s11130-004-0047-9

Bertazzoni EC, Damasceno-Júnior GA (2011) Aspectos da biologia e fenologia de Oryza latifolia Desv. (Poaceae) no Pantanal sul-mato-grossense. Acta Bot Bras 25:476–486. https://doi.org/10.1590/S0102-33062011000200023

Biazotto KR, de Souza Mesquita LM, Neves BV, Braga ARC, Tangerina MMP, Vilegas W, Mercadante AZ, De Rosso VV (2019) Biodiversity fruits: discovering bioactive compounds from underexplored sources. J Agric Food Chem 67:1860–1876. https://doi.org/10.1021/acs.jafc.8b05815

Boeri PA (2016) Bioprospección química y propagación de plantas nativas del monte patagónico como estrategias de conservación y uso sustentable. Universidad Nacional de La Plata

Borges CA, Werlw HJS, Rosa DB, Paiva DJ, Moraes EP, Silva LBSM (1997) Geomorfologia. In: BRASIL. Ministério do Meio Ambiente dos RH e da AL (ed) Plano de Conservação da Bacia do Alto Paraguai (Pantanal) – PCBAP. Diagnóstico dos meios físico e biótico. MMA/SEMAM/PNMA, Brasília, pp 73–119

Bortolotto IM, de Mello Amorozo MC, Neto GG, Oldeland J, Damasceno-Junior GA (2015) Knowledge and use of wild edible plants in rural communities along Paraguay River, Pantanal, Brazil. J Ethnobiol Ethnomed 11. https://doi.org/10.1186/s13002-015-0026-2

Bortolotto IM, Hiane PA, Ishii IH, de Souza PR, Campos RP, Juraci Bastos Gomes R, da Silva Farias C, Leme FM, do Carmo de Oliveira Arruda R, de Lima Corrêa da Costa LB, Damasceno-Junior GA (2017) A knowledge network to promote the use and valorization of wild food plants in the Pantanal and Cerrado, Brazil. Reg Environ Chang 17:1329–1341. https://doi.org/10.1007/s10113-016-1088-y

Bortolotto IM, Damasceno-Junior GA, Pott A (2018) Preliminary list of native food plants of Mato Grosso do Sul, Brazil. Iheringia – Ser Bot 73:101–116. https://doi.org/10.21826/2446-8231201873s101

Bortolotto IM, Seleme EP, de Araújo IPP, de Souza Moura S, Sartori ÂLB (2019) Conhecimento local sobre plantas alimentícias nativas no Chaco brasileiro. Oecol Aust 23:764–775

Caldeira SD, Hiane PA, Ramos MILR, Ramos Filho MM (2004) Caracterização físico-química do araçá (Psidium guineense SW.) e do tarumã (Vitex cymosa) do Estado de Mato Grosso do Sul. B Ceppa 22:145–154

Candil RFM, José de Arruda E, Arakaki AH (2007) O Cumbaru (Dipteryx alata Vog.), o desenvolvimento local e a sustentabilidade biológica no assentamento Andalucia, Nioaque/MS. Interações (Campo Gd) 8:75–80. https://doi.org/10.1590/s1518-70122007000100008

Cano ANH, Cano ANH, Suárez ME, Suárez ME (2020) Ethnobiology of algarroba beer, the ancestral fermented beverage of the Wichí people of the Gran Chaco I: a detailed recipe and a thorough analysis of the process. J Ethn Foods 7:1–12. https://doi.org/10.1186/s42779-019-0028-0

Ciconini G, Favaro SP, Roscoe R, Miranda CHB, Tapeti CF, Miyahira MAM, Bearari L, Galvani F, Borsato AV, Colnago LA, Naka MH (2013) Biometry and oil contents of Acrocomia aculeata fruits from the Cerrados and Pantanal biomes in Mato Grosso do Sul, Brazil. Ind Crop Prod 45:208–214. https://doi.org/10.1016/j.indcrop.2012.12.008

Coelho-Ravagnani C de F, Corgosinho FC, Sanches FLFZ, Prado CMM, Laviano A, Mota JF (2020) Dietary recommendations during the COVID-19 pandemic. Nutr Rev:1–14. https://doi.org/10.1093/nutrit/nuaa067
Damasceno-Junior GA, Souza PR, Bortolotto IM, Ramos MIL, Hiane PA, Braga Neto JA, Ishii IH, Costa DC, Ramos-Filho MM, Gomes RJB, Barbosa MM, Rodrigues RB (2010) Sabores do Cerrado & Pantanal: Receitas e boas práticas de aproveitamento. In: Damasceno-Junior GA, Souza PR (eds). Editora UFMS, Campo Grande, p 141
Damasceno-Junior GA, Semir J, Maës Dos Santos FA, Leitão-Filho HDF (2005) Structure, distribution of species and inundation in a riparian forest of Rio Paraguai, Pantanal, Brazil. Flora 200:119–135. https://doi.org/10.1016/j.flora.2004.09.002
Damasceno-Junior GA, Bortolotto IM, Sartori ÂLB, Pott A, Souza PR (2017) Tipos de Vegetação em Mato Grosso do Sul: fontes de plantas alimentícias. UFMS, Campo Grande
Damasceno-Junior GA, da Silva RH, Gris D, de Souza EB, da Rocha M, Pineda DAM, de Amador GA, Oldeland J, Pott A (In press) Monodominant stands in the Pantanal. In: Junior GAD, Pott A (eds) Flora and vegetation of Pantanal Wetland. Springer, New York
Depenthal J, Meitzner Yoder LS (2017) Community use and knowledge of Algarrobo (Prosopis pallida) and implications for Peruvian dry forest conservation. Rev Ciencias Ambient 52:49. https://doi.org/10.15359/rca.52-1.3
Dias FRT, Galvani F (2017) Custeio e Investimento do Sistema Artesanal de Produção da Farinha de Bocaiuva. Doc 147
do Nascimento VT, de Lucena RFP, Maciel MIS, Albuquerque UP (2013) Knowledge and use of wild food plants in areas of dry seasonal forests in Brazil. Ecol Food Nutr 52:317–343 . https://doi.org/10.1080/03670244.2012.707434
Filgueiras TS, Valls JFM, Oliveira RP (2015) Oryza. In: List. Espécies da Flora do Bras
Finco FDBA, Silva IG, de Oliveira RB (2012) Physicochemical characteristcs and antioxidant activity of three native fruits from Brazilian savannah (Cerrado). Aliment e Nutr 23:179–185
Flora do Brasil (2020) Flora do Brasil 2020 em construção. In: Jard. Botânico do Rio Janeiro. http://floradobrasil.jbrj.gov.br/. Accessed 3 Aug 2020
Freitas de Lima F, Traesel GK, Menegati SELT, dos Santos AC, Souza RIC, de Oliveira VS, Sanjinez-Argandoña EJ, Cardoso CAL, Oesterreich SA, do Carmo Vieira M (2017) Acute and subacute oral toxicity assessment of the oil extracted from Attalea phalerata Mart ex Spreng. pulp fruit in rats. Food Res Int 91:11–17. https://doi.org/10.1016/j.foodres.2016.11.019
Freyre M, Astrada E, Blasco C, Baigorria C, Rozycki V, Bernardi C (2003) Valores nutricionales de frutos de Vinal (Prosopis ruscifolia): consumo humano y animal. Cienc y Tecnol Aliment 4:41–46. https://doi.org/10.1080/11358120309487617
GBIF.org (2018) Ocorrência GBIF
Gonzalez CA (2006) Nutrition and cancer: the current epidemiological evidence. Br J Nutr 96:S42–S45. https://doi.org/10.1079/bjn20061699
Gorostegui HA, Orrabalis CJ, Guzmán CA, Calandri EL (2011) Physicochemical properties of fruit, pericarp and seeds from Copernicia alba morong, harvested in the Province of Formosa, Argentina. Nat Neotrop 42:35–43
Greiner T (2013) Vitamin A: moving the food-based approach forwad T. FAO, pp 1–69
Guevara M, Valdés-Silverio LA, Granda-Albuja MG, Iturralde G, Jaramillo-Vivanco T, Giampieri F, Santos-Buelga C, González-Paramás AM, Battino M, Álvarez-Suarez JM (2020) Pechiche (Vitex cymosa Berteo ex Speng), a nontraditional fruit from Ecuador, is a dietary source of phenolic acids and nutrient minerals, in addition to efficiently counteracting the oxidative-induced damage in human dermal fibroblasts. Antioxidants 9. https://doi.org/10.3390/antiox9020109
Hamacek FR, Moreira AVB, Martino HSD, Ribeiro SMR, Pinheiro-Sant'Ana HM (2013) Valor nutricional, caracterização Física e físico-química de jenipapo (Genipa Americana L.) do cerrado de Minas Gerais. Aliment e Nutr 24:73–77
Hartmann T (1967) Nomenclatura Botânica dos Borôro (Materiais para um ensaio etno-botânico). São Paulo

Hart TB (1990) Monospecific dominance in tropical rain forests. Trends Ecol Evol 5:6–11. https://doi.org/10.1016/0169-5347(90)90005-X
Herberts AL (1998) OS Mbayá-Guaikurú: área, assentamento, subsistência e cultura material. Universidade Vale do rio dos Sinos
Hiane PA, Ramos MM, Ramos MIL, Macedo MLR (2005) Óleo da polpa e amêndoa de bocaiúva, Acrocomia aculeata (Jacq.) Lodd. Caracterização e composição em ácido graxos. Braz J Food Technol 8:256–259
Hiane PA, Baldasso PA, Marangoni S, Macedo MLR (2006) Chemical and nutritional evaluation of kernels of bocaiuva. Ciência e Tecnol Aliment Campinas 26:683–689. https://doi.org/10.1590/S0101-20612006000300031
Hiane PA, Ramos Filho MM, Ramos MILR, Braga Neto JA, Barbosa MM, Costa DC (2010) Frutos nativos do Cerrado e Pantanal, Qualidade nutricional e sabor peculiar (ainda) pouco aproveitados. In: Damasceno-Junior GA, de Souza PR (eds) Sabores do Cerrado e Pantanal: receitas e boas práticas de aproveitamento. UFMS, Campo Grande, pp 131–136
IBGE (2010) Censo demográfico 2010. https://censo2010.ibge.gov.br. Accessed 31 July 2020
Kinupp VF, Lorenzzi H (2014) Plantas Alimentícias Não Convencionais (PANC) no Brasil. Instituto Plantarum de Estudos da Flora, Nova Odessa
Lima MC de, Portari GV (2019) Centesimal composition and antioxidant compounds of two fruits from the Cerrado (Brazilian Savannah). Rev Ceres 66:41–44. https://doi.org/10.1590/0034-737X201966010006
Lima AJB de, Corrêa AD, Alves APC, Abreu CMP, Dantas-Barros AM (2008) Caracterização química do fruto jabuticaba (Myrciaria cauliflora Berg) e de suas frações. Arch Latinoam Nutr 58:416–421
Lima FF de, Menegati SELT, Traesel GK, Araújo FHS de, Lescano CH, Peixoto SM, Mao Silva FA, Vieira SCH, Vieira MDC, Oesterreich SA (2016) Study on the cytotoxic, genotoxic and clastogenic potential of Attalea phalerata Mart. Ex Spreng. Oil pulp in vitro and in vivo experimental models. PLoS One 11:1–12. https://doi.org/10.1371/journal.pone.0165258
Lima FF de, Lescano CH, Arrigo JS, Cardoso CAL, Coutinho JP, Bonfá Moslaves IS, Ximenes TV do N, Toffoli Kadri MC, Weber SS, Perdomo RT, Leite Kassuya CA, Vieira MC, Sanjinez-Argandoña EJ (2018) Anti-inflammatory, antiproliferative and cytoprotective potential of the Attalea phalerata Mart. ex Spreng. pulp oil. PLoS One 13:1–17. https://doi.org/10.1371/journal.pone.0195678
Lima DB, Damiani LP, Fujimori E (2018) Deficiência de Vitamina A em crianças brasileiras e variáveis associadas. Rev Paul Pediatr 36:176–185. https://doi.org/10.1590/1984-0462/;2018;36;2;00013
Lorenzi H (2010) Flora brasileira Lorenzi: Arecaceae. Instituto Plantarum, Nova Odessa
Lulekal E, Asfaw Z, Kelbessa E, Van Damme P (2011) Wild edible plants in Ethiopia: a review on their potential to combat food insecurity. Afr Focus 24:71–121. https://doi.org/10.21825/af.v24i2.4998
Mamone G, Sciammaro L, De Caro S, Di Stasio L, Siano F, Picariello G, Puppo MC (2019) Comparative analysis of protein composition and digestibility of Ceratonia siliqua L. and Prosopis spp. seed germ flour. Food Res Int 120:188–195. https://doi.org/10.1016/j.foodres.2019.02.035
Marcelino G, Donadon JR, Caires ARL, Michels FS, Oliveira LCS, Cortes MR, Maldonade IR, Cavalheiro LF, Nazário CED, Maróstica Júnior MR, Santos NMD, Batista PB, Pott A, Nunes ÂA, Figueiredo PS, Hiane PA, Nascimento VA, Guimarães RCA (2018) Characterization and oxidative stability of oils and bioactive compounds of the fruits of Byrsonima cydoniifolia A. Juss. at different ripening stages. J Sci Food Agric 99:2855–2864. https://doi.org/10.1002/jsfa.9498
Marimon BS, de Souza Lima E (2001) Caracterização fitofisionômica e levantamento florístico preliminar no Pantanal dos Rios Mortes-Araguaia, Cocalinho, Mato Grosso, Brasil. Acta Bot Bras 15:213–229. https://doi.org/10.1590/S0102-33062001000200008

Marin AMF (2006) Potencial nutritivo de frutos do Cerrado: composição em minerais e componentes não convencionais. Universidade de Brasília

Martins RC, Filgueiras TDS, Albuquerque UP (2014) Use and diversity of palm (Arecaceae) resources in central Western Brazil. Sci World J 2014. https://doi.org/10.1155/2014/942043

May P, da Vinha VV (2013) Investing in sustainable use of biodiversity for social benefit in Brazil. In: Muradian R, Rival L (eds) Governing the provision of ecosystem services. Springer, pp. 21–47

Mendes M, de Almeida Silva M, da Silva Neves SM, Neves R, Junior SS (2014) A organização e a produção agroindustrial extrativista na fronteira Brasil–Bolívia, na região sudoeste de Mato Grosso. Rev Conex UEPG 10:140–149

Mendoza V, Santos L, Sanjinez-Argadoña E (2016) Elaboração de granola com adição de polpa e castanha de bacuri para consumo com iogurte. Evidência, Joaçaba 16:83–100. https://doi.org/10.18593/eba.v16i2.10382 ARTIGO

Menendez-Baceta G, Aceituno-Mata L, Tardío J, Reyes-García V, Pardo-de-Santayana M (2012) Wild edible plants traditionally gathered in Gorbeialdea (Biscay, Basque Country). Genet Resour Crop Evol 59:1329–1347. https://doi.org/10.1007/s10722-011-9760-z

MMA (2016) Espécies nativas da flora brasileira de valor econômico atual ou potencial : plantas para o futuro : região Centro-Oeste. Brasília

Molina M, Pardo-de-Santayana M, García E, Aceituno-Mata L, Morales R, Tardío J (2012) Exploring the potential of wild food resources in the Mediterranean region: natural yield and gathering pressure of the wild asparagus (*Asparagus acutifolius* L.). Spanish. J Agric Res 10:1090. https://doi.org/10.5424/sjar/2012104-3050

Moraes RM (2014) Palmeras útiles de Bolivia. Editores, Herbario Nacional de Bolibia/Plural, La Paz

Moraes RM, Borchsenius F, Blicher-Mathiesen U (1996) Notes on the biology and uses of the motacú palm (Attalea phalerata, Arecaceae) from Bolivia. Econ Bot 50:423–428. https://doi.org/10.1007/BF02866525

Nesbitt M, McBurney RPH, Broin M, Beentje HJ (2010) Linking biodiversity, food and nutrition: the importance of plant identification and nomenclature. J Food Compos Anal 23:486–498. https://doi.org/10.1016/j.jfca.2009.03.001

Noguchi DK, Nunes GP, Sartori ÂLB (2009) Florística e síndromes de dispersão de espécies arbóreas em remanescentes de Chaco de Porto Murtinho, Mato Grosso do Sul, Brasil. Rodriguésia 60:353–365. https://doi.org/10.1590/2175-7860200960208

Nogueira LMS, Figueiredo PS, Miyagusku L, Campos RP, Hiane PA, de Cássia Avellaneda Guimarães R, Aréco AET (2018) Characterization and sensorial acceptance of frozen yogurt formulated with Laranjinha de pacu (Pouteira glomerata (Miq.) Radlk) and probiotic cultures. Ambiência 14:174–185. https://doi.org/10.5935/ambiencia.2018.12.01

Nyström M, Jouffray JB, Norström AV, Crona B, Søgaard Jørgensen P, Carpenter SR, Bodin O, Galaz V, Folke C (2019) Anatomy and resilience of the global production ecosystem. Nature 575:98–108. https://doi.org/10.1038/s41586-019-1712-3

Oberg K (1949) Terena and the Caduveo of Southern Mato Grosso, Brazil. United States Government Printing Office, Washington

Oliveira CE da S, Zuffo AM, Fonseca WL, Steiner F (2019) Physical characterization of fruits and seeds Inga laurina (Sw.) Willd. (Fabaceae – Mimosaceae). J Neotrop Agric 6:105–111. https://doi.org/10.32404/rean.v6i3.3006

Oliveira JE (1996) Guató – Argonautas do Pantanal. Porto Alegre

Pacheco P, da Paz JG, da Silva CO, Pascoal GB (2014) Centesimal composition, bioactive compounds, physicochemical parameters in in natura jenipapo (*Genipa americana* L.). Demetra 9:1041–1054. https://doi.org/10.12957/demetra.2014.11310

Pardo de Santayana M, Morales R, Aceituno L, Molina M, Tardío J (2012) Etnobiología y biodiversidad: el Inventario Español de los Conocimientos Tradicionales. Ambienta 99:6–24

Peña R (1976) Flora Cruceña. Litografia y Imprentas Unidas, La Paz

Peña-Chocarro M, De Egea J, Vera M, Maturo H, Knapp S (2006) Guía de árboles y arbustos del Chaco Húmedo. The Natural History Museum, Guyra Paraguay, Fundación Moisés Bertoni, y Fundación Hábitat y Desarrollo, Asunción

Pennington TD (1997) The genus Inga botany. The Royal Botanic Gardens, Kew, London

Pereira RJ, Cardoso MDG (2012) Metabólitos secundários vegetais e benefícios antioxidantes vegetable secondary metabolites and antioxidants benefits. J Biotechnol Biodivers 3:146–152

Pinela J, Carvalho AM, Ferreira ICFR (2017) Wild edible plants: nutritional and toxicological characteristics, retrieval strategies and importance for today's society. Food Chem Toxicol 110:165–188. https://doi.org/10.1016/j.fct.2017.10.020

Pott A, Pott VJ (1986) Plantas comestíveis e medicinais da Nhecolândia. Pantanal, Corumbá

Pott A, Pott VJ (1994) Plantas do Pantanal. Embrapa, Corumbá

Pott VJ, Pott A (2000) Plantas Aquáticas do Pantanal. Embrapa. Centro de Pesquisa Agropecuária do Pantanal, Corumbá

Pott A, Pott VJ (In press) Flora of the Pantanal. In: Damasceno-Junior GA, Pott A (eds) Flora and vegetation of Pantanal Wetland. Springer, New York

Pott A, Pott VJ, Sobrinho AAB (2004) Plantas úteis à sobrevivência no Pantanal. IV Simpósio sobre Recur Nat e Sócio-econômico do Pantanal, pp 1–16

Prado DE, Gibbs P, Pott A, Pott VJ (1992) The Chaco – Pantanal transition in southern Mato Grosso, Brazil. In: Furley P, Proctor J (eds) Nature and dynamics of forest savanna boundaries, Chapman & Hill, London, pp 451–470

Prance GT(1989) Manual de Botânica Econômica, 253 p

Prance GT, Schaller GB (1982) Preliminary study of some vegetation types of the Pantanal, Mato Grosso, Brazil. Brittonia 34:228–251. https://doi.org/10.2307/2806383

Prates MFO, Campos RP, da Silva MMB, Macedo MLR, Hiane PA, Ramos Filho MM (2015) Nutritional and antioxidant potential of canjiqueira fruits affected by maturity stage and thermal processing. Ciência Rural 45:399–404. https://doi.org/10.1590/0103-8478cr20131272

Ramos MIL, Siqueira EMA, Isomura CC, Barbosa AMJ, Arruda SF (2007) Bocaiuva (Acrocomia aculeata (Jacq.) Lodd) improved vitamin A status in rats. J Agric Food Chem 55:3186–3190. https://doi.org/10.1021/jf063305r

Ramos MIL, Ramos Filho MM, Hiane PA, Braga Neto JA, de Almeida Siqueira EM (2008) Qualidade nutricional da polpa de bocaiúva Acrocomia aculeata (Jacq.) Lodd. Cienc e Tecnol Aliment 28:90–94. https://doi.org/10.1590/s0101-20612008000500015

Ramos M de LM, Rodrigues G da CG, Soares WRG, Hiane PA, Ramos MIL, Almeida JA de, Sanches FLFZ (2017) Suplementação com amêndoa de bacuri na composição corporal de ratos submetidos ao exercício. Rev Bras Med do Esporte 23:294–299

Siqueira EMA, Rosa FR, Fustinoni AM, de Sant'Ana LP, Arruda SF (2013) Brazilian savanna fruits contain higher bioactive compounds content and higher antioxidant activity relative to the conventional red delicious apple. PLoS One 8:2–8. https://doi.org/10.1371/journal.pone.0072826

Souza-Lima ES, Sinani TR, Pott A, Sartori ÂLB (2017) Mimosoideae (Leguminosae) in the Brazilian Chaco of Porto Murtinho, Mato Grosso do Sul. Rodriguesia 68:263–290. https://doi.org/10.1590/2175-7860201768131

Santos VS dos, Nascimento TV, Felipe JL, Boaretto AG, Damasceno-Junior GA, Silva DB, Toffoli-Kadri MC, Carollo CA (2017) Nutraceutical potential of Byrsonima cydoniifolia fruits based on chemical composition, anti-inflammatory, and antihyperalgesic activities. Food Chem 237:240–246. https://doi.org/10.1016/j.foodchem.2017.05.082

Salis SM, Mattos PP (2009) Floração e Frutificação da Bocaiúva (Acrocomia aculeata) e do Carandá no Pantanal. Comun técnico – EMBRAPA, pp 1–6

Sanjinez-Argandoña EJ, Chuba CAM (2011) Biometrical, physical and chemical characterization of bocaiuva (Acrocomia aculeata (Jacq.) lodd. ex mart) palm fruits. Rev Bras Frutic 33:1023–1028. https://doi.org/10.1590/S0100-29452011000300040

Santos K FD'eN, Silveira s, Martin-Didonet CCG, Brondani C (2013) Storage protein profile and amino acid content in wild rice oryza glumaepatula. Pesqui Agropecu Bras 48:66–72. https://doi.org/10.1590/S0100-204X2013000100009

Sartori ALB, Pott VJ, Pott A, De Carvalho FS (2018) Checklist of angiosperm from the Chaco of Mato Grosso do Sul. Iheringia – Ser Bot 73:22–33. https://doi.org/10.21826/2446-8231201873s22
Schmeda-Hirschmann G (1994) Plant resources used by the Ayoreo of the Paraguayan Chaco. Econ Bot 48:252–258. https://doi.org/10.1007/BF02862325
Schmidt M (1942) Estudos de Etnologia Brasileira. Nacional, Companhia Editorra, São Paulo
Schulz DG, Oro P, Volkweis C, de Matos Malavasi M, Malavasi UC (2014) Physiological maturity and morphometry of Inga laurina (Sw.) willd. seeds. Floresta e Ambient 21:45–51. https://doi.org/10.4322/floram.2014.007
Silva RH (2018) Estabelecimento de protocolo para uso sustentável de Copernicia alba Morong ex Morong & Briton. Universidade Federal de Mato grosso do Sul
Silva MLC, Costa RS, Dos Santos Santana A, Koblitz MGB (2010) Phenolic compounds, carotenoids and antioxidant activity in plant products. Semin Agrar 31:669–682. https://doi.org/10.5433/1679-0359.2010v31n3p669
Silva VM da, Campos RP, Borsato AV, Candido CJ, Donadon JR (2017) Bocaiuva jelly: preparation, physicochemical and sensory evaluation/Geleia de bocaiuva: elaboracao, avaliacao fisico-quimica e sensorial. Rev Bras Frutic 40. http://dx.doi-org.ez51.periodicos.capes.gov.br/10.1590/0100-29452018846 Resumo
Silva Baldivia D da, Sanjinez-Argandonã EJ, Antunes KÁ, Moraes ICF, Dos Santos EL, De Picoli Souza K (2018) The chemical composition and metabolic effects of *Attalea* phalerata nut oil in hyperlipidemic rats induced by a high-fructose diet. Molecules 23. https://doi.org/10.3390/molecules23040960
Silva Filho OJ da, Gomes Júnior NN (2020) O amanhã vai à mesa: abastecimento alimentar e COVID-19. Cad Saude Publica 36:e00095220. https://doi.org/10.1590/0102-311x00095220
Siqueira I, Pelegrin J, Candido C, Santos E, RCA Guimarães, Sanches F (2016) Composição nutricional do bacuri (Scheelea phalerata Mart.) e efeitos de diferentes processos de extração da polpa. In: XXV Congresso Brasileiro de Ciência e Tecnologia de Alimentos
Smith E, Ahmed S, Dupuis V, Running Crane M, Eggers M, Pierre M, Flagg K, Byker Shanks C (2019) Contribution of wild foods to diet, food security, and cultural values amidst climate change. J Agric Food Syst Community Dev 9:1–24. https://doi.org/10.5304/jafscd.2019.09b.011
Teixeira BS, Suzuki RM, Tonin LTD (2020) Capacidade antioxidantes e compostos bioativos dos frutos de Pouteria glomerata (laranjinha-de-pacu). Rev Tecnológica 29:291–308. https://doi.org/10.4025/revtecnol.v29i2.50511
Termote C, Van Damme P, Djailo B D'a (2011) Eating from the wild: Turumbu, Mbole and Bali traditional knowledge on non-cultivated edible plants, District Tshopo, DRCongo. Genet Resour Crop Evol 58:585–618. https://doi.org/10.1007/s10722-010-9602-4
Tomas WM, de Oliveira Roque F, Morato RG, Medici PE, Chiaravalloti RM, Tortato FR, Penha JMF, Izzo TJ, Garcia LC, Lourival RFF, Girard P, Albuquerque NR, Almeida-Gomes M, da Silva Andrade MH, Araujo FAS, Araujo AC, de Arruda EC, Assunção VA, Battirola LD, Benites M, Bolzan FP, Boock JC, Bortolotto IM, da Silva Brasil M, Camilo AR, Campos Z, Carniello MA, Catella AC, Cheida CC, Crawshaw PG, Crispim SMA, Junior GAD, Desbiez ALJ, Dias FA, Eaton DP, Faggioni GP, Farinaccio MA, Fernandes JFA, Ferreira VL, Fischer EA, Fragoso CE, Freitas GO, Galvani F, Garcia AS, Garcia CM, Graciolli G, Guariento RD, Guedes NMR, Guerra A, Herrera HM, Hoogesteijn R, Ikeda SC, Juliano RS, Kantek DLZK, Keuroghlian A, Lacerda ACR, Lacerda ALR, Landeiro VL, Laps RR, Layme V, Leimgruber P, Rocha FL, Mamede S, Marques DKS, Marques MI, Mateus LAF, Moraes RN, Moreira TA, Mourão GM, Nicola RD, Nogueira DG, Nunes AP, da Nunes da Cunha C, Oliveira MD, Oliveira MR, Paggi GM, Pellegrin AO, Pereira GMF, Peres IAHFS, Pinho JB, Pinto JOP, Pott A, Provete DB, dos Reis VDA, dos Reis LK, Renaud PC, Ribeiro DB, Rossetto OC, Sabino J, Rumiz D, Salis SM, Santana DJ, Santos SA, Sartori ÂL, Sato M, Schuchmann KL, Scremin-Dias E, Seixas GHF, Severo-Neto F, Sigrist MR, Silva A, Silva CJ, Siqueira AL, Soriano BMA, Sousa LM, Souza FL, Strussmann C, Sugai LSM, Tocantins N, Urbanetz C, Valente-Neto F, Viana DP, Yanosky A, Junk WJ (2019) Sustainability agenda for the Pantanal Wetland: perspec-

tives on a collaborative Interface for science, policy, and decision-making. Trop Conserv Sci 12. https://doi.org/10.1177/1940082919872634
UNICAMP (2011) Tabela brasileira de composição nutricional de alimentos – TACO, 4th edn. NEPA – UNICAMP, Campinas
Valko M, Rhodes CJ, Moncol J, Izakovic M, Manzur M (2006) Free radicals, metals and antioxidants in oxidative stress-induced câncer. Chemico-Biol Interact 160(1):1–40
Veiga Dias V da, Schultz G, Schuster M da S, Talamini E, Révillion JP (2015) The organic food market: a quantitative and qualitative overview of international publications. Ambient Soc 18:155–174. https://doi.org/10.1590/1809-4422ASOC841V1812015en
Vianna SA, Campos-Rocha A (2020) Acrocomia in Flora do Brasil 2020 em construção
Wittman H (2012) Food sovereignty: a new rights framework for food and nature? Environ Soc 2. https://doi.org/10.3167/ares.2011.020106

Part V
Consumption of Brazilian Food Plants

Biodiversity in Food Consumption Studies

Michelle Cristine Medeiros Jacob, Severina Carla Vieira Cunha Lima, Clélia de Oliveira Lyra, and Dirce Maria Lobo Marchioni

Contents

1 Introduction

In the last few decades, we have seen a tendency toward homogenization in human diets (Khoury et al. 2014). Diets that used to be composed of a wide variety of plants and animals have gradually shifted to include a limited number of species, focusing on very few crops and breeds. It is remarkable that while an estimated 30,000 edible plant species are available to humans, more than half of global food energy is currently met by only four crops: rice, potatoes, wheat, and maize (FAO 2010).

This tendency toward homogenization is problematic since low biodiversity can negatively impact human and environmental health (Willett et al. 2019). Broad evidence shows that food systems which neglect locally available nutrient-rich foods face the effects of simplified energy-dense diets, various forms of malnutrition, and a growing incidence of non-communicable diseases (Khoury et al. 2014; Herforth and Ahmed 2015). Agricultural biodiversity also increases food production resilience, especially in the face of sudden climate changes, disease outbreaks, and

M. C. M. Jacob (✉) · S. C. V. C. Lima · C. de Oliveira Lyra
Department of Nutrition, Federal University of Rio Grande do Norte, Natal, Brazil
e-mail: michellejacob@ufrn.edu.br

D. M. L. Marchioni
Department of Nutrition, Faculty of Public Health, University of São Paulo, São Paulo, Brazil
e-mail: marchioni@usp.br

M. C. M. Jacob, U. P. Albuquerque (eds.), *Local Food Plants of Brazil*, Ethnobiology, https://doi.org/10.1007/978-3-030-69139-4_15

market price fluctuations (Zimmerer 2015). Wild plant diversity, such as in unconventional food plants, is a strong component of food sustainability in many communities worldwide, especially during drought (Powell et al. 2015). The proper management and use of biodiversity can help to restore ecosystems and assist in resolving diet problems, such as micronutrient deficiencies in vulnerable populations (Ruel and Alderman 2013). However, it is worth mentioning that dietary and ecological indicators to assess the complex relationship between food biodiversity and quality of the diet must improve to provide greater accuracy (Lachat et al. 2018).

When assessing the impact of diet on human and environmental health, analysis of what people eat is crucial. In this chapter, our focus is on biodiverse food plants. The definition of biodiverse food plants (BFP) involves the agricultural diversity of conventional crops cultivated locally (e.g., varieties of beans, rice, corn, banana), together with unconventional food plants (UFP), (usually native, neglected, and rarely available in the marketplace) (Jacob and Albuquerque 2020). Despite the potential of these plants to promote sustainable food systems and diets, when we review the literature in dietary assessment studies, we notice a gap in methods for gathering and assessing the intake of these plants. This information gap is even more evident in studies developed for food systems with higher degrees of complexity (e.g., urban).

In this paper, we discuss certain challenges in assessing biodiverse food plants in dietary assessment and present (how to) proposals for step-by-step development. Finally, we summarize the actions needed to bring biodiverse food plants in dietary assessments mainstream. We hope that the debate concerning these limitations can help us to understand the real dietary contribution of these plants to the sustainability of the greater food system.

2 An Overview of the Challenges Involved in Biodiverse Food Plants' Dietary Assessment

To evaluate diet in free-living human beings is a complex task. Incorporating food biodiversity into dietary assessments creates an additional layer of complexity. Through field research, studying people of all life cycles, using the most varied dietary assessments, and analyzing the available literature, we noted the challenges seen in distinct phases of food composition studies that focus on biodiversity. In this topic, we will address these challenges in the three stages of dietary assessments: definition of goals, study design, and analysis.

A first critical step is to define the dietary assessment objectives. Usually, the dietary survey participants, especially in urban settings, are unable to even recognize the different varieties of vegetables they consume. In this sense, including photos of different food items might help the participants to accurately report the food they consume (FAO 2017). A photo book (or other kind of visual aid) containing food (and its varieties of interest), with different local names, might be a useful dietary assessment tool (in the next section, we will explore this in more detail).

However, to map biodiversity is not simple. For this reason, in determining the extent of information on biodiverse foods which might be included in this visual tool, the assessment objectives must be clear. The team should develop a priority list based on the criteria set forming the survey's objectives.

As an example, we bring a pilot study that was performed in ISA research (*Inquérito de Saúde de São Paulo*, in Portuguese) in 2013, where our goal was to assess the diversity of food intake in São Paulo. In this study, we verified that the population presented dietary inadequacy for vitamin A, vitamin C, iron, and folate. Therefore, these were our target nutrients. By consulting the data of a previous survey, we listed all of the fruits, legumes, and vegetables that contribute to the intake of these nutrients consumed by the target population. Subsequently, we crossed the list with data supplied by the Ceagesp warehouse, the biggest supplier of these food items in the city, to identify the species in at least three different cultivars (see Fig. 1).

A second principal step is to define the study's layout. For this task, there are currently a variety of manuals and guidelines on dietary assessment available. However, these same manuals highlight their own cross-cultural limits and the need for personnel with professional capacity in biodiverse food plants, culture, and eating patterns (FAO 2018). In a nutshell, current dietary assessment methods present limits in approaching the nutritional contribution of biodiverse food in diets (FAO 2017). Diets themselves embed cultural and environmental factors. Analysts of diets with a focus on biodiversity must be doubly aware of this. Biodiversity is linked to cultural practices and the given landscape (Kuhnlein 2014). In the same way that ethnography helps to build contextual understanding about a people and their culture, it can support us in building a more robust dietary assessment design (see, e.g., Tumilowicz et al. (2016)).

Fig. 1 Example of three varieties of bananas (*Musa acuminata* Colla x *Musa balbisiana* Colla) and apples (*Malus domestica* Borkh.) Available at the Ceagesp market in São Paulo. From top to bottom, left to right: banana-nanica, banana-ouro, banana-maçã, and maçã-gala, maçã-red-delicious, and maçã-golden. (Source: Ceagesp 2020)

Here we note that an ethnographic diagnosis, when focused on food and nutrition, is an ethnonutritional assessment. Ethnonutrition is the study of diets in the context of food systems of differing peoples and cultures. An ethnonutritional assessment functions as a useful way to rapidly calibrate tools and prototype dietary assessments while considering cultural and environmental peculiarities of the local food system (Jacob and Albuquerque 2020). For example, in a classic ethnonutrition study in Cameroon, de Garine (1972) shows us how dishes and table utensils influence eating practices. People in many traditional societies often eat together from a typical dish, and the amount of food eaten by each person varies according to sex, age, or other situations. Thus, an ethnonutrition assessment can help us to define the best strategy for data-gathering and help to prepare domestic utensil manuals and inventory meals and beverages that are part of local culture. It is also important to identify local culinary processing techniques that may produce nutritional modulation in food composition and nutrient bioavailability. For example, when reviewing studies concerning BFP in the Brazilian Caatinga, one of us identified local consumption of an UFP (mucunã), considered toxic because of chemical composition studies (Grant et al. 1986), (*Dioclea grandiflora* Mart. ex Benth) (Jacob et al. 2020). Mucunã seeds likely contain small molecular weight components that cause its toxicity and which can be eliminated through exhaustive dialysis (Grant et al. 1986). In the community we studied, people empirically knew this. For this reason, they performed specific culinary processing, washing the flour several times (they use the expression in Portuguese *lavar em várias águas*) and eliminating the plant's toxins. By revealing the contextual culinary processing, previous ethnonutrition assessment permitted us to avoid misinterpretations that might lead to wrong conclusions in the dietary assessment.

The third step is analysis. Data on biodiverse food plants are scarce (FAO 2017), especially for plants identified below the species level (e.g., variety, cultivar). This lack of information is problematic, because when considering composition data for a specific variety, it promotes misinterpretation of the nutritional contribution of foods (FAO 2012).

As an example of this, consider the banana varieties *Musa acuminata* Colla x *Musa balbisiana* Colla, both available in the principal food composition table used in Brazil, the *Tabela de Composição de Alimentos* (TACO/Unicamp). Considering the data from this source, we find 32% more potassium in banana-figo than in banana-maçã; we also find 62% more vitamin C in banana-da-terra than in banana-nanica. The genetic diversity of foods affects nutrient profiles. In the past, generic food composition data were considered sufficient for most research purposes, but now the need for cultivar-specific composition data is increasingly being acknowledged (Toledo and Burlingame 2006). When nutrient contents are significantly different in varieties of the same species, the foods should be reported independently in food composition databases and in other printed materials (such as food labels), presenting their unique nutrient profiles (Burlingame et al. 2009). The information should also be shared with the community in health and agriculture programs and food and nutrition education initiatives and incorporated as well into national health and agriculture policies.

Food composition also varies with agro-ecological zone (soil, climate, management techniques) and seasonality (Hunter et al. 2019). For example, some studies show that foods grown in agro-ecological or organic systems present higher absolute levels of micronutrients (Hunter et al. 2011) and bioactive compounds, such as polyphenols, phenolic acids, isoflavones, stilbene, and anthocyanins (Barański et al. 2014).

The lack of food composition data for UFP is another challenge. Recently, we performed a systematic review of biodiverse food plants in the Caatinga (Jacob et al. 2020), and from 15 studies, we built a list of 65 plants. To date, of these 65 plants, only 3 have accessible nutritional data in the Brazilian Food Composition Table – TACO. We have data for another four in the *Tabela Brasileira de Composição de Alimentos* (TBCA/USP), and the nutritional data of six other plants are available at the Information System about the Brazilian Biodiversity (SiBBr, in Portuguese). Yet we still lack nutritional data for 80% of the biodiverse food plants consumed by people in Caatinga.

One explanation for this lack of information is the cost of generating food composition data. During dietary assessment analysis, the process of determining the nutritional value of biodiverse food plants not covered in national food tables commences with a literature review and continues with laboratory studies to generate new data (Nesbitt et al. 2010). We recommend defining priorities and securing reliable identification of food plants before performing the analysis. Food matching strategies that relate data available in the food composition table and other sources lacking food data are an option in cases where proper analysis is not available (FAO 2012).

In addition to these challenges as related to biodiversity in dietary assessments, we bring others. For example, food systems in peri(urban) areas tend to have a higher degree of complexity than in rural areas, considering both (1) the intricate supply chains of retail outlets and urban markets and (2) the higher contribution of processed foods to diets (Lachat et al. 2018). These represent real challenges to mapping biodiverse food chains and to developing the dietary assessment itself. Another difficulty is how to embed this complexity into methods and research presuppositions. We already know that there is a relationship between the number of species consumed (species richness, a biodiversity indicator) and the quality of the diet (mean adequacy ratio, a nutritional indicator) (Lachat et al. 2018). However, we still need to better understand how this diversity translates into health indicators, especially in urban areas, where the dietary matrix changes due to the significant contribution of industrially processed foods. How do these foods interfere with the bioavailability of nutrients in biodiverse foods? We do not as yet know.

The fact is that we need methods of dietary assessment that address the complex set of continuous interrelated variables involved in epidemiological investigations (Sampson 1985; Beaton et al. 1997; Willett 2013; Mumu et al. 2020). Thus, methods normally recognized as the gold standard in clinical research, such as randomized controlled trials (RCT), are not the most suitable to evaluate the effect of dietary intake in epidemiological nutritional outcomes (Zeilstra et al. 2018). For example, one of RCT's presuppositions is uniformity, which in nutrition research in general is difficult to achieve. Consider that the composition of foodstuffs can vary substantially across the research setting, and consider the effect of population non-uniformity (where the non-uniformity might also be unknown). Complex interactions influence nutrient consumption and thus physiological exposure. Bad assumptions can lead to ambiguity in similar experiments, yielding results that are non-reproducible. Here, we are not discussing the RCT method. However, we can see that in nutritional research, the critical step is to verify whether a presupposition can be considered valid. We must understand both the culture and environment as being critical factors for food biodiversity analysis and as potential modifiers of the inquiry's presuppositions.

3 Assessing Biodiverse Food Plants in Dietary Intake Surveys: How to Do It?

The United Nations Food and Agriculture Organization recently released guidelines to assess biodiverse foods in dietary surveys. In Fig. 2, the flowchart summarizes the main steps for dietary assessment with a focus on food biodiversity. In this section, we will address certain critical steps for this task: elaboration of a tool to support dietary inquiries, generally a photo book (step 6), and the selection and adaptation of a dietary intake instrument (steps 7–9).

To elaborate a photo book, the main recommendations are (1) build a multidisciplinary team, composed of an anthropologist, a nutritionist, food composition experts, and a photographer; (2) analyze food intake patterns, nutritional problems in the studied population, and the available biodiversity; (3) define which biodiverse foods to include in the survey; and (4) ensure correct identification of the biodiverse food plants. The final food list in the dietary assessment tool must combine all foods and varieties of interest and all of the different local food names used. With the food list, vernacular names, correct scientific identification, and photos in hand, we have the material to start production of the photo book.

One of us recently developed a photo book to support a dietary assessment in a Caatinga community (the process is summarized in Fig. 3). We started by building a plant list with plants of interest in the semiarid region, which is mainly in the Caatinga Brazilian biome. We performed a systematic review of biodiverse food plants in the region (Jacob et al. 2020), focusing on UFP. To include conventional plants, we consulted data from national research concerning food consumption (Instituto Brasileiro de Geografia e Estatística 2011). And finally, to gather data on agriculture biodiversity, we visited some local markets. With a plant list ready, together with botanists and photographers, we selected photos. At this stage, we only included photos already identified by specialists, and we also performed two additional identification processes, one with local experts and another with two independent botanists to ensure the accuracy of the images. Our final plant list

Fig. 2 How to assess biodiverse food plants in dietary intake surveys? The flowchart summarizes the main steps for dietary assessments with a focus on food biodiversity. (Adapted from FAO (2017))

contained 135 items. We chose a minimal design for our photo book, with the plants organized by their vernacular names to easily register the data during the interviews. With this first version in hands, we went to the field to test our tool in an ethnonutrition assessment.

Through mapping of the biodiverse foods available in our study scenario, we started with a list of 135 plants and yet finished the ethnonutrition assessment with 156 plants. The new plants added were mostly agricultural varieties of beans, manioc, orange, mango, and corn (for correct identification of plants, see Nesbitt et al. 2010). We did not include new UFP species since new species were not mentioned in the assessment. On the contrary, we perceived that the community might well be facing lessened UFP consumption, with an increase in highly processed foods. Further, with the community's feedback, we changed certain pictures (giving greater attention to botanical identification keys such as flowers and fruits), and we updated vernacular names to the more common names used in the community. For example, when we asked if people consumed cumaru (*Amburana cearensis* (Allemão) A.C.Sm.), they told they did not, because they only recognized this plant, the same plant, as amburana-açu. Such vernacular name adjustments were more significant for agricultural varieties than in other plant species. For example, banana-da-terra (*Musa acuminata* Colla x *Musa balbisiana* Colla "terra" (AAB)), one kind of banana reported in the Brazilian national *Pesquisa de Orçamentos Familiares*, is

Fig. 3 Steps to prepare a photographic guide of locally available biodiverse foods. The final product, the photo book with images and names of food resources, is a supportive tool that assists in accurate identification of foods informed during dietary intake surveys. (Source: Authors' elaboration)

known locally not as banana-da-terra but as banana-comprida. Thus, after the ethnonutritional assessment, we had a more precise guide (quality of pictures, varieties of food, and proper ethnotaxonomy) to proceed with the dietary inquiry. In the case of Brazil, even considering the complexity of populational dietary assessments in urban settings, we suggest that a similar approach can be applied to samples in all Brazilian biomes.

Another step during dietary assessments focusing on biodiversity is selection and adaptation of a dietary intake instrument. A common approach used in extensive epidemiologic studies has been the application of two instruments: the Food Frequency Questionnaire (FFQ) and the 24-hour Dietary Recall (applied to a subsample to evaluate and calibrate the FFQ). Calibration consists of adjusting the FFQ to be more like the Recall results and thus (presumably) be less biased. Another example is administering one or more 24-hour Dietary Recall instruments to the full sample, as a principal instrument in combination with the FFQ, a tool that better sheds light on foods consumed less frequently, such as seasonal plants and meals (Thompson et al. 2015).

We can describe a population group's usual intake with the 24-hour Dietary Recall instrument if we apply it more than once. Since it is an open method, it is effective in cross-cultural contexts, allowing the participant to name a mealtime, a food item, or a local recipe (FAO 2017). Nevertheless, this method has weaknesses. It has a low capacity to recognize food patterns over time. For example, if a food item is sparsely eaten, there is a high chance that the item will not be cited if the survey is performed only once or twice. Many survey days might be needed to register the consumption of certain items, making the research costly and increasing the participant's burden.[1] On the other hand, a traditional FFQ, being popular in epidemiological studies, relies on a finite list (covering 80–90% of nutrient intake) of research interest (Mota et al. 2019). The FFQ is used to rank nutrient intake, biocomponents, groups, or food group present in the diet.

Another recommendation for food intake surveys it is to combine the 24-hour Dietary Recall with a Food Propensity Questionnaire (FPQ), with the aid of a photo album or including photos in the platform used to collect the dietary intake data. The FPQ aims to complement information from the 24-hour Dietary Recall instrument to minimize biases related to daily variations in food consumption, which are present if we obtain only a single observation, or the observations are made in short intervals, affecting estimated habitual consumption (Subar et al. 2006).

Finally, we note the process of adapting tools. Adapting 24-hour Dietary Recall involves adding two new columns: one to accurately identify the biodiverse food mentioned by the interviewee and another to match each plant identified with its code in the photo book (FAO 2017). In the case of the FPQ, researchers should use the plant list as elaborated in the photo book development process.

One of us had the opportunity to test both the 24-hour Dietary Recall and FFQ instruments in an ethno-national assessment in a Caatinga community. After testing

[1] Currently, the 24-hour Dietary Recall instrument is supported by digital platforms, with an ample (but not exhaustive) list of foods.

these tools in the field, we decided to add two new columns to the FFQ. The first one concerned seasonality (yes or no). We added it to quickly identify seasonal plants during interviews and to accurately convert the consumption frequency for dietary analysis. We also added a column to identify plants that act as food and medicine at the same time. In our analysis, the participants presented to us certain plants of medicinal value (e.g., jatobá, *Hymenaea courbaril* L.) that they consumed as a tea at breakfast or dinner, even if not being sick, which characterizes a food medicine continuum (Ferreira-Júnior et al. 2018). This (non-binary) system of food classification is different from ours, and our first instinct in the field was to exclude such plants from our food plant list, due to our own biases. However, we re-analyzed our wrong presuppositions and decided to adapt the FFQ. We added a new column to register every plant that possessed a local non-binary identity (food medicine continuum, *yes* or *no*). This measure allowed us to identify natural resources that act as both food and medicine. It also prevented undesirable dietary interpretations that result from excluding plants that might have a nutritional impact on the diet. Depending on each investigation's purposes, the research team may find it necessary to develop new adaptations. In every case, we highly recommend previous ethnonutrition assessment of instruments before use in dietary assessment at the populational level.

4 Fundamental Actions to Mainstream Biodiverse Food Plants in Dietary Assessments

At this point, it is clear that mainstreaming biodiverse food plants in dietary assessments involves several challenges. How to overcome them? Jacob and Albuquerque (2020) enumerate four main gaps that may help us to design an interdisciplinary research agenda for nutritionists, ethnobiologists, anthropologists, and other scientists to consider biodiverse food resources and the cultural and environmental factors attached to them in our dietary assessments.

The first gap is the lack of accessible food biodiversity data. Brazil has an estimated flora of 48,771 species, including algae, angiosperms, bryophytes, fungi, gymnosperms, ferns, and lycophytes (Instituto de Pesquisas Jardim Botânico do Rio de Janeiro 2020), and we still do not know much about the edible potential of this flora. Due to the accelerating process of biodiversity loss that the world faces (Bongaarts 2019), we must act quickly. The second gap is the lack of culinary data in ethnobiology studies. Ethnobiology studies offer nutrition science-relevant data on food biodiversity and its impact on local food systems (Kuhnlein 2014). However, many studies in this area fail to present how food is prepared before being eaten, which we call ethno-culinary data. Actions such as washing, soaking, and peeling, to mill, to heat, to roast, to boil, to infuse, to germinate, to ferment, to cure, to preserve, and to dehydrate, or combinations of these increase nutrient bioavailability and inactivate or reduce anti-nutritional factors. These need to be collected.

Without such registration, we cannot use ethnobiological data to determine whether higher biodiversity consumption correlates with better quality diets and food and nutrition security. The third gap is the scarcity of nutritional data, which we already explored in this chapter. Finally, the authors highlight the need for interdisciplinary methods and interprofessional research teams.

We see the cross-disciplinary gap mentioned by the authors as a starting point for action. Cross-disciplinary approaches favor construction of methods that resonate with the dialectic between nature and culture in food plants. Classic works by nutritionists such as Harriet Kuhnlein (1991), biologists such as Nina Etkin (2006), and anthropologists such as Audrey Richards (1948) remind us that high-level scientific dietary assessment research is not a matter of a single science. Multidisciplinary academic efforts must begin in the undergraduate years (FAO 2014). Nutrition students must think about food as more than just nutrients but as landscapes, as biodiversity, and as cultural products. Ethnobiologists must think about dietary processing and diets. Anthropologists must think about nature, diets, and ecology. Agronomists must think about traditional systems and how their decision-making as professionals affects people and nature (Powell et al. 2015). All of these professionals are invited to understand the complex biocultural factors of diet and to embrace food systems thinking (IPES-Food 2015). Professors need to face the contemporary challenge of fragmentation in science and create the space for interprofessional action to advance biodiverse food plant science.

Finally, we wish to highlight the window of opportunity that "Citizen Science" opens for us to gather data on biodiverse food and its associated traditional knowledge. Citizens can participate actively in the advances and discoveries of science. This participation should be informed, conscious, and voluntary. Some steps toward success in this endeavor will be (1) the enhancement of tools for gathering and processing data, (2) the appreciation of projects with real effects in people's lives, and (3) the creation of research consortiums (Bonney et al. 2014; Heigl et al. 2019). To win the respect of participants, scientists, and policymakers, citizen-based science projects need to be responsibly planned and executed with precision. One example of how this might happen is the CONECT-e initiative,[2] an interactive platform, in wiki format, where people can consult and provide traditional knowledge as related to plants, animals, fungi, and traditional species varieties. The primary mission of CONECT-e in Spain is to preserve and spread traditional ecological knowledge. Similar approaches to map food biodiversity, using participatory methodologies with the support of family farmers, rural communities, and others, can be applied. For instance, the use of geographic information (Fast and Rinner 2018), volunteered from crowd-mapping services, might well be an appropriate solution.[3] We see citizen science as strategic, because it offers new approaches to realize the biodiverse food plant research agenda.

---

[2] See the CONECT-e platform at https://conecte.es/

[3] See, for example, this experience to map food markets using community participation in a municipality from Brazil: www.nutrir.com.vc/mapa

5 Conclusion

In this chapter, we offered various reflections involved in the effort to include cultural and environment variables in nutritional analyses and a set of methods and tools to do so in the field of food consumption studies. We are conscious of the enormous challenges to application in research practice, principally in urban settings and in populational studies. However, we are also aware of how urgently we need to act. The ongoing pandemic outbreak of COVID-19 reinforces the necessity of integrating nutrition research with the environmental challenges imposed by food systems. Within the ecosystem, the nutritional environment is not separate from the biological environment. In other words, human health and environmental health are intrinsically connected, a fact we cannot ignore.

References

Barański M, Rednicka-Tober D, Volakakis N et al (2014) Higher antioxidant and lower cadmium concentrations and lower incidence of pesticide residues in organically grown crops: a systematic literature review and meta-analyses. Br J Nutr 112:794–811

Beaton GH, Burema J, Ritenbaugh C (1997) Errors in the interpretation of dietary assessments. Am J Clin Nutr 65. https://doi.org/10.1093/ajcn/65.4.1100S

Bongaarts J (2019) IPBES, 2019, Summary for policymakers of the global assessment report on biodiversity and ecosystem services of the Intergovernmental Science-Policy Platform on Biodiversity and Ecosystem Services. Popul Dev Rev. https://doi.org/10.1111/padr.12283

Bonney R, Shirk JL, Phillips TB et al (2014) Next steps for citizen science. Science 343:1436–1437. https://doi.org/10.1126/science.1251554

Burlingame B, Charrondiere R, Mouille B (2009) Food composition is fundamental to the cross-cutting initiative on biodiversity for food and nutrition. J Food Compos Anal 22:361–365. https://doi.org/10.1016/j.jfca.2009.05.003

Ceagesp (2020) Ceagesp Guide. http://www.ceagesp.gov.br/guia-ceagesp. Accessed 1 Aug 2020

de Garine I (1972) The socio-cultural aspects of nutrition. Ecol Food Nutr 1:143–163. https://doi.org/10.1080/03670244.1972.9990282

Etkin N (2006) Edible medicines: an ethnopharmacology of food. The University of Arizona Press, Tucson

FAO (2010) Second report on the state of the world's plant genetic resources for food and agriculture. FAO, Rome

FAO (2012) INFOODS: guidelines for food matching 1.2. FAO, Rome

FAO (2014) Second international conference on nutrition. FAO, Rome

FAO (2017) Guidelines on assessing biodiverse foods in dietary intake surveys. FAO, Rome

FAO (2018) Dietary assessment: a resource guide to method selection and application in low resource settings. FAO, Rome

Fast V, Rinner C (2018) Toward a participatory VGI methodology: crowdsourcing information on regional food assets. Int J Geogr Inf Sci 32:2209–2224. https://doi.org/10.1080/13658816.2018.1480784

Ferreira-Júnior W, Campos L, Pieroni A, Albuquerque U (2018) Bases biológicas e culturais do uso de plantas alimentícias e medicinais. In: Albuquerque UP (ed) Etnobiologia: bases ecológicas e evolutivas, 2nd edn. NUPEEA, Recife, pp 173–188

Grant G, McKenzie N, Moreira R, Pusztai A (1986) Dioclea grandiflora and Dioclea sclerocarpa seeds. A nutritional study. Qual Plant Plant Foods Hum Nutr 36:47–61

Heigl F, Kieslinger B, Paul KT et al (2019) Toward an international definition of citizen science. Proc Natl Acad Sci U S A 116:8089–8092. https://doi.org/10.1073/pnas.1903393116

Herforth A, Ahmed S (2015) The food environment, its effects on dietary consumption, and potential for measurement within agriculture-nutrition interventions. Food Sec 7:505–520

Hunter D, Foster M, Mcarthur JO et al (2011) Evaluation of the micronutrient composition of plant foods produced by organic and conventional agricultural methods. Crit Rev Food Sci Nutr 51:571–582. https://doi.org/10.1080/10408391003721701

Hunter D, Borelli T, Beltrame DMO et al (2019) The potential of neglected and underutilized species for improving diets and nutrition. Planta 250:709–729. https://doi.org/10.1007/s00425-019-03169-4

Instituto Brasileiro de Geografia e Estatística (2011) Pesquisa de orçamentos familiares 2008–2009: análise do consumo alimentar pessoal no Brasil. Brasília

Instituto de Pesquisas Jardim Botânico do Rio de Janeiro (2020) Flora do Brasil 2020 – Algas, Fungos e Plantas. http://reflora.jbrj.gov.br/. Accessed 1 Dec 2019

IPES-Food (2015) The new science of sustainable food systems. Belgium

Jacob M, Albuquerque U (2020) Biodiverse food plants: Which gaps do we need to address to promote sustainable diets? Ethnobiol Conserv 9:1–7. https://doi.org/10.15451/ec2020

Jacob M, Medeiros M, Albuquerque U (2020) Biodiverse food plants in the semiarid region of Brazil have unknown potential: a systematic review. PLoS One. https://doi.org/10.1371/journal.pone.0230936

Khoury CK, Bjorkman AD, Dempewolf H et al (2014) Increasing homogeneity in global food supplies and the implications for food security. Proc Natl Acad Sci U S A 111:4001–4006. https://doi.org/10.1073/pnas.1313490111

Kuhnlein H (1991) Traditional plant foods of Canadian indigenous peoples: nutrition, botany and use. Gordon and Breach Publishers, Quebec

Kuhnlein HV (2014) How ethnobiology can contribute to food security. J Ethnobiol 34:12–27. https://doi.org/10.2993/0278-0771-34.1.12

Lachat C, Raneri JE, Smith KW et al (2018) Dietary species richness as a measure of food biodiversity and nutritional quality of diets. Proc Natl Acad Sci 115:127–132. https://doi.org/10.1073/pnas.1709194115

Mota V, Lima S, Lyra C (2019) Questionário de Frequência Alimentar. In: Marchioni DML, Gorgulho BM, Steluti J (eds) Consumo Alimentar: guia para avaliação. Manole, São Paulo, pp 8–23

Mumu SJ, Merom D, Ali L et al (2020) Validation of a food frequency questionnaire as a tool for assessing dietary intake in cardiovascular disease research and surveillance in Bangladesh. Nutr J 19. https://doi.org/10.1186/s12937-020-00563-7

Nesbitt M, Mcburney RPH, Broin M, Beentje HJ (2010) Linking biodiversity, food and nutrition: the importance of plant identification and nomenclature. J Food Compos Anal 23:486–498. https://doi.org/10.1016/j.jfca.2009.03.001

Powell B, Thilsted SH, Ickowitz A et al (2015) Improving diets with wild and cultivated biodiversity from across the landscape. Food Sec 7:535–554. https://doi.org/10.1007/s12571-015-0466-5

Richards A (1948) Hunger and work in a savage tribe. Greenwood Press, Westport

Ruel MT, Alderman H (2013) Nutrition-sensitive interventions and programmes: how can they help to accelerate progress in improving maternal and child nutrition? Lancet 382:536–551. https://doi.org/10.1016/S0140-6736(13)60843-0

Sampson L (1985) Food frequency questionnaire as a research instrument. Clin Nutr 4:171–178

Subar AF, Dodd KW, Guenther PM et al (2006) The food propensity questionnaire: concept, development, and validation for use as a covariate in a model to estimate usual food intake. J Am Diet Assoc 106:1556–1563. https://doi.org/10.1016/j.jada.2006.07.002

Thompson FE, Kirkpatrick SI, Subar AF et al (2015) The National Cancer Institute's dietary assessment primer: a resource for diet research. J Acad Nutr Diet 115:1986–1995. https://doi.org/10.1016/j.jand.2015.08.016

Toledo A, Burlingame B (2006) Biodiversity and nutrition: a common path toward global food security and sustainable development. J Food Compos Anal 19:477–483

Tumilowicz A, Neufeld L, Pelto GH (2016) Using ethnography in implementation research to improve nutrition interventions in populations. Matern Child Nutr 11:55–72

Willett W (2013) Nutritional epidemiology, 3rd edn. Oxford University Press, New York

Willett W, Rockström J, Loken B et al (2019) Food in the Anthropocene: the EAT–Lancet Commission on healthy diets from sustainable food systems. Lancet 393:447–492. https://doi.org/10.1016/S0140-6736(18)31788-4

Zeilstra D, Younes JA, Brummer RJ, Kleerebezem M (2018) Fundamental limitations of the randomized controlled trial method in nutritional research: the example of probiotics. Adv Nutr. https://doi.org/10.1093/ADVANCES/NMY046

Zimmerer KS (2015) Understanding agrobiodiversity and the rise of resilience: analytic category, conceptual boundary object or meta-level transition? Resilience. https://doi.org/10.1080/21693293.2015.1072311

Wild Food Plant Popularization and Biocultural Conservation: Challenges and Perspectives

Patrícia Muniz de Medeiros, Déborah Monteiro Barbosa, Gabriela Maria Cota dos Santos, and Rafael Ricardo Vasconcelos da Silva

Contents

1 Introduction

The world is facing a continuous process of food homogenization and simplification, as only three crops (corn, wheat, and rice) account for 51% of the world's plant food production (FAO 2019). Such a process may lead to nutrient deficiency and excessive energy consumption (Johns and Eyzaguirre 2006).

For this reason, current policies need to focus on the diversification of food systems, and the popularization of wild food plants (WFPs) can be a promising option to achieve this purpose. This group of plants is widely consumed in many rural communities throughout the world and may play a crucial role in providing food security for these populations (Neudeck et al. 2012). Moreover, its dissemination to novel consumers, for example, in urban areas, could have enormous potential to account for many of the Sustainable Development Goals proposed by the United Nations (UN 2015).

First, studies have shown that wild food plants play an important role in meeting people's nutritional needs (Turan et al. 2003; Flyman and Afolayan 2006);

P. M. de Medeiros (✉) · D. M. Barbosa · G. M. C. dos Santos · R. R. V. da Silva
Laboratory of Biocultural Ecology, Conservation and Evolution (LECEB), Campus of Engineering and Agricultural Sciences, Federal University of Alagoas, Rio Largo, AL, Brazil
e-mail: patricia.medeiros@ceca.ufal.br; rafael.vasconcelos@ceca.ufal.br

M. C. M. Jacob, U. P. Albuquerque (eds.), *Local Food Plants of Brazil*, Ethnobiology, https://doi.org/10.1007/978-3-030-69139-4_16

therefore, their consumption could improve the nutritional quality of diets in urban areas. Second, as they grow spontaneously in anthropic, agricultural, forest, and other environments, WFPs are usually free from conventional agricultural management based on the use of pesticides and fertilizers. This issue is especially important in countries that still strongly rely on unsustainable food production, such as Brazil (Soares and Porto 2009). Finally, the spread of WFP consumption can increase the income of small producers and extractors around the world (Delang 2006), as they are placed near these resources and have knowledge about WFP management techniques.

Several investigations have demonstrated the WFP potential for sustainable harvest (see, e.g., Holm et al. (2008) and Emanuel et al. (2005)) and human well-being (Kusters et al. 2006). However, this expected contribution may prove wrong under some circumstances, especially in contexts of overharvesting and harmful management techniques.

By adopting a biocultural conservation approach, we discuss the main arguments favorable to the maintenance and increase of WFP harvesting, as well as threats to their sustainable use. We present some examples that are specific to WFP use and others that consider the broader concept of non-timber forest products (NTFP). We also examine the case of the Brazilian *açaí* to discuss challenges to WFP popularization.

2 Wild Food Plants and Biocultural Conservation

Ethnobotanical literature has provided arguments favorable to the maintenance of WFP collection practices, especially in forest environments. These arguments are aligned with the concept of biocultural conservation, which has well-established principles and approaches (Gavin et al. 2015). Its central premise is to consider not only biodiversity maintenance but also the whole body of knowledge, practices, and symbols associated with it (Gama et al. 2018).

The most common hypothesis is that WFP (or, in most cases, general NTFP) harvesting by local communities may contribute to forest conservation since its domestic or commercial importance prevents the suppression of forested areas (Ros-Tonen 2000; Lowore 2020). The central idea is that the practical and economic values generated by the extraction of such resources are more relevant than the benefits that could be produced with deforestation and land-use change. Some studies use the term "conservation by commercialization" to refer to this hypothesis (Evans 1993; Lowore 2020). The main challenge related to this hypothesis is how to effectively test it, which is why most studies can only make inferences or provide anecdotal evidence on the subject.

Another source of relevant information concerning the importance of wild food plants comes from domestication studies. Whether dealing with current management strategies or trying to recover domestication paths of cultivated plants, some investigations have suggested that people may either tolerate plants with desirable characteristics (e.g., edible species) during deforestation, promote them in other areas, or protect them from herbivores, parasites, and competing plants (Casas et al. 1997). This means that the importance of a species for food purposes may contribute to its own maintenance in the environment. However, little is known regarding the protective effect that food use may have on other (more harmful) plant uses (e.g., wood uses).

In this regard, there is evidence of the protective effects of other non-timber forest products. A study developed with medicinal plants in a semiarid area of northeast Brazil has shown that controlling for the species' availability and perceived efficiency, the most popular medicinal plants are less used for wood purposes (Silva et al. 2021). This phenomenon is even stronger for the top five medicinal species. It is possible that the domestic and commercial importance of wild food plants also prevents species from being harvested for wood purposes, but this hypothesis needs to be formally tested. To date, a previously established association between use categories in another community from the Brazilian semiarid region has not found a correlation between the importance for food and wood purposes and that for medicinal and wood purposes (Barbosa et al. 2020). However, this study was not planned to test the protection hypothesis but rather to find associations between use categories and local perceptions of the decline in plant populations (mostly based on multinomial variables from participatory workshops).

An eventual protective effect of food uses on wood uses, together with strategies to achieve the sustainable use of wild food plants, would contribute to the maintenance of WFP populations. Moreover, the continuity of harvesting practices may also have a protective effect on the whole set of forest species, including those with little or no edible importance. This may be the case since studies have suggested that the abandonment of WFP harvesting may negatively affect income generation in local populations, thereby leading to a higher dependence on other forest resources, which is often more harmful to plant populations than the food use itself (Delang 2006).

In addition to the arguments concerning a protective effect of WFP harvesting, these species may play a fundamental role in reducing the vulnerability of social-ecological systems to climate change. This hypothesis is based on evidence that agricultural and livestock practices, especially those related to small-scale agriculture, are being particularly affected by scenarios of climatic uncertainties since small farmers often rely exclusively on rain-fed agriculture (see, e.g., Gbetibouo et al. 2010). Several wild food plants, in turn, typically occur in natural environments, especially native forests, which are far more efficient than agroecosystems in regard to retaining water (Perry 1994). However, as is the case for the previously introduced hypothesis, although some studies make inferences about the potential contribution of wild food plants in contexts of climate change (Gradé 2012), we need to employ proper research designs to test this hypothesis.

Concerning the human dimension, the popular claim that WFP commercialization may be responsible for positive social, financial, and human assets has been proven correct for several social-ecological contexts. By analyzing 61 case studies (15 with food plants/spices), Kusters et al. (2006) found that, for almost all of them, trade has a positive impact on local livelihoods. The main challenge in this case is to ensure that significant increases in the popularity of certain WFPs also revert to socioeconomic improvements.

Another common argument favorable to the dissemination of WFP knowledge is that it may contribute to the cultural valuation of the societies that possess knowledge about their management and preparation techniques. This may be even more important under certain circumstances, considering that, in several communities, the consumption of WFP is associated with "tribalness," poverty, and difficult times (Narayanan et al. 2004; Nascimento et al. 2012; Cruz-Garcia and Howard 2013).

3 The Other Side: When WFP Consumption and Popularization Do Not Achieve Sustainable Development Goals

Although WFP harvesting has the potential to conciliate biodiversity conservation, income generation, and human well-being, under several circumstances, overharvesting, together with harmful harvesting patterns, may lead to structural alterations in plant populations (see, e.g., Sinha and Bawa 2002). A study developed in the Brazilian state of Pará showed that chronic anthropogenic disturbance in forest areas could be as (or even more) harmful as deforestation in terms of biodiversity loss (Barlow et al. 2016). Therefore, the naive general notion that NTFP harvesting is always sustainable must be avoided, and proper diagnostics of social-ecological systems need to be conducted to aid sustainable harvesting.

The conciliation between biodiversity conservation and human well-being may not always take place. The previously mentioned investigation of NTFP performed by Kusters et al. (2006) using 61 case studies (15 with food plants/spices) found that, under certain circumstances, higher livelihood outcomes are associated with lower environmental outcomes. The authors also found that the commercial extraction of NTFP may lead to resource depletion, especially in cases of natural areas without further management strategies (e.g., seed propagation, protection of the target species from competitors, etc.).

The obvious conclusion is that WFP harvesting needs to be accompanied by target species management in natural areas to help achieve sustainability, especially in contexts of commercial use. However, a "management dilemma" may occur. On the one hand, management strategies for the target WFP could help maintain its population. On the other hand, when management is intense toward a single (economically important) species, this may lead to the restructuring of the other components of the community (e.g., floristic and functional simplification of natural environments). We will further use the case of *açaí* as an example of this phenomenon. Such dilemmas indicate that finding the best strategy may be more challenging than what is commonly stated in studies with WFP.

Furthermore, population maintenance itself may not always mean that harvesting strategies are sustainable. When species have dual reproductive strategies (clonal and with seeds), intense fruit harvesting may decrease recruitment and cause clonal reproduction to gain importance. Gaoue et al. (2018), for example, found that the proportion of clonal offspring was higher for moderately and heavily harvested populations of *Pentadesma butyracea* Sabine (Clusiaceae) when compared to those in more lightly harvested areas. Although the population is being maintained, the predominance of clonal reproduction may lead to a decrease in genetic diversity and, consequently, affect the population's ability to cope with environmental change.

In terms of the socioeconomic benefits of WFP extraction, a critical constraint has to do with land use and tenure. Data from communities in 24 developing countries concerning wild foods (including animals and mushrooms) showed, for example, that state land is the primary source of income derived from wild food harvesting in forests (Hickey et al. 2016). Therefore, as they are not private lands, local harvesters may face challenges such as competition with "outsiders" for resource collection and a lack of control under the harvesting area, especially when wild foods are gaining in popularity.

Moreover, some researchers have challenged the general idea that WFP harvesting may increase resilience in social-ecological systems and decrease vulnerability to climate change. A study developed in two South African villages, for example, found that, according to local perceptions, wild food availability suffers seasonal variations and decreases with extreme events (Paumgarten et al. 2018). However, the authors also indicated that some specific fruit species are drought-resistant and perceived as essential when coping with extreme events.

Finally, many studies on NTFP have pointed out challenges for commercialization beyond the local level. Belcher and Schreckenberg (2007), for example, discussed that (1) production is often spread, and markets are poorly developed; (2) markets are faddish, and some "fashion" products may gain and then lose space; (3) sustaining supply could be a problem since volumes are typically small; (4) sometimes high technology is needed to benefit certain products; (5) each destination industry has its own marketing requirements, and it may be difficult to meet all of them; (6) certification (e.g., environmental, health, etc.) requires a high level of social organization, as well as technical sophistication, which may be costly and difficult to acquire; and (7) there is a growing concern with intellectual property rights (e.g., how to compensate the holders of local knowledge about a product in case they are not involved in the commercialization).

4 Challenges for WFP Biocultural Conservation: The Case of *Açaí* (*Euterpe oleracea* Mart)

Açaí (*Euterpe oleracea* Mart.) is an outstanding example of how a wild food plant can reach overseas popularity. *Açaí* is a palm species endemic to South America. Brazil is the country that possesses most of the *açaí* populations, and the Brazilian state of Pará is the leading producer in the country. Although the systematic cultivation of *açaí* is increasing, approximately 80% of fruit production comes from its natural ecosystems (Vaz and Nabout 2016), with different management intensities.

Açaí has only recently reached national and international markets. In terms of consumption spread, it began as an indigenous food in pre-Colombian times. It then became a rural staple food, an urban staple food (in regional urban centers), and an urban fashion food (nationally spread) and finally reached the status of an international fashion food (Brondízio et al. 2002). It is currently the most collected NTFP in Brazil (IBGE 2017).

The species' popularization was strongly influenced by mediatization, and it was one of the first examples of how an Internet-based market can help to spread local products on a global scale (Heinrich et al. 2011). The national spread of *açaí* has also been related to the history of Brazilian Jiu-Jitsu and the Gracie clan, a family of practitioners (and the creators) of this sport (Vanni 2018). As Brazilian Jiu-Jitsu became internationally famous, the association between *açaí* consumption and sports practicing also spread outside the country. The species boom in North American markets also had to do with the mention of its claimed nutritional properties in a popular TV program, The Oprah Winfrey Show (Marcason 2009; Heinrich et al. 2011).

As the fruit is highly perishable, it cannot travel significant distances if traded *in natura*. Therefore, frozen fruit pulp has gained market value. Moreover, *açaí* has been popularized in a mixture with guarana syrup as part of a strategy to overcome food neophobia (aversion to eating unknown foods) by adding popular ingredients to novel foods. This mixture leads to a sweet preparation and removes the "earthy, dense, slightly metallic flavor of *açaí*" (Vanni 2018), thereby increasing its acceptance among new audiences.

Such a process raises the first issue in terms of biocultural conservation: a loss of connection between *açaí* and its cultural inheritance and significance. Therefore, the rise of the *açaí* national and international markets did not contribute much to the valuation of local cultural practices, as new consumers hardly know its biocultural origins (e.g., that it is traditionally consumed with tapioca flour and fish (Oliveira and Schwartz 2018)).

Açaí is also not the best example of how WFP popularization may contribute to sustainable development goals. Current management systems are located at some point between wild gathering and systematic cultivation. Several studies have found that species management intensification has had negative impacts on Amazonian floodplain ecosystems. The studies have suggested, for example, that management practices (e.g., *açaí* planting in natural environments, the suppression of competing species, and stem cutting to increase fruit production) are converting native floodplain forests into *açaí*-dominated forests that are quite similar to systematic plantations (Weinstein and Moegenburg 2012). Specific processes include the loss of species diversity, as well as functional and phylogenetic diversity (Freitas et al. 2015; Freitas 2019).

In terms of local livelihoods, *açaí* harvesting intensification has given local dwellers (known as *ribeirinhos*) access to the market economy and increased penetration in the urban-industrial society. Local communities in the state of Pará, for example, have experienced several changes in their livelihoods as a response to the increased market interest in *açaí*. The amplification of their economic gains has led to significant changes in their house and boat structures, thereby making them more modern (Silva 2019).

Although this process has had several positive outcomes, especially regarding poverty reduction, there have been some negative consequences of *açaí* popularization on local livelihoods. Remarkably, the commercialization of a product whose main marketing argument has to do with health claims has led to unhealthier eating behaviors in some contexts. As the *açaí* economy became predominant in several local communities, people abandoned or reduced the cultivation of vegetables that were once important for their subsistence (Silva 2019). Instead, they use their increased economic power to buy industrialized products, which can have negative implications in terms of food and nutrition security and sovereignty.

5 Final Considerations

The different and sometimes contrasting views of whether WFP harvesting can help to achieve sustainable development goals indicate that success and failures are context-dependent. Therefore, the outcomes may be influenced by a set of variables (e.g., natural stocks, biological profile of the target species, ecosystem characteristics, management practices, demand and market dynamics, characteristics of the value chains, land tenure, social organization, public policies). All these factors need to be deeply studied and considered to diagnose whether popularization strategies toward certain WFPs are welcome.

Additionally, to better understand the social-ecological role of WFP, the central hypothesis related to the benefits of these species needs to be addressed with proper research designs. Such analysis would help to strengthen arguments favorable to WFP harvesting or even point out which ways not to pursue.

Finally, most of the concerns regarding WFP popularization have to do with the eventual incorporation of these species into conventional (and unsustainable) agri-food systems, which, in some cases, may have negative environmental, cultural, or socioeconomic outputs. Therefore, we believe that attempts to popularize WFP may not be unrelated to strategies for changing our relationship with food, agriculture, and consumption in ways that help us search for ecological-based alternatives and socially just food economies.

Acknowledgments The authors thank L'Oréal and UNESCO for the grants given to PMM (Brazilian "For Women in Science" and "International Rising Talents" Awards) and CNPq for the productivity grant given to PMM (302786/2016-3).

References

Barbosa EC, Cruz RS, Santos GMC et al (2020) Use categories and local perception of decline in plant populations: a case study of woody medicinal plants in Northeastern Brazil. Econ Bot. https://doi.org/10.1007/s12231-020-09502-3

Barlow J, Lennox GD, Ferreira J et al (2016) Anthropogenic disturbance in tropical forests can double biodiversity loss from deforestation. Nature 535:144–147

Belcher B, Schreckenberg K (2007) Commercialisation of non-timber forest products: a reality check. Dev Policy Rev 25:355–377

Brondízio E, Safar C, Siqueira A (2002) The urban market of Açaí fruit (Euterpe oleracea Mart.) and rural land use change: ethnographic insights into the role of price and land tenure constraining agricultural choices in the Amazon estuary. Urban Ecosyst 6:67–97

Casas A, Caballero J, Mapes C, Zárate S (1997) Manejo de la vegetación, domesticación de plantas y origen de la agricultura en Mesoamérica. Boletín la Soc Botánica México 61:31–47

Cruz-Garcia GS, Howard PL (2013) "I used to be ashamed". The influence of an educational program on tribal and non-tribal children's knowledge and valuation of wild food plants. Learn Individ Differ 27:234–240

Delang CO (2006) The role of wild food plants in poverty alleviation and biodiversity conservation in tropical countries. Prog Dev Stud 4:275–286

Emanuel PL, Shackleton CM, Baxter JS (2005) Modelling the sustainable harvest of Sclerocarya birrea subsp. caffra fruits in the South African lowveld. For Ecol Manag 214:91–103

Evans MI (1993) Conservation by commercialization. In: Hladik CM, Hladik A, Linares OF, Pagezy H, Semple A, Hadley M (eds) Tropical forests, people and food: biocultural interactions and applications to development. UNESCO, Paris and Parthenon Publishing Group, Carnforth, pp 815–822

FAO (2019) Voluntary guidelines for the conservation and sustainable use of farmers' varieties/landraces. Food and Agriculture Organization of the United Nations, Rome

Flyman MV, Afolayan AJ (2006) The suitability of wild vegetables for alleviating human dietary deficiencies. South African J Bot 72:492–497

Freitas MAB (2019) O extrativismo de açaí (Euterpe oleracea mart.) e a natureza das assembleias de árvores em várzea amazônica. Universidade Federal de Pernambuco

Freitas MAB, Vieira ICG, Albernaz ALKM et al (2015) Floristic impoverishment of Amazonian floodplain forests managed for açaí fruit production. For Ecol Manag 351:20–27

Gama ADS, de Paula M, da Silva RRV et al (2018) Exotic species as models to understand biocultural adaptation: challenges to mainstream views of human-nature relations. PLoS One 13:e0196091

Gaoue OG, Gado C, Natta AK, Kouagou M (2018) Recurrent fruit harvesting reduces seedling density but increases the frequency of clonal reproduction in a tropical tree. Biotropica 50:69–73

Gavin MC, McCarter J, Mead A et al (2015) Defining biocultural approaches to conservation. Trends Ecol Evol 30:140–145

Gbetibouo GA, Ringler C, Hassan R (2010) Vulnerability of the South African farming sector to climate change and variability: an indicator approach. Nat Resour Forum 34:175–187

Gradé JT (2012) Karamojon (Uganda) pastoralists' use of wild edible plants: a traditional coping mechanism towards climate change. In: Berhe MG, Butera J-B (eds) Climate change and pastoralism: traditional coping mechanisms and conflict in the horn of Africa, 1st edn. Institute for Peace and Security Studies and University for Peace, Addis Ababa, pp 153–176

Heinrich M, Dhanji T, Casselman I (2011) Aai (Euterpe oleracea Mart.) – a phytochemical and pharmacological assessment of the species' health claims. Phytochem Lett 4:10–21
Hickey GM, Pouliot M, Smith-Hall C et al (2016) Quantifying the economic contribution of wild food harvests to rural livelihoods: a global-comparative analysis. Food Policy 62:122–132
Holm JA, Miller CJ, Cropper WP (2008) Population dynamics of the dioecious Amazonian palm Mauritia flexuosa: simulation analysis of sustainable harvesting. Biotropica 40:550–558
IBGE (2017) Tabela 289 – Quantidade produzida e valor da produção na extração vegetal, por tipo de produto extrativo. Retrieved from http://www.sidra.ibge.gov.br/bda/tabela/protabl.asp?c=289&z=p&o=30&i=P
Johns T, Eyzaguirre PB (2006) Linking biodiversity, diet and health in policy and practice. Proc Nutr Soc 65:182–189
Kusters K, Achdiawan R, Belcher B, Pérez MR (2006) Balancing development and conservation? An assessment of livelihood and environmental outcomes of nontimber forest product trade in Asia, Africa, and Latin America. Ecol Soc 11(2):20
Lowore J (2020) Understanding the livelihood implications of reliable honey trade in the Miombo woodlands in Zambia. Front For Glob Chang 3:28
Marcason W (2009) What is the Açaí berry and are there health benefits? J Am Diet Assoc 109:1968
Narayanan MKR, Swapna MP, Kumar NA (2004) Gender dimensions of wild food management in Wayanad, Kerala. Community Agrobiodiversity Centre, M.S.Swaminathan Research Foundation and Uttara Devi Resource Centre for Gender and Development, Kerala
Nascimento VT, Vasconcelos MAS, Maciel MIS, Albuquerque UP (2012) Famine foods of Brazil 's seasonal dry forests: ethnobotanical and nutritional aspects. Econ Bot 66:22–34
Neudeck L, Avelino L, Bareetseng P et al (2012) The contribution of edible wild plants to food security, dietary diversity and income of households in Shorobe Village. Northern Botswana Ethnobot Res Appl 10:449–462
Oliveira MSP, Schwartz G (2018) Açaí— Euterpe oleracea. Exotic Fruits:1–5
Paumgarten F, Locatelli B, Witkowski ETF (2018) Wild foods: safety net or poverty trap? A south African case study. Hum Ecol 46:183–195
Perry DA (1994) Forest ecosystems. The Johns Hopkins University Press, Baltimore
Ros-Tonen MAF (2000) The role of non-timber forest products in sustainable tropical forest management. Holz als Roh – und Werkst 58:196–201
Silva SM (2019) Do consumo local ao mercado de alimentos saudáveis: como a crescente demanda por açaí (Euterpe oleracea mart.) afeta os meios de vida de comunidades ribeirinhas? Universidade Federal de Viçosa
Silva JPCS, Gonçalves PH, Albuquerque UP, Silva RRV, Medeiros PM (2021) Can medicinal use protect plant species from wood uses? Evidence from Northeastern Brazil. J Environ Manage 279: 111800
Sinha A, Bawa KS (2002) Harvesting techniques, hemiparasites and fruit production in two non-timber forest tree species in South India. For Ecol Manag 168:289–300
Soares WL, Porto MF d S (2009) Estimating the social cost of pesticide use: an assessment from acute poisoning in Brazil. Ecol Econ 68:2721–2728
Turan M, Kordali S, Zengin H et al (2003) Macro and micro mineral content of some wild edible leaves consumed in Eastern Anatolia. Acta Agric Scand Sect B Soil Plant Sci 53:129–137
UN (2015) Transforming our world: the 2030 Agenda for Sustainable Development
Vanni MBS (2018) Brazilian Açaí Berry and non-timber forest product value chains as determinants of development from a global perspective. London School of Economics, London
Vaz ÚL, Nabout JC (2016) Using ecological niche models to predict the impact of global climate change on the geographical distribution and productivity of Euterpe oleracea Mart. (Arecaceae) in the Amazon. Acta Bot Brasilica 30:290–295
Weinstein S, Moegenburg S (2012) Açaí palm management in the Amazon Estuary: course for conservation or passage to plantations? Conserv Soc 2:315–346

Potentials of Value Chains of Unconventional Food Plants in Brazil

Rafael Ricardo Vasconcelos da Silva, Patrícia Muniz de Medeiros, and Danúbia Lins Gomes

Contents

1 Introduction

The sale of wild food plants, among other biodiversity products, is a topic of multidisciplinary interest. In the field of ethnobiology, for example, efforts have been made to understand the influence of commercial demands on the collection behaviors of local human populations as well as their impacts on biocultural conservation (Albuquerque et al. 2019).

Undoubtedly, commercial relations around biodiversity products are an exciting topic, mainly for their practical applications (Belcher and Schreckenberg 2007). Existing experiences usually indicate the potential for promoting these commercial relations as a strategy to reconcile the conservation of biodiversity with the generation of income for local populations (Angelsen et al. 2014). However, in order to understand this potential, greater efforts are still needed in studies that address these commercial relationships in the broader perspective of the value chain.

The relevance of studies on the value chain of wild products is acknowledged (Booker et al. 2012; Sardeshpande and Shackleton 2019). By analyzing all the stages involved in the production of a good or service, from its production to final consumption, these studies have helped to reveal recurring weaknesses in the chains of wild food plants (Gomes et al. 2010; Ingram et al. 2012; da Silva et al. 2017).

R. R. V. da Silva (✉) · P. M. de Medeiros · D. L. Gomes
Laboratory of Biocultural Ecology, Conservation and Evolution (LECEB), Campus of Engineering and Agricultural Sciences, Federal University of Alagoas, Rio Largo, AL, Brazil
e-mail: rafael.vasconcelos@ceca.ufal.br; patricia.medeiros@ceca.ufal.br

M. C. M. Jacob, U. P. Albuquerque (eds.), *Local Food Plants of Brazil*, Ethnobiology, https://doi.org/10.1007/978-3-030-69139-4_17

These weaknesses refer to organizational and/or institutional characteristics that represent obstacles to sustainable production and marketing and therefore have social, economic, and environmental implications.

In general, the organizational characteristics of a production chain refer to the circulation of information, associations, agreements, management, and production techniques among other factors. The institutional characteristics include the legal apparatus, control bodies, customs, and traditions among others. Together, these characteristics interfere in the relationships between the actors involved and, consequently, in the functioning of the chain (da Silva et al. 2014). In this sense, a central issue in studies on the value chain of wild products consists of the power relationships between the different members of the chain and between them and the production environment (Booker et al. 2012). From this perspective, the vast majority of value chain investigations focus on well-known products or products with a consolidated market – characteristics that do not apply to most of the so-called unconventional food plants (UFP).

In Brazil, the acronym UFP (PANC in Portuguese) became popular when mentioning wild food plants or the food use of unusual parts of widely consumed plants (Kinupp and Lorenzi 2014). In this sense, UFP can vary from herbaceous species to arboreal and palm trees, as they comprise food consumption both of whole plants and of fruits, tubers, and leaves among other edible parts of plant species.

Although the acronym UFP is open to criticism for its subliminal message of ethnocentrism, it is under this label that many species of wild food plants are gaining notoriety and arousing market interest in the Brazilian context. Despite this, little attention has been given to their value chains, as they are generally little known, as well as short and unstable.

Evidently, the consumption and commercialization of UFP in local geographic contexts makes it difficult to identify their value chains and analyze their potential. However, some of these species have proven their value and aroused the interest of the markets (Steward 2013; Guariguata et al. 2017; Gomes et al. 2020). This is probably a path of no return, which leads to scenarios of risks and opportunities.

In this chapter, we will discuss the potential of the production chains of unconventional food plants in Brazil. We understand this terminological and geographic scope as interesting due to the emerging behavior of its market, which allows us to establish relationships with the broader experiences of selling wild food plants in different parts of the world.

2 Evolution of Ideas About the Strategic Role of Value Chains for Wild Food Plants

Conceptually, the sale of wild products always constitutes value chains. However, these chains can vary immensely in terms of geographic scope, volumes sold, values, number of links and institutions involved, and technological development among other relevant and dynamic aspects, both temporal and geographically. In this way, ideas about the strategic role of value chains for wild food plants have been changing over time and in the geographic space where they are inserted.

For example, in Brazil, the strategic view on the value chains of wild plants has been strongly influenced by the logic of commercial relations and ideas prevalent worldwide. For centuries, the consumption and commercial exploitation of wild plants has played an important role in the country's colonial culture and economy (Dean 1991; Medeiros and Albuquerque 2014; Tomchinsky and Ming 2019). It can be argued that during this long period, the prevailing idea in Brazil and in the world about the strategic role of chains of wild plants was predominantly exploratory. In other words, the commercialized products were extracted from natural environments with the objective of generating profits for explorers, leaving local populations relegated to the role of sources of information and labor, and natural environments as mere stocks of resources for extraction.

Over time, ideas about the economic relevance of the value chains of wild food plants and other biodiversity products matured. The insights came from broader discussions about environmental problems, such as pollution and deforestation, which started to be guided by international debates. In this context, the first United Nations Conference on the Human Environment, held in 1972 in Stockholm, is considered an important milestone.

During this conference, the document "Stockholm Declaration on the Human Environment" was prepared, which established a set of criteria and principles to guide peoples and countries in their relations with the environment. The fourth principle listed in this document highlights "(...) when planning economic development, importance must be given to the conservation of nature, including wild flora and fauna." In this sense, from the 1970s onward, actions aimed at the preservation of charismatic species, usually animals, and the delimitation of protected areas, usually due to their scenic beauty, gained strength. In this period, local human populations and the components of biodiversity consumed and/or traded by them did not yet assume a prominent position in conservation policies.

From the 1980s onward, the economic importance of biodiversity products gained greater political notoriety, due to the rapid technological development and increased biopiracy, patenting, and privatization of biodiversity components. The vulnerability of local populations in poor countries was evident. In light of this, the pressure and the articulation of different sectors of society and governments grew in favor of mechanisms that would ensure the conservation of biodiversity and the rights of traditional populations.

The articulations culminated in the Convention on Biological Diversity (CBD), signed during the United Nations Conference on Environment and Development, which took place in 1992 in the city of Rio de Janeiro, Brazil. The CBD defined the main conditions and legal bases for regulating the sustainable use of biodiversity, which would influence government programs. The Brazilian Federal Senate approved the final text of the CBD in 1994. Since then, more and more biodiversity products have come to be seen as a strategy to generate income for local populations.

Although there is no unanimity, there is a widespread understanding that the economic relevance of value chains for biodiversity products, such as wild food plants, grows on smaller geographic scales. That is, even if they have a secondary or irrelevant role in the global economy, at the local level, situations in which entire communities depend heavily on the marketing of these products to guarantee their livelihood are recurrent (da Silva et al. 2017). This understanding has a strong influence on ideas about the potential of UFP as a strategy to generate income, conserve biodiversity, and strengthen food security.

Within Brazilian policies, some UFP can be identified as a product of sociobiodiversity, an expression adopted by a public policy implemented in Brazil in 2009 (Brasil 2009). The concept of sociobiodiversity represented an advance, as it recognized the social and economic importance that certain products of biodiversity have in Brazil. This allowed these products, as well as the human populations economically and culturally associated with them, to receive special attention from government programs to strengthen their value chains. Examples of policies associated with sociobiodiversity products include access to the Food Acquisition Program (PAA in Portuguese), which allowed the government to invest in these products for inclusion in social programs to fight hunger (e.g., school meals, popular restaurants). Another example is the National Minimum Price Guarantee Policy, which opened the possibility of a public subsidy to extractive communities that may be affected by the drop in the price of sociobiodiversity products below the minimum value defined by the government control body. However, these policies favored the most consumed and known sociobiodiversity products, leaving out many UFP whose value chains are not well known or are not sufficiently developed.

This lack of visibility of many UFP with economic potential, as well as the organizational limitations of many extractive communities, paved the way for another idea of the strategic role of biodiversity-based value chains. This aims to generate commodities from biodiversity and seeks to promote these value chains inspired by the agribusiness model, with a focus on the final product and the optimization of its production process. The biocultural complex in which these "raw materials" are inserted is not taken into account or is used only as a marketing strategy. Under the logic of agribusiness, extractive communities tend to be relegated to the role of labor of companies interested in raw materials for their products.

This demonstrates that a relevant feature of the strategies that seek to promote the UFP value chains should be, in addition to the popularization of these products, the empowerment of the local human populations associated with them, whether extractive, agroextractive, or traditional farmers. This empowerment is one of the most challenging goals for the promotion policies of these chains, as it means the role of these communities in defining objectives, rules for the collection, processing, and negotiation of products. In addition, it is necessary that these strategies ensure land security and resource control for communities, which often represents one of the greatest institutional obstacles. In addition, many communities demand support to develop articulation and negotiation skills with the government, companies, or final consumer.

3 Contributions and Gaps of Value Chain Studies to Income Generation and Biodiversity Conservation Strategies

One of the main goals of many studies on biodiversity-based value chains, such as the UFP, is to suggest strategies to promote and strengthen these chains on a sustainable basis. For this, many studies seek to emphasize the social, economic, and/or environmental weaknesses in the dynamics of production and commercialization of these products.

The situations of social fragilities are verified in products that are not able to consolidate themselves in the market because they do not have institutional support, even though they present a notable commercial potential. In many cases, the lack of promotion policies, contracts, and regulations hinders production and access to markets willing to pay a fair price for the product, causing the chain to operate in a precarious context of informality (da Silva et al. 2015). This is the case of the pequi (*Caryocar coriaceum* Wittm.), a wild tree species whose fruits are commercialized in the south of the state of Ceará, northeast of Brazil. In this context, da Silva et al. (2015) found that pequi collection represented the most profitable productive activity carried out by local extractivists. However, the commercialization of the fruits occurred in a logic of competitive entrepreneurship among the families. This scenario, according to the authors, stimulated the adoption of inadequate collection practices, that is, in which families aimed to increase the amount of fruit extracted, resulting in damage to the natural populations of plants. In the search for greater individual return, extractivists obtained collective losses.

Economic flaws are highlighted in situations where there is a marked asymmetry in profits along the chain, where extractivists are generally the least benefited (Gomes et al. 2010). Among the examples is the case of *Schinus terebinthifolius* Raddi (aroeira), a tree species very abundant in the region of Baixo São Francisco (Northeast Brazil), where its food use is still little known and unconventional. According to de Jesus and Gomes (2012), the demand for aroeira fruits, marketed under the label of pink peppercorn, came from processing industries that are located distant from the production sites. After processing, the industries exported pink peppercorn to several countries in the European Union, the United States, Canada, and Argentina. According to the authors, the commercialization of "pink peppercorn" for the international market reached an export value of up to U$18.00/kg (2008). In the same period, the processing industries paid extractivists up to U$0.95/kg at the collection sites.

On the other hand, environmental weaknesses are related to situations in which risks of negative impacts on the structure of populations of exploited species are present due to overexploitation pressures (Ingram et al. 2012). In the Congo basin, for example, the leaves of wild lianas of the genus *Gnetum* are consumed and marketed for food purposes. Ingram et al. (2012) found that more than half of the production of *Gnetum* spp. is collected unsustainably in forest areas. According to the authors, the lack of adequate regulation and increased collection pressure led to a decrease in the availability of leaves and increased prices, resulting in unsustainable trade.

Although urgent and highly relevant, the emphasis of many studies on identifying flaws may not be the most appropriate for many UFP chains, whose potential is still poorly understood. In view of this, it is necessary to direct efforts toward understanding the potential of UFP' value chains.

4 Potentialities to Promote and Strengthen Sustainable UFP Value Chains

In Brazil, countless species that are considered unconventional food plants (Kinupp and Lorenzi 2014) had their production chains expanded due to the increase in demand. Açaí (Steward 2013) and Brazil nut (Guariguata et al. 2017) are examples. These Amazonian products generate income for local communities and contribute to forest conservation. To have an idea, in 2018, the production of açaí yielded R$592 million (U$ 152,774,193.55) and Brazil nuts R$130.9 million (U$ 33,780,645.16) (IBGE 2018).

It is likely that there are species of great economic potential in the different Brazilian biomes. In the northeastern region of Brazil, for example, scientific evidence indicated various native UFP tree species (e.g., *Genipa americana* L, *Myrciaria floribunda* (H. West ex Willd.) O. Berg, *Attalea funifera* Mart., *Syagrus coronata* Becc., *Schinus terebinthifolia* Raddi, *Spondias purpurea* L) with high commercial potential in areas of coastal vegetation called *restinga* (Gomes et al. 2020). Also according to this study, although these UFP do not constitute the nutritional base of local communities, they are widely consumed locally due to the pleasant "taste." This evidence suggests that these products have the potential to please the palate of different consumer markets.

However, most UFP are represented by wild species acquired in an extractive manner. This reality has implications for the conservation of its natural populations. One of the ideas that circulate about extraction indicates a fateful trend that the growth in demand generates a disorderly increase in collection, resulting in the depletion of resources (Homma 2000). However, the accumulated experiences have indicated that the sustainability of extractive activities results from the interaction of several factors of an ecological, social, economic, and technological nature (Fig. 1).

In some cases, easily accessible information such as the part of the plant used is essential for assessing the sustainability potential of extractivism. There is a tendency that the collection of certain parts of the plant that cause no lethality to the plant, such as fruits, causes less damage to the conservation of wild populations (Emanuel et al. 2005). It is evident that, even in these situations, extraction requires monitoring and adequate management as a way to prevent damage in the near term (Murali et al. 1996; Ticktin 2004; Ndangalasi et al. 2007).

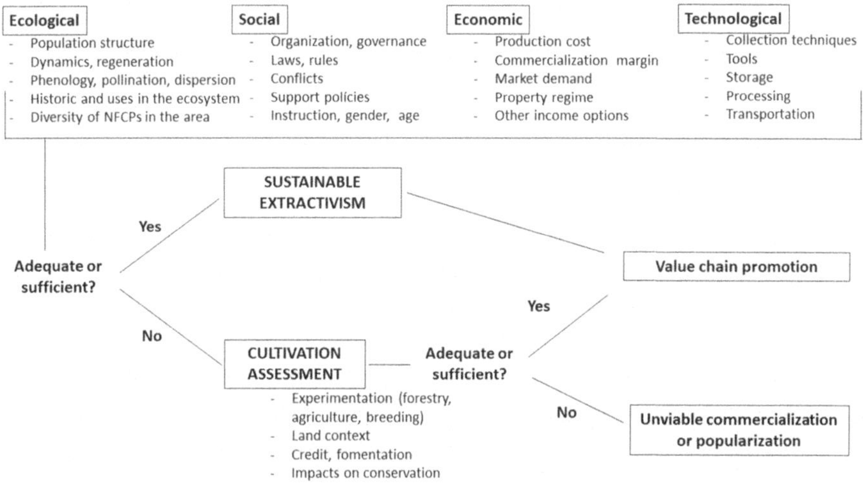

Fig. 1 Flow of basic information to promote UFP productive chains on a sustainable basis

Many UFP species are fruitful (Kinupp and Lorenzi 2014). If, on the one hand, this generates positive expectations regarding the sustainability potential of fruit extraction, on the other hand, it results in serious problems in the production process. Fruits are highly perishable food products that require careful storage, handling, and rapid transportation to the processing or fresh consumption site. This means that the extraction of UFP must be limited to sufficient volumes for economic viability of the stages after collection, avoiding losses. Therefore, the extractive production of perishable food products must be of low intensity, considering not only the conservation of wild populations but also the quality of the products.

In cases where the extraction of wild UFP populations is not sufficient or adequate to promote the value chain, it is recommended to evaluate the viability of the cultivation (Fig. 1). One of the first steps in this direction is the identification of the best propagation techniques for the species, which requires investments in agronomic and forestry experiments. Following this, other obstacles will need to be overcome, such as access to credit and technical assistance for incorporating the species in the cultivated areas. However, it should be noted that cultivation alone does not always represent a strategy capable of reducing extraction and contributing to the conservation of wild populations (Trauernicht and Ticktin 2005; Williams et al. 2014).

The fundamental difference between cultivation and extractive production is the need for the existence and conservation of wild areas in the extractive production process. In this perspective, depending on the context of production, the promotion of a UFP value chain may be associated with the strategy of reconciling income generation with biodiversity conservation or with the generation of income from new agricultural commodities.

Regardless of the motivation of the production base driven by commercialization, the necessary relationship between products and the market represents the point of convergence between extractive and cultivated production. It can even be considered that the success of the UFP value chains on a sustainable basis largely depends on the balance of this relationship (Belcher and Schreckenberg 2007). In general, various storage, processing, and transportation strategies are verified, between and within extractive and cultivated production systems. These variations occur due to the production environment, the intrinsic characteristics of the product, the degree of processing, and the profiles of the producer and the consumer among other aspects. This reinforces the need to understand the relationship with the market from a broad perspective of the value chain, in order to involve the entire UFP commercialization route, from the production environment to the final consumer.

5 Final Considerations

The experiences with value chains of biodiversity products make it possible to enumerate three main goals for reaching its potential: (1) to generate income for the populations that produce UFP, (2) to conserve biodiversity, and (3) to empower populations that produce UFP. It is possible to expect excellent results whenever these three goals can be achieved together.

The vast majority of value chain investigations focused on well-known products with an established market, which does not apply to most UFP. Generally, these investigations highlight the relations of power between the actors and institutions involved in the chain, addressing their weaknesses. Despite the great relevance of this approach, studies on UFP production chains require methodological adjustments in order to direct the focus toward the identification of the potential for promoting these chains on a sustainable basis.

References

Albuquerque UP, Nascimento ALB, Soldati GT et al (2019) Ten important questions/issues for ethnobotanical research. Acta Bot Brasilica 33:376–385. https://doi.org/10.1590/0102-33062018abb0331

Angelsen A, Jagger P, Babigumira R et al (2014) Environmental income and rural livelihoods: a global-comparative analysis. World Dev 64:S12–S28. https://doi.org/10.1016/j.worlddev.2014.03.006

Belcher B, Schreckenberg K (2007) Commercialisation of non-timber forest products: a reality check. Dev Policy Rev 25:363

Booker A, Johnston D, Heinrich M (2012) Value chains of herbal medicines – research needs and key challenges in the context of ethnopharmacology. J Ethnopharmacol 140:624–633. https://doi.org/10.1016/j.jep.2012.01.039

Brasil (2009) Plano nacional de promoção das cadeias de produtos da sociobiodiversidade. MDA, MMA, MDS, Brasília

da Silva RRV, Gomes LJ, Albuquerque UP (2014) Methods and techniques for research on the supply chains of biodiversity products. In: Albuquerque UP, Lucena RFP, Cunha LVFC (eds)

Methods and techniques in ethnobiology and Ethnoecology. Springer Protocols Handbooks, New York, pp 335–347
da Silva RRV, Gomes LJ, Albuquerque UP (2015) Plant extractivism in light of game theory: a case study in northeastern Brazil. Ethnobiol Ethnomedicine 11:1–7. https://doi.org/10.1186/1746-4269-11-6
da Silva RRV, Gomes LJ, Albuquerque UP (2017) What are the socioeconomic implications of the value chain of biodiversity products? A case study in Northeastern Brazil. Environ Monit Assess 189:1–11. https://doi.org/10.1007/s10661-017-5772-2
de Jesus NB, Gomes LJ (2012) Conflitos socioambientais no extrativismo da aroeira (schinus terebinthifolius raddi), baixo são francisco – sergipe/alagoas. Ambient e Soc 15:55–73. https://doi.org/10.1590/S1414-753X2012000300005
Dean W (1991) A botânica e a política imperial: a introdução e a domesticação de plantas no Brasil. Rev Estud Históricos 4:216–228
Emanuel PL, Shackleton CM, Baxter JS (2005) Modelling the sustainable harvest of Sclerocarya birrea subsp. caffra fruits in the South African lowveld. For Ecol Manag 214:91–103. https://doi.org/10.1016/j.foreco.2005.03.066
Gomes JL, Gomes MAO, Jesus NB (2010) Aspectos socioambientais da atividade extrativista de produtos florestais não-madeireiros: os casos da fava-d'anta (Dimorphandra sp.) e da aroeira-da-praia (Schinus terebinthifolius Raddi). In: Albuquerque UP HN (ed) Árvores de valor e o valor das árvores: pontos de conexão, 1st edn. NUPEEA, Recife, pp 61–106
Gomes DL, dos Santos Ferreira RP, da Costa Santos ÉM et al (2020) Local criteria for the selection of wild food plants for consumption and sale: a case study in rural settlements in Alagoas, Brazil. Ethnobiol Conserv 10:1–15. https://doi.org/10.15451/ec2020-04-9.10-1-15
Guariguata MR, Cronkleton P, Duchelle AE, Zuidema PA (2017) Revisiting the 'cornerstone of Amazonian conservation': a socioecological assessment of Brazil nut exploitation. Biodivers Conserv 26:2007–2027. https://doi.org/10.1007/s10531-017-1355-3
Homma AKO (2000) Amazônia: os limites da opção extrativa. Ciência Hoje 27:70–73
IBGE (2018) Produção da extração vegetal e da silvicultura. Instituto Brasileiro de Geografia e Estatística, Rio de Janeiro
Ingram V, Ndumbe LN, Ewane ME (2012) Small scale, high value: Gnetum africanum and buchholzianum value chains in Cameroon. Small-scale For 11:539–556. https://doi.org/10.1007/s11842-012-9200-8
Kinupp VF, Lorenzi H (2014) Plantas Alimentícias Não Convencionais (PANC) no Brasil, 1st edn. Instituto Plantarum de Estudos, Nova Odessa
Medeiros MFT, Albuquerque UP (2014) Food flora in 17th century northeast region of Brazil in Historia Naturalis Brasiliae. J Ethnobiol Ethnomed 10:1–20. https://doi.org/10.1186/1746-4269-10-50
Murali KS, Shankar U, Shaanker RU et al (1996) Extraction or non-timber forest products in the forests of Biligiri Rangan Hills, India. 6. Fuelwood pressure and management options. Econ Bot 52:320–336. https://doi.org/10.1007/BF02862151
Ndangalasi HJ, Bitariho R, Dovie DBK (2007) Harvesting of non-timber forest products and implications for conservation in two montane forests of East Africa. Biol Conserv 134:242–250. https://doi.org/10.1016/j.biocon.2006.06.020
Sardeshpande M, Shackleton C (2019) Wild edible fruits: a systematic review of an underresearched multifunctional NTFP (non-timber forest product). Forests 10:1–24. https://doi.org/10.3390/f10060467
Steward A (2013) Reconfiguring agrobiodiversity in the Amazon estuary: market integration, the Açaí trade and smallholders' management practices in Amapá, Brazil. Hum Ecol 41:827–840. https://doi.org/10.1007/s10745-013-9608-6
Ticktin T (2004) The ecological implications of harvesting non-timber forest. J Appl Ecol 41:11–21. https://doi.org/10.1111/j.1365-2664.2004.00859.x
Tomchinsky B, Ming LC (2019) As plantas comestíveis no Brasil dos séculos XVI e XVII segundo relatos de época. Rodriguésia 70:1–16. https://doi.org/10.1590/2175-7860201970040

Trauernicht C, Ticktin T (2005) The effects of non-timber forest product cultivation on the plant community structure and composition of a humid tropical forest in southern Mexico. For Ecol Manag 219:269–278. https://doi.org/10.1016/j.foreco.2005.09.005
Williams SJ, Jones JPG, Annewandter R, Gibbons JM (2014) Cultivation can increase harvesting pressure on overexploited plant populations. Ecol Appl 24:2050–2062. https://doi.org/10.1890/13-2264.1

Including Biodiversity Food in the Brazilian School Feeding: A Strategy to Ensure Food and Nutritional Security in Childhood

Camila Pia Delgado da Silva, Ana Laura Benevenuto de Amorim, Daniel Henrique Bandoni, and Veridiana Vera de Rosso

Contents

C. P. D. da Silva
Graduate Program in Food, Nutrition and Health, Universidade Federal de São Paulo (UNIFESP), Santos, SP, Brazil

A. L. B. de Amorim
Interdisciplinary Graduate Program in Health Sciences, Universidade Federal de São Paulo (UNIFESP), Santos, SP, Brazil

D. H. Bandoni
Nutrition and Food Service Research Center/Department of Health, Clinic and Institutions, Universidade Federal de São Paulo (UNIFESP), Santos, SP, Brazil

V. V. de Rosso (✉)
Nutrition and Food Service Research Center, Universidade Federal de São Paulo (UNIFESP), Santos, SP, Brazil
e-mail: veridiana.rosso@unifesp.br

M. C. M. Jacob, U. P. Albuquerque (eds.), *Local Food Plants of Brazil*, Ethnobiology, https://doi.org/10.1007/978-3-030-69139-4_18

1 Introduction

Modern agri-food systems have been based on favoring monocultures and increasing productivity, under the hegemony of transnational companies that trade seeds and inputs. Such productive model demands high energy consumption and excessive use of agricultural inputs and agrochemicals and stimulates deforestation, erosion, and soil, water, and air pollution and, above all, loss of biodiversity (Beltrame et al. 2016; Hunter et al. 2016; Tutwiler et al. 2017).

The concern with the environment and quality of life of the population has stood out in discussions on sovereignty and food and nutrition security, human right to adequate food, and foodstuff productive systems (Tutwiler et al. 2017). Food system activities appear as one of the main factors responsible for climate change, with data showing that between 2010 and 2050, the increasing world population and income will cause a 50–90% raise in adverse environmental effects as a result of the current food system (Springmann et al. 2018). Several options for reducing the environmental effects of the food system have been pointed out, including shifts from the current diet to healthier, more plant-based ones, improvements in technology and management, and reductions in food loss and waste (Springmann et al. 2018). Springmann et al. (2018) found that no single measure is sufficient to maintain these effects within all planetary limits simultaneously and that a synergistic combination of measures will be needed to sufficiently mitigate the projected increase in environmental pressures.

In this way, redirecting the agricultural productive system toward more sustainable activities, rescuing the use of biodiversity foods, and reconnecting the rural and urban communities become a priority to boost the commercialization and consumption of these products (Ferigollo et al. 2017; Félix et al. 2018).

Generating discussion and actions on the topic of organic and agroecological production and consumption of biodiversity foods in the scope within public policies can help mitigate the impacts of monoculture, favoring the sustainable use of natural resources and the current or potential value of biodiversity foods (Scott 2017). As such, it would be possible to stimulate the production and consumption of healthy and nutritious foods, minimizing deleterious effects on the environment (Scott 2017). Biodiversity is the lifeblood of what we eat. Biodiversity – both wild and cultivated – underpins the sustainability of agricultural production by providing the genetic diversity and material needed to drive innovation and adaptation, as well as essential ecosystem services and processes (Fanzo et al. 2013).

The Brazil National School Feeding Program (PNAE), through regulations, encouraged the use of organic and agroecological foods by instituting that at least 30% of the National Education Development Fund (FNDE) be invested in the acquisition of products from family entrepreneurs and/or family farming, prioritizing organic and agroecological foods (Brasil 2009). The PNAE also encourages acquiring a diversity of foods (Brasil 2018a), obtained by planting or extractivism, either fresh or as derivatives. This strategy for promoting natural resources comes as a facilitator for a differentiated market, since it enables commercializing these foods at a national level, serving around 42.6 million basic education students and young people and adults in Brazil (Brasil 2013).

The encouragement by public policies for sustainable production, the use of Brazilian biodiversity, and the social inclusion of family farmers can alleviate the impacts of globalization and hegemony of the agri-food system, as well as the standardization of the menus offered to schoolchildren, guaranteeing food and nutrition security in childhood (De Paiva et al. 2012; de Sousa et al. 2015; Beltrame et al. 2016).

However, implementing the use of these foods in the Brazilian school menus can face numerous challenges, such as the regional particularities of these foods, which require developing an entire distribution chain and recipes so that food handlers know how to take advantage of the nutritional value of each biodiversity ingredient. Reason why, we believe, biodiversity foods are rarely present in Brazilian school meals.

Considering, therefore, the biological and genetic diversity existing in Brazil and the policies that encourage trade and consumption of biodiversity foods as a key to mitigate environmental, social, and nutritional problems, this study aims to evaluate the use of food species from the Brazilian biodiversity in school feeding and to discuss the main challenges for expanding the use of such foods as a strategy to guarantee student food and nutrition security.

2 Methods

This is a cross-sectional study developed in two stages: first, the analysis of the presence of biodiversity foods in school feeding meals from all Brazilian regions and, second, interviews with nutritionists (technical managers of the municipalities of Southeast Brazil) about acquisition of biodiversity foods.

2.1 Analysis of the Presence of Biodiversity Foods in School Feeding Menus from All Brazilian Regions

For analyzing the presence of biodiversity foods in school feeding menus from all Brazilian regions, we used the Efficient Manager of School Lunch Award database. This manager award, created and coordinated by the Non-Governmental Organization (NGO) Zero Hunger Action, aims to evaluate and classify how executing agencies (municipal governments) managed school meals in Brazil (Zero 2012). Launched in 2004, the last year of the award was 2013, referring to data on school feeding management in 2012.

Our study included the municipalities enrolled in the award, which sent the school feeding menus and a description of the food used. Using the 2011 School

Feed Manager Award database, we compiled the foods mentioned in the menus sent and identified the presence of organic, regional, and biodiversity foods.

Food was classified as organic when this information appeared on the menu itself, as regional when included on the Ministry of Health list of regional foods (Brasil 2015), and as biodiversity foods when included on the Ministerial Ordinance No. 284 (Brasil 2018a), which lists species that can be acquired for school feeding as fresh or derived products.

Additionally, data on the acquisition of these foodstuffs were compiled from family farms, and we analyzed the relationship between purchase from family farming and purchase of organic, regional, and biodiversity foods.

Analyses also related city size of the cities, region of Brazil in which the city is, and use of organic, regional, and biodiversity foods. Cities were grouped according to the number of inhabitants: small size 1 (up to 20,000 inhabitants), small size 2 (from 20,001 to 50,000 inhabitants), medium size (from 50,001 to 100,000 inhabitants), and large size (over 100,001 inhabitants), adapting the Brazilian Institute of Geography and Statistics (IBGE) classification (IBGE 2018).

Data appear as relative and absolute frequencies. In order to evaluate the relationship between the variables, the chi-square was used, considering $p < 0.05$ as significant. Analyses were performed in SPSS software version 17.0.

2.2 *Interviews with Nutritionists About the Acquisition of Biodiversity Food Products*

Data collection occurred from August 2014 to December 2015, involving municipalities of the four states of Southeast Brazil. All the cities of the states of São Paulo (n = 645), Rio de Janeiro (n = 92), Espírito Santo (n = 78), and Minas Gerais (n = 853) were included in the sample, totaling 1668 municipalities.

The questionnaires were sent electronically to nutritionists and/or the school feeding department of each municipality using data from the Nutritionists Registration System (SINUTRI) of the Ministry of Education/National Education Fund (FNDE).

This semi-structured questionnaire addressed general topics about the acquisition of products for composing school feeding menus in the municipality using FNDE resources and also specific topics related to the acquisition of biodiversity foods.

The variables analyzed in this stage were number of municipalities that purchased food from family farming, reasons for not buying from family farming, agricultural mapping (municipal census on agricultural production), factors prioritized by the nutritionists when preparing menus for school feeding (regional dietary habits, variety of foods in the menu, presence of biodiversity foods, and production sustainability of foods included in the menu), and description of biodiversity foods included in the school feeding menu and the types of preparations served using biodiversity foods.

The study was approved by the Research Ethics Committee under opinion number 1,069,621/2015.

3 Results

3.1 Database Analysis: Presence of Biodiversity Foods in School Feeding Menus

Of the 1082 Brazilian cities that registered to compete for the 2011 Efficient Manager of School Lunch Award, 229 submitted their menu and were included in this study. Of these, we excluded eight cities for lacking information on food purchase from family farming, thus totaling 221 cities.

Table 1 Regional distribution and classification according to city size considering the purchase of family farming foodstuffs and the presence of organic, regional, and biodiversity foods in school feeding menus. Brazil, 2011

		Family farm purchase						
Variables		Yes – *n* (%)		No – *n* (%)		Total – *n* (%)		*p**
Regions of Brazil								0.003
	North	3	42.86	4	57.14	7	3.17	
	Northeast	21	63.64	12	36.36	33	14.93	
	Midwest	10	62.50	6	37.50	16	7.24	
	Southeast	76	65.52	40	34.48	116	52.49	
	South	45	91.84	4	8.16	49	22.17	
City size								0.009
	Small 1	79	76.70	24	23.30	103	46.61	
	Small 2	41	68.33	19	31.67	60	27.15	
	Medium	22	75.86	7	24.14	29	13.12	
	Large	13	44.83	16	55.17	29	13.12	
Presence of organic foods								0.008
	Yes	13	100.00	0	0.00	13	5.88	
	No	142	68.27	66	31.73	208	94.12	
Presence of biodiversity foods								0.151
	Yes	50	75.76	16	24.24	66	29.86	
	No	105	67.74	50	32.26	155	70.14	
Presence of regional foods								0.072
	Yes	150	71.43	60	28.57	210	95.02	
	No	5	45.45	6	54.55	11	4.98	
Total		**155**	**70.14**	**66**	**29.86**	**221**	**100**	

*$p < 0.05$ (Chi-square)

Table 1 shows the regional distribution of cities according to food purchase from family farming, classification according to city size, and presence of organic, regional, and biodiversity foods in school feeding menus. We found a significant difference when analyzing purchase from family farming by region. Thirteen (5.88%) cities reported buying organic food from family farming, a notably low number considering the relation between family farming purchase and presence of biodiversity and regional foods, especially when obtained by extractivism. It should be noted that the menu of 210 (95.02%) cities mentioned regional foods, while only 66 (29.86%) menus reported using biodiversity foods.

Table 2 describes the regional foods on school menus by region, with the Southeast presenting a greater diversity of foods, followed by the Northeast and the South. The Brazilian biodiversity species mentioned in the school menus were cashew (*Anacardium occidentale L.*), *cupuaçu* (*Theobroma grandiflorum K. Schum*), *yerba mate* (*Ilex paraguariensis A.St.-Hil.*), passion fruit (*Passiflora edulis Sims*), *taioba* (*Xanthosoma taioba* E.G. Gonç.), and *umbu* (*Spondias tuberosa* Arruda).

Table 3 summarizes the data on the distribution of organic, biodiversity, and regional foods according to the cities' region. We found a significant difference between the Brazilian regions on the presence of organic and regional foods, with the South region having 11 (21.57%) cities that purchased organic foodstuff, whereas in the North, Northeast, and Midwest regions, no city purchased such foods; moreover, all cities from the South region bought regional foods.

3.2 Interviews with Nutritionists' Results About Acquisition and Use of Biodiversity Food

We contacted all cities ($n = 1668$) from the Southeast region, and 180 participated in this study – 62.78% ($n = 113$), 31.68% ($n = 57$), 2.78%, and 1.67% ($n = 3$) located in the states of São Paulo, Minas Gerais, Espírito Santo, and Rio de Janeiro, respectively.

Table 2 Regional foods present in the respective regions. Brazil, 2011

Brazilian regions				
North	Northeast	Midwest	Southeast	South
Banana (*pacova*), cupuaçu (2)	Barbados cherry, dwarf banana, cashew, papaya, passion fruit, umbu, pumpkin, beans, string beans, cassava, cassava flour, *tapioca* flour, chives (14)	Passion fruit, pumpkin, cabbage, manioc, green corn, parsley, onion (07)	Avocado, persimmon, guava, orange, mango, zucchini, watercress, chayote, kale, spinach, bell pepper, okra, cabbage, arugula, *taioba*, pod, arracacha, corn, parsley (19)	Banana, fig, apple, tangerine (bergamot), grape, chicory, beet, cabbage, tomato, lentil, potato, sweet potato, cinnamon, clove (14)

Table 3 Distribution of organic, biodiversity, and regional foods according to the regions of the country. Brazil, 2011

Regions of Brazil		North		Northeast		Midwest		Southeast		South		Total		p*
		n	%	n	%	n	%	n	%	n	%	n	%	
Presence of organic foods														
	Yes	0	0	0	0	0	0	3	2.46	11	21.57	14	6.11	0.000
	Not	7	100	33	100	16	100	119	97.54	40	78.43	215	93.89	
Presence of biodiversity foods														
	Yes	2	28.57	14	42.42	7	43.75	32	26.23	12	23.53	67	29.26	0.220
	Not	5	71.43	19	57.58	9	56.25	90	73.77	39	76.47	162	70.74	
Presence of regional foods														
	Yes	1	14.29	32	96.97	16	100	118	96.72	51	100	218	95.20	0.000
	Not	6	85.71	1	3.03	0	0	4	3279	0	0	11	4.80	
Total		**7**	**3.06**	**33**	**14.41**	**16**	**6.99**	**122**	**53.28**	**51**	**22.27**	**229**	**100**	

*$p < 0.05$ (Chi-square)

Of the contacted cities, 78.33% ($n = 141$) reported acquiring food directly from family agriculture, showing low compliance with Federal Law No. 11,947 (Brasil 2009). The cities reported spending between 3% and 76.9% of the funds earmarked by the FNDE for acquiring foodstuffs from family farming.

Agricultural mapping is an efficient tool for aiding PNAE municipal managers to find suppliers of local products (Monego et al. 2013) and should include a description of local products, quantity of production, and time of harvest and/or collection. Our study found a significant difference when analyzing the relation between purchase from family farming and agricultural mapping: 58.8% ($n = 106$) of the cities have access to agricultural mapping, of which 15 reported not buying food from family farming. Lack of information on what is regionally produced may be one of the main constraints to acquiring local production.

Nutritionists attributed the 30% non-purchase of food from family farms to the farmers' lack of interest in selling their products for school feeding, difficulty in regularizing the necessary documents, the logistical operation of delivery directly to schools, and lack of dialogue with other sectors of the city council directly involved with family farms.

Sixteen cities (8.89%), 5 of them from the state of Minas Gerais and 11 from the state of São Paulo, reported purchasing the following biodiversity foods: "cajú," cashew apple (*Anacardium occidentale* Linn); "maracujá," passion fruit (*Passiflora edulis* Sims); "pequi" (*Caryocar brasiliense* Cambess); "pitanga," Brazilian cherry (*Eugenia uniflora* Linn); "mangaba" (*Hancornia speciosa* Gomes); "jabuticaba," Brazilian grape tree (*Plinia cauliflora* Kausel); "juçara" (*Euterpe edulis* Martius); "cupuaçu" (*Theobroma grandiflorum* Willd. ex Spreng. K. Schum); "pupunha," peach palm (*Bactris gasipaes* Kunth); and "urucum," achiote (*Bixa Orellana* Linn).

The current National School Feeding Program (Brasil 2013) norms require that the responsible nutritionist considers regional habits, variety, biodiversity, and sustainability while developing the menu. Table 4 shows how nutritionists observe this

Table 4 Priority factors listed by nutritionists for planning the menus offered to schoolchildren. Brazil, 2015

Priority factors in menu development	Yes (% – *n*)	No (% – *n*)	Did not answer (% – *n*)
Regional eating habits	90.0% (*n* = 162)	1.1% (*n* = 2)	8.9% (*n* = 16)
Variety of food	88.9% (*n* = 160)	2.2% (*n* = 4)	8.9% (*n* = 16)
Biodiversity	38.9% (*n* = 70)	52.2% (*n* = 94)	8.9% (*n* = 16)
Sustainability	60.0% (*n* = 108)	31.1% (*n* = 56)	8.9% (*n* = 16)

requirement. Although biodiversity foods can participate in regional habits, thus contributing to a varied menu, nutritionists interpreted these concepts as disassociated.

All biodiversity native fruits listed in the questionnaire feature in the Interministerial Ordinance No. 284 (Brasil 2018a), but other species could be included in the questionnaire, provided the city was making the purchase. The cities reported acquiring these fruits directly from family farming, with eight claiming to have access to agricultural mapping; thus, having access to the mapping tool allows nutritionists to know the location of family farmers and buy from them.

Nutritionists pointed out that biodiversity foods were offered on school menus as simple preparations, or only in fresh form, reducing the possibility of frequent inclusion by its use in creative and tasty recipes based on regional cuisine (de Santiago et al. 2012). Among preparations cited by nutritionists, we highlight rice with *pequi* and *carne de sol* (sun-cured meat), a traditional dish in cities of Minas Gerais (Monego et al. 2013) but also the use of some little-known fruits such as *juçara*, offered as a juice with banana and lemon, whose pulp is often consumed in several recipes in *caiçara* cuisine, typical of fishing communities (IPEMA 2012; Beltrame et al. 2016). The information on the frequency of purchase of food from the biodiversity and the supply of preparations for schoolchildren reported by study participants was incomplete.

Some municipalities offered *PANCs* non-conventional food plants (Plantas Alimentícias Não Convencionais) and exotic fruits (Figs. 1 and 2, respectively). PANCs have limited distribution, restricted to certain localities or regions, exerting great influence on the food and culture of traditional populations, with no productive chain organization, unlike conventional vegetables (tomato, cabbage, lettuce, etc.), not arousing commercial interest by seed, fertilizer, or agrochemical companies (Brasil 2010).

Fig. 1 Organic vegetable plantation including PANCs – non-conventional food plants

4 Discussion

As evidenced in the literature, the present study found a prevalence of purchase from family farming in the South region, as well as smaller municipalities (Saraiva et al. 2013; De Amorim et al. 2016), while the purchase of organic food had a positive relationship with the purchase from family farming, and thus organic food was also more present in the South region. However, there was no difference between the presence of biodiversity foods and Brazilian regions.

Despite the incentives in Brazilian legislation for school feeding regarding the purchase and use of biodiversity foods and the biomes rich in plant species of the Southeast region (REFLORA 2020), such as the Atlantic Forest and the Cerrado, our results reinforce the lack of knowledge about native regional products, reaffirming the tendency to standardize eating habits and low varied menu without local reference (de Sousa et al. 2015). This last data concerns the dissociation observed

Fig. 2 Exotic fruit, pitaya – *Hylocereus undatus* Britton & Rose

on the nutritionist's understanding of biodiversity foods, regional foods, sustainability, and varied menu.

Another limitation often pointed out by nutritionists for not including biodiversity foods in the menus is the low acceptance by schoolchildren; but (da Cunha et al. 2014) have shown that all preparations developed with regional foods, including carambola cake, green soup, jackfruit preserve, and persimmon jelly, had an acceptance of over 85% in both urban and rural schools.

The Organic Law for Food and Nutritional Security (LOSAN) proposes expanding access to food, highlighting traditional and family farming as contributors to job creation and income redistribution, identified as participatory and sustainable strategies (Brasil 2006). In parallel, PNAE requires that food be purchased from family farmers and encourages organic, agroecological, and biodiversity foods, which represents a step forward in consolidating sustainable food systems, considering the premises of the National Policy on Food and Nutritional Security (PNSAN). Moreover, it has contributed to increase the diversity and supply of quality food and to ensure the lowest transportation cost while respecting local culture and food tradition (De Sousa et al. 2009; Ferigollo et al. 2017; Schwartzman et al. 2017). The use of biodiversity foods has also proved to be efficient for sustainable food production in other parts of the world, such as the semiarid West Africa, where local management offers opportunities to collect a vast diversity of nutrient-rich species year round, supporting domestic nutrition associated with crop diversification (Félix et al. 2018).

Food purchases from family farmers have increased over the years, but a number of acquisition bottlenecks still subsist, such as bureaucratic barriers, late payment for farmers, limitations imposed by health, and tax legislation (Bandoni et al. 2014; Gonçalves et al. 2015). Gonçalves et al. (2015) reported that 74.1% of the cities studied in the state of São Paulo bought foodstuffs from family farmers in 2010 and 2011. We see, thus, a slow progress on procurement of family farming products for school meals between 2011 and 2015, highlighting the need to adopt new strategies such as supplying food of the biodiversity and improving the articulation between PNAE actors.

Studies addressing the acquisition and use of biodiversity foods in school meals are virtually non-existent. Amorim et al. (2016) reported that the analysis of 122 public calls made in the state of São Paulo in 2013 allowed to identify only one biodiversity species – *cambuci* (*Campomanesia phaea* Bergen).

Large cities and those located in urban centers have encountered additional difficulties regarding the logistics of receiving, storing, and distributing these foods, especially fresh produce; additionally, larger cities require larger amounts of food, which only big cooperatives can offer (de Sousa et al. 2015; De Amorim et al. 2016).

To meet future food security and sustainability needs of the world's population, food production must grow substantially; however, losses from agricultural intensification have been shown in many organisms and ecosystems, leading to their depletion and/or species extinction. As food production is the economic activity that most contributes to climate change (Springmann et al. 2018), society pushes for the consumption of food produced in systems with less environmental and human health impacts. Thus, organic and agroecological agriculture has gained space, since they supposedly reduce the negative effects of traditional agriculture, being based on the dynamic interaction between soil, plants, animals, and people, the ecosystem, and the environment (Foley et al. 2011; Winqvist et al. 2012). The correlation between family farming and organic foods corroborates other literature findings, highlighting how the inclusion of family farming in public policies involved with food security is beneficial and might serve as a model for other countries. The PNAE can strengthen food systems based on agro-ecology principles and this should be valued and referenced (Guerra et al. 2018).

Family farming food production is essential for basic food crops, representing 83% of the production of cassava, 70% of beans, 58% of milk, and 45% of corn, which are strategic for the local economy but do not contribute to increasing biodiversity food consumption (Maluf et al. 2015) since the production/extractivism of such foods is low when compared with the basic food crops. To overcome this scenario, the Ministry of Social Development and Hunger Control launched the National Plan for the Promotion of Socio-Biodiversity Product Chains in 2009, aiming to develop integrated actions for promoting and strengthening biodiversity product chains, with added value and consolidation of sustainable markets (Brasil 2017). One of its main actions was the development of the Minimum Price Guarantee Policy for Socio-Biodiversity under the coordination of the Ministry of Agriculture, Livestock and Food Supply, aimed to reduce income oscillations and ensure a minimum remuneration for producers by a direct subsidy, acting as a marker of product supply. Between 2009 and 2017, the plan invested 47 million *reais* (US$12 million) in subsidies for extractive producers throughout Brazil.

Interministerial Ordinance No. 284 (Brasil 2018a) instituted a list of biodiversity species, for marketing as fresh or derivate products, in operations conducted by the Food Acquisition Program (PAA) and the National School Feeding Program (PNAE). This facilitates purchasing food biodiversity, following the specific rules of each program. The published list of species is an important guideline for PNAE managers who often claim lack of knowledge to justify noncompliance with the norm requiring that 30% of the resources passed on to school feeding by the federal government be used to purchase family farming products.

The nutritionists often cited lack of knowledge of species and ways of preparing biodiversity foods as a crucial factor limiting their offer in the school meals served. To fill this gap, the Ministry of Environment published a cookbook named *Biodiversidade Brasileira: Sabores e Aromas* (*Brazilian Biodiversity: Flavors and Aromas*), which presents biodiversity species as ingredients of preparations and can significantly assist nutritionists in planning the school menu (de Santiago and Coradin 2018). Given this context, training for nutritionists, school food handlers, and school feeding managers is fundamental for the introduction and/or improvement of all actors about the importance of buying family farming foodstuffs, as well as acquiring organic, agroecological, and biodiversity foods.

Another difficulty reported by nutritionists is access to data on the nutritional composition of biodiversity food in traditional nutritional tables, which is another obstacle to the dietary calculation of menus. In 2018, the government released a digital database called "Biodiversidade & Nutrição" (Biodiversity and Nutrition) (Brasil 2018b), with data on the nutritional composition of Brazilian biodiversity foods, as well as recipes using these foods as ingredients. Indeed, facilitating the access to knowledge about nutritional composition is a strategy to encourage the consumption and use of such foods.

5 Conclusion

Although most cities buy foodstuffs from family farming, including regional foods, biodiversity and organic foods are still rare in some regions, suggesting a needed incentive to increase consuming such foods.

Brazilian ministries and universities jointly conducted institutional actions to fill gaps that could represent limitations for using food biodiversity in school meals. These actions, implemented by laws, financial subsidies, and the development of teaching materials, allow all actors involved with the Feeding Policy and School Meals to play their roles in ensuring adequate and healthy food for Brazil's 41 million schoolchildren. Like links in a sustainable chain, providing healthier food and preserving the environment are effective ways of guaranteeing the food and nutritional security of this population group.

To strengthen institutional actions, we must carry out interventions aimed at improving the knowledge of nutritionists and school feeding managers, the nutritional value of biodiversity species, and the different preparations in which these foods can be served to students.

Acknowledgments The authors would like to thank the Biodiversity for Food and Nutrition (BFN) project, developed in Brazil, with support from the United Nations Environment Program (UNEP) and the Food and Agriculture Organization of the United Nations (FAO) and additional support from the CGIAR Research Program on Agriculture for Nutrition and Health (A4NH).

Conflict of Interest The authors declared no conflict of interest.

References

Bandoni DH, Stedefeldt E, de Amorim ALB et al (2014) Desafios da regulação sanitária para a segurança dos aliments adquiridos da agricultura familiar para a alimentação escolar. Rev Visa em debate 2:107–114

Beltrame DMO, Oliveira CNS, Borelli T et al (2016) Diversifying institutional food procurement – opportunities and barriers for integrating biodiversity for food and nutrition in Brazil. Raizes 36:55–69

Brasil (2006) Lei n° 11.346. Presidência da República. Cria o Sistema Nacional de Segurança Alimentar e Nutricional – SISAN com vistas em assegurar o direito humano à alimentação adequada e dá outras providências. Brasil

Brasil (2009) Lei n° 11.947. Presidência da República. Dispõe sobre o atendimento da alimentação escolar. Brasil

Brasil (2010) Manual de hortaliças não-convencionais. Ministério da Agricultura, Pecuária e Abastecimento. Secretaria de Desenvolvimento Agropecuário e Cooperativismo, Brasilia

Brasil (2013) Resolução n° 26. Ministério da Educação. Fundo Nacional de Desenvolvimento da Educação. Brasil

Brasil (2015) Alimentos Regionais Brasileitos, 2nd edn. Ministério da Saúde. Secretaria de Atenção à Saúde. Departamento de Atenção Básica, Brasil

Brasil (2017) Arranjos produtivos locais : APLs de produtos da sociobiodiversidade. Ministério do Meio Ambiente, Brasil

Brasil (2018a) Portaria Interministerial n° 284. Ministério do Meio Ambiente e do Desenvolvimento Social, Brasil

Brasil (2018b) Sistema de Informação sobre a biodiversidade brasileira. Biodiversidade & Nutrição Composição nutricional e receitas. In: Ministério do Meio Ambietnte. https://ferramentas.sibbr.gov.br/ficha/bin/view/FN/

da Cunha DT, Gonçalves HVB, de Lima AFA et al (2014) Regional food dishes in the Brazilian national school food program: acceptability and nutritional composition. Rev Nutr 27:423–434. https://doi.org/10.1590/1415-52732014000400004

De Amorim ALB, De Rosso VV, Bandoni DH (2016) Acquisition of family farm foods for school meals: analysis of public procurements within rural family farming published by the cities of São Paulo state. Rev Nutr 29:297–306. https://doi.org/10.1590/1678-98652016000200012

De Paiva JB, De Freitas MDCS, Da Silva Santos LA (2012) Hábitos alimentares regionais no Programa Nacional de Alimentação Escolar: Um estudo qualitativo em um município do sertão da Bahia. Brasil Rev Nutr 25:191–202. https://doi.org/10.1590/S1415-52732012000200002

de Santiago RAC, Coradin L (eds) (2018) Biodiversidade brasileira Sabores e Aromas. Ministério do Meio Ambiente, Secretaria de Biodiversidade

de Santiago RAC, Monego ET, de Assis JN, Zago MFC (2012) Alimentação saudável na culinária regional Alimentação saudável na culinária regional, 1 3d. Goiania

De Sousa LM, Martins KA, Cordeiro MDM et al (2009) Alimentação escolar nas comunidades quilombolas : desafios e potencialidades School nutrition in ' quilombola ' communities : challenges and opportunities. Cien Saude Colet 18:987–992

de Sousa AA, da Silva APF, de Azevedo E, Ramos MO (2015) Cardápios e sustentabilidade: Ensaio sobre as diretrizes do programa nacional de alimentação escolar. Rev Nutr 28:217–229. https://doi.org/10.1590/1415-52732015000200010

Fanzo J, Hunter D, Borelli T, Mattei F (2013) Diversifying food and diets: using agricultural biodiversity to improve nutrition and health, 1st edn. Routledge, New York.

Félix GF, Diedhiou I, Le Garff M et al (2018) Use and management of biodiversity by smallholder farmers in semi-arid West Africa. Glob Food Sec 18:76–85. https://doi.org/10.1016/j.gfs.2018.08.005

Ferigollo D, Kirsten VR, Heckler D et al (2017) Products purchased from family farming for school meals in the cities of Rio Grande do Sul. Rev Saude Publica 51(6). https://doi.org/10.1590/S1518-8787.2017051006648

Foley JA, Ramankutty N, Brauman KA et al (2011) Solutions for a cultivated planet. Nature 478:337–342

Gonçalves HVB, da Cunha DT, Stedefeldt E, de Rosso VV (2015) Family farming products on menus in school feeding: a partnership for promoting healthy eating. Ciência Rural 45:2267–2273. https://doi.org/10.1590/0103-8478cr20150214

Guerra J, Blesh J, Filho ALS, Wittman H (2018) Caminhos para o Manejo Agroecológico através de participação no Programa Nacional de Alimentação Escolar (PNAE) em Santa Catatina, Brasil. Cad Agroecol – An do VI CLAA, X CBA e V SEMDF 13

Hunter D, Özkan I, Moura de Oliveira Beltrame D et al (2016) Enabled or disabled: is the environment right for using biodiversity to improve nutrition? Front Nutr 3:1–6. https://doi.org/10.3389/fnut.2016.00014

IBGE (2018) Consulta Classes de tamanho da população dos municípios. In: IBGE, Inst. Bras. Geogr. e Estatística –. https://ww2.ibge.gov.br/home/estatistica/populacao. Accessed 20 Aug 2018

IPEMA (2012) Culinária Juçaneira. Instituto de Permacultura e Ecovilas da Mata Atlântica (Ipema), Associação para Cultura, Meio Ambiente Cidadani (Akarui) e Petrobrás, Ubatuba

Maluf RS, Burlandy L, Santarelli M et al (2015) Agricultura sensível à nutrição e a promoçâo da soberanía e da segurança alimentar e nutricional no Brasil. Cienc e Saude Coletiva 20:2303–2312. https://doi.org/10.1590/1413-81232015208.14032014

Monego ET, Alexandre VP, de Sousa LM et al (2013) Produção e potencial agrícolas de alimentos destinados à alimentação escolar em Goiás e no Distrito Federal, na Região Centro-Oeste do Brasil Agricultural production and potential and foods intended for schools in Goiás and Distrito Federal, located in the Midwest Region of Brazil. Rev Nutr 26:233–241

REFLORA (2020) Plantas do Brasil: Resgate Histórico e Herbário Virtual para o Conhecimento e Conservação da Flora Brasileira. http://floradobrasil.jbrj.gov.br/reflora/PrincipalUC/PrincipalUC.do

Saraiva EB, da Silva APF, de Sousa AA et al (2013) Panorama da compra de alimentos da agricultura familiar Para o Programa Nacional de Alimentação escolar panorama of purchasing food products from family farmers for the Brazilian School Nutrition Program. Cien Saude Colet 18:927–935. https://doi.org/10.1590/S1413-81232013001000004

Schwartzman F, Mora CAR, Bogus CM, Villar BS (2017) Antecedentes e elementos da vinculação do programa de alimentação escolar do Brasil com a agricultura familiar. Cad Saude Publica 33:1–2. https://doi.org/10.1590/0102-311X00099816

Scott P (2017) Global panel on agriculture and food systems for nutrition: food systems and diets: facing the challenges of the 21st century. Food Secur Sci Sociol Econ Food Prod Access to Food 9:653–654

Springmann M, Clark M, Mason-D'Croz D et al (2018) Options for keeping the food system within environmental limits. Nature 562:519–525

Tutwiler A, Padulosi S, Hunter D (2017) Securing sustainable and nutritious food systems through mainstreaming agricultural biodiversity: an interdisciplinary study. Lancet 389:S22. https://doi.org/10.1016/s0140-6736(17)31134-0

Winqvist C, Ahnström J, Bengtsson J (2012) Effects of organic farming on biodiversity and ecosystem services: taking landscape complexity into account. Ann N Y Acad Sci 1249:191–203. https://doi.org/10.1111/j.1749-6632.2011.06413.x

Zero AF (2012) Gestão Eficiente da Merenda Escolar – Histórias Gostosas de Ler e Boas de Copiar

Part VI
Learning and Teaching Brazilian Food Plants

Plant Identification Using Artificial Intelligence: Innovative Strategies for Teaching Food Biodiversity

Elias Jacob de Menezes Neto, Djackson Garcia de Lima, Ivanilda Soares Feitosa, Sávio Marcelino Gomes, and Michelle Cristine Medeiros Jacob

Contents

1 Introduction

"People care about what they know" (Balmford 2002). For this reason, species literacy, which involves species identification skills, is a pillar of projects engaged in biodiversity conservation (Aldhebiani 2018). There is evidence that species knowledge is associated with positive attitudes toward both fauna and flora conservation (Hooykaas et al. 2019). Within the biophilia hypothesis of Edward

E. J. de Menezes Neto (✉) · D. G. de Lima · S. M. Gomes · M. C. M. Jacob
Universidade Federal do Rio Grande do Norte, Natal, Brazil
e-mail: eliasjacob@ufrn.edu.br; djackson7@ufrn.edu.br; saviogomes@ufrn.edu.br; michellejacob@ufrn.edu.br

I. S. Feitosa (Deceased)
Universidade Federal de Pernambuco, Laboratório de Ecologia e Evolução de Sistemas Socioecológicos, Recife, Brazil

M. C. M. Jacob, U. P. Albuquerque (eds.), *Local Food Plants of Brazil*, Ethnobiology, https://doi.org/10.1007/978-3-030-69139-4_19

Osborne Wilson, biodiversity teaching projects find a good argument: human beings have an innate desire to catalog, understand, and spend time with other forms of life (Wilson 1984). However, studies in many countries where biodiversity education has been in place for decades (e.g., Finland, Sweden) reveal that people, even when involved in such projects, do not always perceive plants well or identify species (Randler 2008; Salatino and Buckeridge 2016; Yli-Panula et al. 2018). A probable explanation is that as industrialization and urbanization reduce our direct interactions with nature, our species' biophilic potential is redirected toward human artifacts (Balmford 2002). Technological devices, for example, are seen as reducing interactions between humanity and the environment, with the resulting consequence of little or no species literacy (Neves et al. 2019).

Some authors argue that technology can be a solution (instead of a problem) in biodiversity education projects (Rogers et al. 2004; Pfeiffer et al. 2009; March 2012; Sung et al. 2016). They understand technology as a means of teaching biodiversity-related topics and arousing the interest of generations more engaged with technology. One application in education technology is "mobile learning" or "M-learning," a modality of e-learning in which learning occurs through mobile devices, such as cell phones or tablets (Pfeiffer et al. 2009). In these cases, mobile devices are seen as tools which support the learning process, reducing the gap between the real world and classroom experiences, by transporting the information used in the room to the outside world and vice versa. M-learning initiatives, such as the "Encyclopedia of Life,"[1] can involve new audiences outside the scientific community and allow them to contemplate the richness, beauty, and relevance of biodiversity in our lives (Wilson 2003).

M-learning has proven to be more successful than traditional learning methods in producing educational outcomes (Sung et al. 2016). Some researchers have already tested M-learning in the field of biodiversity education. As an example, Yvonne Rogers et al. (2004) designed an outdoor learning experience, in the forest with teenagers, to encourage contextualized scientific biodiversity research. The study confirmed that this type of exploration promotes interpretation and reflection and has better results than traditional methods. Technologies involving direct interaction between subjects (e.g., Crowdmap, using volunteered geographic information) enable construction of complex collaboration networks between individuals while simultaneously stimulating connectivity and cooperation (Hemingway et al. 2015).

Tools that combine the use of mobile devices with artificial intelligence (AI) can also be powerful instruments in biodiversity education. "AI is a branch of computer science, and is concerned with construction and deployment of intelligent agents as computer programs and with understanding the behavior of these artifacts" (Feldman 2001, p. 792). AI already has several applications in education, such as adaptive platforms (e.g., Khan Academy, Geek, among others), that recognize student learning patterns and use them to improve strategies that increase student motivation

[1] See at: https://eol.org/

and engagement (Gatti 2019). For biodiversity, in addition to engaging young audiences with nature, AI apps address practical problems related to biodiversity conservation, such as rapidly identifying species menaced by extinction (Kwok 2019). The data processing potential of AI is crucial in responding to the current biodiversity loss crisis, guiding on-time decision-making.

This chapter discusses our development of a web-based app, trained with artificial intelligence to identify nine unconventional food plants (UFP) in a community garden in Northeastern Brazil. During their activities in the garden, participants were encouraged to visit a website, choose a plant, take a picture with their mobile devices, and submit them for identification to learn more about the species;and then to evaluate the classification provided by our model. We named this app NEIDE (NEural IDEntifier) in honor of Neide Rigo (also an author in this book), one of the Brazilian women who inspired us with her vast knowledge of UFP.[2] To build NEIDE, we used a convolutional neural network (CNN), an architecture that has successfully applied AI in a wide range of image classification tasks (including plant image identification), performing in state-of-the-art models, as well as human experts. Before presenting the elaboration of NEIDE, we will provide a narrative summary of the main ideas that supported our product's idea. Finally, we will finish our chapter by discussing the implications of using innovative products in education projects – in both formal and informal learning spaces (e.g., schools, school gardens, and community gardens).

2 Why Did We Create NEIDE?

2.1 Biodiversity Decline and Lack of Species Literacy

The current data concerning biodiversity losses are alarming. A recent report by the "Intergovernmental Science-Policy Platform on Biodiversity and Ecosystem Services" (2019) has revealed that biodiversity is declining more rapidly than at any other time in history, with the average abundance of native species in most major terrestrial habitats dropping by at least 20% since 1900. The report also shows that more than 1,000,000 species are currently under threat of extinction. It is unlikely that scholars will be able to describe even a small part of our biodiversity before it is extinguished. In Brazil, the leading causes of this loss are intensive agricultural activity (da Silva 2018) and unsustainable extraction of forest products (Lima et al. 2012; Nunes et al. 2012; Feitosa et al. 2017). In the *Cerrado* biome, agricultural degradation increases biodiversity loss and food insecurity (Cunha et al. 2008; Lima et al. 2012). The *Caatinga* (after the Atlantic Forest and the *Cerrado*) is considered the third most degraded biome in Brazil, with 80% of its vegetation modified by deforestation (Souza et al. 2015).

[2] Code and data available at https://github.com/eliasjacob/neide

The lack of interactions between people and nature is both *cause* and *consequence* of this rapid decline in biodiversity (Soga and Gaston 2016): *cause*, because when we interact little with biodiversity, we learn little about it (Hooykaas et al. 2019), and thus, the greater tendency toward alienation concerning the problem of environmental degradation and its consequences (Rozzi 2013), and *consequence* because the reduction of green spaces caused by deforestation reduces people's opportunities to interact with plant, animal, and fungi species and even with other human beings (in parks, for instance) (Celis-Diez et al. 2017).

Biodiversity education projects aim to break this harmful cycle. A primary objective of many initiatives of this nature is to increase species literacy. Species literacy involves a broad and deep knowledge about species, starting with identification skills (Aldhebiani 2018). Yet, species literacy also involves awareness of species diversity (species richness), positioning in the ecological food chain (trophic level), natural life environment (habitat), lifecycle knowledge (e.g., insect egg-larva-adult metamorphosis), knowledge of origins (whether native, exotic, or cultivated), and conservation status (abundant, threatened, or extinct) (Hooykaas et al. 2019).

2.2 The Need for a Better Approach in Biodiversity Education

As already mentioned, even in countries where biodiversity education has been a topic for decades, species literacy lacks much. The EU has had legislation on biodiversity conservation and education since the 1970s. Yet in the "Attitudes Towards Biodiversity" (2013) survey, assessing the population's knowledge of biodiversity, the European Commission interviewed roughly 25,000 citizens over the age of 15 across the European Union (EU), and the results revealed that after more than 40 years of effort, 44% of Europeans did not even know what the term "biodiversity" meant. Three out of ten knew the term, but did not know what it meant, and a quarter of the interviewees had never even heard of biodiversity. Other research (Bebbington 2005) has demonstrated that even among A-level students in biology, the ability to recognize and name common wildflowers, for example, tends to be low; the students thought that naming organisms was not an important ability and showed little interest in learning identification skills.

Teaching methods are a contributing factor for low engagement in developing biodiversity skills. Teachers who coordinate biodiversity projects frequently simply reproduce the model they know from their own training yet without contextualizing the topic to the student's reality (Silva 2013). Subjects involving botany, for example, are often taught in a mnemonic and annoying way, where the students do not recognize applications in their daily lives (da Fonseca and Ramos 2019). Consequences for a society that does not recognize its plants are drastic, and thus there is an urgent need to break this vicious cycle in science education. Ignorance of the importance of trees in rural and urban environments leads a society to stop caring about its environment and thus toward destruction of its own ecosystem (Salatino and Buckeridge 2016).

In a literature review, Navarro-Perez and Tidball (2011) identified inadequate communication as a principal obstacle toward achieving biodiversity educational outcomes. Textbooks, still considered the primary tool in the teaching and learning process, reveals this low capacity to communicate and attract the public. The content presented in such materials does little to contribute to a comprehensive understanding of biodiversity at either global or local levels, as well as not stimulating a critical understanding of nature conservation (de Fonseca 2007). As a result, student engagement is low, and despite growing public concern, environmental issues still rank below many other problems, such as terrorism, health, economics, and family values (Novacek 2008).

2.3 *Use of Technology and Innovative Methods*

The United Nations recommends development of platforms that allow young people to positively connect with biodiversity conservation (United Nations Environment Programme 2019). They argue that in biodiversity projects, the use of technology may expand young people's interaction with nature and, consequently, their engagement and learning outcomes.

Costa Rica, for example, internationally recognized for its biological diversity, is a reference country in promoting educational strategies for biodiversity conservation. Jiménez et al. (2017) report that significant progress toward biodiversity conservation has been achieved through programs that implement innovative teaching methodologies, where technology plays an important role. Virtual learning communities, such as the project "Cibercolmenas" (CyberHive, in English), promote innovative use of science and technology in the classroom, the field, and cyberspace, aiming to generate practical experiences and projects that stimulate student learning of biodiversity in both primary and secondary schools (Zamora and Calvo 2012). Another innovative way to teach biodiversity is through the use of robotics. Ruiz Vicente et al. (2020) developed an application using robotics as a teaching tool in a project involving "sustainability," which presented promising educational outcomes and demonstrated how technology is a potentially active learning tool. In the context of undergraduate education, the tool "CROPVIEW" (Comprehensive Resources for Observing Plants in a Visual Interactive Enhancement Window) through interaction and simulation on a website has the potential to increase students' understanding of global foods, production systems, agricultural practices, plant biology, geography, and climatology. Evaluating the tool, Lori Snyder et al. (2012) argued that "CROPVIEW" used to complement the teaching of environmental sciences appears promising.

The use of innovative methodologies in education is not a new topic. In the 1970s, the behaviorist B. F. Skinner defended what he called "progress in teaching" based on the complementary use of mechanical and electrical artifacts. These artifacts – called "teaching machines" – contained carefully planned material with adaptive characteristics in which the next learning problem would depend on the previous answer (Skinner 1972). The constructivist Seymour Papert also defended the computer as a teaching and learning resource. However, differently from Skinner, he argued that in place of a *computer-aided instruction* methodology, the learner should be the one to program the computer (Papert 1985). He presented a futuristic view of education, believing in the 1980s that we should be taught to communicate with computers (via programming language) to transform the way we learn. Papert was a pioneer in the field of artificial intelligence in education (AIEd).

In the past 30 years, researchers in this field of knowledge have investigated teaching and learning environments along three principal lines of research: (a) intelligent tutoring systems, which provide adaptive and individualized instruction to the learner (e.g., Duolingo); (b) intelligent support for collaborative learning, which supports and enhances group learning processes (e.g., eBird, see below); and (c) intelligent virtual reality, which generates 3D environments to make education more efficient (e.g., Blippar) (Luckin et al. 2016).

AIEd knowledge has also been explored in biodiversity projects that use human and computer learning networks (HCLNs). Such projects leverage networks of volunteers who act as intelligent sensors to collect immediate observations. These are filtered and improved by AI processes to compose aggregated databases capable of providing immediate feedback on the accuracy of the information collected. This in turn contributes to the advancement of the observer's experience, and simultaneously, as the quality of the observer data improves, the training data in which the AI processes make their decisions also improves. The idea is to have robots (by receiving help from a human collaboration network) to perform their tasks, to learn, and to become more autonomous. The eBird initiative, for instance, is a citizen science project that takes advantage of human observational capacity and machine learning methods in a biodiversity conservation and research HCLN (Kelling et al. 2013). eBird engages a global network of bird watchers to identify birds and report their observations to a centralized database. The eBird database content, together with that of other AI-based biodiversity conservation projects, provides scientists, students, and amateur naturalists with data concerning species distribution and abundance across varying spatial-temporal extensions, raising both awareness concerning biodiversity and the potential to increase species literacy.

2.4 The Idea to Create NEIDE

Some of us are part of an open-air laboratory, *LabNutrir*,[3] which provides a community garden project based at a Brazilian public university. In this space, we develop biodiversity education activities focused on unconventional food plants (UFP) through guided visits, joint projects, and culinary workshops. Each semester we receive up to 500 visitors (primary schools and higher-education students, teachers, and nutritionists among others) interested in learning more about our more than 50 UFP in the LabNutrir.

[3] See further information at http://www.nutrir.com.vc/

Our primary purpose is for the laboratory to serve as an education space. The great majority of the signalization (sign posts) within it identifies plant species, which is significant. At the first moment, we fabricated signposts from wood to identify the UFP but, due to the exposure to weather (rain and sun), they did not lasted. So, wanting to boost our teaching practices with "tech," we engaged our students and visitors using active methods such as M-learning. And thus, inspired by the ideas presented in this section and by our practical need, NEIDE was born.

3 How Did We Elaborate a Web-Based App to Identify Unconventional Food Plants?

3.1 The Dataset

Our entire dataset consists of 2561 photos of 9 species (*Bidens pilosa* L., *Bixa orellana* L., *Clitoria ternatea* L., *Commelina erecta* L., *Costus spiralis* (Jacq.) Roscoe, *Oxalis regnelli var. triangularis* Miq., *Peperomia pellucida* (L.) Kunth, *Plectranthus ornatus* Codd., and *Portulaca oleracea* L.), which were split into 2040 (training) and 521 (validation) labeled photos.

Initially, we took 997 photos for our training dataset. To avoid overfitting (i.e., when a model memorizes a dataset, instead of learning its general characteristics), we designed our training dataset to capture as many plant features as possible. We took photos on 28 different days to capture the plants' attributes during their lifecycles (i.e., with and without flowers) and under several lighting conditions (i.e., cloudy and sunny days, at noon, and sunset) and with different angles and zoom. When collecting our data, we tried to simulate (as much as possible) the final users' behavior, making the model results more robust for small differences in pictures taken by different people, with varied equipment.

Even though we collected photos with differing characteristics, they were naturally similar to one another. To mitigate this problem and to add more diversity to our training dataset, we manually collected and labeled additional 1043 photos from various sources on the Internet. We decided not to use the traditional training/validation random data splits, since this might result in a validation dataset that was artificially similar to our training dataset and undermine our ability to assess our model generalization capabilities accurately. To tackle this issue, we waited 6 months from the first collection to create our validation dataset (composed of 521 labeled images). We hoped that by waiting (due to seasonality), the general landscape and plants would change enough to make our validation dataset substantially different from our training dataset.

We conducted all experiments using the fast.ai library (Howard and Gugger 2020), which provided us with many data augmentation techniques, a critical step toward regularizing computer vision models, since simple changes in the images can result in poor deep neural network performance (Engstrom et al. 2019). With these image transformations, we managed to artificially increase our model's exposure to differing target representations by applying several affine transformations and light changes. The following image transformations were randomly applied to our training dataset:

- Horizontal and vertical flipping
- Rotation changes, up to 45 degrees (both to right and left)
- Zoom level changes, up to 1.3x
- Symmetric warping; i.e., tilting the image on all four directions (left, right, top and bottom)
- Contrast and brightness changes
- Resizing and cropping only parts of the full image

All image transformations were applied in tandem, with a probability of 80%. Afterward, we resized them for each training to fit a square, which was then gradually upsampled, allowing our model to capture more information about the target.

3.2 Architectures Used and Results

We tested the ResNet (He et al. 2016) and VGG (Simonyan and Zisserman 2015) architectures at varying depths. ResNet is arguably the most used architecture, and newer models generally present the same basic ideas (residual connections between convolutional layers). We decided to use CNNs because they are broadly used for flower (Thanh et al. 2016) and plant identification in the natural environment (Sun et al. 2017). Bodhwani et al. (2019) have successfully used ResNet to identify plants from the LeafSnap dataset, providing a reasonable feasibility baseline.

The models were pre-trained on the ImageNet dataset (Deng et al. 2009), starting with a model already trained for another image classification task and then fine-tuning its weights to our desired goal. This approach is called transfer learning, and it results in better models with far less data than would be required if we had started with randomly initialized weights (Shaha and Pawar 2018). Although Mehdipour Ghazi et al. (2017) have used transfer learning between similar datasets (from one plant identification task to another), we show that one can apply this technique even when the target classification task is intrinsically different from the pre-trained model.

Based on previous tests on other datasets, we added an extra batch normalization to all models, just before the output layer; this slightly improved their accuracy. We also set the learning rate of each parameter group in accordance with the one-cycle learning rate policy described by Smith and Topin (2017). This policy leads to what they describe as super-convergence, which allows neural networks to train much faster than with standard learning rate schedules, boosting the performance in cases like ours, where labeled data is limited.

Our best model achieved an accuracy of 96.35%, with an apparent correlation (R^2 = 0.56) between the model's capacity (expressed as its number of trainable

Table 1 Final error rate for each model architecture, considering the number of trainable parameters (better model in bold)

Model architecture/depth	Error rate (%)	# of trainable parameters
ResNet 18	10.75	11,709,019
ResNet 34	4.60	21,817,179
ResNet 50	8.83	25,619,547
ResNet 152	**3.65**	**60,255,323**
VGG 11	8.06	9,758,473
VGG 13	9.79	9,943,369
VGG 16	7.49	15,255,643
VGG 19	7.87	20,567,881

parameters) and performance, where larger models (in general) achieved better results. This is evident when we compare the error rate of the smallest ResNet (10.75%) with the largest (3.65%). In Table 1, we present the final error rate for each architecture/size.

Despite our good results, we were dissatisfied working with a black-box model alone, especially considering our educational goals. This led us to look deeper into the model and understand its decisions better. From the work of Selvaraju et al. (2020), which uses gradient-weighted class activation mapping (Grad-CAM) to create visual explanations for a model's output, i.e., the gradient of a label to map important regions of an image, or what is influencing its final decision. We randomly sampled examples from our validation dataset, which enabled us to verify that our model (to classify plant images) was correctly targeting flower and leaf structures. We present two examples below, where relevant areas are seen as brighter spots (Fig. 1).

3.3 *Safety Considerations for Model Deployment*

Our model is readily available for anyone who visits our website, where anyone can submit a sample image for analysis.[4] Currently, no machine learning model can provide results for a category it is not trained to predict: it will always transform any input into one previously known label. In simpler terms, the model cannot say by itself "I do not know." Any implementation must take this into account.

The model's final layer output (logit scores) consists of an array with one raw score for each class. A softmax function converts these scores into numbers between zero and one (and which sum to one). This can conveniently be interpreted as the probability distribution of potential outcomes for a given input. For the predicted result, the model will choose the label with the highest probability. The limitations

[4] See at http://nutrir.com.vc/neide

Fig. 1 Mapping the regions of the image used by AI to identify plants. (**a**) Grad-CAM visualization for *C. ternatea*, highlighting its leaves and blue flowers. (**b**) Grad-CAM visualization for *P. pellucida*, highlighting its leaves and inflorescences

pose a challenge for the deployment phase, especially for the user interface, which needs to account for user out-of-context submissions. As an example, in one case, a child took a selfie and submitted it for analysis. A plant label was assigned to her photo.

During our beta test, we implemented three measures to mitigate such problems. First, we informed users regarding how the model is trained to identify only listed species. Second, a threshold level before showing results for final use was selected, such that the user interface refrains from showing any results if the maximum probability for a given class is below the threshold. Finally, we provided differing interpretations for each probability interval, which was easily achieved with color-coded results. In this case, we decided to hide results with a $p < =0.6$, display progress bars in red ($0.6 > p > 0.7$), yellow ($0.7 > p > 0.85$), and green ($p > = 0.85$), and always provide a warning concerning the intrinsic uncertainty of the model. This caution is especially important since we are working with food plants, and we need to avoid any risk of undesirable consumption.

Any final implementation must also possess other security checks, like collecting GPS metadata from the photo itself, to ensure it comes from a plant inside the garden. Alternatively, if a WiFi connection were available in the garden, the service could be rendered inaccessible to those outside of the local network. Such security measures may reduce misuse and guide users to understand the tool's limitations better.

4 Educational Implications of NEIDE and Other Innovative Methods

All considered, educators seeking to foster species literacy in their own projects will value tools like ours since they may stimulate spontaneous and autonomous learning concerning plants. If applied in proper pedagogical proposals, people of different age groups can benefit from projects using apps like NEIDE, whether individually or collectively.

Community and school gardens are environments that create unique opportunities to increase environmental awareness and biocultural diversity (Vandebroek et al. 2020). For this reason, they are also strategic spaces for developing and using plant identifiers with pedagogical purposes. We highly recommend that educators use tools as trained for in controlled spaces, where they can count on a well-known plant inventory. A plant list can increase the educators' ability to prepare and invest their students (and other users) with species literacy. Tech is a tool to facilitate the learning process. The educator's role remains irreplaceable: to foster debates, to provide context, and to enhance user's and students' capacity toward thinking in broader terms, particularly about the presence or absence of UFP in local markets.

During the web-based app activities, the user can collaborate to improve the model even as the tool allows active information searches. If the tool is used only locally, data generated concerning improvement of the model may also foster both discussions and the students' potential to instruct the model. Students can also be stimulated to take notes and/or print-screens of the processing results, or of other observations made in the field to debate in class. Details in pictures, for example, can serve as kick-starters for discussing various science education topics, from botany to tech.

In Brazil, biodiversity is one of the themes included in the "National Common Curricular Base" (Marconces 2018). One of the main challenges for implementing technological innovations in addressing the theme is the scarcity of physical resources and qualified personnel in schools. According to data from PISA, the "Program for International Student Assessment," Brazilian schools average 22.1 students per computer. The country is second only to Turkey and Tunisia, respectively, with 44.9 and 53.1, and is well behind the average of the countries in the "Organization for Economic Cooperation and Development" (OECD 2015), which is 4.7. The same report has demonstrated that one in four Brazilian teachers reports needing professional training in information technology to improve their teaching skills. Finally, data from the 2018 "Brazilian School Census" reveal that only 44% of schools possess a science laboratory (Instituto Nacional de Estudos e Pesquisas Educacionais Anísio Teixeira 2019), an essential space for experimentation, hypothesis testing, and consolidation of ideas (Krasilchik 2008; Alexander and Mar 2017). Our proposal, which uses mobile devices and community gardens, may thus be adequate for the country's scenario. The problem regarding teachers' low capacity to deal with information technology can be addressed through multiprofessional initiatives, involving support of extension projects from the university.

5 Final Considerations

In this chapter, we argued that using technology and innovative methods may provide a better approach to biodiversity education. We also revealed a practical how-to example of developing a web-based app with AI to identify UFP, which may be useful to guide those supporting biodiversity projects in education. We have also demonstrated that there is no need for large datasets when training deep CNNs. This is due to the ease with which one can apply completely unrelated classification tasks through transfer learning.

The use of NEIDE, or another similar tool, may open opportunities to build knowledge, not just in biodiversity but also in AI. We believe that even data collection (the most involved arduous step) can become part of the educational activities, where participants engage with the environment to better understand how computers themselves learn. Educating people to comprehend the basics in AI will also increase the likelihood of a more diversified societal debate – including diverse gender, social class, and ethnicity. As in biology, diversity in AI will be critical to building a more comprehensive portrait of our plants (including the neglected ones), animals, landscapes, ethnicities, and cultural diversity.

Acknowledgments We dedicate this chapter to our beloved friend Ivanilda Feitosa who suddenly left us. We feel thankful for all friendship and knowledge shared.

References

Aldhebiani AY (2018) Species concept and speciation. Saudi J Biol Sci 25:437–440. https://doi.org/10.1016/j.sjbs.2017.04.013

Alexander Y, Mar O (2017) Enseñanza y Aprendizaje de las Ciencias the teaching of biodiversity: trends and challenges in pedagogical experiences. Revista Góndola 12:173–185. https://doi.org/10.14483/23464712.11599

Balmford A (2002) Why conservationists should heed Pokemon. Science 295:22367. https://doi.org/10.1126/science.295.5564.2367b

Bebbington A (2005) The ability of A-level students to name plants. J Biol Educ 39:63–67. https://doi.org/10.1080/00219266.2005.9655963

Bodhwani V, Acharjya DP, Bodhwani U (2019) Deep residual networks for plant identification. Proc Comput Sci 152:186–194. https://doi.org/10.1016/j.procs.2019.05.042

Celis-Diez JL, Muñoz CE, Abades S et al (2017) Biocultural homogenization in urban settings: public knowledge of birds in city parks of Santiago, Chile. Sustainability (Switzerland) 9:485–490. https://doi.org/10.3390/su9040485

Cunha NR, Lima JE, Gomes MF, Braga MJ (2008) A intensidade da exploração agropecuária como indicador da degradação ambiental na região dos cerrados, Brasil. Rev Econ Sociol Rural 46:291–323. https://doi.org/10.1590/S0103-20032008000200002

da Fonseca LR, Ramos P (2019) Ensino de botânica na licenciatura em Ciências Biológicas em uma universidade pública do Rio de Janeiro: contribuições dos professores do ensino superior. Ensaio Pesquisa em Educação em Ciências (Belo Horizonte) 20:1–23. https://doi.org/10.1590/1983-21172018200l026

da Silva CM (2018) Between fenix and ceres: the great acceleration and the agricultural frontier in the brazilian cerrado. Varia Historia 34:409–444. https://doi.org/10.1590/0104-87752018000200006

de Fonseca MJ (2007) A biodiversidade e o desenvolvimento sustentável nas escolas do ensino médio de Belém (PA), Brasil. Educacao e Pesquisa 33:63–79. https://doi.org/10.1590/s1517-97022007000100005

Deng J, Dong W, Socher R et al (2009) ImageNet: a large-scale hierarchical image database. In: 2009 IEEE conference on computer vision and pattern recognition. IEEE, pp 248–255

Engstrom L, Tran B, Tsipras D et al (2019) Exploring the landscape of spatial robustness. In: proceedings of the 36th international conference on Machine Learning, pp 1802–1811

European Commission (2013) Attitudes towards biodiversity. Flash Eurobarometer 379:1–145

Feitosa IS, Sobral A, Monteiro JM et al (2017) Impact of collection on bark regeneration from Stryphnodendron rotundifolium Mart in northeastern Brazil. Environ Monit Assess 189:1–11. https://doi.org/10.1007/s10661-017-5908-4

Feldman J (2001) Artificial intelligence in cognitive science. In: International Encyclopedia of the Social & Behavioral Sciences. Elsevier, Oxford pp 792–796

Gatti FN (2019) Educação básica e inteligência artificial: perspectivas, contribuições e desafios. Dissertation, Pontifícia Universidade Católica

He K, Zhang X, Ren S, Sun J (2016) Deep residual learning for image recognition. In: 2016 IEEE conference on computer vision and pattern recognition (CVPR). IEEE, pp 770–778

Hemingway C, Adams C, Stuhlsatz M (2015) Digital collaborative learning: identifying what students value. F1000 Res 74:1–13

Hooykaas MJD, Schilthuizen M, Aten C et al (2019) Identification skills in biodiversity professionals and laypeople: a gap in species literacy. Biol Conserv 238:1–10. https://doi.org/10.1016/j.biocon.2019.108202

Howard J, Gugger S (2020) Fastai—a layered API for deep learning. Information 11:1–27

Instituto Nacional de Estudos e Pesquisas Educacionais Anísio Teixeira (2019) Noventa e cinco por cento das escolas de ensino médio têm acesso à internet, mas apenas 44% têm laboratório de ciências. In: INEP. http://portal.inep.gov.br/artigo/-/asset_publisher/B4AQV9zFY7Bv/content/dados-do-censo-escolar-noventa-e-cinco-por-cento-das-escolas-de-ensino-medio-tem-acesso-a-internet-mas-apenas-44-tem-laboratorio-de-ciencias/21206%0D. Accessed 28 Jan 2020

Intergovernmental Science-Policy Platform on Biodiversity and Ecosystem Services (2019) The global assessment report on biodiversity and ecosystem services – summary for policy makers. IPBES, pp 1–56

Jiménez A, Monroe MC, Zamora N, Benayas J (2017) Trends in environmental education for biodiversity conservation in Costa Rica. Environ Dev Sustain 19:221–238. https://doi.org/10.1007/s10668-015-9734-y

Kelling S et al (2013) eBird: a human/computer learning network to improve biodiversity conservation and research. AI Magazine, Spring 10–21

Krasilchik M (2008) Prática de Ensino de Biologia. 6th edn. Edusp, São Paulo

Kwok R (2019) AI empowers conservation biology. Nature 567:133–134. https://doi.org/10.1038/d41586-019-00746-1

Lima ILP, Scariot A, de Medeiros MB, Sevilha AC (2012) Diversidade e uso de plantas do cerrado em comunidade de geraizeiros no norte do estado de Minas Gerais, Brasil. Acta Botanica Brasilica 26:675–684. https://doi.org/10.1590/S0102-33062012000300017

Luckin R, Holmes W, Griffiths M, Forcier L (2016) Intelligence unleashed: an argument for AI in education. Pearson, London

March KA (2012) Backyard botany: using GPS technology in the science classroom. Am Biol Teach 74:172–177

Marconces M (2018) As Ciências da Natureza nas 1ª e 2ª versões da Base Nacional Comum Curricular. Estudos Avançados 32:269–284. https://doi.org/10.1590/s0103-40142018.3294.0018

Mehdipour Ghazi M, Yanikoglu B, Aptoula E (2017) Plant identification using deep neural networks via optimization of transfer learning parameters. Neurocomputing 235:228–235. https://doi.org/10.1016/j.neucom.2017.01.018

Navarro-Perez M, Tidball KG (2011) Challenges of biodiversity education: a review of education strategies for biodiversity education. Int Electr J Enviro Educ 2:13–30. https://doi.org/10.18497/iejee-green.65475

Neves A, Bündchen M, Lisboa CP (2019) Cegueira botânica: é possível superá-la a partir da Educação? Ciência & Educação (Bauru) 25:745–762. https://doi.org/10.1590/1516-731320190030009

Novacek MJ (2008) Engaging the public in biodiversity issues. In: Proceedings of the National Academy of Sciences of the United States of America, pp 11571–11578

Nunes AT, Nascimento VT, Feitosa IS et al (2012) Caatinga plants with nutritional potential: a review from the work "contribution to the study of the Flora from Pernambuco, Brazil" (1954) by Dárdano de Andrade Lima. Ethnobiol Conserv 5:1–18. https://doi.org/10.15451/ec2012-8-1.5-1-18

OECD (2015) Students, computers and learning: making the connection. PISA, OECD, Paris

Papert S (1985) LOGO: Computadores e educação. Editora Brasiliense, São Paulo

Pfeiffer VDI, Gemballa S, Jarodzka H et al (2009) Situated learning in the mobile age: mobile devices on a field trip to the sea. Alt-J 17:187–199. https://doi.org/10.1080/09687760903247666

Randler C (2008) Teaching species identification – a prerequisite for learning biodiversity and understanding ecology. Eurasia J Math Sci Technol Educ 4:223–231. https://doi.org/10.12973/ejmste/75344

Rogers Y, Price S, Fitzpatrick G et al (2004) Ambient wood: designing new forms of digital augmentation for learning outdoors. Proceedings of the 2004 Conference on Interaction Design and Children: Building a Community, IDC 2004 3–10. https://doi.org/10.1145/1017833.1017834

Rozzi R (2013) Biocultural ethics: from biocultural homogenization toward biocultural conservation. In: Linking ecology and ethics for a changing world: values, philosophy, and action. Springer, Dordrecht, Netherlands, pp 9–32

Ruiz Vicente F, Zapatera A, Montes N, Rosillo N (2020) Steam robotic puzzles to teach in primary school: a sustainable city project case. In: Merdan M, Lepuschitz W, Koppensteiner G, Balogh R, Obdržálek D (eds) Advances in intelligent systems and computing. Springer, New York, pp 65–76

Salatino A, Buckeridge M (2016) Mas de que te serve saber botânica? Estudos Avancados 30:177–196. https://doi.org/10.1590/S0103-40142016.30870011

Selvaraju RR, Cogswell M, Das A et al (2020) Grad-CAM: visual explanations from deep networks via gradient-based localization. Int J Comput Vis 128:336–359. https://doi.org/10.1007/s11263-019-01228-7
Shaha M, Pawar M (2018) Transfer learning for image classification. In: 2018 second international conference on electronics, communication and aerospace technology (ICECA). IEEE, pp 656–660
Silva JR (2013) Concepções dos professores de botânica sobre ensino e formação de professores. Thesis, Universidade de São Paulo
Simonyan K, Zisserman A (2015) Very deep convolutional networks for large-scale image recognition. In: Published as a conference paper at ICLR 2015, pp 1–14
Skinner BF (1972) Tecnologia do ensino. Ed. da Universidade de São Paulo, São Paulo
Smith LN, Topin N (2017) Super-convergence: very fast training of neural networks using large learning rates. In: Pham T (ed) Proceedings of Artificial Intelligence and Machine Learning for Multi-Domain Operations Applications, Spie Defense, Baltimore. https://doi.org/10.1117/12.2520589
Snyder LJU, Cathey SE, Quesenberry K et al (2012) Evaluation of CROPVIEW as a crop science teaching resource for post-secondary educators. NACTA J 56:64–69
Soga M, Gaston KJ (2016) Extinction of experience: the loss of human-nature interactions. Front Ecol Environ 14:94–101
Souza B, Artigas R, Lima E (2015) Caatinga and desertification. 131–150. https://doi.org/10.4215/RM2015.1401
Sun Y, Liu Y, Wang G, Zhang H (2017) Deep learning for plant identification in natural environment. Comput Intell Neurosci 2017:1–7. https://doi.org/10.1155/2017/7361042
Sung YT, Chang KE, Liu TC (2016) The effects of integrating mobile devices with teaching and learning on students' learning performance: a meta-analysis and research synthesis. Comput Educ 94:252–275. https://doi.org/10.1016/j.compedu.2015.11.008
Thanh T, Nguyen N, Le VT et al (2016) Flower species identification using deep convolutional neural networks. In: Regional Conference on Computer and Information Engineering
United Nations Environment Programme (2019) Gaming for the planet: how green tech can help us care. In: UN. https://www.unenvironment.org/news-and-stories/story/gaming-planet-how-green-tech-can-help-us-care. Accessed 27 Jan 2020
Vandebroek I, Pieroni A, Stepp JR et al (2020) Reshaping the future of ethnobiology research after the COVID-19 pandemic. Nat Plants 6:723–730. https://doi.org/10.1038/s41477-020-0691-6
Wilson EO (1984) Biophilia. Harvard University Press, Cambridge, MA
Wilson EO (2003) The encyclopedia of life. Trends Ecol Evol 18:77–80. https://doi.org/10.1016/S0169-5347(02)00040-X
Yli-Panula E, Jeronen E, Lemmetty P, Pauna A (2018) Teaching methods in biology promoting biodiversity education. Sustainability (Switzerland) 10:1–18. https://doi.org/10.3390/su10103812
Zamora N, Calvo A (2012) Cyberhives: virtual communities of learning on biodiversity Ar. Biol Int 50:5

An Educational Experience with a Focus on Plants: Capacity Building in Nutrition Students in a Brazilian Laboratory-Vegetable Garden

Joana Yasmin Melo De Araujo, Laura Brito Porciuncula, and Thiago Perez Jorge

Contents

1 Introduction

The global agenda for sustainable development (emphasizing the United Nations Agenda for 2030, UN 2017) points to the need for investment in training of professionals ready in their practice to recognize and address the impacts of food systems[1] (Jacob and Araújo 2019). Among such professionals, the nutritionist stands out, since his practice (the task of thinking about diet) integrates relevant aspects of human and environmental health (Lang and Barling 2013). In view of

[1] Food systems deal with all actors and their interrelated activities in "production, aggregation, processing, distribution, consumption, and disposal of food." They consist of several subsystems, such as agriculture and food supply, and, with the multiplicity of participants and dimensions involved in the systems, have impacts on the social, economic, and environmental scales on a global scale (FAO 2013).

J. Y. M. De Araujo
Departamento de Nutrição, Universidade Federal do Rio Grande do Norte, Natal, Brazil

L. B. Porciuncula · T. P. Jorge (✉)
Nutrition Department, Federal University of Rio Grande do Norte, Natal, Brazil
e-mail: lauraporciunc@ufrn.edu.br

M. C. M. Jacob, U. P. Albuquerque (eds.), *Local Food Plants of Brazil*, Ethnobiology, https://doi.org/10.1007/978-3-030-69139-4_20

this, by working toward sustainable diets, the nutritionist might lead in the construction of sustainable food systems. According to the Food and Agriculture Organization of the United Nations (FAO), sustainable diets present low environmental impact: they contribute to a healthy life for present and future generations, are protective and respectful of biodiversity and ecosystems, are culturally acceptable and accessible, are economically fair and accessible, are nutritionally adequate and safe, and optimize natural resources (Burlingame and Dernini 2012).

In addition to that, if on one hand, excessive fragmentation of knowledge in nutritionist training, in most cases, favors an approach focused only on nutrients (Azevedo and Rigon 2016), neglecting social, environmental, and economic contexts, on the other hand, it has also presented a slow evolution toward incorporation (into curricular structures) of new concepts and questions concerning sustainable development. The official Brazilian document in force for higher education in Nutrition dates back to 2001 and when dealing with the contents to be covered during training explains environmental themes only once, as part of understanding "ecological" determinants (Brasil 2001).

However, the 2013 "Consensus document concerning skills and competency of nutritionists in the field of public health," being relevant to the entire field, incorporates the dimensions of sustainability and biodiversity, as being based on four professional competencies that must be contained in training courses, being intrinsic to delineation of the new professional:

1. Identify and value aspects related to sustainability and food consumption and incorporate them into Food and Nutrition Education practices;
2. Promotion of a sustainable food supply;
3. Identify and analyze the relationship between behavior, eating habits, culture, territoriality, sustainability, and food diversity.
4. Promotion of water quality in the context of the Human Right to Adequate Food. (Recine and Mortoza 2013)

It is with this formative concern that the Community Garden/Horta (LabNutrir) emerges, seeking to incorporate and strengthen the topic of sustainability in nutrition training. The vegetable garden laboratory brings an agroecological perspective to growing food, of teaching, and of research and extension aimed at learning about food and nutrition with a focus on the biodiversity of unconventional food plants (UFP). The applied methodology is garden-based learning (GBL), which uses vegetable gardens as teaching tools (Desmond et al. 2004).

In this chapter, we will present the experience at LabNutrir to exemplify how UFP can be approached based on GBL as we share instruments and ideas we developed with our students. Also, we will explore the perception of Nutrition students with respect to the potential knowledge, skills, and attitudes that they observe and experienced in GBL strategy considering their living in LabNutrir.

2 Materials and Methods

This study is qualitative and describes the GBL method, analyzing the effects of teaching nutrition with plants; and we will use the case study approach (Creswell 2014).

In this case, we intended to evaluate the learning process at LabNutrir with the starting question: how can teaching and learning of sustainability and biodiversity be developed though the GBL method? This starting point is the motivation for us to present the outcomes of using UFP as a learning and teaching tool with undergraduate students of Nutrition and their perspectives concerning knowledge, abilities, and attitudes developed in LabNutrir. It should be noted that the authors are members of the study site/locale, which is why the text takes the form of an experience report. The data were obtained through two combinations:

1. Collection and examination of data sheets produced during 2019, on the eight non-conventional food plants studied, being available on the Project website.[2] These data sheets systematize information concerning the UFP – scientific name of the plant, photography, vernacular names, origin, occurrence biome, food uses, indicators of nutritional composition (macronutrients, micronutrients, and bioactive compounds), and references. In this way, we choose this tool to systematize UFP information based on the idea that it materializes the transdisciplinary approach applied through GBL method, as the students involved in this data collection have to connect production and consumption of the species thought ethnobotanical interviews and research, and thus it supports for the education development in food and nutrition, focused especially on biodiversity during the conception of it, and the product is a tool of easy, accessible, and democratic scientific dissemination.
2. Semi-structured interviews carried out with four students by telephone. In 2019, of the ten students directly involved in the project, four had joined this year. Thus, it was decided to choose these students to assess the potential of GBL method in a short period of time in comparison to other students of the project. The interviews were conducted in late 2019 by telephone. The conversations were recorded, and their content was transcribed. A floating reading allowed first contact and organization of the material. Students' speeches will be used in both sessions, firstly, as basis to emphasize benefits from the GBL method. For categorical aggregation in the three items of capacity building (knowledge, skills and attitudes), we opted for direct interpretation of students' responses (Creswell 2014).

Conceptually, competencies are structured cognition modalities, that is, they refer to the relationships between people, objects, situations, and phenomena that one desires to understand. Skills, on the other hand, stem from the competencies acquired by the individual and refer to immediate "know-how." Thus, through

[2] See at www.nutrir.com.vc/blog/biodiversidade/

actions and operations, skills improve and articulate themselves, aiming at a new (re)organization of competences (Brasil 2000). Strictly speaking, capacity building is not carried out in isolation. For the future professional, thinking, doing, living, and being are intrinsically linked to competencies and skills development. Thus, in each answer from the academics, all from the nutrition course, we obtain an example necessary for construction of categories of analysis with respect to the development of knowledge, skills, and attitudes.

Thus, it is of extreme importance to say that it is not possible to "teach" competencies. According to Perrenoud (2008), from the perspective of the subject of education, instead of a "teaching" process, there is a mobilization of his mind from resources, skills, and knowledge with the starting point of any demand (i.e., needs and new or unforeseen situations). Thereby, capacity building emphasizes the pedagogical method and the learning situation, overcoming the false theory-practice dichotomy.

3 The GBL Proposal as Developed in LabNutrir

The Community Garden-Based Laboratory (GBL) (LabNutrir) is located on the central campus of the Federal University of Rio Grande do Norte, in the city of Natal, Northeastern Brazil. Natal is the capital of Rio Grande do Norte and has the largest population in the state, with more than 800,000 inhabitants.

LabNutrir covers an area of 1200 m^2 divided into different growing beds, scattered plantings, an orchard, a pollinator garden, compost space, seedling nursery, bee houses, two circular areas for group conversations, and a space for tool storage. In accordance with agro-ecological practices, no chemical-based fertilizers or pesticides are used. We have more than 131 edible plant species in LabNutrir,[3] distributed over 55 different botanical families; half of them are UFP.

Opened in November 2017 as the Community Nutrition Garden (HCN, in Portuguese), as a result of the need to incorporate local biodiversity into the curriculum, it was institutionalized in April 2019 as a laboratory of the Department of Nutrition of the Federal University of Rio Grande do Norte (UFRN) (Fig. 1).

Institutionalization of the first open-air laboratory in Brazil guaranteed the maintenance of the space, greater fundraising, and autonomy, with recognition of the legitimacy of the garden as a structure of excellence in teaching and learning, within a perspective that links biodiversity to nutritionist training in both content and skills.

Thus, an approach was sought that dealt with differing fields of knowledge in a transversal way, an active and transdisciplinary methodology where students are invited to deal critically and reflectively with the problems and solutions involved in the themes worked experientially, from the physical space, and in management of a vegetable garden (Desmond et al. 2004; Jacob 2020).

---

[3] Check our inventarium at http://nutrir.com.vc/horta/Completo.pdf

Fig. 1 The LabNutrir physical space. (*Source*: LabNutrir)

Thus, to put the GBL method into practice, the creators of the then HCN sought a theme (to guide their pedagogy) that reconciled the integration of curricular components of different teaching departments at UFRN: Agronomy, Botany, Ecology, and Nutrition. In this way, we came to *the topic of UFP, the main axis of the GBL method*. UFP are characterized as edible plants, exotic or native, which are often hard to identify or inaccessible to a certain population to buy. Their definition varies in time and space, depending on whether these plants are part of the usual diet of a given community, and thus, it offers the potential to contribute with food and nutrition security, health, generation of income, and ecological integrity (Jacob 2020).

At LabNutrir, planning each semester, the UFP that will be studied in the following term are selected in a wide debate involving the Garden/Horta members. In 2019, the laboratory-garden had 5 teachers, 3 assistants, 10 student monitors, and 20–30 volunteers, including students from various courses at UFRN and residents of the metropolitan region of Natal-RN.

The following sections present our results and discussions. We start presenting the result of a data collection instrument developed at LabNutrir to study the UFP during the year of 2019 among our students, as a training strategy for the promotion of environmental sustainability based on the valorization of local biodiversity. In the next section, we explore the perceptions of impacts on the teaching and learning process based on the potential of acquisition of knowledge, skills, and attitudes involved in the GBL method with their experience in the garden lab.

4 Unconventional Food Plants: An Axis in Education to Promote Biodiversity and Sustainable Diets

It is known that Brazil as a country presents the greatest biological diversity in the world; its flora represents more than 20% of the total species existent on the planet (Martinelli and Moraes 2013). However, loss of food biodiversity is a global reality and affects large reservoirs of biological diversity on the planet, noting that more than half of the energy currently consumed globally comes from only four crops: rice, potatoes, wheat, and corn (FAO 2010).

In this panorama, there is a huge amount of edible plants' species underexploited, and thus they are classified as unconventional. Even so, their use in traditional communities is rescued, thinking cultural heritage and local ethnoculinary are distant from and less influenced by the urban globalization process. Also, these plants are of great ecological and economic importance to local communities (Kinupp and De Barros 2008; Kinupp and Lorenzi 2014; UN 2017). To this matter, UFP are a medium for several lines of study, including Ethnobotany, Agroecology, Biodiversity, Sustainability, and Nutrition.

UFP can well function as a tool for multidisciplinary studies with biocultural perspectives. They fit in as plants which strengthen food biodiversity, being essential for construction of sustainable diets. The sustainable diet has been characterized according to Willett et al. (2019) and Forouhi and Unwin (2019), as one based on plant diversity and, occasionally, moderate amounts of animal protein. To fill gaps involving (1) accessibility of data on food biodiversity, (2) lack of data on culinary use in ethnobotanical studies, and (3) lack of data on nutritional composition, Jacob and Albuquerque (2020) reveal the need for this perspective.

Elaboration of sustainable menus or diets by nutritional professionals has been limited, this considering that the principal tools involving nutritional assessment do not consider culture and, for the most part, do not contain data that relate biodiversity between species and subspecies (Fao 2017; Jacob and Albuquerque 2020).

LabNutrir seeks to overcome these limitations with the implementation of activities for the development of UFP. The data sheets serve as instruments of scientific dissemination that allow simultaneous analysis of the plants, the people, and their diets, in addition to corroborating the dialogue between nature and food plant culture. The instrument assists in design activities involving Food and Nutrition

Education within the community, whether in practical classes, culinary workshops, or agricultural management (Etkin 2006; Tumilowicz et al. 2016; Jacob 2020).

For this, the data sheets were developed: for compiling data related to food biodiversity, assisting in the propagation of their content, and bringing knowledge of the nutritional and cultural nature of the plants. They serve as training instruments for nutrition professionals and contribute to an education that overcomes the barriers of nutritionism.[4]

Each semester, four UFP are chosen for research among students, project team, and community. Favorable conditions for plant development, functionality for recipes, and seasonality are considered when choosing which UFP to work with. The variables selected for collection are representative photos of the species, popular names and uses, morphological characteristics, origin, ecology of the species, propagation and cultivation, food uses, nutritional information, toxicity, and a recipe, once they are the main gaps for access and recognition of these species by the population. This activity is associated as main research of the four curricular components associated with LabNutrir: Socio-anthropological Aspects of Human Food, Food and Nutrition Education, Principles of Agroecology, and Sustainable Food Systems. In this way, students of this component carry out research in scientific databases, by culinary practice and through ethnobotanical surveys carried out with the local population.

After their collection and interviews, the information is compiled and edited by student monitors linked to the project, who structures the final product of the data sheets. The teachers perform a review of the material, possibly indicating more in-depth searches and aligning the information. Finally, the sheets are made available in the Project website,[5] as a tool for scientific dissemination of easy access. This workflow of how the data sheets are made is being represented in Fig. 2.

By 2019, 131 species of plants were identified in the garden space, where 64 were characterized as unconventional and/or medicinal and 35 as native, such as Jurubeba (*Solanum paniculatum* L.), Cariru (*Talinum triangulare* (Jacq.) Willd.), Taioba (*Xanthosoma taioba* E.G. Gonç), and Couvinha (*Porophyllum ruderale* (Jacq.) Cass), among others. In 2019, eight species plants were cataloged. Table 1 presents the data for scientific name, origin, distribution/occurrence, and food uses of the eight plants cataloged by LabNutrir in 2019, represented in Fig. 3.

With the aim of demonstrating education outcomes specially related to sustainable food systems, food biodiversity, and food and nutrition education achieved through the elaboration of UFP data sheets, we have made explicit here some speeches from students included in the project that denote in a clear way how this process is operationalized in practice.

To begin, the following sentence of Student 2 "I've learned about UFP and non UFP, discovered that there are specific ways to photograph plants, about plants'

[4] Refers to ideological discourses employed by food industry, based on nutrition dietary rationality that reduces the dimensions of food, disconnecting the act of eating of food, diet, environment, and the world (Scrinis 2008).

[5] UFP data sheets are available at http://nutrir.com.vc/blog/biodiversidade/

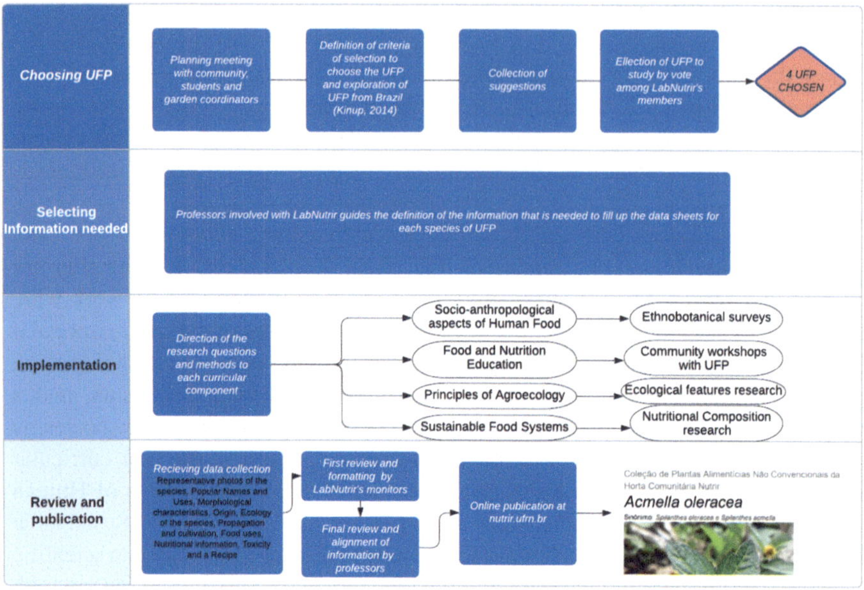

Fig. 2 Workflow for UFP data sheets. (*Source*: Authors)

families (…)" illustrates the importance of this notion. In this way, it is possible to infer that the recognition of plants' family and common characteristics between species, such as the photography that allows distinguishing details, facilitates, for instance, the comprehension of similar preparation techniques among those species of the same family.

Moreover, when it comes to plants, it is commonplace that there are confusions in the moment of recognition of species, particularly with popular names. With that in mind, acquiring the knowledge about origin, place of occurrence, and scientific name of species helps the understanding about food utilization and proper handling. To this matter, Student 1 says that "I did not have notion about the amount of food available in an easy, economic and biodiverse way," which can be related with Food Uses column of Table 1. This statement exposes a gap around preparation techniques and uses of UFP, once the student reveals that his experience in LabNutrir promoted an amplification of his food perspective and thus made it possible to see new possibilities of preparation and consumption of plants that can enrich local market and contributes with healthier and sustainable food habits, since they encompass biodiversity through UFP.

Furthermore, experiential education obtained through the process of making UFP data sheets, illustrated by Fig. 1 (flowchart), allows us to recognize LabNutrir's importance of broadening the comprehension of Nutrition as social-environmental science, once the inclusion of outcomes from the promotion of ethnobotanical surveys with locals brings up the biocultural commonly neglected in this field of knowledge. In this case, food biodiversity comprehension requires

Table 1 Data from the UFP studied at the Garden/*Horta Comunitária Nutrir* (HCN) Laboratory in 2019. Natal-RN, Brazil

Scientific name	Origin	Distribution/occurrence	Food uses
Moringa oleifera Lamarck	India	Africa, Central America, South America, Sri Lanka, India, Mexico, Malaysia, and the Philippines	Flowers and leaves, raw or cooked. Fruits, cooked
Eugenia pyriformis Cambess	Brazil	Cerrado and Atlantic Forest	Fruits and leaves, raw or cooked
Cyperus rotundus L.	Brazil	Wide distribution in temperate tropical and subtropical climate. Cerrado, Amazon, Atlantic Forest, and Caatinga	Rhizome, raw, cooked, or roasted
Momordica charantia L. "Goya"	East India and South China	All regions of Brazil	Fruits, blanched, brine, preserved, fried, or cooked. Raw flowers and leaves
Acmella oleracea (L.) R. K. Jansen	India or South America[a]	Tropical regions close to the Equator in Africa, Asia, and South America Atlantic Forest and Amazon	Branches, leaves, and flowers, raw, cooked, sautéed, or roasted
Clitoria ternatea L.	Indonesia, with reports in Asia, South America, and the Caribbean[a]	All tropical areas of the globe. America, Asia, and Australia	Flowers, raw. Pods and leaves, cooked
Coccinia grandis (L.) Voigt	East Africa	Tropical areas India, Philippines, Cambodia, China, Indonesia, Malaysia, Myanmar, Thailand, Vietnam, eastern Papua New Guinea, Australia, and the Americas	Cooked fruits, leaves and stem
Oxalis regnellii	South America	Tropical zones of South America	Leaves, stem, flowers and bulbs, raw or cooked and bleached

Source: Authors

[a]Possibly originating from these locations, as there is no consensus in the literature on this information. The most recurring locations are listed as being origins

the understanding that plant species are incorporated via appropriation of uses and customs attributed to historical moments that transmits the culture of traditional communities and people. This idea is reflected by Student 3:

> Inside the Garden you are able to see food as a whole, from his genesis to end (...), the principal is related to Food Biodiversity, which is very well placed in practice (...), both inside the gardens' discourse and in task force with community, such as in curricular components associated with it (…), it is also exercised the thing about contact with the earth. I used to see Nutrition in a more closed way.

Fig. 3 UFP studied in 2019. (*Source*: LabNutrir collection)

That speech justifies that education with plants inside Nutrition Science widens up closed visions to sustainability, biodiversity, and soil care. In this manner, from the point of view of Nutrition formation, the data from Table 1 represents the culmination of an experiential learning process with studied plants and, to that point, points out proper attention by students to an integrated sight between food and ethnobotanical dimensions, which can be strategic to fight against nutritionism ideology.

Also, there is to say that LabNutrir is the first open sky laboratory in Brazil, and it enlightens a unique perspective for those who are involved with it. In this way, this aspect can be measured by Student 4's: "I believe that all of us from Nutrition had to pass by the Garden because it is an experience that they would not have in a conventional lab. There are primordial knowledge that can only me acquired with this method."

Thereby, LabNutrir promotes necessary dialogue between Nutrition, Ethnobiology, and Anthropology with the elaboration of UFP data sheets. Besides that, it enables the improvement and reformulation of food assessment tools, complement studies in Ethnobiology and compiles data for inclusion of these plants in food and nutritional security within the local economy and biodiversity conservation (Reyes-García et al. 2019; Kasper-Pakosz et al. 2016; O'Neill et al. 2017). However, there are still limitations. Food composition data reveals significant differences related to agroecological zone, management techniques, soils, seasonality, and genetic diversity (Hunter et al. 2019).

On top of that, the data sheets encourage nutritional composition studies with these plants and combat harmful notions attributed to the condition of resistance and spontaneity of such plants, such as "scrubs", or weeds or even "famine foods." The data sheets also prioritize the spread of knowledge of the plant's nutritional character in a reliable manner (Jacob and Albuquerque 2020; do Nascimento et al. 2012).

The negative denotation "famine foods" refers to the memory of hunger and/or extreme poverty, in which ethnoculinary resources were essential to enable consumption of wild species during drought; this was revealed in research carried out on *Pereskia aculeata*, known as "the poor people's meat" in northeastern Brazil.[6] Thus, culinary use is analyzed and compiled to enable an understanding of what are the difficulties and practical potentials of working with the species (do Nascimento et al. 2012). However, over time, the practices of collecting and preparing these foods have fallen into disuse, in a scenario of privilege given to the ultra-processed fruits of the industry lobby and UFP consumption associated only to the memory of hunger (Lang and Barling 2013).

To overcome the negative meaning of "famine foods" and to value plants from the local biome, LabNutrir's GBL method, by linking curricular components to the production and dissemination of UFP data sheets, brings these subjects closer to the student, in an effective theory-practice relationship based on the singular work involving each component and its problematization set.

In this scenario, the focus of working with plants develops contextualized learning, in which theoretical contents are experienced (agroecological management, botanical classification of plants, seasonality and plant development, methods of culinary elaboration of plant parts) and finally systematized (scientific name, origin, distribution and occurrence, nutritional information, and food uses).

In effect, the data sheets become instruments for scientific dissemination and support and development of educational activities in the community in free classes, cooking workshops and agricultural management (Jacob 2020).

We suggest that the active GBL methodology applied in LabNutrir is a positive strategy for incorporating knowledge of ethnobotany as a pillar of Nutrition, which, in addition to being biological, is socio-environmental. In turn, the GBL methodology mitigates botanical blindness (the inability to perceive plants in the environment)

[6] For the data *P. aculeata*, see in http://nutrir.com.vc/horta/Ficha_Pereskia.pdf

(Salatino and Buckeridge 2016). Rediscovered and valued plants become new references for the promotion of biodiversity and sustainable diets, a reconciliation between nature and culture.

5 An Answer for the Future: GBL for Professional Training of Nutritionists

Reformulation of Nutrition curricular structures in Brazil, contemplating the complexities of knowledge and practices of SFS, is much needed. In view of the window found in interdisciplinary and interprofessional training of Nutrition professionals in knowledge, skills, and attitudes for the development of SFS, Jacob and Araújo (2019) proposed summarizing the conclusions of Brazilian experts concerning gaps in nutrition education and outlined aspects that limit the professional's performance, formalizing desires not yet addressed in the curricular structures of most courses. Among the recommendations for the proposed training paradigm, attention has been drawn to "pedagogical gardens structured as laboratories." It is in this sense that LabNutrir is placed. Recognizing the urgency of implementing a holistic approach to food, its practical scenario aims to strengthen the topic of biodiversity in Nutrition training (Jacob 2020).

Table 2 presents a systematization of the students' responses. We prioritized those promoting dialogue both for sustainable food systems and for food biodiversity.

These fragments from students' interviews are reflections and potential positive outcome indicators for the implementation of GBL method supporting capacity building toward sustainable food aystems.

Applying the lens of capacity building in education with respect to the demands of global agendas for achievement of Sustainable Development Goals, authors have been presenting *capacity building* as a critical point in training systematization. *Capacity building* refers to the process in which individuals strengthen and maintain the attitudes, skills, and knowledge needed to develop their goals over time (Baillie et al. 2009; Delisle et al. 2017).

The cognitive structure for knowing or for "learning to know" refers to the domain of *knowledge acquisition instruments*, whose foundation is the pleasure of understanding (Delors 2013).

As an example of that, Student 4 sayings allowed us to infer that "Composting" is a content of knowledge developed in our laboratory garden. In 2019, as part of the training activities, a course was being held in which one of the modules involved the proper environmental management of organic waste, with a focus on composting. The process was conducted by professors from UFRN, specialists in soil management and conservation, and included implementation of composting in our vegetable garden. In the words of our student, the "most sensational" experience refers to the "use of the composter."

Table 2 Mastery of skills for the interviewed students

Capacity building	Fragments of interest
Knowledge Composting Sustainable diets and/or menus	"I understand the impact in relation to waste that would be dumped in nature and the importance of composting." (Student 4) "I think that for me the [experience] that was most sensational, and I was very interested in learning is the use of the composter. So much that we had there in the garden, that we always did the 'turning' and using leftover food from the laboratory and from home as well." (Student 4) "I learned what UFP is. [...] benefits and methods of preparation. I would never imagine that a certain plant, like purslane - [Portulaca oleracea], could make a souffle" (Student 2)
Skills Identify plants, develop sustainable diets and menus with UFP	"I discovered that there are specific ways to take pictures of plants"; "Now I know how to pass on the street and identify" (Student 2)
Attitudes Awareness of food biodiversity	"There are seedlings here at home that I brought from the garden, purslane [P. oleracea] and ora-pro-nobis [Pereskia aculeata], I [...] have it here at home and always use it in my food [...]" (Student 4) "Now I see that places in disuse can become urban gardens." (STUDENT 1) "Close to my house, I see that many people take out the chananas [Turnera subulata], today I even asked them not to start. Whenever I can and I see people destroying for lack of knowledge, I [...] try to talk and talk a little more about them." (Student 2) "[The vegetable garden] mainly changed the way I looked at the plants. Before, I just looked at the color, [...] fruit. I see today with different eyes, I observe more specific details. I have a new view on plants [...] now it is mainly the role of the nutritionist to instruct on [UFP] [...]. It is a differential for nutritionists who know and those who do not know the project." (Student 2)

Source: Authors

These excerpts suggest the importance of the strategy, using gardens as a teaching method for acquisition of practical knowledge. Perhaps, the concrete experience mediated by the vegetable garden is the motivation necessary to learn a given subject, situation, or phenomenon with pleasure. A pleasurable form of learning, in addition to memorization, is situated as contextualized and meaningful learning, taken home from the institutional garden, as revealed by Student 4 above.

The construction of meaningful knowledge, through concrete experiences, echoes in the reports of students from another project, previous to ours, and very close to our ideals, the Garden/Horta of the Faculty of Public Health (FSP) of the University of São Paulo (USP):

> After my participation in the project, I became even more concerned with the issue of the environment, with the reuse of organic waste, building a compost bin at home, and helping with the maintenance and care of plants
>
> It was in the vegetable garden that I learned what a composter is, and an earthworm, and the importance of both in preserving the environment, mainly related to the issue of accumulating garbage (Higino et al. 2018, p. 27–9)

The history of the FSP/USP Garden/Horta begins in 2014. As a teaching laboratory, it aims to be a space for observation, research, and teaching with the conviction that production and consumption of food from the garden contributes to healthy eating practices, promotes interaction with the environment, and encourages interdisciplinary studies (Higino et al. p.8).

The two Brazilian vegetable garden/laboratories present other similarities, such as acquisition of knowledge concerning UFP:

> There was a UFP tasting, like Peixinho tea and mint leaves, it was very tasty. It was also really cool to be able to plant seedlings, have this contact with the land, and all of this is part of our involvement with food, isn't it? (Higino et al. 2018, p. 21)

In the previous section of this chapter, if our case study presented and discussed the relevance of lessons learned from data sheets, in which the UFP are studied in the biome in which they are located, for sustainable and socio-environmental nutrition, then the responses of project students provide evidence for the importance of rediscovering and valuing plants, making them new references for promotion of biodiversity and sustainable diets: "[...] I learned that [...] that we can insert UFP at times with more benefits than traditional [food]. Now I know that UFP open up a new range of options in the form of consumption" (Student 1, from LabNutrir).

In the wake of these experiences sharing an expanded proposal for a laboratory garden, a stage for contextualized learning, we are able to argue in favor of building skills and abilities for the nutritionist to act in the pursuit of sustainable food systems (SFS).

But, how does learning about composting relate to SFS? Our student replies: "Today, I understand how food can impact sustainable development, from when we produce, to when we will discard" (Student 1). Learning about composting makes it possible to concretize the notion of complexity in food systems, where care for each constituent element (environment, people, inputs, processes, infrastructure, institutions, markets, and commerce) is fundamental for the balance of the whole of its related activities (production, processing, distribution, and consumption of food) and the results of these activities (including social, economic, and environmental results).

Boff (2012, p.153) teaches us that one does not learn inside classrooms or closed into laboratory libraries, or even on Internet search sites. Students "must be led to experience nature, to discover biodiversity, to learn about the history of these landscapes [...] it is a plunge into the real world, to find Mother Earth."

The experience of learning mediated by mother nature "alive" in vegetable gardens reveals that in the immediate plain of "know-how," the acquisition and improvement of skills (identifying plants, preparing diets, and sustainable menus with UFP) are renewed in capacity building, necessary to become a professional capable of working in sustainable food systems.

Worldwide, other experiences have demonstrated the formative importance of this change toward sustainability of the food system.

At Appalachian State University, in the North Carolina, USA, university students from the Nutrition and Dietetics course ($n = 12$) participated in a program providing practical learning and service opportunities in vegetable gardens. The results relate, among the main benefits, greater vegetable intake, local food purchasing, and an increase in vegetable literacy (Gartman et al. 2016).

At the Sunshine Coast University, Queensland, Australia, a study with eight students from the Nutrition and Dietetics course in the universe surveyed (n = 14), revealed that, of the motivations of these university students to remain in volunteer work at the university community vegetable garden, the most common factor was the desire to learn and gain knowledge, particularly in themes related to vegetable gardens, sustainability, and healthy eating (Anderson et al. 2018).

Certainly the most powerful initiative is in Canada. The traditional University of British Columbia (UBC) has enjoyed the UBC Food System Project (UBCFSP) since 2001, with a series of courses at various levels of training, whose teaching-applied learning initiative aims to promote knowledge, skills, and experience in students so that they can serve as agents of change, as community leaders and as responsible global citizens. Despite its short time in existence, we can say that LabNutrir's experience also seeks to "offer students opportunities for applied learning, research, and professional development, to bring a positive impact to ecological and human health" (UBCFSP 2020).

The importance of the work articulated, between the dimensions of knowledge and doing, in capacity building for Sustainable and Socio-Environmental Nutrition, is realized. This is made possible by the fact that the GBL learning method transcends formal teaching and takes place through a unique experience in which the construction of knowledge and the development of skills are linked in a contextualized way, which is fundamental to the pursuit of Sustainable Development Goals/SDGs, leaving the rhetorical plane and facing in the garden, a starting point for problems of a transdisciplinary nature, and stimulating new learning paths.

As to development of attitudes and thoughts linked to ways of acting, feeling, and positioning oneself in relation to tasks in social life, the responses obtained suggest a certain degree of internalization of the concepts, behaviors, and facts, derived from the pedagogical experiences in the LabNutrir vegetable garden.

> [The vegetable garden] mainly changed the way I look at plants. Before I just looked at the color, the fruit. I see today with different eyes, I observe more specific details. I have a new view on plants [...] now it is mainly the role of the nutritionist to teach on [UFP]. [...]. You can take your knowledge about UFP with you. It is a differential between nutritionists who know and those who do not know the project. (Student 2)

The professional training that acts for the transition of SFS necessarily passes through an awareness of food biodiversity (Table 2). As "learning to be," it becomes part of the person's total development – spirit and body, intelligence, sensitivity, aesthetic sense, personal responsibility, and spirituality (Delors 2013).

An inseparable development in the cognitive and "knowing how to do" dimensions, by changing "my worldview" (Student 1), can change the local context: "Now I see that places in disuse can become urban gardens." (Student 1); "Near my house, I see that many people remove the chananas [*T. subulata*], today

I even asked them not to. Whenever I can and I see people destroying... for lack of knowledge, I [...] try to talk and say a little more about them." (Student 2).

Biodiversity becomes an incorporated attitude. Learning that promotes decongestion of botanical blindness. It is discovery that leads to attitudes supported by an awareness of food biodiversity: "There are seedlings here at home that I brought from the garden, purslane [*P. oleracea*] and ora-pro-nobis [*P. aculeata*], I [...] have it here at home and I always use it in my food [...]" (Student 4).

Although the number of students is not significant when considering the universe of Nutrition students at UFRN, it should be noted that all attended the project for 1 year. The results show that teaching and learning experiences focused on plants promote capacity building in food biodiversity. GBL is an answer that deals with the questions of the present and, in passing, strengthens bridges of the future for sustainable environments and food systems.

6 Final Considerations

The pedagogical method employed by LabNutrir, insofar as it establishes a contextualized teaching-practice-reflection relationship with food plants in the biome of their territory, advances the training of nutritionists able to promote sustainable diets. We understand that studying, disseminating, and enhancing UFP are an important strategy to expand and reframe food plants that are culturally acceptable and accessible, economically fair and accessible, and nutritionally adequate, safe, and healthy; and it is also fundamental to protecting and respecting biodiversity and ecosystems while optimizing natural resources. Basing *capacity building* in science that sees itself as sustainable and socio-environmental, and in which the teaching of food biodiversity can be performed from theory to practice in gardens, becomes an important pillar for the new professional profile, trained to work in sustainable food systems: a sensitive and proactive view from ethnobotanical and biocultural perspectives.

Although the universe of our case report is limited, we understand that our results and their implications contribute to the debate on training human resources in Nutrition toward having the competence to promote sustainable diets, in which biodiversity plays a central role.

Acknowledgments We would like to say thank you to the Pró-Reitoria de Extensão and to our Pró-Reitoria de Graduação, whose contribution of financial resources and quota of scholarships allow us to continue the actions of projects linked to LabNutrir with excellence. To the Pró-Reitoria de Administração (PROAD), who made it possible to translate this text. Also, we would express our gratitude to the members of our garden laboratory, whose dedication keeps LabNutrir's ideals enlightened. Finally, a special thank you to Michelle Cristine Medeiros Jacob, idealizer and creator of this open sky living lab, who has contributed to the first version of this manuscript with dedicated reading.

References

Anderson C, Maher J, Wright H (2018) Building sustainable university-based community gardens: Volunteer perceptions of enablers and barriers to engagement and benefits received from volunteering in the Moving Feast. Cogent Soc Sci 4:1–19. https://doi.org/10.1080/23311886.2018.1488211

Azevedo E, Rigon S (2016) Sistema Alimentar com base no conceito de sustentabilidade. In: Taddei J et al.(2nd ed) Nutrição em Saúde Pública. Rubio, São Paulo, p 467–478

Baillie E, Bjarnholt C, Gruber M, Hughes R (2009) A capacity-building conceptual framework for public health nutrition practice. Public Health Nutr 12:1031–1038. https://doi.org/10.1017/S1368980008003078

Boff L (2012) Sustentabilidade: o que é o que não é, Petrópolis, RJ

Burlingame BA, Dernini S (2012) Sustainable diets and biodiversity. In: Food and Agriculture Organizations of the United Nations. International Scientific Symposium Biodiversity and Sustainable Diets United against Hunger (2010: Rome, Italy). Rome

BRASIL. INEP. Instituto Nacional de Estudos e Pesquisas Educacionais (2000) Exame Nacional do Ensino Médio: documento básico. Brasília

Brasil (2001) Institui diretrizes curriculares nacionais do curso de graduação em nutrição. Resolução CNE/CES 5, de 7 de novembro de 2001. Diário Oficial da União; Nov 9, Seção 1, p. 39. DCN/MEC/2001. Available at: http://portal.mec.gov.br/cne/arquivos/pdf/CES05.pdf. Accessed 29 Mar 2020

Creswell JW (2014) Investigação Qualitativa e projeto de pesquisa: escolhendo entre cinco abordagens. Edição em português: da Rosa SM. Porto Alegre

Delisle H, Shrimpton R, Blaney S et al (2017) Capacity-building for a strong public health nutrition workforce in lowresource countries. Bull World Health Organ 95:385–388. https://doi.org/10.2471/BLT.16.174912

Delors J (2013) The treasure within: Learning to know, learning to do, learning to live together and learning to be. What is the value of that treasure 15 years after its publication? Int Rev Educ 59:319–330. https://doi.org/10.1007/s11159-013-9350-8

Desmond D, Grieshop J, Subramaniam A (2004) Revisiting garden-based learning in basic education. FAO, Rome

do Nascimento VT, da Silva Vasconcelos MA, Maciel MIS, Albuquerque UP (2012) Famine Foods of Brazil's Seasonal Dry Forests: Ethnobotanical and Nutritional Aspects. Econ Bot 66:22–34. https://doi.org/10.1007/s12231-012-9187-2

Etkin N (2006) Edible medicines: an Ethnopharmacology of food. University of Arizona Press, Tucson

FAO (2010) The second report on the state of the world's plant genetic resources for food and agriculture. Journal of Medical Entomology, Rome

FAO (2013) Sustainable food systems concept and framework. Rome, Italy

FAO (2017) Guidelines on assessing biodiverse foods in dietary surveys. Rome, Italy

Forouhi NG, Unwin N (2019) Global diet and health: old questions, fresh evidence, and new horizons. Lancet 393:1916–1918. https://doi.org/10.1016/S0140-6736(19)30500-8

Gartman M, Ball L, Gutschall M et al (2016) Design and implementation of a garden-based service learning program for nutrition and dietetic students: a CBPR approach. J Nutr Educ Behav 48:S60. https://doi.org/10.1016/j.jneb.2016.04.160

Higino AF, Prager AC, Bertolini AM et al (2018) Histórias da horta da Faculdade de Saúde Pública. In: Faculdade Saúde Pública, USP. São Paulo

Hunter D, Borelli T, Beltrame DMO et al (2019) The potential of neglected and underutilized species for improving diets and nutrition. Planta 250:709–729. https://doi.org/10.1007/s00425-019-03169-4

Jacob MCM (2020) Biodiversidade De Plantas Alimentícias Não Convencionais Em Uma Horta Comunitária Com Fins Educativos. DEMETRA Aliment Nutr Saúde 15:e44037. https://doi.org/10.12957/demetra.2020.44037

Jacob MCM, Albuquerque UP (2020) Biodiverse food plants: Which gaps do we need to address to promote sustainable diets? Ethnobiol Conserv 9:1–6. https://doi.org/10.15451/ec2020-04-9.09-1-6
Jacob MCM, Araújo FR (2019) Desenvolvimento de competências para Nutrição no contexto de Sistemas Alimentares Sustentáveis Capacity Building in Nutrition in the context of Sustainable Food Systems. a 1–18
Kasper-Pakosz R, Pietras M, Łuczaj Ł (2016) Wild and native plants and mushrooms sold in the open-air markets of south-eastern Poland. J Ethnobiol Ethnomed 12. https://doi.org/10.1186/s13002-016-0117-8
Kinupp VF, De Barros IBI (2008) Protein and mineral contents of native species, potential vegetables, and fruits. Cienc e Tecnol Aliment 28:846–857. https://doi.org/10.1590/s0101-20612008000400013
Kinupp VF, Lorenzi H (2014) Plantas alimentícias não convencionais (PANC) no Brasil: guia de identificação, aspectos nutricionais e receitas ilustradas. Instituto Plantarum de Estudos da Flora, São Paulo
Lang T, Barling D (2013) Nutrition and sustainability: An emerging food policy discourse. In: Proceedings of the Nutrition Society
Martinelli G, Moraes M (2013) Red book of flora of Brazil. Botanical Garden, Rio de Janeiro
O'Neill AR, Badola HK, Dhyani PP, Rana SK (2017) Integrating ethnobiological knowledge into biodiversity conservation in the Eastern Himalayas. J Ethnobiol Ethnomed 13:1–14. https://doi.org/10.1186/s13002-017-0148-9
Perrenoud P (2008) Dez novas competências para ensinar. Porto Alegre
Recine E, Mortoza AS (2013) Consenso sobre habilidades e competências do nutricionista no âmbito da saúde coletiva. In: Observatório de Políticas de Segurança e Nutrição. Brasília
Reyes-García V, Powell B, Díaz-Reviriego I et al (2019) Dietary transitions among three contemporary hunter-gatherers across the tropics. Food Secur 11:109–122. https://doi.org/10.1007/s12571-018-0882-4
Salatino A, Buckeridge M (2016) Mas de que te serve saber botânica? Estud Avancados 30:177–196. https://doi.org/10.1590/S0103-40142016.30870011
Scrinis G (2008) On the Ideology of Nutritionism. Gastronomica 8:39–48. https://doi.org/10.1525/gfc.2008.8.1.39
Tumilowicz A, Neufeld LM, Pelto GH (2016) Using ethnography in implementation research to improve nutrition interventions in populations. Matern Child Nutr 11:55–72. https://doi.org/10.1111/mcn.12246
UBC Sustainability (2020) UBC Food System Project. https://sustain.ubc.ca/teaching-applied-learning/seeds-sustainability-program/ubc-food-system-project. Accessed 29 Mai 2020
United Nations (2017) Transforming our world: the 2030 Agenda for Sustainable Development. New York
Willett W, Rockström J, Loken B et al (2019) Food in the Anthropocene: the EAT–Lancet Commission on healthy diets from sustainable food systems. Lancet 393:447–492. https://doi.org/10.1016/S0140-6736(18)31788-4

Teaching and Learning About Unconventional Food Plants in an Edible Urban Landscape: A Brief Report

Neide Rigo

Contents

My interest in edible plants is an old one, and I have always dedicated myself to being aware of the edible landscape around me. I like taking walks with the purpose of discovering edible plants in the urban space, either planted or spontaneously sprouted. In 2012, I wanted to do something special for December 10, when the *Slow Food* movement celebrates *Terra Madre Day* internationally. So, I decided it would be pleasant and productive to share my experience with other people. I gathered friends and strangers through an invitation on my blog,[1] and equipped with bags, scissors, and paraphernalia for gathering plants, we were 12 people roaming around the streets of my neighborhood. We walked for about 3 h, and the result could not have been any better. We brought home fruits, edible flowers, leaves, and roots, as well as aromatic plants, which were all prepared collectively for lunch. When eating, each with their own plate on their lap, the feeling was of enchantment for understanding oneself as part of a system surrounded by good and free food that no one notices in the rush of everyday life.

I repeated the experience a few more times, whenever I had the time. Many of the people who took part in these sporadic foraging events were insistently asking for more gatherings, and I myself found the activity much more exciting when I had the company of others who shared the same interest, so I decided that I could turn these outings into a job. In 2016, *PANC na City* was officially born as a guided foraging tour to observe, identify, and possibly harvest every kind of unconventional food

[1] See at www.come-se.blogspot.com. See also my Instagram account @neiderigo.

N. Rigo (✉)
Autonomous Researcher, São Paulo, SP, Brazil

M. C. M. Jacob, U. P. Albuquerque (eds.), *Local Food Plants of Brazil*, Ethnobiology, https://doi.org/10.1007/978-3-030-69139-4_21

plants (UFP) that we found along the way. It happened in the City Lapa neighborhood, located in the western part of São Paulo. It was initially planned to be a monthly event, and for a long time, I was able to do so. Over the past 2 years (since 2018), I tried to keep the project alive with at least six tours a year. Forages were always preceded by a talk and an exposition of dozens of forgotten, unknown, or neglected edible species. After that, our lunch was prepared using several of the plants observed, some found in the streets, other cultivated in my yard. Having different people signing up for each forage is a source of enjoyment, but besides that, it is great to have the opportunity to observe the changing of the species throughout the different seasons in the year, a fact rarely noticed in a world used to just a few mainstream vegetables artificially available at all times in the market. One of the stops of the tour is at *Horta* (kitchen garden, in Portuguese) *City Lapa*, a space full of perennial species, either edible, aromatic, or medicinal.

Our community garden was born in 2014 from a group of neighbors outraged by the sad fate of a small piece of land of about 170 square meters (Fig. 1). This corner looks like it was a remnant of the block, and although sitting between two wide streets full of trees, it was used as a disposal point for rubbish, pruning remains and broken furniture. All this waste was thrown amid high grass, lacking proper maintenance by the neighborhood council, and because of these features, the place was considered dangerous, and residents would avoid passing by.

Trying to transform that idle and poorly maintained space, one other neighbor and I started to plant useful species and organize ideas and actions to make that space more friendly. We set up a meeting and invited as many people from the neighborhood as we could, but only the two of us attended. So, without much support from the community in that first moment, we went to the neighborhood council to ask for help to clean the space. In that same week, a truck of workers removed the garbage and mowed the overgrown grass. We preserved the larger plants and left everything ready for planting.

At first, we started working on it ourselves. Overtime, other neighbors joined us, and soon we won the trust of some who shared our desire for recovering and revamping that space. There was no planning on what to cultivate, but as I had many aromatic seedlings at home, these were the first species to be planted, clove basil

Fig. 1 Horta City Lapa in 2014 and in 2019. (Photos by Neide Rigo)

(*Ocimum gratissimum* L.), manjericão-anis (*Ocimum carnosum* (Spreng.) Link & Otto ex Benth), ruby leaf (*Alternanthera brasiliana* (L.) Kuntze), lemon grass (*Cymbopogon citratus* (DC.) Stapf), and sweet-scented geranium (*Pelargonium graveolens* L'Hér.), in addition to more common species such as rosemary, kale, etc.

Some of the plants proved unsuitable for the space we had, so we gradually replaced those. We came to that conclusion during our first winter. Summer rains were beneficial to our kitchen garden, as we never had any water source, but when the cold dry winter arrived, the garden suffered and lost a good portion of the species. Another decisive factor for the choice of plants that we would grow there was the interest aroused by some species. The garden is open and located in a place with considerable pedestrian traffic, so many of the plants were uprooted and taken away.

A natural selection slowly took place and defined the current configuration of the garden. At the same time, we had to deal with angry neighbors who did not want a public garden close to their homes. Their arguments were flawed and full of prejudice, but it was their best shot. They said that the work groups for the garden would attract strange people to our area; that the place had bad soil; that people who came and went to the nearby train station would take everything away; that what we were growing was not a garden as they knew it; that it was something of such bad taste it could not even be called a garden; that we did not have authorization from the city government to mess with public gardening; that people who eventually ate something grown there were at great risk; that we did not have technical knowledge to work with plants; and that we destroyed the sidewalk (which has never existed, it used to be just high grass), among other empty accusations. We were then called to the neighborhood council, where authorities tried to convince us to formally care for the garden (sort of an adoption). Our group was larger at that point, and we agreed to remain informal. After much nuisance and dried plants by the drought, summer rains calmed the spirits and outlined what would remain in our garden. By that time, our space received a plaque with a name honoring the neighborhood: "Horta City Lapa." Figure 2 shows some gardeners of our group.

It was then that we realized that although we did not plan it, the plants that today are part of the garden are those we call UFPs. They were the ones who could resist all bad weather, both meteorological conditions properly said and bad tempers, and they keep on firm and strong, come rain or sunshine.

As I write, we are still in quarantine due to COVID-19 pandemic and in the midst of a winter drought. Still we have perennial and rustic species that cross the seasons and also some very resistant annuals harvested in winter. In addition to the two types of *Ocimum* sp. already mentioned and various spontaneous seasonal species, we have chaya (*Cnidoscolus aconitifolius* (Mill.) I.M.Johnst.), moringa (*Moringa oleifera* L.), cinnamon (*Cinnamomum verum* J. Presl), bay leaves (*Laurus nobilis* L.), curry leaves (*Murraya koenigii* L. Spreng), banana (*Musa* sp.), elderberry (*Sambucus nigra* L.), Turk's turban (*Malvaviscus arboreus* Dill. Ex Cav.), leaf cactus (*Pereskia aculeata* Mill.), spiked spiralflag ginger (*Costus spicatus* (Jacq.) Sw.), boldão (*Plectranthus grandis* (L.H. Cramer) R.H. Willemse), boldinho (*Plectranthus ornatus* Codd.), arrowroot (*Maranta arundinacea* L.), turmeric (*Curcuma longa* L.), and sweet potato (Ipomoea batatas (L.) Lam.), among others.

Fig. 2 Gardeners of *Horta City Lapa* after planting sweet potatoes. (Photos by Neide Rigo)

After all these years, the vegetable garden has become a reference on our street, and during normal times, it is a pleasant place to take children and maybe pick some herbs. I know that our garden is not an example of an urban agriculture with vast food production, but it certainly is an example of how a degraded space can be transformed into a useful garden that, in addition to providing a little food and aromatic and medicinal plants, also serves the purpose of reconnecting people with the ground and the cycles of nature. It also connects the local community, uniting them around a common will to make cities more humane, safe, and friendly.

The few neighbors who were unfavorable to our initiative are still around, but their voices were silenced in the face of the unquestionable benefits that a public garden brings to the neighborhood. Nowadays, daily care is given spontaneously by community volunteers and more intensely during task force days, when we join neighbors and people from other areas who are interested in helping.

I believe that the success of these experiences strengthens the idea that citizens occupying the public space make any city a safer and more pleasant place to live in, nourishing everyone's physical and mental health. And that any idle space can be transformed into a useful and low maintenance garden, if only we choose rustic and appropriately adapted species, the majority of which will surely be among the UFPs group.

Finally, below I provide a list of references that can be useful as education resources to people interested in starting community-based projects with Brazilian UFP (Brasil 2019; Kinupp and Lorenzi 2014; Ranieri 2017, 2018; Rigo 2017). These references provide helpful guidance in identifying and cooking these plant species.

References

Brasil (2019) Amazônia à mesa: Receitas com produtos da sociobiodiversidade para a alimentação escolar. MAPA/SAF, Brasília/DF
Kinupp V, Lorenzi H (2014) Plantas Alimentícias Não Convencionais (Panc) no Brasil. Editora Plantarum, São Paulo
Ranieri G (2017) Guia prático de plantas alimentícias não convencionais. Instituto Kairós, São Paulo
Ranieri G (2018) Guia prático de plantas alimentícias não convencionais para escolas. Instituto Kairós, São Paulo
Rigo N (2017) Mesa farta no semiárido. COOPERCUC, Uauá/BA

The manufacturer's authorised representative in the EU is Springer Nature Customer Service Centre GmbH, Europaplatz 3, 69115 Heidelberg, Germany. If you have any concerns regarding our products, please contact ProductSafety@springernature.com

Printed and bound by CPI Group (UK) Ltd, Croydon, CR0 4YY
30/09/2026
02225506-0009